Interdisciplinary Evolution Research

Volume 2

About the Series

The time when only biologists studied evolution has long since passed. Accepting evolution requires us to come to terms with the fact that everything that exists must be the outcome of evolutionary processes. Today, a wide variety of academic disciplines are therefore confronted with evolutionary problems, ranging from physics and medicine, to linguistics, anthropology and sociology. Solving evolutionary problems also necessitates an inter- and transdisciplinary approach, which is why the Modern Synthesis is currently extended to include drift theory, symbiogenesis, lateral gene transfer, hybridization, epigenetics and punctuated equilibria theory.

The series Interdisciplinary Evolution Research aims to provide a scholarly platform for the growing demand to examine specific evolutionary problems from the perspectives of multiple disciplines. It does not adhere to one specific academic field, one specific school of thought, or one specific evolutionary theory. Rather, books in the series thematically analyze how a variety of evolutionary fields and evolutionary theories provide insights into specific, well-defined evolutionary problems of life and the socio-cultural domain.

Editors-in-chief of the series are Nathalie Gontier and Olga Pombo. The Series is edited from within the Applied Evolutionary Epistemology Lab, more information on the lab is available at http://appeel.fc.ul.pt.

More information about this series at http://www.springer.com/series/13109

Emanuele Serrelli · Nathalie Gontier
Editors

Macroevolution

Explanation, Interpretation and Evidence

AppEEL
Applied Evolutionary Epistemology Lab

Springer

Editors
Emanuele Serrelli
"Riccardo Massa" Department
of Human Sciences
University of Milano-Bicocca
Milan
Italy

Nathalie Gontier
AppEEL, Centre for Philosophy of Science
Campo Grande
University of Lisbon
Lisbon
Portugal

ISSN 2199-3068 ISSN 2199-3076 (electronic)
Interdisciplinary Evolution Research
ISBN 978-3-319-15044-4 ISBN 978-3-319-15045-1 (eBook)
DOI 10.1007/978-3-319-15045-1

Library of Congress Control Number: 2014959855

Springer Cham Heidelberg New York Dordrecht London

Printed on acid-free paper

Springer International Publishing AG Switzerland is part of Springer Science+Business Media (www.springer.com)

Preface and Acknowledgments

In the early 1980s of the twentieth century, which is now a little less than 40 years ago, and in response to the uproar that the introduction of Punctuated Equilibria theory had caused within evolutionary biology, John Maynard Smith invited paleontologists to join what he called, the "high table" of evolutionary theory. In the 1970s, Niles Eldredge and Stephen J. Gould provided convincing theoretical arguments as well as empirical data that the Modern Synthesis was incomplete because its founders had not adequately incorporated the study of life's evolution above and beyond the differential distribution of genes across populations through time.

They defined macroevolutionary research by arguing that species are real biological individuals that have a clear beginning, lifespan, and ending in time, and furthermore proved that the fossil record evidences that species demonstrate long periods of stasis where no apparent morphological change occurs, which are interspersed by rapid speciation events that follow a branching pattern. When the fossil record does not give proof for intermediate life forms, then instead of assuming the incompleteness of the fossil record as Darwin and Neodarwinians did, they suggested to understand the gaps as data, and to postulate that no intermediates existed when none are found. Because species are understood as real biological individuals, speciation and extinction events are also understood as real phenomena that require scientific investigation.

Researching morphological stasis, species and above species phenomena such as (mass) extinctions and speciation events, large-scale evolutionary trends, and major transitions across all domains of life; or mapping the various units, levels, and mechanisms whereby life evolves as well as the hierarchical nature there exists between these units, levels, and mechanisms; require an epistemic incorporation of fields such as developmental biology, ecology, systematics, and biophysics. Most of all, it requires a change in scholarly research attitudes, a willingness to transcend classic field-specific disciplinary boundaries that remain focused on reductionist, gene-centered theoretical accounts, in favor of complexity-focused, holistic epistemic stances.

In recent years, the heated aura that surrounded these often polemic debates that associated with the introduction of micro- versus macroevolutionary stances appears to have cooled down a bit. The explosive entrance of macroevolutionary areas of research has cleared room for what we can characterize as an almost silent integration of the major claims first put forth by macroevolutionary scholars into standard evolutionary research. One can safely say that evolutionary biologists in both micro- as well as macro-oriented fields today accept that species are real biological entities with an important evolutionary role to play, and research on the evolutionary and abiotic causes that underlie constraints, stasis, extinction, and speciation has never ranged so "high" at the "evolutionary table". Indeed, these topics defined the talks of many of the lectures of the speakers that were invited to present their work at the Darwin 2009 bicentennial that was organized at the University of Chicago.

One of us, Nathalie Gontier, attended that conference as an audience participant. What amazed me was that the topics introduced by macroevolutionary scholars ranged so high, while at the same time, little attention and credit was given to where these ideas stemmed from and in which context they had originated. During that same period, and with a grant from the European Marie Curie fellowships, I held a one-year visiting research position at the American Museum of Natural History in New York, under the supervision of Niles Eldredge, with the goal to perform theoretical research on punctuated equilibria theory in particular, and how macroevolutionary theory in general is applicable within the sociocultural sciences. For many years, the Museum has set a crucial scene for the development of evolutionary theory, both in what regards its micro- as well as macroevolutionary stances, not in the least by employing scholars such as Niles Eldredge and Stephen J. Gould as well as Ernst Mayr.

At that moment in time, by on the one hand reading these scholars' important works that lie at the basis of micro- and macroevolutionary stances as well as on the other hand having the experience to attend the excellently organized Darwin commemoration where I had the chance to discuss many of these ideas with my contemporaries, I felt scholars from my generation had lost the roots of these significant ideas. From there grew the idea to edit a book on the matter, as well as to provide dissemination and outreach activities on the specificities of macroevolutionary research outside the classic paleontological and biogeographical disciplines where they were first introduced, and to highlight how macroevolutionary-oriented scholars have contributed to a richer conceptualization as well as demonstration of life's evolution.

To obtain these goals, and back in Belgium, I started with writing out these ideas in the form of a grant proposal which I submitted to the John Templeton Foundation. The proposal, that had as goal to investigate the importance of both macroevolution as well as reticulate evolution for the Extended Synthesis within the field of evolutionary biology, as well as the impact these new areas of research have on the growing sociocultural evolutionary sciences, was accepted by the Foundation and was successfully executed at the Applied Evolutionary Epistemology Lab (http://appeel.fc.ul.pt), at the Portuguese University of Lisbon in 2012–2013. I hereby want to thank again the staff of the Templeton foundation and, in particular, Paul Wason as well as Kevin Arnold and Drew Rick-Miller, my program officers. I am grateful to the project's team, including Olga Pombo as well as Márcia Belchior, Francisco Carrapiço, Luís Correia, Larissa Mendoza Straffon, Marco Pina, and Emanuele Serrelli; and I want to express my gratitude toward the Portuguese Gulbenkian Foundation and Ciência Viva Agency for their enthusiasm and efforts in helping me to bring my ideas on outreach and knowledge dissemination into action during what turned out to be a crazily busy year.

Together, we organized a session on how macroevolutionary theory transcends the Modern Synthesis for the 2013-meeting for the American Association for the Advancement of Science (http://appeel.fc.ul.pt/sub/eve/dir/aaas/aaas2013.html) that among the speakers had Alycia Stigall who also contributed to this volume; a session for the 2012-meeting of the American Anthropological Association that partially focused on cultural macroevolutionary theories and methodologies (http://appeel.fc.ul.pt/sub/eve/dir/aaa/aaa2012.html); two international evolution schools for pre- and postdoctoral university students with modules on macroevolution (http://evolutionschool.fc.ul.pt) taught, among others, by Ilya Tëmkin and Folmer Bokma who also contributed to this volume as well as the conference on Micro and Macro, Horizontal and Vertical Evolutionary Patterns (http://evolutionarypatterns.fc.ul.pt); a public conference on evolution with special sessions for teenagers (http://evolutionconference.fc.ul.pt); we conducted several video interviews with the scholars who participated in the activities (https://www.youtube.com/user/appeellisboa); and we concretized my multiple book ideas further, including this one on macroevolution. I am grateful to Emanuele for having accepted my invitation to co-edit the book.

We are very much indebted and grateful to the numerous scholars we encountered along our paths, who gave us advice and guidance, encouragement, and critique. In particular, we want to thank Wallace Arthur, Quentin Atkinson, Michael Benton, Luis Borda de Agua, Rituparna Bose, Dan Brooks, Mike Bruford, Jorge Carneiro, Maurizio Casiraghi, Lounes Chikhi, Alex de Voogt, Charbel Niño El-Hani, Claus Emmeche, Douglas H. Erwin, Frietson Galis, Root Gorelick, Ronald Jenner, Fiona Jordan, Bruce Lieberman, Paulo Madruga, Octávio Mateus, Daniel McShea, Richard P. Meisel, Alex Mesoudi, Sérgio Moço, Ana Noronha, Samir Okasha, Mark Pagel, Octávio Paulo, Telmo Pievani, Anna Marie Prentiss, Luís Paulo N. Rebelo, Luis Mateus Rocha, Michael Ruse, Ana Sanchez, David Sepkoski, Artur Santos Silva, Rui Silveira, Andrej Spiridonov, Ian Tattersall, Inês Tomé, Derek Turner, Rosalia Vargas, Davide Vecchi, Emílio Rui Vilar, Tyler Volk, Matthew Joseph Walsh, Richard Watson, and Jon Umerez.

We also want to cordially thank the whole Springer team, and in particular Anette Lindqvist and Sabine Schwarz.

This book on macroevolution is neither a handbook for beginning scholars for which there exist numerous excellent works written by paleontologists, nor a work aimed toward philosophers and historians of science to exclusively highlight the history and epistemology of macroevolutionary ideas. This book highlights some of the most important research topics that macroevolutionary scholars have introduced into the evolutionary sciences, including debates on how microevolution differs from macroevolution; what the nature is of evolutionary stasis, extinction, speciation; how we can define and measure evolutionary rates; how we can model biological hierarchies; how biophysics, ecology, evo-devo, genetics, and systematics shed new light on life's major patterns, trends and transitions, its origination, extinction and diversity; how macroevolutionary theory transcends biology and is applied within the sociocultural sciences; and how all the latter requires us to reconceptualize the very nature of evolutionary research in light of an extended synthesis.

We are enormously grateful toward the authors who found the time to contribute a chapter for the book, as well as the scholars who, behind the scenes, were willing to peer-review the chapters. We hope the reader will find as much enlightenment on the subject of macroevolution as we did editing the volume.

Nathalie Gontier
Emanuele Serrelli

Contents

Macroevolutionary Issues and Approaches in Evolutionary Biology 1
Emanuele Serrelli and Nathalie Gontier

Part I Macroevolutionary Explanations and Interpretations

Can Modern Evolutionary Theory Explain Macroevolution? 29
Douglas J. Futuyma

Evolution as a Largely Autonomous Process 87
Folmer Bokma

Visualizing Macroevolution: From Adaptive Landscapes to Compositions of Multiple Spaces 113
Emanuele Serrelli

Toward a Natural Philosophy of Macroevolution 163
Stanley N. Salthe

Networks and Hierarchies: Approaching Complexity in Evolutionary Theory 183
Ilya Tëmkin and Niles Eldredge

Uniting Micro- with Macroevolution into an Extended Synthesis: Reintegrating Life's Natural History into Evolution Studies 227
Nathalie Gontier

Part II Evidencing Macroevolution with Case Studies

Using Microevolution to Explain the Macroevolutionary Observations for the Evolution of Sex 279
Lutz Becks and Yasaman Alavi

Speciation: Expanding the Role of Biogeography and Niche Breadth in Macroevolutionary Theory 301
Alycia L. Stigall

Morphological Misfits and the Architecture of Development 329
Alessandro Minelli

Macroevolution in and Around the Hominin Clade 345
Bernard Wood and Mark Grabowski

Facing the Big Sixth: From Prioritizing Species to Conserving Biodiversity 377
Elena Casetta and Jorge Marques da Silva

Contributors

Yasaman Alavi Department of Zoology, The University of Melbourne, Melbourne, VIC, Australia

Lutz Becks Max Planck Institute for Evolutionary Biology, D-Plön, Germany

Folmer Bokma IceLab and Department of Ecology and Environmental Science, Umeå University, Umeå, Sweden

Elena Casetta CFCUL, Centre for Philosophy of Science, University of Lisbon, Lisbon, Portugal; LabOnt, Laboratory for Ontology, University of Turin, Turin, Italy

Niles Eldredge American Museum of Natural History, New York, NY, USA

Douglas J. Futuyma Department of Ecology and Evolution, Stony Brook University, NY, USA

Nathalie Gontier Applied Evolutionary Epistemology Lab, Faculty of Science, Centre for Philosophy of Science, University of Lisbon, Lisbon, Portugal

Mark Grabowski Center for Advanced Study of Hominid Paleobiology, George Washington University, Washington, D.C., USA

Jorge Marques da Silva Faculty of Sciences, Department of Plant Biology and Center for Biodiversity, Functional and Integrative Genomics, University of Lisbon, Lisbon, Portugal

Alessandro Minelli Department of Biology, University of Padua, Padua, Italy

Stanley N. Salthe Brooklyn College, City University of New York, New York City, USA

Emanuele Serrelli "Riccardo Massa" Department of Human Sciences, University of Milano Bicocca, Milan, Italy

Alycia L. Stigall Department of Geological Sciences and Ohio Center for Ecology and Evolutionary Studies, Ohio University, Athens, OH, USA

Ilya Tëmkin Biology Department, Northern Virginia Community College, Smithsonian National Museum of Natural History, Washington, D.C., USA; Department of Invertebrate Zoology, Smithsonian National Museum of Natural History, Washington, D.C., USA

Bernard Wood Center for Advanced Study of Hominid Paleobiology, George Washington University, Washington, D.C., USA

Macroevolutionary Issues and Approaches in Evolutionary Biology

Emanuele Serrelli and Nathalie Gontier

Abstract Many fields and approaches evidence, quantify, and analyze macroevolution. From biogeography to paleontology, from ecology to phylogenetics, and from biophysics to philosophy of biology, macroevolution elicits definitions and theoretical problems related to concepts such as species, lineage, ecology, niches, and extinction, which are relevant for general evolutionary biology. Macroevolutionary theories provide new epistemic frameworks to explain evolution in deep time, and macroevolution is also a phenomenon exemplified by myriads of real life-history case studies. This volume Macroevolution: Interpretation, Evidence and Explanation samples the rich reservoir of macroevolutionary knowledge, and evidences the macroevolutionary phenomenon in various episodes in time.

Keywords Macroevolution · Evolutionary biology · Speciation · Extinction · Deep time · Fossil record

Outlining the table of contents and writing an introduction to the various chapters of a book volume always comes with a reflection on the sequence in which we present the topics discussed by the authors, a sequence that in turn associates with the reasons we invited the scholars to contribute. Macroevolution on the one hand associates with theory formation and the methodological means by which we can interpret and explain evolution in deep time and above the species level. On the other hand, macroevolution is itself a phenomenon that can be evidenced by actual

E. Serrelli (✉)
"Riccardo Massa" Department of Human Sciences, University of Milano Bicocca, Milan, Italy
e-mail: emanuele.serrelli@epistemologia.eu

N. Gontier (✉)
Applied Evolutionary Epistemology Lab, Centre for Philosophy of Science, University of Lisbon, Lisbon, Portugal
e-mail: nlgontier@fc.ul.pt

E. Serrelli and N. Gontier (eds.), *Macroevolution*,
Interdisciplinary Evolution Research 2, DOI 10.1007/978-3-319-15045-1_1

cases in life's history. For that reason, we have divided the book into two parts, one that focuses on theory formation, and one that evidences macroevolution.

1 Introduction to Part 1: Macroevolutionary Explanations and Interpretations

For the first part, Macroevolutionary Explanations and Interpretations, we invited our contributors to focus on the theoretical, methodological, and epistemological aspects of macroevolution, defined as a scientific area of research that endorses specific scientific practices. Evolutionary scholars today continue to disagree on the nature and scope of evolutionary theory. Is there such a single field as "evolutionary biology" or is evolution a phenomenon studied by a variety of scientific disciplines? How does the field of macroevolution relate to microevolutionary biological areas of research? Is the Modern Synthesis complete, and can it adequately explain macroevolutionary problems above the population level such as speciation and extinction, evolutionary trends, major transitions, biological hierarchies, or species sorting? Does macroevolution delineate one or multiple distinct area(s) of research, or does it merely complement microevolutionary theory and practice? Answers to these questions not only vary, but also they continue to raise significant debate between micro- as well as macroevolutionary scholars. We have sought out both the controversies and agreements, and we have invited our contributors to write on how they, from within their specific disciplines, understand and define macroevolutionary epistemology, and how they see their theoretical frameworks fit or dissociate from the standard evolutionary paradigm both in theory and practice. We have favored quality over quantity and invited a selected group of scholars to provide extensive review chapters instead of aiming for shorter, more concise position papers.

For **Douglas J. Futuyma**, a leader in evolutionary biology (Futuyma 2013), macroevolutionary theory primarily associates with on the one hand ideas on developmental constraints as introduced by adherents of punctuated equilibria, and on the other hand, with the role speciation plays in bringing forth biodiversity at an ecological and biogeographical level. He opens the first part by asking "**Can Modern Evolutionary Theory Explain Macroevolution**?" As one of the scholars who has long recognized the importance of the issues raised by macroevolutionary scholars, and who in his career has focused on reconciling aspects of punctuated equilibria theory with population genetics, he answers the question mostly in the affirmative. In his chapter, the author provides a rich contextualization of both the origin of the synthetic theory and how its architects tried to explain macroevolutionary above-population phenomena, as well as the challenges that evidence and hypotheses on developmental constraints and stasis, among others, pose to the synthetic theory. Futuyma provides historical insight into how post-synthetic evolutionary biologists have been reconciling these ideas into standard evolutionary theory, and he especially points toward the rising disciplines of evo-devo and evolutionary ecology as the means by which such reconciliation is possible. Eco-evo-devo presents a more evolved and richer synthetic evolutionary view, and the continuously evolving and expanding framework therefore remains valid.

That evolutionary theory is valid and is undisputed by macroevolutionary scholars (Ridley 2003). Nonetheless, many macroevolutionary scholars are less optimistic about how well microevolutionary theory can explain macroevolutionary problems or predict evolutionary outcomes. How do macroevolutionary scholars differentiate their research agenda from microevolutionary biology, and how do they define the microevolutionary fields they oppose? Macroevolutionary areas of research today associate with fields such as paleobiology, ecology, systematics, and biophysics, and scholars that form part of these fields in general think that microevolutionists place too much emphasis on genetic selection at the expense of other principles, such as physical and ecological ones, that equally contribute to our understanding of evolution.

Folmer Bokma tests some of the microevolutionary predictions on speciation, extinction as well as the mode and tempo of evolution in his chapter "**Evolution as a Largely Autonomous Process**." He gives a series of examples wherein he demonstrates the means by which microevolutionary scholars provide explanations and make predictions on the evolutionary fates of species as well as how they interact with other species (flowers and their pollinators, for example), and weighs them against the actual evolutionary history that those life forms undergo through time, which he in turn deduces from molecular phylogenetic analyses. His examples demonstrate an epistemic ambivalence and duality in the works of microevolutionists, which he characterizes as "ascribing change to natural selection when it occurs, but failing to account for the frequent cases where no evolution is observed" though it is predicted. To explain periods of stasis as well as rapid speciation, he turns to punctuated equilibria theory and agrees with many of the founders of macroevolutionary thought that in real-life history events, natural selection, in and of itself, cannot account for speciation, extinction, or stasis. Again, he firmly grounds these conclusions upon the incoming results of molecular phylogenetics, a rising field that today forms a bridge between fossils and genes, and where Bokma is a leading and pioneering expert. He furthermore turns to epigenetics and evo-devo to explain evolution as a largely autonomous process.

In his chapter "**Visualizing Macroevolution: From Adaptive Landscapes to Compositions of Multiple Spaces**," **Emanuele Serrelli** details how macroevolutionists have visualized life's evolution. By taking classic models of evolutionary change as depicted in adaptive landscapes as point of departure, Serrelli demonstrates that the original population geneticists understood evolutionary change mostly as the various distributions of genes within populations through time, while macroevolutionary schools of thought understand evolution as the outcomes of adaptations to environmental conditions, and thus favor a more spatial, ecological approach. He furthermore demonstrates how new visualizations of evolution conceived as occurring in multiple spaces, such as morphospaces, geographical, and ecological spaces, as well as diversity diagrams and distribution maps provide new methodological tools to deduce the major patterns and trends of life's evolution.

Stanley Salthe, one of the architects of the macroevolutionary discipline, understands biological evolution as part of a larger, cosmic evolutionary process that both transcends and influences the evolution of life. In his "**Toward a Natural Philosophy of Macroevolution**," he explains how on a cosmic scale a distinction can be made between the physical, chemical, biological, and social realm. Besides by biological principles such as natural selection, life evolves according to thermodynamic and overall physical principles that act within life, both in what regard its development as well as its survival, expansion, and extinction in ecological settings. Salthe emphasizes that taking on a macroevolutionary perspective implies a return to natural history research as well as natural philosophy. In such a framework, all the natural sciences, including physics and chemistry, are put to use in explaining just how it is that life originates, diversifies, and dies. He demonstrates how such a naturalistic, ecological approach to life necessitates hierarchical thinking and explains how especially biophysics is able to account for life at a grander scale.

Hierarchy theory is also the topic of **Ilya Tëmkin and Niles Eldredge**'s chapter, "**Networks and Hierarchies: Approaching Complexity in Evolutionary Theory**." One of the claims made by macroevolutionary scholars is that species and higher taxa are real entities or biological systems that evolve in an equally real biotic and abiotic environment, respectively, by proximal evolutionary processes and abiotic drivers. Such a stance, for Tëmkin and Eldredge, requires an ontological investigation into the multiple levels of the genealogical and economic (ecological) hierarchy. By understanding biological entities as hierarchically nested, complex emerging systems that occupy an equally hierarchical and multilayered economy, they demonstrate how hierarchy thinking provides new means to delineate and identify the underlying patterns and processes of evolution.

Ontological and epistemological hierarchy thinking also forms the topic of **Nathalie Gontier's** chapter "**Uniting Micro- with Macro- evolution into an Extended Synthesis: Reintegrating Life's Natural History into Evolution Studies**." She demonstrates how the modern synthesis defined evolution at a meso-level, and details how microevolutionary and macroevolutionary research schools necessitate a reconceptualization of older hierarchical levels such as the inorganic, organic, and superorganic. She furthermore details how both the micro- as well as macroschools have evolved different scientific practices and epistemic frameworks to understand life's evolution. Macroevolutionary scholars understand evolution as the outcome of natural history, while microevolutionary scholars understand evolution as the result of a causal mechanism (i.e., natural selection), and Gontier points out that the various epistemic stances underlie radically different concepts on matter, space, and most of all, time. Finally, she details how macroevolutionary thought currently extends the biological sciences and is successfully applied within the sociocultural domain.

2 Introduction to Part 2: Evidencing Macroevolution with Case Studies

Macroevolution does not merely define a specific scientific agenda; it also delineates a phenomenon. In the second part of this volume, Evidencing Macroevolution with Case Studies, we invited scholars to contribute with specific topics and examples that explain the specificity of macroevolution as a phenomenon. In our selection, we have of course had to make choices, because not all case studies and examples can fit one book volume. We have therefore focused on some of the most important macroevolutionary phenomena typical of eukaryotic evolution, namely the origin of eukaryotic sex, the evolution of distinct body plans, hominid evolution, speciation and extinction, and biodiversity.

Lutz Becks and Yasaman Alavi contribute with a chapter "**Using Microevolution to Explain the Macroevolutionary Observations for the Evolution of Sex**." The origin and evolution of sex in eukaryotic organisms poses one of the biggest enigmas for evolutionary theory, and sex is rightfully characterized as one of the major transitions of life. Becks and Alavi understand the emergence of sex as a macroevolutionary phenomenon, which they define as an observation, and explain how traditional microevolutionary theories can explain its evolutionary emergence. More specifically, they understand sexual reproduction as an evolutionary pattern that associates with various geographic and phylogenetic distributions not found in asexual organisms and demonstrate how microevolutionary processes can explain the macroevolutionary observations.

Because species are considered real biological entities, macroevolutionary scholars also understand speciations as events resulting from processes often distinct from genetic selection. In the chapter titled "**Speciation: Expanding the Role of Biogeography and Niche Breath in Macroevolutionary Theory**," **Alycia Stigall** discusses how a multiplicity of abiotic and biotic, external factors including, among others, climate change and plate tectonics as well as niche occupation and breath, and species invasions of ecological niches, underlies speciation events. She avers for understanding speciations as the outcome of multiple factors that often lead to speciation as combined factors that together bring forth evolutionary change. Stigall reviews some of the vicariant speciation events that occurred in bivalves and brachiopods (which are both shelled marine animals) that lived through the Late Devonian Biodiversity Crisis, a period that is designated as a crisis because the ecosystem underwent radical changes while the marine animals form an anomaly to these extinction events; as well as the Late Ordovician Richmondian Invasion, another crisis period characterized by fluctuating sea levels and associated invasions of foreign species that before did not occupy these niches, where the marine benthos fared less well in comparison with the period before they had to share their habitat. She ends with detailing the speciation events of North American horses of the Neogene (the second period of the Cenozoic), this time during a radiation period associated with favorable climate change and subsequent abundant food availability (grasslands).

Prokaryotes come in distinct shapes and sizes, and although they can form morphologically complex colonies, and most certainly contribute to the anatomical form of eukaryotic beings, they are mostly unicellular organisms that often contain organelle-like structures. The evolution and diversification of anatomically distinct body plans, organs, and organelles is typical of eukaryotic organisms. Macroevolutionary theory has played a significant role in reviving and reintegrating embryological and overall developmental biology into standard evolutionary theory.

In his chapter "**Morphological Misfits and the Architecture of Development**," **Alessandro Minelli** examines morphological "misfits," i.e., taxa whose morphology diverges from the conventional structural body plan of the major clade they belong to. A correct phylogenetic positioning of morphological misfits is therefore the first obligate step toward a tentative interpretation of their evolution. Some misfits are "systemic": homologies between them and their relatives are hard to find, while homologous structures are what enables morphological comparisons. Macroevolutionary research on body plan formation helps to make the degree of "evolutionary freedom" of a structure visible and enables insight into character evolvability. Divergent structures 'behave as evolutionarily independent modules,' because their independence is structural, and often results from the largely independent genetic control of their development, which is the case in arthropod segments for example. The neck of the giraffe, on the other hand, although very divergent in a phylogenetic context where necks are rather monotonous, is not a module, because it involves several body units. For Minelli, rapidly evolved misfits deserve detailed studies to estimate the time of their divergence from "normal" relatives.

Macroevolutionary phenomena also impact our own history. **Bernard Wood and Mark Grabowski** document "**Macroevolution In and Around the Hominin Clade**." The authors begin their work by delineating how paleontologists, informed by both micro- and macroevolutionary theory, classify fossil finds into species and genera to build taxonomies. They exemplify by listing the reasons why recent fossil finds such as *Sahelanthropus tchadensis*, *Orrorin tugenensis*, and the two *Ardipithecus* species (*Ardipithecus kadabba* and *Ardipithecus ramidus*) can, and for the authors cannot, be categorized as members of the hominin clade (which is a subclade of the hominids). Until more evidence is available, they suggest instead to designated them as "possible hominins." They go on to demonstrate how a variety of morphological and developmental features, including body growth, sexual maturation, and reduction in teeth size, so typical of the actual hominin clade, enable deductions on the tempo and mode of evolution. They exemplify how the evolution of *Australopithecus afarensis* and *Paranthropus boisei* is characterized by stasis, and they demonstrate how macroevolutionary morphological trends, such as increase in brain size and the overall morphological differences between *Homo neanderthalensis* and *Homo sapiens*, can be explained by drift. Finally, the authors end their chapter with an account of the difficulties posed by homoplasies, morphological traits present in sister taxa but not in their most recent common ancestor.

Finally, macroevolutionary theory not merely associates with the fossil record or the study of the dead, it also associates with the living and those who are threatened with extinction. The final case study, written by **Elena Casetta and Jorge Marques da Silva**, therefore focusses on bioconservation efforts in a chapter titled "**Facing the Big Sixth: From Prioritizing Species to Conserving Biodiversity**." Scholars in general distinguish between five large extinction events but Eldredge distinguishes a sixth that is imminent. Casetta and Marques da Silva first investigate the criteria by which species are declared extinct from an environmental ethics point of view, and subsequently focus on how species can be prioritized for conservation. They provide a rich overview and analysis of the distinct means by which the United States of America and the European Union delineate their conservation policies and give suggestions on how they can be improved.

3 Alternative Ways to Read the Volume

The division of the book into two parts is only one way in which scholars can read the book. Macroevolutionary research is also characterized by a set of specific research questions, and these questions present alternative ways in which we suggest the reader to digest the various chapters.

3.1 Macroevolutionary Fields and Approaches

Macroevolutionary research fields and approaches to macroevolutionary phenomena are remarkably many, and distinguishing between a research area or an approach is not always easy. Biogeography, for example, delineates a particular field of study, but in its approach it also demarcates a particular dimension that can be incorporated and put to use in other research fields. Consolidated areas of research that deal with macroevolutionary issues include paleontology, systematics, geology, ecology, phylogenetics, evolutionary developmental biology, population genetics, conservation biology, theoretical biology, biophysics, and philosophy of evolutionary sciences. Each field has developed a particular series of methodologies and theoretical frameworks, whereby macroevolution can be evidenced, quantified, and analyzed. This book samples and surveys a good deal of them.

Wood and Grabowski's chapter is very informative of how macroevolutionary inferences are made in *paleontology* (Benton 2004; Foote and Miller 2006) and in particular in *paleoanthropology* (Wood 2011; Begun 2013) which is a subbranch of both anthropology as well as paleontology. The hominin features they focus on, such as teeth and brain size, are much more durable than any systematic classification of our kind into genera and species which in many details remain unresolved due to the fragmentary nature of the fossil record. Because paleoanthropology sets out to reconstruct our own evolutionary past, the authors also note that our demand for conclusive resolution is disproportionate, and instead highlight why evolutionary inferences of the fossil record are by necessity only one means in which we can approach the problem of phylogenetic reconstructions.

As evidenced by **Bokma**, *molecular phylogenetics* provides a complementary means to find answers on macroevolutionary issues of deep time, and our increasing possibility to quantify, model, and simulate large-scale datasets, links fossils to genes (Prothero 2003). Bokma's work on the temporal patterns and rates of evolutionary change (the "tempo" of evolution) is based on pioneering work by Avise and Ayala (1975) where correlations between species richness and quantity of evolutionary change across clades evidence a central thesis of punctuated equilibria, i.e., evolutionary change is concentrated in speciation events.

Taxonomies that are developed within the field of *systematics* are often taken for granted, but the amount of theorizing and quantitative research that precedes any systematic reconstruction cannot be underestimated. As **Gontier** points out, systematics started out as a logical and philosophical discipline, but today, the classification of natural phenomena that are bounded in space and time into various taxa, and their division into subspecies, species, genera, kingdoms, and domains, necessarily precedes any evolutionary investigation into a specific group. Systematic classification impacts all the evolutionary sciences, and systematic classification is in turn subject to advances made in molecular genetics, evo-devo, biogeography, paleontology, and ecology, work that often necessitates taxonomic revisions (Minelli 1993; Wilkins and Ebach 2013).

Even the intuitively clear-cut distinction between sexually and asexually reproducing taxa turns out to be anything less than straightforward, as **Becks and Alavi** reveal. **Minelli** devotes some reflection on taxonomy in his exemplification of morphological misfits, case study anomalies that can lead to the introduction of novel taxonomical units. **Wood and Grabowski** understand fossil genera as different from "neontological" genera: they are "grades." A grade is an informal grouping united by a level of morphological or physiological complexity supposedly reflecting adaptation, without the strict phylogenetic requirements implied by "clades" (i.e., monophyletic groups) that fossil hominins cannot meet. Furthermore, inferences of function from morphology, such as the inference of bipedalism from a few skeletal characters, are, for Wood and Grabowski, particularly frail. Wood and Grabowski are "splitters" and think the "lumpers" underestimate the number of species there existed in the hominin clade, because morphological differences across the geographical range of a putative "species" may actually hide speciation events and many groups probably went extinct without fossilizing.

Biogeography is another fundamental macroevolutionary field. To understand the evolution of eukaryotic sex, **Becks and Alavi** compare the distributional range of asexuals and sexuals: asexuals tend to range to higher latitudes and altitudes, and they tend to colonize previously glaciated and devastated areas. Shallow waters can be coevolutionary "hot spots" for the evolution of sex.

In macroevolutionary time scales, biogeography inevitably flows into *geology*, *paleogeography*, and *paleoecology*. There is a very intimate relationship between *stratigraphy* and living beings, dating back to the seventeenth century, as shown for example by the principle of faunal succession: Fossilized flora and fauna succeed each other vertically in a specific, reliable order that can be identified over wide horizontal distances, and they are used as time markers for rocks. In macroevolutionary stratigraphy, geological layers are moreover tagged with the biotic evolutionary processes that are ongoing in the correspondent era at the appropriate time scale, normally spanning a few million years.

Stigall's approach to paleontology combines biogeographic methods with phylogenetics, bringing to light the speciation patterns of single genera. It also integrates *niche evolution analysis*, to study the macroevolutionary dynamics of generalists and specialists (Peterson et al. 2011). Niche evolution in a taxon can be measured by extrapolating from geographical distribution the niche parameters at one time slice, then by extrapolating from niche parameters the expected geographical distribution at a second time slice, and then by comparing the expectations with the actual distribution found in the fossil record. The concept of niche is also explained by **Serrelli** and used by **Salthe** in his comprehensive view of Earth's history.

Ecology is slowly but steadily entering the list of disciplines that are considered indispensable to understand macroevolution (Allmon and Bottjer 2001; Price 2003; Loreau 2010). Ecology is, as **Salthe** describes it, "the study of energy flow relations taking place on Earth between the influx of solar radiation and its reradiation into space," a unifying science that studies the world in terms of thermodynamics and hierarchical structures. **Becks and Alavi** show that community ecology may combine the available theoretical and experimental knowledge on the evolution of sex and accommodate pluralistic explanations combining the accumulation of deleterious mutations and Red Queen dynamics (Van Valen 1973). Some branches of ecology seem well prepared to face the increasing complexity revealed in macroevolution. For **Tëmkin and Eldredge**, ecology is not only a good half of the macroevolutionary story, but also the locus of the triggering causes of evolutionary change all the way through the biological hierarchies (Eldredge 1989, 1999).

Many essays in the book emphasize how *phylogenetic methods* are now able to process both morphological and molecular evidences and even take into account within-species diversity. The progress of phylogenetic techniques has greatly improved the inferential, predictive, and testing possibilities of evolutionary hypotheses (Wiley and Lieberman 2011).

Comparative disciplines such as morphology, physiology, cytology, and genetics remain fundamental; indeed, they are boosted thanks to the "skeleton" of relationships provided by phylogeny. *Evolutionary morphology*, in particular, studies the existence of clusters of species sharing similar anatomies. Morphological classifications are often a useful first organization of knowledge, as **Minelli** argues in his chapter. The categorization of morphological misfits in three morphological kinds—divergent by reduction, by building blocks, or by synorganization—guides the search for specific developmental pathways and steps along which their developmental schedules may have evolved.

"Steps" is correctly understood, like many other words in our book (e.g., "intermediate" or "transitional" form), in the context of macroevolutionary *trends*, an extremely important concept. Trends are recognizable long-term, large-scale changes that involve great numbers of species, genera, and even higher taxa (Turner 2011). **Wood and Grabowski** need the demonstration of trends in cranial, dental, and postcranial morphology. They cannot rely on single diagnostic traits, because traits come and go in phylogenies, and what is diagnostic for one period is not automatically for another one. Not all the differences between *H. sapiens* and Panini (chimpanzees and bonobo) may be used to distinguish ancestral Panini from ancestral *Homo*. If, on the one hand, *H. sapiens*' hypodontism came about as a reversion after million years of megadontism, on the other hand there are more reliable trends in the clade: An incipient *Homo*-like cranium will be diagnostic all the way back to stem hominins to tell basal hominins from basal Panini. Corroborated trends allow for back-tracing and provide criteria to distinguish taxa that lived close to their divergence time.

One of the most exciting fields approaching macroevolution today is *evolutionary developmental biology*, abbreviated EDB by **Futuyma**, and evo-devo by **Minelli** (Arthur 2002, 2011; Minelli and Fusco 2008), while others link evo-devo to ecology under the heading eco-evo-devo (see Gilbert 2012 and Ledón-Retting and Pfennig 2011 for an overview). For **Futuyma**, evo-devo inspires an improved theory of variation. **Minelli**, following a classification by Arthur (2000, 2002), emphasizes how most macroevolutionary differences seem to consist in modifications of the temporal, positional, and quantitative (metric or meristic) aspects of the ontogenetic production of individual body parts, and how important developmental trends are in macroevolution. Developmental modularity is evident in those misfits that are odd, not systemically, but only with respect to a few well-circumscribed body parts. Other misfits have uncommon life cycles, with oddly shaped stages or rearranged sequences of stages.

Knowledge and understanding of these macroevolutionary phenomena have exploded after the advent of developmental genetics. Molecular methods, for example, can now be used to determine that only some developmental modules are also structural modules, characterized by a more well-defined genetic control. Developmental genetics in macroevolutionary context is a fervent field, very rewarding but also very difficult for its struggle with "the intricacies of the genotype → phenotype map." **Bokma** refers to the evo-devo schools to explain developmental constraints, selection plateaus and stasis and demonstrates how, partly because of developmental mechanisms, evolution is a largely autonomous process. Stigall also turns to eco-evo-devo to explain phenomena such as phenotypic plasticity and niche construction as well as habitat tracking, migration, and niche invasion, because all present behavioral and thus anatomically underlain responses to ecological settings.

Futuyma gives his own reconciliation of *population genetics* with punctuated equilibria. The claim of punctuated equilibria (Eldredge and Gould 1972, based on Mayr 1942) that rapid evolutionary change is coupled with bottlenecks, is, for Futuyma, "surely wrong" in the light of available evidence. Instead, the claim that rapid evolutionary change is coupled with biological speciation is plausible. Futuyma's solution focuses on the niche heterogeneity that is expected in widespread species and highlights the role of reproductive isolation in protecting local adaptations from dissolution.

As **Serrelli** narrates in his chapter, population genetics is essentially a powerful mathematical theory of factors such as selection, mutation, drift, population size, and allele frequencies. The synthetic work of Sewall Wright in the 1930s already hinted to "the way in which both speciation and extinction can flow mechanically from the processes of modulation of variation," in the words of Lewontin (1980: 61). But today, macroevolution does not get forced into population genetics models. *Evolutionary quantitative genetics* addresses the complex dynamics of phenotypes and their genetic underpinnings under different regimes of selection and other conditions (Falconer and Mackay 1996; Roff 1997). Selection can be detected in the covariation of character complexes, instead of being assumed as an ad hoc explanation for postulated adaptive traits.

The same holds for developmental correlations that present another crucial role in explaining morphological evolution. **Wood and Grabowski** are confident that quantitative genetics can now be applied to detect the relative weight of natural selection and drift even in hominin fossils. **Becks and Alavi's** chapter illustrate how *population biology and genetics* (Hanski and Giaggiotti 2004; Hartl and Clark 2007) may be put to use to explain innovation or, at the opposite, persistence of macroevolutionary distributions. Why does sex evolve? And why are sexual populations or taxa not invaded and replaced by asexual taxa when sex seems to be a bad choice as it comes at high costs? The case of sex is particularly interesting because the evolving trait is also one that shapes variance and, consequently, microevolutionary mechanisms. The "short- and long-term effects of sexual reproduction" need to be studied mathematically: It is assumed that sex always increases variation, but mathematical models demonstrate that it is not always the case and identify boundary conditions. Other questions may be asked, such as could sex accelerate adaptation to new environments?

Gontier approaches research on sex, so typical of microevolutionary fields, from within *history and philosophy of science*, which are also fields that can contribute to macroevolutionary research. She details how epistemic approaches to the origin and evolution of eukaryotic sex as well as the asexual behavior characteristic of prokaryotes relate to sociocultural and political ideas on the common goods of society as well as the rise of liberal as social thought in the nineteenth century. Is sex a social behavior or is it an adaptive outcome that enables the fit to survive and reproduce successfully?

Also from within philosophy of science, **Serrelli** hones in on the different kinds of *modeling* that map macroevolution as the realization of actual life forms in the huge domain of possible virtual alternatives. Many modeling approaches use biologically unrealistic exploration mechanisms to study general properties of low-dimensional "spaces of possibilities" that imitate the biological possibilities. High-dimensional spaces, which require a holistic, probabilistic kind of mathematics with a strong role of statistics, are used to understand how whole genomes or sets of phenotypic traits should behave in macroevolution. Some models, such as sequence spaces of RNA and proteins for example, can even guide empirical work.

Although of course macroevolution is not liable to direct experimentation, *experimental evolution* is sometimes used in relation to macroevolution (Garland and Rose 2009). **Becks and Alavi** provide the example of the few experimental tests of mechanistic causes for the evolution of sex, limited by the use of model systems whose sexuals and asexuals do not actually coexist under natural conditions.

As **Casetta and Marques da Silva** demonstrate with their essay, *conservation biology* entered an age in which it needs to deal with macroevolutionary knowledge and tools (Louys et al. 2012). Importantly, however, empirical knowledge will never be enough to settle issues such as how species should be prioritized, which conservation targets deserve focus, and how biodiversity conservation should be ethically justified; rather, the authors note that "Conventions and procedures have to be agreed upon and established … facing the Big Sixth is not a matter for biologists only." Following Soulé the authors characterize conservation biology as a mission- or crisis-oriented discipline, and its relation to biology has been compared to that of surgery to physiology, or of war to political science. So, ethics—and *philosophy of science* more generally—joins the table of disciplines that deal with macroevolution.

Because macroevolutionary thought redefines the cosmic scale, hierarchy theory, an intrinsic topic of *philosophy*, plays a crucial role in theory formation. *Hierarchy theory* is discussed in the works of Salthe, Tëmkin and Eldredge, and Gontier. **Gontier** highlights how many macroevolutionary scholars define themselves as naturalists, because of the emphasis they put on historical narration of past events. Both Salthe and Gontier trace hierarchy theory and macroevolutionary thought back to natural philosophy as it developed in the nineteenth century, and as a naturalist, **Salthe** legitimizes the approach. For him, natural philosophy is "an attempt to construct a scientifically based 'Big Picture' understanding of the world" whose goal is to find the different realms of reality. His picture is ecological and Salthe criticizes 'idealistic' interpretations of ecology that are based on an attempt to unify energy and information. Instead, he chooses a "materialist" position that concentrates on energy connections and energy flows.

Hierarchy theory also presents scholars with a means to *unify* knowledge on macroevolution. Advised by Marjorie Grene, Eldredge (1985, 1986, 1989) first turned to *Hierarchy Theory* in the 1980s. **Tëmkin and Eldredge** present an up-to-date version of hierarchy theory tightly integrated with network theory. Biological systems are arranged hierarchically, with smaller units forming the components of larger systems: trees in a forest, cells in a body, organisms in a population, trophic groups in ecosystems, genes within chromosomes, within cells, within organs, within organisms, within populations, within species, within ecosystems. Hierarchy theory of evolution is a theory of how biological systems are hierarchically organized, how they function, and how evolution takes place through them over time. "Hierarchies" is a plural term in hierarchy theory, not only because every system (organism, local ecosystem) is an instance thereof (a hierarchical entity) but also because there are distinct kinds of hierarchies. The hierarchy theory the authors present in their work identifies two main kinds that they delineate as the only two that are really important in evolution. One is named "economic" or "ecological," and it is based on matter–energy transfers and ecological processes. The other is named "genealogical" or "evolutionary," and it is based on information and replication. The two hierarchies overlap at some point. For example, an organism is both an economic entity and a replicator, being part of ecological processes and also a fundamental element of a lineage. At other points, the two hierarchies are clearly distinct, for example a species is a genealogical entity that does not participate, as such, in any economic system. Even when the two hierarchies overlap, hierarchy theory sees them as ontologically distinct, they interact in complex ways, and it is their interaction that brings about the patterns of evolution at all scales.

Finally, **Salthe** and **Gontier** point out the important role the rising field of *biophysics* has in contributing to a richer understanding of macroevolutionary phenomena. **Salthe's** ecological perspective is very much inspired by thermodynamics. And **Gontier** reviews how abiogenesis, traditionally a field associated with physics, exo- and astrobiology, can shed light on the origin of life, a theme that remains understudied from within classic Neodarwinian theory. Complex adaptive systems theory is relevant for understanding the spontaneous generation of catalytic and autocatalytic biochemical systems. She also points out that naturalists adhere to a different notion of time that she associates with relativity theory and quantum physics which she distinguishes from Newtonian mechanical worldviews she sees as underlying research on cause and effect as well as the evolution of purposeful behavior in Neodarwinian frameworks.

3.2 Macroevolution Defined as Evolution at and Above the Species Level

Wood and Grabowski point out that macroevolution as a concept was first used in 1934, in the work of the Russian geneticist Filipčenko (see Sepkoski 2012). The authors further note that the most accepted technical definition of macroevolution is the one presented by Simpson (1944) and Hallam (1989) that define macroevolution as "evolution at and above the species level." Such a characterization of macroevolution is also adhered to by **Futuyma**, **Bokma**, **Casetta and Marques da Silva**, and **Gontier**. **Wood and Grabowski** note that "if the species is the rubicon that divides macroevolution from microevolution, then the type of taxonomic hypothesis that is adopted will have profound implications for what is included."

Defining macroevolution as evolution at and above the species level therefore first and foremost requires a good definition of species (Wilkins 2011), and such a definition impacts how scholars understand speciation and extinction events. **Wood and Grabowski**, following Smith (2009), consider five species concepts and divide them into two classes: process-related and pattern-related species concepts. Process-related species concepts include the biological species concept (BSC), the evolutionary species concept (ESC), and the species-mate recognition species concept (RSC). Pattern-related species concepts are the phenetic species concept (PeSC) and the phylogenetic species concept (PySC). Paleoanthropologists often uses the PySC and tries to identify the smallest clusters of populations that are recognizable from the available set of characters. But the taxonomical and, most of all, classificatory controversies surrounding the hominin clade lead the authors to adopt a dataset-relative definition of macroevolution as "what you can learn from the fossil record."

Gontier points out that Mayr's BSC defines species based upon "sexual exclusivity and geographical accessibility" and especially the former implies an annihilation of the sexual individual in favor of a higher-order classification: the group or the species. **Stigall**, who endorses the BSC, considers species as "groups of organisms that maintain genetic continuity by interbreeding among members of the group but that are distinct from other reproductive groups" and underlines that in the fossil record this condition by necessity has to be inferred from morphology. Stigall understands speciation as "the separation of a set of organisms into a newly isolated reproductive unit that is discrete from the ancestral species," typically happening in "less than ten thousand years." Stigall is interested in a geographical classification of speciation modes: most frequently speciation begins in allopatry, i.e., in geographical separation, although sympatric and parapatric speciation is considered possible, and, under an integrated view of macroevolution inclusive of the ecological environment; it is possible to say that modes of speciation themselves shift in prevalence over evolutionary time (Coyne and Orr 2004). As Stigall explains, allopatric speciation comes in two kinds: vicariance and dispersal. In vicariance, the geographic separation is due to the formation of barriers. In dispersal, it is due to active migration. While speciation by vicariance appears predominant in modern taxa and in the fossil record (for example in trilobites), in the Devonian Stigall finds an opposite prevalence of speciation by dispersal, in other words, an anomalously low proportion of vicariant speciation.

In **Tëmkin and Eldredge's** hierarchy theory, species are lineages demarcated by origin through lineage-splitting (or speciation) events and by eventual demise through extinction, although the temporal boundaries of species become less distinct at smaller time scales due to a gradual process of divergence that appears instantaneous at geological time scales. **Casetta and Marques da Silva** observe that when describing the Big Sixth, species concepts break down: several animals hybridize, either spontaneously or by human intervention, and hybridizing also becomes more likely as selective pressures increase. **Casetta and Marques da Silva** describe the ongoing mass extinction as partially caused by our own species and its activities (Wilson 2002). A symbol of the Big Sixth is the Amur leopard, which today, mainly due to habitat alterations caused by the exploitation of forests, only consists of about twenty individuals living in southwestern Russia. In the USA alone, the list of endangered species is huge, from the Grey Wolf to the Puget Sound Killer Whale. Casetta and Marques da Silva explain why a species-based approach to conservation will not work and point out the necessity of integrating macroevolutionary mechanisms and logics into conservation biology: "the aim is not just to save endangered species, but also to allow the continued production of novel diversity, i.e., to improve the conditions and mechanisms that help the diversity to be generated."

If anthropogenic pressure depresses speciation rates, how can extinction be compensated? **Salthe's** perspective embraces the recent socio-technological evolution of *H. sapiens*, seen as a fervent cooperator to universal energy dissipation. Salthe's interpretation of macroevolution leads him to affirm that we should not be surprised at all if "disorder threatens everywhere, and we must work harder than seems reasonable to achieve anything." Salthe holds a more conventionalist view of species: Species do not have a role in macroevolution since they are not ecological actors. The status of populations is not granted either, because although populations do play ecological roles, they actually do as aggregations of individuals. Yet, Salthe grants populations at least an indirect influence in regulating the abundance of individuals through their reproduction network. For Salthe, speciation is a by-product of natural selection that rewards the ability to switch to new energy sources as those being utilized become locally depleted.

Tëmkin and Eldredge, whose hierarchy theory sees species as genealogical entities, focus on the reproductive aspect: Species replication is the process of speciation, or cladogenesis, that results from the perturbation and partitioning of demic networks. In fact, the hierarchical framework proposed by thc authors fixes the upper bound of evolution at the level of the species. Even though supraspecific entities (monophyletic taxa) do show patterns of differential survival, they lack the capacity to replicate: Evolution above the species level is an epiphenomenon that results from processes that occur at lower levels. Moreover, species are not interactors (Hull 1980, 1988). Interactors are those entities that interact with their environments in such a way as to make replication differential, and they belong in the economic hierarchy. The most extensive population-level interactor corresponds to a metapopulation, which is a geographic and ecological mosaic of contemporaneous avatars and, as such, a synchronous subset of a species lineage. Interactors are affected by physical perturbations. According to the sloshing bucket model (Eldredge 2003), the higher the level of perturbation, the higher the level in the economic hierarchy at which its effects will be expressed and, consequently, the higher the level of the genealogical hierarchy at which the evolutionary pattern of change in diversity and disparity will be recorded. Intermediate levels of environmental disturbance yield the maximum speciation rates, because they maintain a balance between population fragmentation and establishment of favorable conditions for isolate persistence.

Leaving aside the particular focus on levels and the species boundary, **Serrelli**, relying on current scientific practice, describes macroevolution as a simultaneous exploration of morphospaces, geographical spaces, ecological possibilities, and genealogical outgrowths. This exploration of multiple interrelated spaces may be represented by means of compositions of multiple graphical representations. Sometimes these visual representations take the form of landscapes: geographical landscapes (based on spatial distribution), genealogical landscapes (based on relatedness), morphological landscapes (based on shared combinations characters), and ecological landscapes (based on the combinations of environmental variables that are suitable for the considered organisms). The spaces of possibilities are connected in complex ways in macroevolution. For example, descent relationships may not go along with morphological resemblance, generating visualization conflicts. Knowledge of macroevolution must be creatively connected by means of composite pictures and, more importantly, composite and interdisciplinary studies.

Finally, several chapters approach macroevolution from the point of view of contemporary species. In particular, **Bokma** notices that with respect to fossil species, contemporary species are interesting because they allow biologists to observe more characters, at a molecular level, with their respective evolutionary rates. But the present also poses specific problems: declaring extinction, measuring biodiversity, quantifying and modifying our own impacts and effects for ethical reasons.

3.3 From Ediacara to the Grey Wolf: Embracing All of Life's History

Macroevolution is evolution on a large scale. It is therefore natural for macroevolution to span long periods of time and to embrace broad and diverse parts of the living world. In fact, not only the examples in this volume have such characteristic broad scope but they are also drawn from the actual history of life. Another way to read the volume is by following the geological age, the chronological order in which these phenomena first appeared.

In **Serrelli's** chapter, we get to know the earliest known complex of multicellular organisms: The worldwide Ediacaran biota, which existed from 575 to 541 Mya (million years ago), just preceding the more famous Cambrian explosion (Erwin and Valentine 2013). Most Ediacaran life forms left only indirect traces ("fossil traces"), but some of them had carbonate structures, and their fossils are found in Australia, Canada and Namibia (although, when they lived, their locations had completely different geographical coordinates than today, in a completely different configuration of continents).

Several chapters in our book refer to the Big Five, the five largest mass extinctions in the history of life on Earth: the Ordovician-Silurian, the Late Devonian, the Permian-Triassic, the End Triassic, and the Cretaceous-Tertiary (Raup 1991). But there is reason to believe that there are at least 7 major mass extinction events. The first major extinction event probably involves the mass extinction of prokaryotic anaerobe life forms that must have followed the great oxygenation event, the other, often dubbed the sixth extinction event is the one that faces us now and that is discussed by **Casetta and Marques da Silva**. The authors remind us that 99 % of the species that have ever lived on Earth are now extinct. Besides these Big Seven, extinction events occur constantly during life's evolution, on less grand scales, and the events are variable in intensity (Lawton and May 1995).

Stigall's chapter brings us back circa 450 Mya (Late Ordovician), in an area that today is near Cincinnati, Ohio. There, a cyclical pattern of sea level rise and fall determined periodical basin invasions by bivalves and other components of marine benthos. The author analyzes one of these invasions, the Richmondian, and studied how biodiversity and geographical differentiation fell down, speciation rate was depressed, and broad ranging species, during a moment in time, had prevailed on species with narrower ranges. Many millions of years later, precisely 375 Mya, the world went through the "Late Devonian biodiversity crisis." Again, the geographical context is characterized by sea level rises that, by connecting previously separated water basins, facilitated geographical invasion and limited the uprise of new physical barriers. The fossil record of brachiopods, bivalves, and other shallow sea organisms shows relatively high extinction rates as well as relatively low speciation rates. Of the two anomalous rates, the most determinant was the low speciation rate, because even clades with normal extinction rate had declined in number of species. There is a fundamental pattern connecting the Devonian biodiversity crisis with the Cincinnati invasions, granted some differences due to the global versus local scale: Physical events bring about invasions that destabilize ecological niches, generalists prevail, and speciation "by dispersal" overwhelms speciation by vicariance (see above).

Much more recently, between ca. 8 and ca. 5 Mya, in Africa, there lived the so far unidentified group that would become the common ancestor to both *H. sapiens* and chimpanzees/bonobos. **Wood and Grabowski** group this ancestor, along with all its descendants in a subfamily: Homininae. Later in the same continent, among Homininae, a tribe originated: the hominins (Hominini). This tribe would give birth to genera like *Australopithecus* and *Homo*. The current consensus, based mostly on molecular data, considers chimpanzees and bonobos as the "outgroup" for comparing hominins to each other, but within the hominin clade the phylogenetic context is all but clear: Fossils are rare and incomplete, taxa lack obvious ancestors, and there are many sources of error (Wood 2011). One of the reasons why these problems arise right here is that scientists and outsiders are uncommonly curious and demanding toward the hominin clade, looking for levels of detail that we would never ask to analog fossils of animals that are more distantly related to us.

While the ancestors of hominins lived in Africa, in North America, between the Miocene and the Early Pliocene (ca. 5.3 Mya), horse species (subfamily Equinae) went through an event traditionally referred to as an "adaptive radiation." **Stigall** demonstrates that in horses—which are vagile and migratory organisms—speciation by dispersal normally prevails on vicariant speciation. On land, as opposed to what happens in water, climate change may have led to geographical fragmentation, increase of vicariant speciation, and radiation, not necessarily adaptive to varying environmental conditions. From **Serrelli's** chapter, we learn that this phase of the evolution of horses was also well studied by one of the founders of the Modern Synthesis: George Gaylord Simpson. We see how Simpson (1944) described the phases by which a lineage of browser mammals, Hyracotheriinae, split into browsers and grazers under the effect of environmental change combined with correlation among some of their characters.

And **Gontier** demonstrates how macroevolutionary questions are today becoming the dominant mode by which we can approach sociocultural and linguistic evolution. She notes that on a grand cosmic scale, Julian Huxley already characterized evolution as going from the physical to the biological and psychosocial, and traces these ideas back to classic cosmic lineups that go from the inorganic to the organic and superorganic. Such classification has profoundly impacted our division of the sciences, that go from the physical and astronomical, to the natural and biological, and sociocultural sciences. Today, the sociocultural sciences and classic humanities are embracing an evolutionary approach to the study of human behavioral traits, languages, and cultures. And in these studies, a transition is taking place from studying isolated sociocultural and linguistic traits synchronically to understanding them diachronically, by tracing the genealogical origins of these traits across cultures in space and time. Such an approach by necessity takes on a macroevolutionary perspective, and she highlights how in particular the jargon associated with punctuated equilibria is finding its use in these new macrocultural fields.

3.4 A Rich Research Agenda

Many more fascinating topics are addressed in the dense chapters of this book. We end this introduction with listing some of the more open-ended research questions that are currently associated with macroevolutionary research fields.

What is the temporal pattern—the "tempo"—of macroevolution? **Tëmkin and Eldredge**, **Stigall**, and **Bokma**, all emphasize stability, and frame macroevolutionary change as coincident with the achievement of new equilibria after disturbance. The roots of contemporary approaches to the tempo of evolution have to be traced back to the idea of punctuated equilibria (Eldredge and Gould 1972) and to the lively methodological debate that followed, and still goes on today (Gould and Eldredge 1977, 1993; Lieberman and Eldredge 2014). **Wood and Grabowski** notice how the quest for detecting punctuated equilibria is translated, in paleontology, into four research questions on: (1) the relative importance of gradualist versus punctuated evolution, (2) the role of speciation events versus within-lineage evolution, (3) adaptive versus neutral processes, and (4) the operation and incidence of 'species selection.' But the authors point out that none of these questions can be answered definitely for the hominin clade. Significant improvements will be enabled by new technologies and methodologies, much more than by the accumulation of more and more fossils.

How is stasis explained? **Stigall** emphasizes geographical and ecological relationships; **Bokma** emphasizes the stability of development and the selection pressures exerted by traits on each other, while **Futuyma** concentrates on demic structure and interconnectedness. All these aspects, and many others, are integrated into **Tëmkin and Eldredge's** hierarchy theory of Evolution, for which "The overall stability of biological systems across levels of organization is not surprising given the general tendency of complex systems to acquire complex network architectures that ensure relative insensitivity to external perturbations."

What's the role of the environment in macroevolution? The biogeography of invasions and radiations demonstrates, for **Stigall**, environmental "control" over macroevolution, i.e., a direct causation of climatic and geologic events upon the periodic restructuring of biodiversity in the history of life. Mass extinctions, climate change, and continental drift all show the tremendous importance of physical factors in shaping the history of life. The environment has multiple causal roles also in hierarchy theory: "Environmental perturbations contribute to both generating variation at the molecular level and facilitate the expression of hidden phenotypic variance by compromising evolutionary capacitance. The spread and fixation of novel genotypes, ultimately responsible for producing taxic evolutionary patterns, are enabled by temporary removal of control over population dynamics brought about by environmentally triggered disruption or destabilization of ecological networks at the level of biocenosis." On the other hand, "the stochasticity and non-linear dynamics characterizing the processes of the biocenosis in flux" for Stigall, are another important contribution "to the evolutionary contingency of life's history."

But does such importance of the environment mean that macroevolution is adaptive? As **Serrelli** notices, macroevolution and adaptation are peacefully decoupled epistemologically, although, of course, still related in complex ways. Macroevolutionary phenomena such as speciation, diversity, and disparity, with their peculiar patterns, do not necessarily constitute adaptive "peak climbing," although, for instance, patterns of adaptation such as niche breadth are integral part of macroevolutionary explanations. Paleoenvironments were not a major factor controlling the extent of Ediacara morphospace. **Futuyma** addresses the issue of constraints: for him, constraints are demonstrated (e.g., by extinction) and relevant to evolution, although they do not explain stasis, nor do they jeopardize optimality and adaptation. **Minelli** considers the possibility that morphological uniqueness be a symptom of peculiar adaptations, which may in turn suggest adaptive dead ends. More radically, for **Bokma,** macroevolution is "a largely autonomous process." Bokma does not downplay the most representative and fascinating studies that historically convinced the scientific community of the agency and efficacy of natural selection, but Fisher's idea that complex traits have additive genetic basis is at best inaccurate, and its persistence is due to theoretical inertia in face of theoretical alternatives and empirical issues, such as missing heritability and the persistent abundant variation in ecology-related traits. The autonomy advocated by Bokma is not immediately in conflict with the environmental drive demonstrated by **Stigall**. In fact, **Bokma** focuses on the rate of adaptive change, which is autonomous from environmental change, whereas **Stigall** focuses on rates of speciation and extinction, which crucially depend on ecological events. Climatological influence does not necessarily mean *adaptive* drive, as shown in the case of horses, traditionally and hastily called an "adaptive radiation."

Is macroevolutionary change essentially concentrated during speciation events? All authors seem to agree upon a positive answer to this question. **Stigall** relies on the many studies in evolutionary biology—from Ernst Mayr, to Eldredge and Gould's (1972) punctuated equilibria, to contemporary followers and revisers—that have identified speciation as the fundamental locus of evolutionary change. In fact, in her framework, macroevolutionary "events" are *defined as* those critical periods *in which* evolution proceeded above the species level, with uncommon rates of speciation and/or extinction. For **Bokma,** there is little doubt that the number of speciation events is much more predictive of evolutionary divergence than the amount of environmental change. For **Futuyma**, according to the available evidence, biological speciation is indeed plausibly the typical context where rapid evolutionary change takes place. But, in the available evidence, Futuyma sees that speciation does not require bottlenecking, as reproductive isolation can arise in a widespread species that spans across heterogeneous niches (protecting, in turn, adaptations from dissolving into the species).

Are there any exquisitely macroevolutionary processes? **Stigall's** paleoenvironmental distribution data form the Late Devonian crisis demonstrate, for her, sorting in favor of broadly distributed, generalist, invasive species *versus* narrowly distributed, ecologically specialized species. Natural selection is a sorting process that happens among individual organisms in a population. Species sorting would be *another* sorting process, not completely analogous, of course, to natural selection. The peculiarly cosmopolitan Late Devonian fauna might be a result of species sorting, i.e., the preferential survival—in sustained large-scale environmental conditions—of invasive ecological generalists with low speciation rates. This hierarchical expansion of processes is one of the cores of hierarchy theory, presented orderly and carefully by **Tëmkin and Eldredge**. And if we think about the concept of *evolutionary potential* that for **Casetta and Marques da Silva** should inform conservation efforts, we may imagine conservation as a process of artificial deme or species selection, something that may or may not take place in nature, and that would yield trends of increasing evolutionary potential through time, visible at levels above the individual.

What is the relationship between micro- and macroevolution and how do both relate to evolutionary theory? Well, as **Bokma** tells us, the topic was considered an "old" question already by George Gaylord Simpson in 1944. **Gontier** confirms and explains how it relates to philosophical cosmologies, worldviews that delineate the cosmological hierarchy that have traditionally been brought to us in the form of Greek Chains of Beings, Medieval *Scala Naturae* and Far Eastern Wheels of Time. The question might be old, but the associated problems are all but resolved and all essays in this book demand for the composition of an articulate picture of micro- and macroevolution and how both define the field of evolutionary biology, as well as what the scope of evolutionary theory is in general.

What are the requirements on evolutionary theory with regard to prediction? **Gontier** discusses how especially Mayr, as one of the founders of the Modern Synthesis, was also involved in delineating a specific area of research for evolutionary biologists within academia. Classic physics is traditionally conceived as the queen science because it enables predictions on the future, while biologists cannot predict the future path of life. It remains impossible to predict speciation and extinction events. Nonetheless, the evolutionary sciences are on the rise and have successfully incorporated the sociocultural sciences that increasingly approach their subject areas from within an overall evolutionary framework. She also points toward different scientific practices among micro- and macro- evolutionary scholars. Microevolutionary scholars focus on the causal explanations of evolution and endorse uniformitarian epistemic stances that enable them to tackle questions on teleology and goal-directed behavior of living organisms, while macroevolutionary scholars take on more relativistic stances that she associates with epistemic pluralism that does not require evolutionary scholars to predict the future, but rather to narrate the past from within a series of referential frameworks that necessitate a comparative approach. **Futuyma** also writes interesting lines on prediction. With respect to macroevolutionary diversification, he compares evolutionary theory to meteorological theory that is able to provide explanation without large-scale prediction. On the other hand, Futuyma points out the innumerable successful predictions of DNA sequences and adaptive associations between phenotypic traits and environmental conditions. **Bokma** agrees that the existing evolutionary framework is not predictive over longer periods of time. In fact, he argues that the constant confirmation of the ubiquity of adaptation as a driver of evolution is largely an artifact of an epistemological bias. Other predictions, such as quantitative predictions about relationships between environmental change, genetic change, and speciation rates, can offer more sound evaluations and bring forward better evolutionary hypotheses.

How is human cultural evolution to be understood in macroevolutionary terms? **Salthe** mentions how cultural evolution—especially technological innovation—has been a crucial continuation of the tendency of evolution toward entropy. Evolution—cosmic, biological, and technological—has produced systems that depend upon, and produce, energy flows of greater and greater intensity through them. **Gontier** goes into the fine-grained epistemological nuances of studying cultures and languages as "beings" or biological individuals that blur the distinction between the "living" and the "dead" and shows the dance of ontological partitioning that has been going on for centuries in the definition of disciplines and their domains of study, a dynamic evolution that still goes on today with fascinating developments.

How does macroevolution relate to the Modern and Extended Synthesis? Macroevolution is the title of a collection in honor of Stephen Jay Gould (Vrba and Eldredge 2005). Gould, in his scientific testament, had written: "For some reasons still unclear to me, I always found the theory of how evolution works more fascinating than the realized pageant of its paleontological results, and my major interest therefore always focused upon principles of macroevolution" (Gould 2002, p. 38). Vrba, Stanley, Eldredge, and Gould were part of the paleobiological revolution that, in the seventies, brought forth theory and practice of macroevolutionary research. Decades have passed, and macroevolution not only remained a sustained and growing field of research but it also became entangled with more and more fields of biology, demanding for their specific empirical and theoretical contributions.

Several essays of the book address the relationship between macroevolution and the Modern Synthesis (MS). Although the MS, as a concept used to designate a shared theoretical framework, is itself a problematic object, Mayr and Provine (1980) describe the origin of the MS as an outcome of two distinct phases in time. The first phase, ranging from 1910–1920 to 1940 encompasses the reconciliation of Darwin's theory of natural selection with Mendel's theory of inheritance as well as with aspects of de Vries' and others' mutation theory. The 'second synthesis,' that took place from 1930 to 1940, was characterized by the incorporation of previously independent fields, such as systematics and taxonomy, zoology and botany, paleontology and morphology, and embryology. The 1980 account edited by Mayr and Provine was paralleled and immediately followed by pleas for an extension of the modern synthesis. One of those pleas came from within the field of paleobiology, where in the 1970s a 'revolution' took place (Sepkoski and Ruse 2009; Sepkoski 2012). Since Darwin, paleontology and its subject field, the fossil record, had been occupying a paradoxical position: Fossils were, on the one hand, necessary evidence, and, on the other hand, a place of embarrassing failures of prediction. Work by Simpson (1944), Eldredge and Gould (1972), and Stanley (1979) proved that the fossil record can indeed exhibit law-like patterns and regularities of its own. Subsequently, macroevolutionary scholars started to develop field-specific methods, but scholars on both micro- and macro- fields continue to disagree on whether macroevolution in and of itself requires a radical reconceptualization of Neodarwinism. **Gontier**, for example, points out that Eldredge repeatedly characterizes himself as a "knee-jerked Neodarwinian," while she attributes more revolutionary élan to Gould.

In his magisterial chapter, **Futuyma** provides an informative background on the foundation of the MS, which he designates as the Evolutionary Synthesis and focusses on the contributions made by scholars from a variety of countries who, although well-known in evolutionary circles, are not routinely listed as 'architects' of the discipline. Futuyma explains how founders of the evolutionary synthesis never really advocated that natural selection is the "sole cause" of evolution, and they merely provided evidence in favor of gradual evolution which they used to explain aspects of macroevolution. Scholars such as Bernhard Rensch, for example, provided macroevolutionary explanations for apparent orthogenetic trends. **Futuyma** exhorts critics to consider the particular challenges to which the Evolutionary Synthesis has responded successfully, and he downplays the idea of a 'hardening' of the late ES, emphasizing instead its flexibility. **Bokma**, on the other hand, in his focus on stasis and punctuations in macroevolution, points out that stasis was uncritically considered as fully compatible with basic evolutionary theory, and even denied as an observational fact. He agrees with Gould (1983, 2002) that the MS went through a "hardening" during the twentieth century, perhaps exacerbated by a defensive attitude.

Serrelli presents Ernst Mayr's description of the MS as a process engaging separate fields of research by virtue of communication, reciprocal awareness and familiarity, producing a shared vocabulary that is still largely used today in macroevolutionary studies. Serrelli explores how graphical representations may have worked as a major bridge between fields that were mutually independent, sometimes suspicious of each other, and locked.

4 Conclusion

In this introduction, we have given spots and keys to reading this rich volume on the interpretation, evidence, and explanation of macroevolution. The reader will be able to find more threads and much more knowledge running through the pages by our excellent contributors and will get the feeling of an exciting field of research that is going to grow even more and surprise us in the next years. While this book is being published, the debate on extending the Modern Synthesis is carried over on major scientific journals (e.g., Laland et al. 2014), as well as by other media. The interested reader will follow these developments to see how this debate will be settled, at least temporarily, in the next years, and the contributors of this book as well as we at the Applied Evolutionary Epistemology Laboratory are happy to have contributed to the overall debate.

Acknowledgments This work was written with the support of the John Templeton Foundation (grant ID 36288).

References

Allmon W, Bottjer DJ (2001) Evolutionary paleoecology: the ecological context of macroevolutionary change. Columbia University Press, New York

Arthur W (2000) The concept of developmental reprogramming and the quest for an inclusive theory of evolutionary mechanisms. Evol Dev 2:49–57

Arthur W (2002) The emerging conceptual framework of evolutionary developmental biology. Nature 415:757–764

Arthur W (2011) Evolution: a developmental approach. Wiley, Chichester

Avise JC, Ayala FJ (1975) Genetic change and rates of cladogenesis. Genetics 81:757–773

Begun DR (2013) A companion to paleoanthropology. Wiley, New Jersey

Benton MJ (2004) Vertebrate palaeontology, 3rd edn. Wiley, Oxford

Coyne JA, Orr HA (2004) Speciation. Sinauer Associates, Sunderland

Eldredge N (1985) Unfinished synthesis. Biological hierarchies and modern evolutionary thought. Oxford University Press, Oxford

Eldredge N (1986) Information, economics, and evolution. Ann Rev Ecol Syst 17:351–369

Eldredge N (1989) Macroevolutionary dynamics: species, niches and adaptive peaks. McGraw-Hill, New York

Eldredge N (1999) The pattern of evolution. W.H. Freeman & Co., New York

Eldredge N (2003) The sloshing bucket: how the physical realm controls evolution. In: Crutchfield JP, Schuster P (eds) Evolutionary dynamics: exploring the interplay of selection, accident, neutrality, and function. Oxford University Press, Oxford, pp 3–32

Eldredge N, Gould SJ (1972) Punctuated equilibria: an alternative to phyletic gradualism. In: Schopf TJM (ed) Models in paleobiology. Freeman, Cooper & Co., San Francisco, pp 82–115

Erwin D, Valentine J (2013) The Cambrian explosion: the construction of animal biodiversity. Roberts and Company Publishers, Colorado
Falconer DS, Mackay TFC, Cummings B (1996) Introduction to quantitative genetics, 4th edn. Benjamin Cummings, New York
Foote M, Miller A (2006) Principles of paleontology. W.H. Freeman & Co, New York
Futuyma DJ (2011) Expand or revise? The evolutionary synthesis today. Q Rev Biol 86(3):203–208
Futuyma D (2013). Evolutionary Biology, 3rd edn. Sinauer, Sunderland, MA
Garland T, Rose MR (eds) (2009) Experimental evolution. Concepts, methods, and applications of selection experiments. University of California Press, Berkeley
Gilbert SF (2012) Ecological developmental biology: environmental signals for normal animal development. Evolution and development 14(1):20–28
Gould SJ (1983) The hardening of the modern synthesis. In: Grene M (ed) Dimensions of darwinism. Cambridge University Press, Cambridge, pp 71–93
Gould SJ (2002) The structure of evolutionary theory. The Belknap Press of Harvard University Press, Cambridge
Gould SJ, Eldredge N (1977) Punctuated equilibria: the tempo and mode of evolution reconsidered. Paleobiology 3(2):115–151
Gould SJ, Eldredge N (1993) Punctuated equilibrium comes of age. Nature 366(6452):223–227
Hallam A (1989) What can the fossil record tell us about macroevolution? In: Hecht MK (ed) Evolutionary biology at the crossroads. Queen's College Press, Flushing, pp 59–73
Hanski IA, Giaggiotti OE (2004) Ecology, genetics and evolution in metapopulations. Academic Press, San Diego
Hartl DL, Clark AG (2007) Principles of population genetics, 4th edn. Sinauer Associates, Sunderland
Hull DL (1980) Individuality and selection. Annu Rev Ecol Syst 11:311–322
Hull DL (1988) Interactors versus vehicles. In: Plotkin HC (ed) The role of behavior in evolution. MIT Press, Cambridge, pp 19–50
Laland K et al (2014) Does evolutionary theory need a rethink? Nature 514(7521):161–164
Lawton JH, May RM (1995) Extinction rates. Oxford University Press, Oxford
Ledón-Retting CC, Pfennig DW (2011) Emerging model systems in eco-evo-devo: the environmentally responsive spadefoot toad. Evol Dev 13(4):391–400
Lewontin RC (1980) Theoretical population genetics in the evolutionary synthesis. In: Mayr E, Provine WB (eds) The evolutionary synthesis: perspectives on the unification of biology. Harvard University Press, Cambridge, pp 58–68
Lieberman BS, Eldredge N (2014) What is punctuated equilibrium? What is macroevolution? A response to Pennell et al. Trends Ecol Evol 29(4):185–186
Loreau M (2010) From populations to ecosystems. Theoretical foundations for a new ecological synthesis. Princeton University Press, Princeton
Louys J, Wilkinson DM, Bishop LC (2012) Ecology needs a paleontological perspective. In: Louys J (ed) Paleontology in ecology and conservation. Earth system sciences series. Springer, Berlin, pp 23–37
Mayr E (1942) Systematics and the origin of species. Columbia University Press, New York
Mayr E, Provine WB (eds) (1980) The evolutionary synthesis: perspectives on the unification of biology. Harvard University Press, Cambridge
Minelli A (1993) Biological systematics. The state of the art. Chapman & Hall, London
Minelli A, Fusco G (eds) (2008) Evolving pathways: key themes in evolutionary developmental biology. Cambridge University Press, Cambridge
Peterson AT, Soberón J, Pearson RG, Anderson RP, Martínez-Meyer E, Nakamura M, Araújo MB (2011) Ecological niches and geographic distributions. Princeton University Press, Princeton
Price PW (2003) Macroevolutionary theory on macroecological patterns. Cambridge University Press, Cambridge
Prothero DR (2003) Bringing fossils to life: an introduction to paleobiology. McGraw-Hill, Boston

Raup DM (1991) Extinction: bad genes or bad luck?. W.W. Norton & Co., New York
Ridley M (2003) Evolution, 3rd edn. Blackwell Scientific Publishing, Malden
Roff D (1997) Evolutionary quantitative genetics. Springer, London
Sepkoski D (2012) Rereading the fossil record. The growth of paleobiology as an evolutionary discipline. The University of Chicago Press, Chicago
Sepkoski D, Ruse M (eds) (2009) The paleobiological revolution: essays on the growth of modern paleontology. University of Chicago Press, Chicago
Simpson GG (1944) Tempo and mode in evolution. Columbia University Press, New York
Smith AB (2009) Systematics and the fossil record: documenting evolutionary patterns. Wiley, New York
Stanley SM (1979) Macroevolution: pattern and process. W.H. Freeman, San Francisco
Turner D (2011) Paleontology: a philosophical introduction. Cambridge University Press, Cambridge
Van Valen L (1973) A new evolutionary law. Evol Theor 1:1–30
Vrba ES, Eldredge N (eds) (2005) Macroevolution: diversity, disparity, contingency: essays in honor of Stephen Jay Gould. Paleontological Society, Lawrence
Wiley EO, Lieberman BS (2011) Phylogenetics: theory and practice of phylogenetics systematics. Wiley, Hoboken
Wilkins JS (2011) Species: a history of the idea. University of California Press, California
Wilkins JS, Ebach MC (2013) The nature of classification: relationships and kinds in the natural sciences. Palgrave MacMillan, New York
Wilson EO (2002) The future of life. Alfred A. Knopf, New York
Wood B (ed) (2011) Wiley-Blackwell encyclopedia of human evolution, vol 2. Wiley, New Jersey

Part I
Macroevolutionary Explanations and Interpretations

Can Modern Evolutionary Theory Explain Macroevolution?

Douglas J. Futuyma

Abstract Ever since the Evolutionary Synthesis of the 1930s and 1940s, some biologists have expressed doubt that the Synthetic Theory, based principally on mutation, genetic variation, and natural selection, adequately accounts for macroevolution, or evolution above the species level. Some questions pertain to the history of biological diversity, but the greatest argument has concerned the evolution of major changes in organisms' form and function. Such changes have been the subject of debate on the nature and phenotypic effect of mutations (especially the role of "macromutations" or saltations), the role of developmental mechanisms and processes, and the importance of internal constraints on adaptive evolution. Bridging the two major macroevolutionary themes, the hypothesis of punctuated equilibria invoked constraints on phenotypic evolution and the role of speciation in both diversification and the evolution of form. This chapter describes the Evolutionary Synthesis and the challenges to it and addresses the extent to which the modern formulation of the Synthetic Theory (ST) adequately addresses the observations that have prompted skeptical challenge. I conclude that although several proposed extensions and seemingly unorthodox ideas have some merit, the observations they purport to explain can mostly be interpreted within the framework of the Synthetic Theory.

Keywords Evolutionary synthesis • Punctuated equilibria • Evolutionary developmental Biology • Epigenetics • Genetic constraints • Adaptation

1 Introduction

From Darwin's time to the present day, biologists have debated the question of whether or not the evolutionary theory of the time suffices to explain "macroevolution." Before the Evolutionary Synthesis (ES), extending from about 1930 to 1950, few

---

D.J. Futuyma (✉)
Department of Ecology and Evolution, Stony Brook University, Stony Brook, NY, USA
e-mail: futuyma@life.bio.sunysb.edu

E. Serrelli and N. Gontier (eds.), *Macroevolution*,
Interdisciplinary Evolution Research 2, DOI 10.1007/978-3-319-15045-1_2

biologists, and almost no paleontologists, thought it did. According to the architects of the Synthetic Theory (ST) that emerged during the ES from the union of ideas and evidence from genetics, systematics, natural history, and paleontology, the processes of evolution within species, accumulated over time, explain the origin and diversification of higher taxa (Futuyma 1979, Mayr and Provine 1980). For about 20 years, rather little disagreement was audible, but challenges arose in the 1970s and 1980s that lay bare deep differences within evolutionary biology. For example, disagreements among some parties at a major conference on macroevolution in 1980 were so intense that proceedings of the conference were never published (Levinton and Futuyma 1982). Discontent at that time arose chiefly within paleontology and systematics, disciplines that have since achieved greater rapprochement with the Synthetic Theory, although the arguments raised by paleontologists have not been entirely settled. In the last decade or so, new calls for extension, reconsideration, or even repudiation of the ST have been issued, this time largely from developmental biology. In this essay, I will address several of the major challenges to the Synthetic Theory, ranging from the 1970s to the present. I will conclude that many of these challenges have had a positive impact on evolutionary biology, but that the fundamental principles of the ST remain valid, and can explain known evolutionary phenomena with only modest extension.

It would be useful to define "macroevolution," but definitions vary. Simpson (1944, p. 97) wrote that "Micro-evolution involves mainly changes within potentially continuous populations…[whereas] Macro-evolution involves the rise and divergence of discontinuous groups." In *Evolution Above the Species Level*, Rensch (1959, p. 1) objected to the lack of a clear borderline between "larger" and "smaller" events (and to the hybridization of Greek and Latin roots) and referred instead to "infraspecific" (referring to processes that occur within a species or lead to a new species) and "transspecific" evolution (referring to processes that "lead to new genera, families, and lesser divisions, and thus to new constructional types"). Rensch thus focuses on the evolution of characters of individual organisms that distinguish taxa above the species level. (Levinton (2001) is among modern authors who adhere to this usage.) For many authors, however, "evolution above the species level" also includes patterns and causes of diversification of higher taxa, such as variation in diversity, speciation rates, and extinction among clades or geographic regions or geological periods.

I must at this point emphasize that I am neither a historian nor a philosopher and cannot address many questions that arise in those contexts. For example, I am hesitant to say whether or not the ST *explains* macroevolution, because I do not know what "explain" means. By "explanation," I usually mean consistency of *explananda* with a set of postulated, sufficient causal processes. Others may require that an explanation enables prediction of the *explananda*, such as prediction of macroevolutionary diversification from a theory of mutation and natural selection. Current evolutionary theory cannot provide so grand a prediction, but it often can predict patterns (e.g., that mitochondrial mutations are more harmful to males than females, on average Innocenti et al. 2011) or very short-term responses to selection. By way of analogy, all meteorological phenomena are manifestations of physical principles, but you will be disappointed if you expect physics to predict the weather in your location a month from now.

1.1 Background: The Evolutionary Synthesis and Its Aftermath

In order to appreciate discussion of the sufficiency of today's evolutionary theory, we must be familiar with the Evolutionary Synthesis, which familiarity itself requires a glance further back. Many historians recognize three major stages in the development of evolutionary theory: Darwinism (from 1859 until about 1898) in which natural selection among "random" variations (meaning undirected with respect to need) was urged as the most important but not sole cause of evolution (for some, inheritance of acquired characters was allowed); neo-Darwinism (from about 1898), referring to August Weismann's and Alfred Russel Wallace's complete rejection of Lamarckian inheritance in favor of selection as the sole cause of evolution; and the Synthetic Theory, which in my view extends from about 1930 (with the publication of Fisher's *The Genetical Theory of Natural Selection*) to about 1950 (with Stebbins's *Variation and Evolution in Plants*). (The definition, temporal extent, major players, and content of the Evolutionary Synthesis, or Modern Synthesis are all debated by historians.)

The major elements of the ST, which remain major elements of evolutionary theory today, include (1) the units of evolution are populations of organisms, not types or single organisms ("population thinking"); (2) evolution is based on mutations that are random with respect to the adaptive needs of the organism (but not necessarily random in other respects), resulting in inherited variation that may be amplified by recombination; (3) natural selection (at the level of individual organisms), acting on inherited variation, is the major cause of evolution of adaptive characteristics; (4) changes in the genetic composition of populations can also result from random genetic drift, especially in small populations; (5) new species are formed by divergence between populations of an ancestral species, owing to factors that reduce or prevent gene flow between populations that undergo different evolutionary changes; (6) gradual accumulation of changes by these same factors results in character differences that distinguish higher taxa, i.e., macroevolution (Reif et al. 2000; Kutschera and Niklas 2004). In particular, as embodied in the equations of theoretical population genetics, the theory was cast in very general terms. "Selection" is not identified with any specific mode or agent (and so could include ecological sources of selection, sexual selection, the "internal selection" stemming from functional interactions among characters Schmalhausen 1949, and genic selection owing to factors such as meiotic drive). "Mutations" are any kind of reasonably stable alternatives ("allelomorphs") to a prevailing unit of heredity; the equations for the dynamics of mutations in populations apply equally to what we now identify as single-base pair substitutions (whether in structural or regulatory sequences), chromosome inversions, polyploids, and even epigenetic "mutations." These broad concepts lack mechanistic content; empirical data are needed to describe real instances of evolution, such as the agents of selection and the molecular and developmental basis of phenotypic variants. Thus, the conception of causes of evolution embodied in the Synthetic Theory, i.e., gene frequency change, is quite different from the causes of differences in morphology, physiology, or behavior that are commonly envisioned by mechanistic developmental biologists, physiologists, or neurobiologists (cf. Amundson 2005).

The Evolutionary Synthesis was both a synthesis (especially of genetics and natural selection) and a “constriction” (Provine 2001). The seeming exclusivity of the ES can be understood (and excused, if deemed necessary) only by appreciating the state of evolutionary discourse in the early twentieth century (see Simpson 1944; Rensch 1959; Bowler 1983; Reif et al. 2000). Darwinism was in “eclipse” (Huxley 1942; Bowler 1983), in that almost no biologists accepted natural selection as a significant agent of evolution. (The exceptions were chiefly some of the naturalists.) Almost nobody had attempted to measure selection in natural populations, so it simply had not been documented. Many biologists doubted that organisms’ characteristics are adaptive; Robson and Richards (1936), for instance, devoted much of their book to the thesis that differences between related species are nonadaptive. Selection was thought of as a “random,” undirected process, so “orderly” phenomena such as trends and parallel or convergent evolution were thought to refute evolution by natural selection. After centuries of a theological world view that included divine design and purpose, many morphologists were “idealists” who held a Platonic interpretation of each species’ form as “an element in the overall pattern imposed by Mind upon the material world” (Bowler 1983, p. 47; Winsor 2006; Amundson 2005; and others disagree). Moreover, to those who thought in terms of purpose, Darwinian selection was far less appealing than theories that did not include struggle for survival, and in which organisms could be viewed as active agents, directing their own evolution (Bowler 1983, p. 15). Among these was “neo-Lamarckism,” especially popular and long lasting among paleontologists, even after geneticists had refuted and abandoned “soft inheritance.” Lamarckism, in which organisms direct their evolution by use and disuse of certain organs, was related to Haeckel’s recapitulation theory (for ontogeny displays “progress” toward the “goal” of the adult organism), and both of these to orthogenesis, the belief (again persistent among paleontologists) that evolution is driven by irresistible internal factors in specific directions; in some versions, the drive is inexorable progress, while in others it involves momentum that carries the species into maladaptive degeneration and extinction. One might imagine that the geneticists, having disposed of two arguments against the efficacy of Darwinian selection (Lamarckian inheritance and blending inheritance), would have been staunch Darwinians, but Hugo de Vries and Thomas Hunt Morgan, founders of genetics, instead interpreted mutations as a sufficient cause of evolution. Early in his career, Morgan thought that species arise simply as mutations; natural selection simply eliminated mutations that were unfit. If selection explained anything, it was adaptation, not the origin of species—but he denied that most characteristics were adaptations (Bowler 1983, p. 198). A more extreme mutationism was voiced by some paleontologists, such as (Schindewolf 1950; cited by Simpson 1953a, b), and most notoriously by the (otherwise respected) geneticist Richard Goldschmidt (1940), who considered gene mutations and selection instrumental within species, but argued that species and higher taxa originate by an entirely different process, involving a major reconfiguration of the genome. Such a “macromutation” would often, perhaps usually, yield a hopelessly dysfunctional organism, but occasionally a coherent, adapted "hopeful monster" instead. Thus, Goldschmidt proposed evolution by saltation, i.e., a "large" discontinuous change in one or more characteristics that arises in a single generation.

Those who today disparage the Evolutionary Synthesis as a constrained, dogmatic assertion that evolution consists only of natural selection on random genetic mutations within species must recognize that the authors of the Synthesis were responding to an almost complete repudiation of natural selection, adaptation, and coherent connection of macroevolution to these processes. Macroevolution, in particular, was explained by Lamarckian modification, orthogenesis (for which no mechanism was ever articulated), and saltation (mutationism). It is instructive, then, to glance at some of the main arguments presented by the contributors to the Evolutionary Synthesis.

1.2 The Content and Authors of the Evolutionary Synthesis

The best known contributors to, or "architects" of, the Evolutionary Synthesis (sensu lato) are R.A. Fisher, J.B. Haldane, Sewall Wright, Theodosius Dobzhansky, Julian Huxley, Ernst Mayr, George Gaylord Simpson, and G. Ledyard Stebbins. Bernhard Rensch is rightfully placed in this company by those who know his work (see below), but that number is regrettably dwindling. A considerable number of other authors should be credited with major conceptual or empirical contributions, especially in Germany and Russia (Adams 1980; Reif et al. 2000). To mention only a few, in Russia, Sergei Chetverikov was a founder of population genetics, and I.I. Schmalhausen integrated natural selection with genetics and some aspects of development. Nikolai Timofeeff-Ressovsky did pioneering, insightful work on genetic variation in natural populations in Russia and later in Germany. Experimental population genetics was initiated by Georges Teissier and Philippe L'Héritier in France. To the well-known names in England should be added, at least, the cytogeneticist C.D. Darlington, author of *The Evolution of Genetic Systems* (1939), E.B. Ford, who with his students created ecological genetics and applied Fisher's theory to real genetic data, and Gavin de Beer, who in his many books (e.g., *Embryos and Ancestors*, 1940) used comparative embryology to dismantle Haeckel's recapitulation theory (the "biogenetic law") and interpret macroevolutionary changes in form. In the United States, major contributions to the genetic aspects of evolution came from Herman Muller and from the botanists Edgar Anderson (author of *Introgressive Hybridization*, 1949), E.B. Babcock (Smocovitis 2010), and the famous trio of Clausen et al. (1948). Non-Darwinian views of evolution in Germany were countered by many adherents to Darwinism, informed by genetics (Reif et al. 2000). Erwin Baur, Max Hartmann, Wilhelm Ludwig, and Alfred Kühn, among others, developed arguments for evolution by natural selection of genetic variants that conformed fully to the ST as it developed in England and the United States. As early as 1930, in *Die Phylogenie der Pflanzen, ein Überblick über Tatsachen und Probleme*, the botanist Walter Zimmermann argued against idealistic morphology, typology, Lamarckism, and saltation, interpreted plant characters as adaptations formed by natural selection, and "single-handedly accomplished a synthesis years before other synthesists" (Reif et al. 2000). (Zimmermann had almost no impact, even during his lifetime, because of academic politics and because he embedded his arguments in scientific philosophy that Simpson (1949), for one, found hard to read.) The broadly trained zoologist Gerhard Heberer edited a book (*Die Evolution der Organismen*, 1943) in which he and most of the 18 other authors argued for the gradual evolution of higher taxa, and against saltation, Lamarckism, and orthogenesis.

Most of the well-known "architects" of the ES addressed aspects of macroevolution to at least some extent. Huxley (1942) sketched the newly forming theory most broadly in the book that gave the Evolutionary Synthesis its name; his most significant personal contribution was probably his formulation of allometric growth (unequal growth rates of different features or dimensions), which was used, by Haldane among others (Haldane 1932b), to explain some apparently nonadaptive characters—and which, incidentally, illustrates an awareness of the importance of development in evolution. Wright (1932) intended his quite abstract "shifting balance theory," with the "adaptive landscape" as its metaphor, as a theory of long-term progressive evolution, and his landscape metaphor was adopted by Dobzhansky, Simpson, and subsequently many others. Dobzhansky (in *Genetics and the Origin of Species*, 1937) said almost nothing about macroevolution, but drew attention to Wright's and Fisher's theoretical arguments, including the efficacy of even very weak selection. He marshaled most of the existing evidence of the operation of natural selection, and it is striking to read how few his examples were—and how many more he could cite 14 years later (Dobzhansky 1951). Among the major themes in *Systematics and the Origin of Species* (Mayr 1942) is geographic variation: its nature, adaptive significance, and the evidence it provides of the gradual evolution of species. Mayr emphasized the uncertain borders of many genera as evidence of continuity of divergence and cited many examples of species that are clearly closely related but were assigned to different genera on the basis of one or a few character differences. In a final chapter on "The higher categories and evolution," he listed seven factors that "deprive the macroevolutionary processes of much of their former mysteriousness," including "the smallness and frequency of mutations," pleiotropy, the polygenic basis of traits, allometric growth, and the power of selection (citing Fisher). He closed the book by stating, "all the processes and phenomena of macroevolution and the origin of the higher categories can be traced back to intraspecific variation, even though the first steps of such processes are usually very minute." Complementing Mayr's book that was written "from the viewpoint of a zoologist," Stebbins (in *Variation and Evolution in Plants*, 1950) described macroevolutionary patterns in plants (e.g., fusion of flower parts) and interpreted them both in terms of likely adaptive value and developmental mechanism, a dual approach that he revisited in *Flowering Plants: Evolution Above the Species Level* (1974). In the later book, he listed many instances in which a diagnostic feature of a higher taxon is found as a difference between congeneric species in other plant families, illustrating that evolutionary changes at different taxonomic levels do not differ in kind.

The most widely known treatment of macroevolution during the Synthesis is *Tempo and Mode in Evolution* (1944), by the vertebrate paleontologist George Gaylord Simpson. By interpreting patterns in the fossil record in terms of genetics, Simpson achieved a remarkable union of widely disparate disciplines. He cited both genetic data (e.g., the adequacy of mutation rates to account for rates of phenotypic evolution in the fossil record, the polygenic basis of most phenotypic traits) and geological data to explain apparent saltations in fossil series, but agreed that the common absence of forms transitional to high-level taxa (e.g., orders, classes) requires a special explanation. He postulated "quantum evolution," a forerunner of punctuated equilibria, in which new forms evolve very rapidly as they adapt to very different habitats or ways of life. Intermediate fossils would not be found "if the animals involved in the transitions were relatively few in number and if they were evolving at unusually high rates" (p. 117). This was not a saltationist hypothesis, for "in general the genetic processes involved do not permit making the step with a single leap" (p. 210). Having dispatched saltationism, Simpson addresses "inertia, trend, and momentum" in a multifaceted attack on orthogenesis. Rectilinear evolution does occur, but is far from universal, for most clades show a pattern of branching, diversifying in different directions. Mutation can be biased in certain directions, but appears not to coincide "with the direction in which the group is really evolving." Progressive, rectilinear change is most consistent with persistent natural selection (for example, increasing tooth height in grazing horses). Apparent momentum can be produced by selection in many ways, such as the effect of selection on two correlated characters that reach their optima at different times. Simpson does not claim to have demonstrated that particular evolutionary events had these causes, only that they are realistic possibilities, consistent with genetic data and theory, in contrast to Lamarckian inheritance or the undefined, almost mystical factors invoked by supporters of orthogenesis.

Bernhard Rensch, in my opinion, is the great unsung hero of the Evolutionary Synthesis; it is a great misfortune that he is so poorly known, especially in English-speaking countries. Although he was a Lamarckian early in his career, he soon became a neo-Darwinian, did extensive research on geographic variation in land snails, lizards, birds, and mammals, and formulated the well-known Bergmann's, Gloger's, and Allen's "rules" that provided important evidence for adaptation and natural selection. Anticipating much later research, he experimentally altered the color pattern of birds' eggs, found that more markedly altered eggs elicited more frequent rejection by the parents, and interpreted the egg polymorphism of the brood-parasitic cuckoo (*Cuculus canorus*) as an adaptive response to egg rejection. He published major papers in 1939 on *Typen der Artbildung* (kinds of speciation) and in 1943 on *Die paläontologischen Evolutionsregeln in zoologischer Betrachtung* (paleontological rules of evolution from a zoological viewpoint). He worked on his book, *Neuere Probleme der Abstammungslehre* (recent problems of the theory of evolution), during World War II, but was unable to publish it until 1947; only after correcting proofs did he see for the first time the books by Huxley (1942), Mayr (1942), and Simpson (1944). He took these into account in the second edition that was published in German in 1954 and finally appeared in English, as *Evolution Above the Species Level*, in 1959.

Rensch's treatment of macroevolution is, I think, more impressive, and certainly more multifaceted, than Simpson's. He not only counters Goldschmidt's saltationism by citing abundant gradations from geographic races to species to genera, but he also provides deeply insightful analyses of the major problems of macroevolution (arguing against saltation, orthogenesis, and neo-Lamarckism) by summarizing paleontological studies and especially evolutionary patterns revealed by comparative morphology and embryology—an approach that Simpson did not take. In reviewing the first edition of Rensch's book, Simpson (1949) lavished praise. Although Rensch is not a paleontologist, Simpson wrote, he provides interpretations that may be "commended to paleontologists as examples of how…to understand the facts of their subject." Simpson praised the book for "an extraordinary richness of pertinent examples and for clearly reasoned interpretation, … so packed with well-integrated information that summary is impossible."

Rensch interprets the temporal course of clade diversification (cited by paleontologists as an inherent "life cycle") as adaptive proliferation when a lineage adapts to new habitats that are relatively free of competitors; the rate of diversification may last for more than 100 million years (not a brief, vigorous "adolescence"), and declines, he suggests, because competition increases with the number of species. Dollo's "law" of irreversibility (a mainstay of orthogenesis) has many exceptions, but there is seldom reversal to exactly the ancestral character, because during the interim, the "whole organism of the animal has undergone change:" any reversal has to be functionally integrated with the entire system. Examining development and the "mechanisms of construction" shows that many possibly nonadaptive features may be ascribed to allometry or to the multiple effects of hormones. Increases in body size carry with them changes in such features as the number of retinal cells or brain cells, which may support new functions and even be the basis of selection on size. Parallel evolution may arise from similar hereditary factors and development, as seen in lepidopteran wing patterns, or from similarity of natural selection, as in the longer wings of diverse migratory bird species compared to nonmigratory relatives. Apparent orthogenesis, as exemplified by Cope's rule of increase in body size, can readily be caused by selection or by correlated growth (especially allometry). Adaptively novel clades (e.g., of arboreal mammals) must originate from small, unspecialized ancestors, not giant forms. (This argument anticipates Stanley's 1973 explanation of Cope's rule of size increase.) Alterations of embryonic development (described in a 27-page passage that draws on comparative and experimental embryology) show that ontogeny can be altered in so many ways that the direction of evolution cannot be set by autonomous factors. "Jumps" in the fossil record of the origin of new "structural types" are explained by failure to fossilize (the then recently discovered living coelacanth *Latimeria* strikingly shows the incompleteness of the fossil record[1]), by geographic shifts in the distribution of new species that evolved elsewhere, and by the accelerated evolution and strong directional selection of lineages that adapt to new habitats or ways of life. (Cf. Simpson's "quantum evolution" and Eldredge and Gould's "punctuated equilibria.") New organs are usually modifications of features that evolved long before (e.g., mammalian ear ossicles, derived from jaw bones), and probably evolved by successive small changes (since most "large" mutations are harmful in *Drosophila* and other species). Rensch cautioned that "I do not wish to deny the possibility that some day further evidence of the evolutionary effects of macro-mutation may come to light" (p. 106), but he concludes (p. 358) that the wealth of forms that compose a single, giant tree of life "is the result of continuous, undirected mutation and is patterned by the respective conditions of selection," and that "if there are some special problems to which we can only say 'Ignoramus' [we do not know], we need not add 'Ignorabimus' [we shall not know]."

In my view, the major contributors to the Synthesis marshaled available evidence (which on some points, such as the prevalence of natural selection, was strikingly sparse) logically and effectively in support of gradual evolution, chiefly by natural selection acting on undirected mutations of mostly small magnitude. The Synthesis "architects" successfully banished orthogenesis and, together with geneticists, Lamarckian mechanisms of change. They replaced mutationism with "population thinking," although they did not (and could not, as the quotation from Rensch admits) demonstrate that "macromutations" never contribute to the evolution of major change in form. They did not address all evolutionary phenomena, by any means; they said little about patterns and causes of extinction, for example, and Mayr (1960) noted, in a famous essay on evolutionary novelties, that "the problem of the emergence of evolutionary novelties has undoubtedly been greatly neglected during the past two or three decades." As is widely recognized, ecology, morphology, developmental biology, and phylogeny received little attention during the Synthesis, relative to genetics.

Empirical evolutionary research in the 1950s and 1960s greatly increased information on genetic variation in natural populations, the seeming ubiquity of natural selection (Endler 1986, Table 5.1 lists at least 85 studies of "natural selection in the wild" from these decades), and speciation. Major theoretical advances included the articulation of kin selection, the distinction between individual selection and group selection (the latter still a contentious issue), and the consequent development of theory, based on individual and kin selection, to explain classes of characteristics such as life history traits and social behaviors. These developments may well have led to "hardening" of evolutionary thinking

[1] The group of lobe-finned fishes known as coelacanths was thought to have become extinct in the Cretaceous until a living species, named *Latimeria chalumnae*, was described from the Indian Ocean in 1938.

around selection as an almost exclusive factor of evolution, as I will note later. (Gould's 1983 claim that the Synthesis itself became more exclusively selectionist has not been rigorously analyzed Reif et al. 2000.) But the all-important role of selection was challenged by interpretations of molecular polymorphism and evolution in neutralist terms (King and Jukes 1969; Kimura 1968, 1983). Students of phenotypic evolution nevertheless tended to remain convinced that the features they studied evolved mostly by natural selection, and Kimura himself agreed that this is likely the case. (Remarkably, Wright disavowed the role of genetic drift in any but very small populations and was little interested in Kimura's theory, because "the condition that gives the maximum amount of such drift is that of complete neutrality and hence of no evolutionary significance" Provine 1986, p. 472.)

Evolutionary biology since about 1970 has seen immense growth and integration with other areas of biology (e.g., ecology, behavior, physiology, developmental biology, and especially molecular biology). The ST has proven flexible because it was cast in general terms that could be easily honed to describe specific, newly discovered phenomena such as codon bias and transposable elements. Massive evidence of selection and adaptation was revealed not only by demographic studies of the kind that Endler (1986) and Kingsolver et al. (2001) summarized, but also by "signatures" of selection in DNA sequences, experimental evolution (chiefly in laboratory cultures of microorganisms), the revival and documentation of sexual selection, and the frequent fit of data to adaptive models of life history, behavioral, physiological, and morphological traits. Phylogenetic inference became increasingly rigorous and reliable and is now a major element in evolutionary biology, appreciated not only as a reconstruction of some aspects of evolutionary history, but also as an analytical approach to inferring some evolutionary processes. Evolutionary studies became increasingly quantitative and increasingly compared data against neutral (random) null models. Evolutionary biologists became increasingly cognizant of mechanistic biology: It is necessary to know some molecular biology in order to interpret molecular data. At the same time, there has been a resurgence of challenges to the ST (Depew and Weber 2013), with calls to expand the ST (e.g., by recognizing selection at different levels), to extend it (by incorporating other processes and other fields of study), or to replace it. Paleontologists have been most conspicuous in challenging the ST, but developmental biologists and a few "neontological" evolutionary biologists have also issued calls for change, in some instances echoing the paleontologists. Stephen Jay Gould, the most incessant and articulate critic of the ST, played all three roles. Most of these calls for change explore or advocate explanations other than individual selection within populations, which is commonly viewed as too exclusive a theory of evolutionary process. In the remainder of this essay, I will comment on four contentious issues: (1) alternatives to gradualism, (2) internal constraints on adaptation by natural selection, (3) challenges from developmental biology, and (4) long-term changes in diversity.

2 Alternatives to Gradualism

Two major alternatives to gradualism have been posed since 1970, both led by Gould: a revived dalliance with saltation (Gould 1980) and the model of punctuated equilibria introduced by Eldredge and Gould (1972) and elaborated and defended, especially by Gould, in many later publications (see Gould 2002). These are entirely different propositions.

2.1 Macromutation

Gould (1980) envisioned a discontinuity between intraspecific evolution and the origin of new species (which he named the "Goldschmidt break") and advocated a more favorable reconsideration of a role for discontinuous, macromutational changes in the evolution of major character changes. In an introduction to a 1982 reprint of Goldschmidt's notorious *The Material Basis of Evolution*, Gould wrote "I find [Goldschmidt] not victorious, but weighted equally with his self-proclaimed Darwinian opponents." (The book was reviewed in *Paleobiology* by four reviewers, among whom Charlesworth (1982) and Templeton (1982) wrote scathing criticisms of Goldschmidt that included highly unflattering comments on Gould's advocacy.)

"Macromutation" has been used with a variety of very different meanings: For some, such as Goldschmidt, it is manifested as the origin of a radically altered character or set of characters (or a major morphological remodeling, as expressed by Schindewolf's (1950) famous speculation that the first bird emerged from a reptile's egg). For other authors, a macromutation merely causes a discrete difference in a character, the magnitude of which need not be specified. An evolutionary role for single discrete mutational changes of single characters, of substantial magnitude, has been admitted from the beginnings of the Synthesis. Haldane (e.g., 1932a), for example, suggested that evolutionary "jumps" could arise by a variety of processes, such as hybridization, polyploidy, and the substitution of fairly "large" mutations, followed by modifier alleles with small effects. Fisher (1930) suggested that the latter model would account for data on inheritance of mimetic color patterns in butterflies: A major "switch" gene decides between two alternatives, each of which may be modified later in evolution by other substitutions. This model was adopted and supported by mimicry researchers such as Philip Sheppard (a student of E.B. Ford) (Sheppard et al. 1985) and by the population geneticist Charlesworth (1980), a vocal defender of the ST.

It is true that by the early 1980s, it was widely thought that almost all the allele substitutions underlying variation in polygenic traits had very small effects, but on closer examination, it became clear that many character differences between closely related species are based on fewer gene differences, of larger effect, than previously supposed (Gottlieb 1984; Orr and Coyne 1992). Nevertheless, quite a few genes contribute to such differences. For example, a bee-pollinated and a hummingbird-pollinated species of *Mimulus* have very different flowers that differ in at least 12 features. Analysis of interspecific crosses documented one to six quantitative trait loci (QTL) contributing to each trait. In nine traits, at least one QTL accounted for more than 25 % (but always less than 50 %) of the variance (Bradshaw et al. 1998). The authors interpreted the data as meaning that genes of large effect can contribute to speciation. However, the traits were affected by different QTL, a total of 47 QTL were detected, each QTL might well be a cluster of genes rather than a single gene, and the considerable unexplained variance surely is attributable, in part, to many loci with effects too small to be detected by the rather coarse genetic analysis. This is a far cry from a macromutation that might be imagined to underlie a major change in flower form. To be sure, it had been recognized since before the Synthesis that evolution of major differences in form could sometimes arise from changes in development that ensured coherent, integrated change in multiple traits. The chief example was paedomorphosis, as exemplified by salamanders (*Ambystoma*), known as axolotls, that retain larval features in the reproductive adult stage.[2] The difference between the metamorphic and paedomorphic life cycles is closely associated with a major gene that affects delay of metamorphosis, but other genes clearly contribute to the threshold that determines which developmental mode is expressed (Voss and Smith 2005). Of course, paedomorphosis is merely a change in timing of development of features that are the product of a very long history of, possibly, entirely gradual evolution.

Whether or not major, discontinuous single-step changes in phenotype have occurred in evolution is an empirical question. No mathematical theory excludes the possibility; Fisher's (1930) famous geometrical model, often cited as an argument against macromutation, is a metaphor, not comparable to, say, models of the conditions for stable polymorphism. The manifold effects of polyploidy (Levin 1983; Ramsey 2011) might be considered macromutational; possibly newly established endosymbioses will likewise have large but beneficial effects. The most intriguing possibilities are raised by developmental genetics, in which major regulatory genes have "coopted" different developmental pathways, or have different spatial expression and so are associated with major morphological differences among taxa. (For example, the somites that develop into different classes of vertebrae differ among major vertebrate taxa, apparently caused by differences in the expression domains of certain *Hox* genes.) It is conceivable that a single mutational change in the association between a regulatory gene and a developmental pathway accounts for such cases, but it is also possible that it happened incrementally.

---

[2] This example loomed large in Gould's first book, *Ontogeny and Phylogeny* (1977). It is with some amusement that I read in my review (entitled, with some slight pretension, "The axolotl as Parsifal") that when evolutionary biology embraces developmental biology, "it is then most likely to prove that slight changes in the regulatory systems of development—micromutations, no doubt, at the molecular level—can be amplified by exponential growth and allometric relations, and by the profusion of developmental effects that we call pleiotropy, into major phenotypic changes, some of which will seem discontinuous" (Futuyma 1978, p. 43). Both of these anticipated results, especially discontinuity, still await discovery.

An instructive example is the gene *shavenbaby*, responsible for a difference between *Drosophila* species in the presence or absence of larval trichomes: The difference is based not on a single mutation, but on a combination of mutations in three different enhancers (Stern 2011). Fisher (1930, p. 164), in the discussion of mimicry that I have already cited, noted that a single gene determines sex in some fishes, but that we would not suppose that the various adaptations of one sex have arisen by a single saltation from the other sex. This example, he wrote, emphasizes "that it is the function of a Mendelian factor to decide between two (or more) alternatives, but that these alternatives may each be modified in the course of evolutionary development, so that the morphological contrast determined by the factor at a late stage may be quite unlike that which it determined at its first appearance." It is certainly possible that instances will be found in which a functionally coherent set of character alterations will be found to have originated by one or a few mutations that affect development. So far, however, it seems as if the bulk of evidence continues to favor the view that phenotypic characters generally evolve more or less independently at different rates (mosaic evolution) and by multiple, polygenic substitutions.

2.2 *Punctuated Equilibria*

The proposition that Eldredge and Gould (1972) dubbed punctuated equilibria (PE) has often been confused with macromutational saltation, but it is entirely different. PE refers both to a pattern that Eldredge and Gould claimed is common in the fossil record, and to a proposed process that would, they said, explain that pattern. The pattern is rapid shift between one long-lasting, virtually constant phenotype and another. The shift is typically not documented by intermediate fossils, but the geological interval during which the shift occurs is typically on the order of thousands of years, long enough for appreciable evolution by standard processes (Stebbins and Ayala 1981; Hunt 2010). Both in 1972 and afterward, both Gould and Eldredge emphasized that character change during the shift (the "punctuation") may well be gradual, i.e., a continuous change in mean character state, caused by natural selection acting on undirected variation. The radical feature of Eldredge and Gould's proposed process is that during the long periods of constancy ("stasis"), populations cannot readily respond to natural selection because of genetic constraints (in the form of epistatic interactions among genes), constraints that might be loosened when a population undergoes a bottleneck in size. Genetic drift might then initiate evolution toward a different genetic equilibrium, which they envisioned to be a different species, reproductively isolated from its more widespread "parent" species. Thus, character evolution occurs chiefly in concert with, and is caused by, speciation—bifurcation of an ancestral lineage into two reproductively isolated descendants. The new, modified "daughter" species originates as a small, geographically localized population, in which the evolutionary transition from one optimal phenotype to another occurs rapidly. Its existence can be preserved and documented in the fossil record only if it eventually expands its geographic range, perhaps supplanting its ancestor (the "parent" species) as it does so. This hypothesis, they suggested, accounts for the paucity of cases of steady transformation of lineages (which they labeled "phyletic gradualism") and for gaps in morphology that have plagued evolutionary biologists from Darwin on. (Note, however, that the phenotypic gaps in this model are small; PE describes not the origin of higher taxa with novel features, but closely related, similar species. PE was born, in large part, from Eldredge's study of the trilobite *Phacops rana*, in which ancestral and descendant forms are distinguished by a small difference in the number of rows of eye lenses.)

Eldredge and Gould argued that if a character evolved only during the origin of a new "daughter" species, and if the direction of character evolution depends only on local selection that is unlikely to be correlated among successive speciation events, widely separated in time and space, individual selection would produce only random fluctuations in the character, averaged over the members of a clade. Eldredge and Gould concluded that long-tern trends in characters should therefore be attributed not to individual selection within species, but selection at the species level: association of character states with rates of speciation or species extinction (the species-level analogs of birth and death). Paleontologist Stanley (1975) phrased the same argument more dramatically, claiming that macroevolution is decoupled from microevolution. This argument epitomized the rebellion against the ST that started in the 1970s.

Eldredge and Gould's hypothesis was not entirely original. As one of several possible reasons for the embarrassing paucity of transitional forms in the fossil record, Darwin (1859, p. 306 of 1979 reprint) wrote: "One other consideration is worth notice: with animals and plants that propagate rapidly and are not highly locomotive, there is reason to suspect, as we have formerly seen, that their varieties are generally at first local; and that such local varieties do not spread widely and supplant their parent-forms until they have been modified and perfected in some considerable degree. According to this view, the chance of discovering in a formation in any one country all the early stages of transition between any two forms, is small, for the successive changes are supposed to have been local or confined to one spot." Rensch (1959, p. 106) wrote that although paleontologists often invoke macromutations to account for "saltatory deviations" in fossils from successive geological horizons, "these 'saltations' are probably due to horizontal shifts of geographic races or closely related species." Eldredge and Gould (1972) stated clearly that their model was an application to the fossil record of Mayr's (1954) founder-effect model of speciation (which Mayr later dubbed "peripatric speciation"). Mayr, in fact, had already made this application explicit, asserting that a locally formed species will invade new areas, and "only then will it become widespread and thus likely to be found in the fossil record. But then it is already too late to record the evolutionary change through which it has gone. All the paleontologist finds is the fact that one widespread numerous species was replaced or succeeded by a rather different species…" (reprinted in Mayr 1976, p. 207).

Mayr (1954, 1963) had suggested that adaptive evolution might be slow in large populations because of what evolutionary geneticists call epistasis for fitness: Many mutations fail to increase because they have deleterious interactions with many of the vast number of genetic backgrounds in which they are distributed in a large, highly polymorphic population. He proposed that changes in allele frequencies due to sampling error (genetic drift) at some loci in a population founded by a few individuals, together with the population's reduced genetic variation, would change the "genetic environment," in which certain alleles would confer high fitness that would not do so in the large "parent" population. This might be a snowballing process (a "genetic revolution"), leading to such great change as to form a new, reproductively isolated species, rapidly and in a localized area. This was envisioned as a process of evolution by natural selection, but the selection was "internal," not necessarily imposed by environmental change. When population geneticists later modeled their interpretation of Mayr's verbal model (and rather similar verbal propositions by Carson and Templeton (1984), they found it almost indistinguishable from Wright's 1932 et seq.) shifting balance theory. Like Wright, they said, Mayr appears to envision a shift between peaks on an adaptive landscape, requiring that selection against departure from the original peak be countered by genetic drift. All the population geneticists who modeled this process agreed that this was very unlikely unless selection is very weak; that is, the "valley" between adaptive peaks is very shallow (Charlesworth and Rouhani 1988; Barton and Charlesworth 1984). But a shallow valley would imply, they said, that the fitness of hybrids between "parent" and "daughter" populations would be quite high; hence, reproductive isolation would be very weak. The majority of population geneticists judged Mayr's peripatric speciation unlikely, and some rejected Wright's shifting balance theory as well (Coyne et al. 1997; but see Wade and Goodnight 2000). Furthermore, they found no genetic evidence, based on allozymes and early DNA data, that speciation is associated with reduced population size. Charlesworth (1984) argued that seemingly unchanging characters actually show substantial fluctuations around a nontrending long-term average, suggesting that they are able to evolve, but are subject to long-term stabilizing selection (see also Gingerich 1983). Thus, evolutionary geneticists criticized the theory on which Eldredge and Gould based punctuated equilibria theory; they rejected genetic constraint as an explanation of stasis; they rejected the proposition that character evolution depends on speciation; and they vigorously defended the ST (Charlesworth et al. 1982).[3]

---

[3] I referred earlier to the intense controversy at a symposium on macroevolution held in Chicago in October 1980. I attended the symposium, and was scheduled to become editor of *Evolution* three months later. When it became evident that no proceedings of the symposium would be published, I invited several participants, representing diverse views, to contribute manuscripts to *Evolution*. This paper, "A neo-Darwinian commentary on macroevolution," as well as a paper by Steven Stanley on macroevolution and the fossil record, a paper on evolution and development by George Oster and Pere Alberch, and a paper by Sewall Wright (based on his plenary address) were the result.

As to evidence, the little evidence from studies of closely related species, based on application of coalescent theory to DNA sequences, generally suggests that speciation has not been associated with bottlenecks in population size (e.g., Rovito 2010; Yeung et al. 2011). Whether or not "punctuational" changes in fossil lineages are associated with biological speciation or are simply episodes of rapid evolution within single, nondividing lineages ("punctuated gradualism") is still unclear. In paleontological taxonomy, "species" are morphologically distinguishable named units, either successive stages in a single evolving lineage ("chronospecies") or reproductively isolated forms (biological species). Temporal overlap between "parent" and "daughter" forms is the best evidence that they represent cladogenesis (biological species) rather than "chronospecies." Although punctuated gradualism has been claimed for some lineages of planktonic Foraminifera, which provide exceptionally complete fossil records (e.g., Malmgren et al. 1983), Gould and Eldredge (1993) claimed that many studies find temporal overlap, supporting their model. A recent comprehensive analysis of 337 Cenozoic "speciation" events in Foraminifera concluded that at most 19 % of Cenozoic events (last 65 million years) and 10 % of Neogene events (last 23 million years) represented change within nondividing lineages: The great majority revealed temporal overlap, and hence biological speciation (Strotz and Allen 2013). Analyses of living species can also shed light on the question. Starting with Avise (1977) and Ricklefs (1980), several authors noted that (controlling for clade age) the total amount of phenotypic variation ("disparity") among species in a clade should be correlated with the number of species according to the PE model, but not if phenotypic evolution is independent of speciation. Statistical methods for testing this hypothesis have been developed only recently (Bokma 2010; Magnuson-Ford and Otto 2012), but have indicated that most of the evolution of body size in mammals (Mattila and Bokma 2008; Monroe and Bokma 2009) and of habitat use in primates (Magnuson-Ford and Otto 2012) is associated with speciation. (See Adams et al. 2009 for a counterexample.) Likewise, DNA sequence divergence seems to have been enhanced by the amount of speciation in some higher taxa, but not others (Venditti and Pagel 2010; Goldie et al. 2011; Duchene and Bromham 2013).

The theoretical criticisms of founder-effect speciation by Charlesworth, Barton, and their coauthors do not necessarily eviscerate Eldredge and Gould's hypothesis that phenotypic evolution is associated with and enabled by speciation. Slatkin (1996) wrote "in defense of founder-flush theories of speciation," noting that relaxation of selection during the exponential population increase that may occur in newly founded populations can enable new advantageous allele combinations to be formed and selected, and Gavrilets (2004) noted that small populations can drift along adaptive ridges in multidimensional genetic landscapes, and achieve reproductively incompatible genetic configurations without having to cross impassably deep fitness valleys. And as explained in the next section, speciation might well promote trait evolution even if it does not proceed by genetic drift and reduced population size: Any mode of speciation might do.

2.3 Reconciling Punctuated Equilibria with Population Genetics

I proposed a simple explanation of why biological speciation (by any mode) is likely to be associated with substantial, long-lasting phenotypic alterations in fossil lineages (Futuyma 1987; see Futuyma 2010). I noted that only a minority of changes in phenotypic characters are advantageous across a broad array of environmental conditions; most advantageous alterations enhance adaptation to particular ecological niches or circumstances. Many herbivorous insects, for example, have the potential for advantageous changes in behavioral or physiological responses to certain plant species, perhaps adding the plant to the insect's diet. Most such adaptations have a polygenic basis, often composed of several functionally interacting components (e.g., recognizing a plant and possessing the enzymes needed to digest it or detoxify its chemical defenses). The geographic distribution of a specific "niche" (e.g., an environmental condition or a resource such as a host plant) is often discontinuous (patchy); moreover, it is likely to change over time, due to climate change, if for no other reason. An adaptation to such a "niche" arises and may be fixed in a local population, but for two reasons, both owing to breakdown of the adaptation by recombination, it may not persist long enough to be registered in the fossil record, much less be inherited by a clade of species. Specifically, the constellation of alleles and component characters associated with an adaptive trait will generally not be maintained intact if the population interbreeds freely with another population (such as the ancestral form) that is adapted to a different niche. Two likely consequences follow. First, if the adaptation does not become widespread, it is unlikely to be documented in the fossil record, and unlikely to persist very long because the natal population will eventually become extinct. But spread of the new adaptation from its birthplace to other patches with the same niche may well be hindered if emigrants are likely to disperse into intervening patches of the ancestral niche, where they will interbreed with ancestral genotypes. Second, environments undergo geographic shifts, dramatically illustrated by Pleistocene glacial and interglacial fluctuations. When this occurs, species commonly "track their niche": They undergo range shift, during which new populations are founded by migrants and some old populations become extinct, and the former geographic structure of the species is broken down and reformed. ("Niche tracking" suggests that dispersal may often be "easier" than adaptation in situ to an environmental change. The several possible reasons include genetic constraints, discussed in Sect. 4.2) The founders of a new population often will be drawn from separate, differentiated populations, causing gene flow on a more massive scale than the "trickle gene flow" that characterizes equilibrium populations (Slatkin 1977; McCauley 1993). Such gene flow, if between differently adapted populations, may break down the differences between them. In both of these scenarios, the evolution of reproductive isolation maintains the locally originated adaptation intact, by preventing free interbreeding with the more widespread, common ancestral genotype. Thus, I concluded, "speciation can facilitate morphological change not by liberating a population from genetic homeostasis or accelerating the response to selection, but by enabling a gene pool to remain subject to consistent selection pressures even as it moves about in space. By isolating gene pools from other gene pools that they encounter as they move about, speciation enables them to retain characters that evolved in a local context…

Although speciation does not accelerate evolution within populations, it provides morphological changes with enough permanence to be registered in the fossil record" (Futuyma 1987, p. 467). In that paper, I emphasized the role of reproductive isolation in protecting adaptations from dissolution during massive changes in geographic range, but I am now inclined to think that its more important effect is in enabling a new adaptation to spread by migrants that do not interbreed with residents of intervening ancestral-type populations. Eldredge et al. (2005) also considered the problem of how to reconcile apparent stasis in fossil lineages with the capacity for rapid evolution, and observed high rates of evolution, in populations of living organisms. (See also Thompson 2013.) Chief among the several factors that they suggested might cause stasis was spatially and temporally heterogeneous selection, owing in part to a "geographic mosaic" of different coevolutionary interactions experienced by different populations of a species (Eldredge 2001; Lieberman and Dudgeon 1995; Thompson 2005). Although individual populations may respond rapidly to local selection, consistent directional selection seldom acts on the species as a whole. Eldredge et al. did not discuss what factors overcome the heterogeneity of selection and enable significant character change, i.e., punctuation. Their hypothesis for stasis is related to mine and can be extended to account for punctuation by postulating, as do I, that evolution of reproductive isolation by one such population enables the phenotype to spread and persist. My model has won modest approval, especially among some paleontologists. Gould (2002, pp. 798–802), in particular, admitted in *The Structure of Evolutionary Theory* that the original explanation of punctuational evolution by founder-effect speciation and "genetic revolution" was untenable, and strongly endorsed my model, writing that "his simple, yet profound, argument has not infused the consciousness of evolutionists because the implied and required hierarchical style of thinking remains so unfamiliar and elusive to most of us" (p. 799). (Well, maybe.) Although this model is not the only one that might account for a correlation between divergence and speciation (Rabosky 2012), the evidence mentioned earlier (e.g., Mattila and Bokma 2008; Venditti and Pagel 2010; Strotz and Allen 2013) is consistent with it. So is evidence suggesting a break between intraspecific evolution and divergence between reproductively isolated populations.

For example, the structure of the phenotypic variance–covariance matrix is much the same among conspecific geographic populations of damselflies, but differs strongly between closely related species, between which divergence has been highly discordant with the intraspecific first principal component of variation (Eroukhmanoff and Svensson 2008). A most intriguing "blunderbuss" pattern of evolution of vertebrate body size has been described by Uyeda et al. (2011), who show that size evolves at a high rate over short time spans, but does not accumulate until lineages have been separated for about a million years or more. That is, the amount of divergence between related lineages is much the same after 10^5 years as at 10^3 years. After 10^6 years, however, the amount of difference mounts steadily and rapidly with time. It is tempting to attribute the million-year break to speciation, which often requires isolation and genetic divergence for about that long (Coyne and Orr 2004).

If Eldredge and Gould (1972) were right in supposing that trait evolution is facilitated by speciation, they were surely wrong about the mechanism, as Gould (2002, p. 796) came to recognize: "I believe that our critics have been correct in this argument, and that Eldredge and I made a major error by advocating, in the original formulation of our theory, a direct acceleration of evolutionary rate by the processes of speciation." It is possible, in my view, that phenotypic evolution and speciation are functionally associated, although more evidence will still be needed before the generality and cause of this pattern can be established. Perhaps Mayr (1963, p. 621) rightly wrote that "without speciation there would be no diversification of the organic world, no adaptive radiation, and very little evolutionary progress. The species, then, is the keystone of evolution."

3 Internal Constraints on Adaptation

Neither Darwin nor major figures in the Evolutionary Synthesis viewed natural selection as the sole important factor of evolution, much less as an omnipotent agent that could always fit organisms optimally to their environment. Darwin made frequent reference to the "mysterious laws of growth," as well as to environmental modifications that he supposed (especially in later editions of *The Origin of Species*) might be inherited. Wright included genetic drift as an important component of his Shifting Balance Theory; Fisher recognized genetic drift (especially as it affects the probability of fixation of a new advantageous mutation), and described "runaway" sexual selection in which female preference evolves not because of an advantage, but because of linkage disequilibrium with the male trait. In his well-known essay "What is an adaptive trait?" Dobzhansky (1956) emphasized the importance of nonadaptive pleiotropic effects of selected genes. Rensch (1959) attributed parallel evolution partly to similarity of hereditary factors, emphasized the role of development and "mechanisms of construction," and explained many characters by character correlation, especially allometric growth. Like Rensch, Mayr (e.g., 1963, p. 608) attributed parallel evolution partly to shared genetic and developmental properties, which also predispose every group of animals "to vary in certain of its structures, and to be amazingly stable in others." Stebbins (1950, 1974) noted that certain traits, such as the number of ovules per carpel, vary in certain taxa and are invariant, both within and among species, in other taxa.

During and after the ES, however, evidence mounted that natural populations are genetically very variable. Lewontin (1974, p. 92) famously wrote that "[t]here appears to be no character—morphogenetic, behavioral, physiological, or cytogenetic—that cannot be selected in Drosophila," and concluded that "there is good reason to suppose that any outbred population or cross between unrelated lines will contain enough variation with respect to almost any character to allow effective selection." This view, still widely held by evolutionary geneticists, supports an optimistic view of species' adaptability, and skepticism that adaptation is often limited or channeled by available genetic variation.

3.1 Adaptation: Critique and Defense

As I noted earlier, evidence for natural selection increased greatly after the Synthesis, and interest grew in explaining the evolution, by natural selection, of classes of characteristics such as life history traits and animal behaviors. Some such literature included plausible, but not well-tested, adaptive interpretations that became disparaged as "just-so stories" by critics, especially Richard Lewontin and Stephen Jay Gould. Much of the literature, though, consisted of optimality models that could be evaluated by empirically testing their assumptions and, especially, by comparing the models' predictions with observations (Maynard Smith 1978). Such models included constraints, or boundary conditions, such as trade-offs among traits. "Adaptationism" came under fire in the 1970s, Gould and Lewontin's (1979) eloquently written paper, "The spandrels of San Marco and the Panglossian paradigm," being by far the most frequently cited critique. Echoing Gould's frequent complaint that the Synthesis had "hardened" around natural selection, and Lewontin's (1977, 1979) critiques of the "adaptationist program" embodied in sociobiology and in *The Selfish Gene* (Dawkins 1976), Gould and Lewontin criticized what they viewed as a practice of atomizing organisms into unitary traits, proposing adaptive explanations of each, and substituting alternative adaptive hypotheses if the first ones fail. Among the many faults they found in the "adaptationist program" was its supposed failure to consider alternatives to natural selection, such as random genetic drift, alternative stable states, and especially nonadaptive by-products of developmental correlation. Maynard Smith (1978), among others, defended optimality theory, noting that the traits usually studied "can hardly be selectively neutral" (e.g., behavior and other traits that affect reproductive success), that the theory does not assume or attempt to show that traits are actually at their optima, and that the models make explicit assumptions about constraints and heredity, but he agreed that it was important to develop adequate methods of testing the models and that the field could benefit from heeding Lewontin's criticisms. Since then, researchers in this field have indeed become more critical, and the literature now includes countless examples of adaptationist hypotheses that were testable, have been tested, and have (usually) provided evidence of adaptation. For example, the inflorescence of wild carrot (*Daucus carota*) consists of an umbrella-like array of many tiny white flowers—with one or a few purple flowers in the center. Maynard Smith (1978) quotes Darwin's passage about this in *The Origin of Species*: "that the modified central flower is of no functional importance to the plant is almost certain," and then writes that, having cited this example in conversation, his companions immediately offered two adaptive hypotheses which, however, struck him as "fanciful." One of these hypotheses was that the dark flower is an "insect mimic" that attracts pollinating insects to the inflorescence. In 2009, Goulson et al. reported experiments, including experimental removal of the dark flower, that showed exactly this effect. As tests of adaptationist hypotheses improved after Gould and Lewontin's (1979) critique, constraints and the possibility of nonadaptive interpretations (especially based on development) became a common theme. A distinction was made between universal constraints (owing to physics and chemistry that affect, for example, the properties of materials) and "phylogenetic" constraints, particular to a clade because the features established in its earlier history can restrict the variety of possible evolutionary paths. (For example, it has long been supposed that the maximum body size of insects is set by the extent to which gas exchange can occur by diffusion through the tracheae.) Constraints might be caused by natural selection ("selective constraints") or by internal factors that restrict or bias the kinds of phenotypic variations that can arise. These "genetic constraints" and "developmental constraints" are closely related and often are much the same thing. Moreover, the distinction between selective and developmental constraints is often unclear, for a phenotypic change may cause death by disrupting development (e.g., failure of proper formation of the embryonic notochord could abort development of vertebrae, which the notochord induces). Smith et al. (1985) offered the most widely used definition of a developmental constraint: "a bias on the production of variant phenotypes caused by the structure, character, composition, or dynamics of the developmental system."

In the now extensive literature on constraints, some authors (e.g., Wake 2009) attributed certain evolutionary patterns, such as toe webbing in some salamanders, to developmental correlation rather than adaptation (as had Rensch and others during the Synthesis). Others provided both theoretical and empirical studies of ways in which the direction and extent of evolutionary change might be biased or limited by genetic variance and, especially, covariance among traits (e.g., Bradshaw 1991; Schluter 1996; Futuyma et al. 1995; Marroig and Cheverud 2005). The broad problem addressed is the extent to which constraints are important in explaining a range of phenomena. These include both existing features (such as toe webbing) and restrictions on adaptation, such as limits on species' geographic range and ecological amplitude (niche width). The following paragraphs summarize my recent review of the importance of genetic constraints, especially as they may limit adaptation (Futuyma 2010). I include under "genetic constraint" both so-called phylogenetic constraint and developmental constraint, which implies strictures set by developmental properties that do not vary, even though they are based at least partly on genetically encoded products (see Sect. 4.1).

3.2 Genetic Constraints

Studies of genetic variation in natural populations, responses to artificial selection, and rapid adaptation to environmental changes have led most population geneticists to conclude that almost every characteristic of most species is so genetically variable that the availability of variation seldom limits the response to selection (e.g., Mather 1955; Barker and Thomas 1987; Barton and Partridge 2000). However, all acknowledge that genetic correlations caused by pleiotropy can greatly retard, or possibly prevent, evolution of a character if there exists antagonistic selection on correlated characters; the strength of this effect grows with the number of correlated characters (Dickerson 1955; Kirkpatrick 2009; Walsh and Blows 2009). Authors past and present (e.g., Schmalhausen 1949; Riedl 1978; Schwenk and Wagner 2004) have emphasized the likely importance of "internal" selection, owing to antagonistic pleiotropy and epistasis, in limiting selection response and evolutionary change.

Several phenomena suggest that genetic constraints may effectively prevent response to selection. (1) The most striking evidence of failure of adaptation is extinction, the fate of the vast majority of species that have existed. Even with plentiful genetic variation, adaptation will lag behind very rapid changes of environment. This was surely the case during certain mass extinctions, but a greater fraction of species have succumbed during periods of "background" extinction. Whether or not the unknown environmental changes that may have caused these extinctions were rapid or slow is not known. (2) Almost all species have limited geographic distribution and habitat occupancy. Why they cannot adapt to often modestly different environments beyond their range is one of the most challenging problems in evolutionary ecology, in my view (see Kirkpatrick and Barton 1997; Holt and Gaines 1992). Bradshaw (1991), an authority on rapid adaptation of plants to metal-contaminated soils, cogently attributed habitat limits, and many other examples of adaptive failure to what he called "genostasis," a lack of selectable genetic variation. (3) Although convergent adaptation to similar selective challenges is common, there are also countless examples of unique, one-off adaptations; many are familiar synapomorphies of higher taxa. No bryophytes are more than about 15 cm tall, because they lack the vascular tissues that evolved only once (as far as known), in the ancestor of tracheophytes; among millions of species of insects, only one lineage (aculeate Hymenoptera) evolved a sting. The quantitative difference between evolving a feature once and not at all is slight, and terrestrial biotas would be very different if vascular plants had not evolved. (4) There are "empty niches," lacunae in the economy of nature, as we see from geographic comparisons (e.g., sea snakes in the Indo-Pacific but not the Atlantic Ocean) and from the replacement of extinct forms by ecological counterparts only after a very long time (e.g., 120 million years between extinction of the first bivalve-drilling gastropods and the evolution of modern oyster drills) or not at all (e.g., sauropod dinosaurs). (5) "Phylogenetic conservatism" is a major feature of life that is largely unexplained. It is hard to envision an adaptive explanation of many morphological synapomorphies that characterize large, old taxa whose species are distributed among many environments, such as certain wing vein patterns that distinguish large families of Diptera and Hymenoptera. Dobzhansky (1956), in ascribing some traits to pleiotropy, cited a diagnostic feature of all of the 600 species of Drosophilidae then known: three orbital bristles, the anterior bristle oriented forward and the others toward the rear. Phylogenetic "niche conservatism," associated with limited variation in physiology, morphology, and behavior, has immense ecological consequences (Wiens and Graham 2005). Many families of herbivorous insects have been associated with a single plant family for more than 70 million years; congeneric species of plants have similar latitudinal distributions and climate associations on different continents, after lengthy opportunity for divergence. Thermal tolerance limits are highly conserved, varying little with latitude, in both lizards and *Drosophila* (Grigg and Buckley 2013; Kellermann et al. 2006).

Genetic evidence of constraints on adaptation is mostly rather indirect. In a few cases, little or no genetic variance could be detected for certain characters in outbred natural populations. Bradshaw's research group found genetic variation for copper tolerance in populations (from uncontaminated areas) of those species of grasses that have evolved copper tolerance in copper-contaminated areas, but no variation at all in other species of grasses that have failed to evolve copper tolerance (Bradshaw 1991; Macnair 1997). Tolerance of desiccation and cold displayed little or no genetic variation in rainforest-dwelling species of *Drosophila* (Kellermann et al. 2006). In a series of tests, my colleagues and I screened four species of *Ophraella* leaf beetles for genetic variation in their willingness to consume, and ability to survive on, species of plants other than their normal host plant; every species failed to display genetic variation in consumption and survival on at least one of the test plants (Futuyma et al. 1995). Moreover, the macroevolutionary pattern of diet evolution in this genus is partly predictable from, and perhaps has been guided by, the abundance or paucity of genetic variation for different responses.

There is considerable evidence that correlations among genetically variable traits may retard response to selection; examples include such traits as sexually selected male features in fishes, crickets, and *Drosophila*, floral traits in *Ipomoea* (morning glories), and tarsus length in flycatchers. In an elegant experiment, Etterson and Shaw (2001) transplanted families from a Minnesota population of *Chamaecrista fasciculata* further south, estimated genetic variance and covariance among several traits, and determined the relationship between trait combinations and fitness in the southern environment—which is expected to prevail in Minnesota about 50 years from now. There was little genetic variance for the trait combinations that would provide the greatest potential enhancement of fitness at that time, suggesting that future adaptation to climate change may be inadequate to ensure population persistence. In several taxa, divergence among species has been along the multivariate axis of greatest intraspecific variation, a pattern that Schluter (1996) called "evolution along genetic lines of least resistance."

The quantitative genetic approach has been used to test whether or not a postulated developmental constraint actually would prevent response to selection. For example, a positive correlation between two characters, perhaps expressed as an allometric relationship, may be postulated to represent developmental constraint.

This can be tested by artificially selecting for a character combination orthogonal to the observed axis of variation (i.e., for increase of one trait and decrease of the other). For example, the features of the several "eye spots" on the wing of the butterfly *Bicyclus anynana* are genetically correlated. Beldade et al. (2002) successfully uncoupled the size of two such eye spots by artificially selecting in different directions, showing that there exists some independent, uncorrelated genetic variation for each trait and that the observed correlation need not constrain response to natural selection, but a similar attempt to decouple their color was unsuccessful (Allen et al. 2008). Theoretically, the likelihood of constraint increases with the number of intercorrelated characters, but all selection experiments to date have addressed bivariate correlations. Nevertheless, both theory and evidence suggest that estimates of genetic variances and correlations generally provide weak evidence on the strength or even existence of genetic and developmental constraints (Conner 2012).

Another kind of evidence, the genetic architecture of a trait, might at least hint at the possibility of constraints or bias on its evolution. The mutational variance of a character, the genetic variance that arises each generation by new mutations, is greater, *ceteris paribus*, if many genes affect its development (Lynch and Walsh 1998). Highly polygenic characters may be expected to display considerable genetic variation. Conversely, if only a few genes affect a character, the origin of a new character state might be a rarer event, and there might be less standing genetic variation, and adaptive evolution might have to wait for suitable new mutations to arise (Houle 1998). If so, the rate and possibly the direction of evolution of the trait might be limited, or at least biased, by mutation (Hartl and Taubes 2008; Stolt Even Simpson 1944), in postulating "quantum evolution," envisioned a rapid shift between peaks in Wright's adaptive landscape, but "in general the genetic processes involved do not permit making the step with a single leap" (p. 210). Stolfus (2006), in contrast to the view that selection generally acts on a non-limiting pool of standing variation. A considerable number of traits, ranging from pesticide resistance in plants and insects to pelvic reduction in stickleback fish, have evolved in diverse species and populations by independent mutations of the same gene (sometimes the same base pair), suggesting that there are few possible genetic avenues to the adaptive phenotype (Wood et al. 2005; Arendt and Reznick 2008; Martin and Orgogozo 2013). The extent to which adaptation is based on standing variation or new mutations is uncertain (Barrett and Schluter 2008), but the evidence of abundant "selective sweeps" in DNA sequences, which occur when new or previously rare mutations increase fitness, suggests that new mutations might play a more important role than traditionally (and still widely) thought.

In summary, several phenomena, among which extinction is most conspicuous, strongly imply that there exist constraints on the rate and direction of adaptation, including genetic/developmental constraints. The great attention to the question of constraint is a major, valuable development in evolutionary biology. Testing the constraint hypothesis in any particular instance, however, is not easy, and the evidence to date does not yet enable us to decide on the importance of internal constraints on adaptation.

4 Challenges from Developmental Biology

As many authors have noted, from Darwin into the early twentieth century, the study of evolution was intimately related to embryology. Starting with Haeckel's recapitulation theory, embryology was viewed as a window into the past, a way of reconstructing ancestral forms. Early leaders of genetics, such as Thomas Hunt Morgan, separated genetics from embryology, which they viewed as speculative; embryology likewise became an experimental science that rejected its speculative past and turned away from evolution, considering it not rigorous enough to be taken seriously (Smocovitis 1996, p. 193; Amundson 2005). However, comparative embryology continued as a parallel discipline: During the 1920s and 1930s, Gavin de Beer, Walter Garstang, and others dethroned recapitulation and described other categories of evolution of development, such as heterochrony (Gould 1977; Love 2003; Love and Raff 2003). The split between genetics and embryology probably affected the formation of the ST (Love 2009), which built on genetic but not developmental mechanisms.

It is often said that development was excluded from the Evolutionary Synthesis, although this claim has been contested by authors like Smocovitis (1996), Amundson (2005), and Love (2009). Mayr claimed that developmental biologists "bitterly resisted the synthesis. They were not left out of the synthesis…they simply did not want to join" (Mayr 1993, p. 32), and the developmental biologist Hamburger (1980, p. 98) noted that leading books on experimental embryology in the 1930s did not treat evolution, and that "the modern synthesis did not receive assistance from contemporary embryologists." The "architects" of the Synthesis were certainly familiar with contemporary comparative embryology. Ford and Huxley (1929) studied the genetics of "rate factors" in a crustacean, Haldane (1932b) wrote on the evolutionary significance of the time of action of genes, and Mayr (1942) alluded to allometry and compensatory growth. Rensch (1947, 1959) treated developmental phenomena in some depth, as I have noted, and Stebbins (1950) gave equal time to developmental and selectionist interpretation of patterns of morphological evolution. Huxley, whose analysis of allometry was his chief conceptual contribution to evolutionary analysis, included de Beer in *The New Systematics* (Huxley 1940), and de Beer included Haldane, Huxley, and Ford in *Evolution: Essays on Aspects of Evolutionary Biology* (de Beer 1938). The only (or at least the only well known) experimental embryologist to address evolutionary processes (and who did not espouse Lamarckism, saltation, or vitalism) was C.H. Waddington, whose experimental studies of canalization and genetic assimilation appeared in the early 1950s, after the Synthesis. Simpson (1953a) expressed some cautious doubt that genetic assimilation is an important factor in evolution, but did not object to it in theory. Dobzhansky (1951) referred very favorably to Schmalhausen's (1949) views on what Waddington called canalization, and in *Genetics of the Evolutionary Process* (1970, the sequel to *Genetics and the Origin of Species*) referred repeatedly to Waddington's and Schmalhausen's concept of canalization. He described Waddington's genetic assimilation experiments, noting that Waddington did not interpret them in Lamarckian terms. One has the impression that the architects of the Synthesis were entirely open to admitting a role for development, but that no one stepped forward to join them as an advocate for development—especially experimental embryology.

Whatever the reasons may have been, development was not effectively assimilated into the Evolutionary Synthesis, which lacked a theory of the origin of phenotypic variation, as many authors have noted. I do wonder what kind of theory of variation could have been derived from developmental biology in the 1940s, when even the physical basis of heredity, much less the mechanisms of development, was unknown. Developmental biologists had phenomenological descriptors, with names such as embryonic induction and prepattern (just as the comparative embryologists had phenomenological descriptors such as heterochrony), but development was a black box. Kirschner and Gerhart (2010, p. 276), who have suggested ingredients of a theory of variation, write that the "Modern Synthesis did not and could not incorporate any understanding of how the phenotype is generated." Certainly, some evolutionary biologists were sensitive to this gap. Early steps toward our growing understanding of developmental mechanisms, especially the models of gene regulation by Jacob and Monod (1961) and Britten and Davidson (1971), informed King and Wilson's (1975) interpretation of molecular differences between chimpanzee and human, and were featured in chapters on macroevolution in the textbooks by Dobzhansky et al. (1977) and Futuyma (1979).[4] Since then, evolutionary developmental biology (EDB) has become (in my opinion) one of the most exciting dimensions of evolutionary biology. Mechanistic understanding of gene action, of regulatory circuits, of the conservation of elements in the "genetic toolkit," and their association with different downstream genes are rapidly deepening our understanding of evolutionary changes in form (Carroll et al. 2005; Kirschner and Gerhart 2005; Stern 2011; Davidson 2011).

4.1 Structuralism: An Alternative to Variation and Selection?

Amundson (2005) places much of modern EDB in the structuralist tradition and contrasts the Synthetic (or "neo-Darwinian") and structuralist concepts of what constitutes the process and "causes" of evolution. Neo-Darwinians, following Dobzhansky (1937), define evolution as change of gene frequencies, and the

[4] As a graduate student, I recognized the gap between developmental biology and evolution, partly because my advisory committee included two "physiological geneticists" (as developmental geneticists were called then): Tahir Rizki (a *Drosophila* geneticist), who had been a student of Dobzhansky, and Morris Foster (a mouse geneticist), who had been one of Sewall Wright's few students. Rizki and his students attempted to (verbally) model mechanisms of gene action, and Foster imparted the importance of Jacob and Monod's work in his course. For some years afterward, it was easy to keep abreast of the little research at the interface. Two chapters in my 1979 textbook prominently featured developmental aspects of evolution.

causes of evolution are therefore the factors that change gene frequencies. For a structuralist concerned with the evolution of form (organisms' bodies), evolution is change in form, which requires change in ontogenies, the mechanical processes by which form develops. For the adaptationist, says Amundson (2005, p. 255), "Individuals don't evolve. Populations do. Populations evolve by natural selection," whereas the structuralist maintains that "Individuals don't evolve. Ontogenies do. Ontogenies evolve by modifications of ontogeny."

The distinction, then, is between change in the frequency of alleles that affect a phenotype, and the material mechanisms by which the phenotype is formed and is altered—a contrast closely related to Mayr's distinction between ultimate and proximal causes. But the distinction between explanation by gene frequency change versus mechanism is not limited to the evolution of ontogenies and form. I noted above (Sect. 1.1) that population genetic theory lacks mechanistic content. The mutations that produce genetic variation have no molecular specification; the trait affected by a mutation is not specified; selection is represented by coefficients that are mute with regard to the ecological or internal sources of selection. Much of evolutionary biology since the 1960s has consisted of applying the abstract theory to real biological systems. A large industry describes the molecular nature of the genes and mutations that affect traits of interest. An even larger industry attempts to identify the sources of selection on life history variables, physiological and biochemical traits, behaviors, and morphological features, often by describing how variation in a trait affects fitness via its interaction with specified environmental factors. Amundson (2005, p. 176) describes a "Causal Completeness Principle," espoused by earlier authors, according to which understanding development is a requirement for understanding evolution. I suggest that understanding developmental mechanisms is just one of the several components of a "causally complete" explanation of the evolution of form.

Practitioners and supporters of EDB are rightly enthusiastic for their subject. Some authors, however, make slightly hyperbolic claims for EDB's revolutionary impact on evolutionary biology, either by claiming a power and prevalence of certain developmental mechanisms well beyond what current evidence supports, or by suggesting that some developmental phenomena can replace genetic variation and natural selection as explanations of the evolution of form. Müller (2010) speaks of a "shift from a predominantly statistical and correlational approach to a causal-mechanistic approach." We can and should applaud a union of these approaches (consider the enormous benefits that have flowed from the union of evolutionary and molecular biology!), but I see no need for a "shift," if that implies lessening the role of the one in favor of the other in explaining evolution. Developmental mechanisms (which count among Mayr's (1961) "proximate" explanations) and population-level processes such as selection ("ultimate" explanations) are, of course, complementary. I will take that position throughout the remainder of this essay, as well as the position that as important as speculative hypotheses are in this as in all fields; a skeptical demand for evidence is also essential.

A strain persists within developmental biology that seems to echo, even if faintly, the idealistic morphology of the nineteenth century that carried over into physicalist or structuralist interpretations of development and evolution. *On Growth and Form*, by the anti-Darwinian Thompson (1917), was intended to show by mathematics that organisms conform to purely formal laws of growth and structure (Bowler 1983, p. 157), such conformation proving (Thompson wrote) "that a comprehensive 'law of growth' has pervaded the whole structure in its integrity, and that some more or less simple and recognizable system of forces has been in control"—by which Thompson presumably meant laws of mechanics. A more recent structuralist interpretation of development and evolution has been provided by Goodwin (e.g., 1984), whose position is appealing to many developmental biologists. Like many other writers, he misinterpreted the "random variation" in neo-Darwinism (i.e., the Synthetic Theory) to imply that "survival is the only constraint," approvingly citing pre-Darwinian rational morphologists who interpreted regularities, such as the segmented body plan of arthropods, as "basic structural constraints." He rejected Darwin's attribution of such similarities among organisms to heredity, and in a vigorous attack on the ability of genetics to explain similarity among organisms concluded that "gene products affecting morphology are to be understood as stimuli which evoke particular categories of response from a structured, self-organizing process which has a limited repertoire of possible responses" (Goodwin 1984, p. 227). The self-organizing processes are the consequences of developmental fields, spatial domains in which "every part has a state determined by neighbouring parts," and which are capable of reconstituting themselves if perturbed. Goodwin illustrated his point with models (e.g., by Oster et al. 1980) that describe developmental events and resulting forms, such as gastrulation and invagination, in terms of the properties of cellular elements such as cytoskeletons. He granted that the "main source of the heritable differences between multicellular organisms" surely resides in DNA (p. 236), even though he maintained, a few pages before, that "there is no way of accounting in causal terms for observed differences of form in organisms by the identification of differences in hereditary factors" (p. 219).

I am baffled by the argument that, on the one hand, genes cannot explain commonality of form among related organisms, and on the other hand that they can explain differences—especially since vast amounts of evidence attest to the role of gene activity in the formation and maintenance of phenotypes, both within and among individual organisms. If mutations of genes cause differences, how is it possible that unchanged genes should not cause unchanged, shared properties, at least in part? But my principal criticism of Goodwin's argument, as I wrote in a review of the book in which it appeared, is that "to provide physicochemical models of developmental events is not to replace genes and selection with a sufficient physicalist theory, as Goodwin believes: obviously the constituents of organisms obey physical laws, but these laws permit innumerable developmental patterns, of which only some are permissible under natural selection" (Futuyma 1984). We see the regularities of development monstrously violated by mutations and environmental teratogens, and we see countless (but not all possible) variations of development and form, even of such supposedly fundamental processes as gastrulation, that are attributable to the action of once-mutated genes that have at least been permitted, if not fixed, by natural selection.

Physical and chemical processes are of course the proximal causes of development, and models of these processes, by Goodwin, Oster et al. and others then and now, are immensely important. They complement not only evolutionary explanations of phenotypes, but also the explanations of development, expressed in terms of gene regulatory pathways and networks (Davidson 2011), that form so much of current developmental biology and describe the genetic "algorithms" or instructions for building concrete features, but not the physical events by which the features are built. Of course, physics sets constraints, but they are broad and do not provide a sufficient account of the origin and evolution of new phenotypes.

Some authors today may disagree with that statement. Perhaps the most thought-provoking structuralist interpretations of evolution today are provided by Müller (e.g., 2007, 2010) and Newman (e.g., 2010).

Newman's approach, in the tradition of Turing's (1952) and Murray's (1981) physicochemical models of animal patterns, is to show in detail that various forms and patterns, both of unicellular and simple multicellular organisms, can arise from known properties of cells and proteins. Newman's models are intriguing and may well be an important step toward understanding the mechanisms by which some phenotypes are produced. (I am not qualified to make that judgment.) But I have deep reservations about Newman's interpretation of the evolutionary scenarios he portrays. For example, in discussing "organismal motifs" (complex multicellular structures, as found in metazoans), he writes (Newman 2010, p. 283) that "the all but inevitable emergence, in this view, of organismal motifs *that were not products of natural selection, but rather serves as its raw material*, raises questions concerning both the necessity and sufficiency of the mechanisms of the neo-Darwinian Modern Synthesis for the origination of ancient multicellular forms" (my italics). The mechanisms ("dynamic patterning modules," or DPMs) he describes include establishment of cell adhesion by cadherins and C-type lectins like those found in choanoflagellates—but note that these must have evolved during the evolution of multicellularity, perhaps in concert with other molecules, since choanoflagellates are not multicellular. Cell clusters then took on different forms, says Newman, via differential adhesion owing to differences in levels of cadherins (resulting in multilayering), lateral inhibition of neighboring cells mediated by the *Notch* transduction pathway (enabling coexistence of multiple cell types), cell polarity mediated by the *Wnt* gene family (enabling lumen formation), and other such changes. From Newman's description of these DPMs, they all appear to involve multigene pathways, or at least regulation of expression level. In other words, they are complex characters that (presumably) did not exist as such in unicellular ancestors of metazoans (perhaps choanoflagellates): They must have arisen during the origin of protometazoans with multiple cell layers, lumens, etc., based partly on gene products in unicellular ancestors. But these gene products required modification if they were to interact in the way the components of the known DPMs do, on which Newman bases his scenario. The only known process by which such modifications can form complex, functional pathways is mutation (and recombination) of genes, coupled with natural selection.

Likewise, I am skeptical of Newman's proposition that new forms emerged abruptly, almost saltationally, and that "since the resulting pattern or form would potentially self-organize in a significant portion of the founding population, there would be no question of a single, isolated individual" (p. 293) establishing a new lineage (one of the criticisms of Goldschmidt's hopeful monsters). A new form might be induced by an environment in many individuals (as in one of the Newman's scenarios), but without genetic specification of the critical components, it will persist only as long as the inducing conditions—unless "genotypes associated with increased reliability of developmental outcome" are selected, "leading to what has been termed genetic assimilation or accommodation" (p. 298). Thus, in addition to what I view as implausible origin of a morphology without the aid of genetic variation and selection, Newman must invoke another quite controversial hypothesis, genetic assimilation (which I treat in the next section). At some point, the concatenation of questionable scenarios or hypotheses should be resisted until sufficient evidence is brought to bear on them.

Müller (2007, 2010) sounds many of the same themes as Newman, mostly in the context of the more familiar realm of the development and evolution of major multicellular clades (specifically, animals). In a thoughtful review (Müller 2007) of EDB, or "evo-devo" in his paper), he analyzes the field's major research programs, themes, and theoretical implications such as evolvability and organization, which includes features such as modularity. I agree with him that, in contrast to the theory of how genetic variation affects population dynamics (i.e., natural selection), evo-devo "does not invalidate the formal framework of the Modern Synthesis, but adds another level of explanation. The reach of evolutionary theory is expanded in that evo-devo accounts not for what kinds of variation are going to be maintained through natural selection, but also what kinds of variation can possibly arise from specific developmental systems" (p. 947). But the evo-devo that I think makes the greatest contribution to understanding evolution is not the one that "assigns much of the explanatory weight to the generative properties of development, with natural selection providing the boundary condition" (p. 947), nor the one that "posits that the causal basis for phenotypic form resides not in population dynamics or, for that matter, in molecular evolution, but instead in the inherent properties of evolving developmental systems" (p. 948). Like Newman, Müller gives lip service to the complementarity of the ST and developmental mechanism, but in effect treats them as alternative explanations.

Müller (2010) provides more concrete examples of his views in his treatment of morphological evolutionary novelties, in which he distinguishes "Type III" novelties, which are major changes of existing characters (e.g., tusks that are modified teeth), from "Type II" innovations, which are "new constructional elements that do not have a homologous counterpart in the ancestral species or in the same organism." (The latter provision excludes serially homologous structures; thus, the paired mouthparts of crustaceans are not novelties because their ancestor had paired biramous locomotory appendages.) Examples of Type II innovations include the carapace of turtles and the patella (knee bone) of mammals. Müller describes developmental mechanisms which, when modeled, produce changes in skeletal patterns as an "emergent consequence of activation-inhibition thresholds in geometrically confined spaces," illustrated by the loss of digits in salamanders and lizards. In his view, such evolutionary changes represent threshold responses to perturbations of "developmental systems that are characterized by cellular self-organization, feedback regulation, and environment dependence" (p. 322). Environmental induction, he says, is a realistic initiating trigger of innovations, via phenotypic plasticity, that are eventually genetically consolidated (or assimilated, or accommodated). Thus, "genetic evolution, while facilitating innovation, serves a consolidating role rather than a generative one, capturing and routinizing morphogenetic templates" (p. 323).

Again, I view this position as an unnecessary concatenation of speculations. I do not understand, for example, why so complex and unsupported a hypothesis is needed to explain the origin of the patella, which arises by osteogenesis in a phylogenetically novel location in the body. Selection of mutations in gene regulation is an alternative, simpler hypothesis. There is plenty of evidence that changes in gene regulation trigger the expression of entire developmental pathways at different times in ontogeny (resulting in heterochrony) or at different locations in the developing body (resulting in heterotopy; Baum and Donoghue 2010). Müller grants that genetic and cellular innovations permitted the formation of such novel mineralized tissues, "but the question of phenotypic novelty is why and how these processes were initiated in specific patterns and at specific locations of the vertebrate body" (p. 313). He does not entertain the hypothesis that mutant heterotopic expression of bone may have occurred a great many times, in various body locations, in diverse vertebrates, and that only those few mutant expression patterns that provided a selective advantage have been retained. The patella is one of many heterotopic bones (cf. the osteoderms of crocodiles, armadillos, and others) that have clear selective value. We must bear in mind, also, that the "inherent properties" of the developmental system themselves can evolve. It has long been known that variation in threshold traits, which display discrete or quasi-discrete states, usually has a polygenic basis, that the position of the threshold can evolve, and that the steepness of transition between states can change under artificial selection. (E.g., Suzuki and Nijhout 2006; Chevin and Lande 2013 provide a review and a model of the evolution of threshold characters from continuous variation). If a simpler hypothesis of genetic variation and natural selection can explain the observation, especially in view of abundant evidence for that hypothesis, those who propose more complex (and more vague) hypotheses should expect to be asked for evidence.

4.2 Genetic Assimilation and Accommodation

Perhaps no developmental theme has had as long an uncomfortable relationship to evolutionary theory as phenotypic plasticity. If defined as "the property of a given genotype to produce different phenotypes in response to distinct environmental conditions" (Pigliucci 2001, p. 1), plasticity can be either adaptive or nonadaptive (as illustrated by malformations and stunted growth if individuals are deprived of key nutrients during growth). The array of phenotypes that a genotype produces is the genotype's norm of reaction. Nobody denies the abundance of adaptive plastic responses (ranging from learning to the different adult morphologies of many organisms that are triggered by environmental stimuli during development); nor does anyone deny that the mean reaction norm can evolve, based on genetic variation in reaction norms. [That is, different genotypes display different reactions to an environmental stimulus or condition, a property called G × E (genotype × environment) interaction.] Under some conditions (such as constant stabilizing selection for a single phenotype), "canalization" may occur: the evolution of a phenotype that is relatively unaffected by environmental (and perhaps also genetic) perturbations. Canalization can sometimes break down in organisms that experience a novel, stressful environment, revealing "cryptic" genetic variation. For instance, body size of marine threespine sticklebacks (*Gasterosteus aculeatus*) increased dramatically when fishes were reared in freshwater, to which many stickleback populations have become adapted (Mc Guigan et al. 2010). The environment, then, may be said not only to exert selection, but also to amplify the variation on which selection can act.

Phenotypic plasticity may affect evolution in a variety of ways (Ghalambor et al. 2007; Wund 2012). For example, many authors have suggested that the expression of a modified phenotype in a newly encountered environment may help populations persist until natural selection improves adaptation to the environment (see Lande 2009). The time-honored idea (e.g., Mayr 1960) that animals' behavior may initiate a shift in ecological niche, leading to morphological and physiological adaptation, provides an important potential role for behavioral plasticity. Aubret et al. (2007) found that young tiger snakes (*Notechis scutatus*), a terrestrial species, could swim faster if reared for 5 months in water than on a solid surface, and they suggested that this plastic response may have facilitated the evolution of fully aquatic snakes, such as sea snakes. Aside from the fairly considerable phylogenetic distance between these taxa, we might ask whether the snakes (like human athletes) might have been trained to become more proficient at any physically possible task to which they might have been set. More studies on the possibility that behavioral plasticity initiates evolutionary change would be desirable.

I am not concerned not with the entire theme of the importance of phenotypic plasticity for evolution, but rather with a single controversial issue: the extent to which a phenotypically plastic response to an environmental stimulus becomes genetically entrained, such that the phenotype develops even in the absence of the stimulus. This is the thrust of several closely related ideas, of which Simpson (1953a, p. 110) wrote "Characters individually acquired by members of a group of organisms may eventually, under the influence of selection, be reinforced or replaced by similar hereditary characters. That is the essence of the evolutionary phenomenon here called 'the Baldwin effect'." Simpson noted that this idea had been independently proposed by Baldwin (in 1896), Lloyd Morgan, H.F. Osborn, and Soviet geneticists whose ideas were promulgated by Schmalhausen (1949). Waddington (1953) introduced "genetic assimilation" to describe the genetic fixation, due to selection, of part of an originally broad reaction norm, a character state that initially required an environmental stimulus. In the most comprehensive treatment of this theme, West-Eberhard (2003) wrote that Baldwin's hypothesis allowed for a broader range of outcomes than Waddington's, and she introduced "genetic accommodation" to mean a variety of genetic changes, caused by selection on genetic variation, in the "regulation, form, or side effects of the novel trait" (p. 140). But the common controversial element in the Baldwin effect, genetic assimilation, and genetic accommodation is precisely what Simpson identified as "the essence" of the Baldwin effect: the evolution from an environmentally triggered individual developmental response to a similar, genetically determined phenotype. I will refer to this specific aspect of the evolution of reaction norms as genetic assimilation.

Simpson (1953a) noted that the postulated process has three elements: (1) Owing to interaction with the environment, at least some individuals develop a nonhereditary character state that is advantageous. (2) The population includes "genetic factors" that produce the same kind of individual modifications (or, as we would say today, affect the reaction norm so as to make the phenotype more likely to develop, independent of environment). (3) These genetic factors (alleles) are favored by natural selection, increase, and make the character state more hereditary. Simpson wrote (p. 113) that each of these processes, viewed individually, does occur and that all may well occur together. Thus, the Baldwin effect may well occur. "Nevertheless two points remain decidedly questionable: whether the Baldwin effect does in fact explain particular instances of evolutionary change, and the extent to which this effect has been involved in evolution or can explain the general phenomenon of adaptation."

By whatever name, genetic assimilation is not a Lamarckian hypothesis, and it is fully compatible with the ST (cf. Lande 2009). After the 1950s, and until very recently, genetic assimilation was the subject of little research and was largely viewed as a "baroque hypothesis" (Orr 1999) that received little attention in most textbooks.[5] Recently, however, it has become a focus of intense interest. (For overviews, see West-Eberhard 2003; Price et al. 2003; Ghalambor et al. 2007; Pfennig et al. 2010; Schwander and Leimar 2011; Moczek et al. 2011; Moczek 2012; Wund 2012.) Genetic assimilation is now very popular with many evolutionary developmental biologists (e.g., Schlichting and Pigliucci 1998; Gilbert and Epel 2009; also authors in Pigliucci and Müller 2010) and is viewed by some as a major extension of, if not replacement for, the Synthetic Theory. The controversy is epitomized (and perhaps partly sparked) by West-Eberhard's (2003) provocative proposal that "most phenotypic evolution begins with environmentally initiated phenotypic change… The leading event is a phenotypic change with particular, sometimes extensive, effects on development. Gene-frequency change follows, as

[5] For example, genetic assimilation is not mentioned in Dobzhansky's *Genetics of the Evolutionary Process* (1970) or in textbooks by Dobzhansky et al. (1977), Freeman and Herron (2001), or Barton et al. (2007). I treated the topic sparingly, in the context of canalization, in all the editions of my textbook.

a response to the developmental change. In this framework, most adaptive evolution is accommodation of developmental-phenotypic change. Genes are followers, not necessarily leaders, in phenotypic evolution" (pp. 157–158).

This proposition is frequently associated with the idea that expression of phenotypic plasticity can help populations persist in a new environment until adaptation evolves by genetic change. That is not controversial, but it is not quite the same as genetic assimilation (or accommodation) of an environmentally induced character state. After all, two different characters might be involved: Animals exposed to unusually high temperature might react via behavioral flexibility, by staying in shady microsites, while natural selection enhances thermal tolerance over the course of generations. I am concerned here only with the question of genetic assimilation: whether or not genes are usually "followers" rather than "leaders" in adaptive evolution. There is abundant evidence that reaction norms evolve, that phenotypic plasticity and canalization can be shaped by natural selection, and that genetic assimilation is possible: It has been demonstrated in artificial selection experiments (e.g., Waddington 1953; Suzuki and Nijhout 2006). The major questions today include Simpson's queries, still unanswered, on whether or not it explains particular instances of evolution and whether it accounts, as West-Eberhard proposed, for "the general phenomenon of adaptation." To address these issues, we should ask how a history of genetic assimilation might be detected and demonstrated and how adequate the evidence is at this time.

West-Eberhard (2003) supported her thesis by describing numerous interesting examples of closely related species or populations that differ in plasticity of one or more traits: Commonly, one species exhibits different adaptive phenotypes under different conditions, and another exhibits a relatively fixed, nonplastic phenotype. However, such examples show only that reaction norms can evolve, which is not disputed. They do not show that genes follow plasticity; on the face of it, plasticity might be the derived trait. What is needed, at the least, is evidence of the direction of change (Moczek et al. 2011): Is plasticity the ancestral condition and genetic fixity the derived state? West-Eberhard's treatment provides little evidence on this point. Evidence on polarity of change is best sought when rapid, recent evolution has been observed or can confidently be reconstructed, or by robust phylogenetic inference (Schwander and Leimar 2011). Only recently has any such evidence been amassed.

One can envision at least three scenarios for genetic assimilation. These are not sharply demarcated.

1. A population with an adaptively plastic trait that experiences a variety of environments becomes subjected to a new selective regime, owing to constant exposure to one of the formerly experienced environmental states. Stabilizing selection now favors one of the previously expressed phenotypic states, resulting in abbreviation, narrowing, of the formerly broad reaction norm. Plasticity, the capacity to produce different phenotypes if exposed to environments that the population no longer experiences, may be lost if it is costly (DeWitt et al. 1998; Snell-Rood et al. 2010), or perhaps, if the population inhabits a constant

environment (Moran 1992; Masel et al. 2007), by mutation and genetic drift that erode the genetic capacity to produce alternative phenotypes. Thus, the phenotype has been canalized around one of the states the ancestral population could express.

2. The population has an adaptively plastic trait and is exposed to a new constant environment in which a quantitatively, but not qualitatively, different adaptive phenotype outside the previous range of observed variation is induced by an environmental stimulus that is simply an extension of, or is similar to, one of the environmental states that selected for plasticity in the ancestor. The expression of this phenotype is later canalized. For example, if the ancestral population had been selected to develop larger size when exposed to lower temperature, the "novel phenotype" might be a still larger size, triggered by an unprecedentedly low temperature or perhaps a novel stimulus. In this case, the novel character state arose by plasticity and the genetic change followed, but the plastic response is an "exaptation," a manifestation of an adaptive reaction norm that had been forged in the past, presumably by selection among genotypes with different reaction norms.
3. The population experiences a qualitatively novel environment that induces a novel phenotypic character state that happens to be advantageous. One possibility is that the new optimal character state is, or is close to, the extension of the ancestral reaction norm, which proves to be "preadaptive" even though the novel environment is not an extension of the range of environmental states the ancestral population experienced. As Ghalambor et al. (2007) have emphasized, however, the ancestral plasticity might well be maladaptive in a new environment: The extended reaction norm might be very different from the new optimum. In this case, plasticity, instead of facilitating adaptation to the new environment, would retard it. Such cases are not uncommon. For example, populations of humans and other vertebrates native to low elevations undergo several maladaptive acclimatization responses to low oxygen availability at high altitude, such as increased hematocrit and decreased affinity of hemoglobin for O_2, in opposition to the genetic adaptations seen in adapted highland populations (Storz et al. 2010).

I consider scenarios 1 and 2 to fit well within the standard Synthetic Theory. In neither case have genes been "followers," and for both cases represent simple modifications—abbreviation or extension—of a reaction norm that had already evolved (according to the Synthetic Theory) by selection of alleles that moved the developmental reaction closer to the optimum for an array of environments experienced by the ancestral population. The claim that genes are "followers" would receive strongest support if scenario 3 proves to be common, i.e., when a fortuitously advantageous expression of phenotype is induced in an ancestral population by a novel environment.

The number of empirical studies in which the polarity of change is known or can be inferred with reasonable confidence is too small to establish any generalizations about how common these several scenarios have been realized. I suspect that environmental induction of novel characters that are not manifestations of adaptive ancestral reaction norms (scenario 3) is likely to be rare (see also Schwander and Leimar 2011). Nevertheless, a few convincing cases have been described. In one of the clearer examples, Lédon-Rettig et al. (2010) showed that a short larval gut, a feature of the carnivorous morph of the spadefoot *Spea multiplicata*, can be induced by an animal diet in a related genus, *Scaphiopus*, which has the ancestral detritivorous habit. Freshwater populations of the stickleback *Gasterosteus aculeatus* have evolved from a marine ancestor and have evolved novel limnetic and benthic-feeding morphologies. Some of their features (body shape and gill raker length) were induced in experimental marine sticklebacks that were reared under conditions of diet and environmental configuration that resembled aspects of those under which the freshwater populations have evolved (Wund et al. 2012).

Some other cases are more difficult to interpret. Aubret et al. (2004) found that an island population of the tiger snake (*Notechis scutatus*), which feeds on larger prey than mainland populations, has a larger head. This is attributable both to a genetic difference and to greater phenotypic plasticity: Young snakes develop larger heads if fed larger prey. The reaction norm of the mainland population, presumably representing the ancestral state, displays similar, but less pronounced, plasticity. This may indicate that natural selection, based on success in prey capture, has shaped adaptive plasticity in the past and that selection in the island population has acted on genetic variation in the reaction norm.

A number of cases illustrate less ambiguously the genetic assimilation of an adaptive ancestral reaction norm (scenario 1). For example, some montane populations of *Daphnia melanica* that inhabit lakes to which fish have recently been introduced have lost the ability to develop pigmentation, a shield against ultraviolet radiation that also makes the animal more conspicuous to visual predators (Scoville and Pfrender 2010). Genetic assimilation has occurred, by abbreviating an adaptive ancestral norm of reaction. Similarly, there are many cases in which species with discrete alternative phenotypes, either genetic polymorphism or developmental polyphenism, have given rise to descendants with a single phenotype (Schwander and Leimar 2011). Since that state implies genetic fixity, the derivation of a monophenic form from a polyphenic ancestor can be considered genetic assimilation. Examples include loss of one of the alternative male mating strategies in insects, of the ability to develop a carnivorous phenotype in some populations of a spadefoot toad (*Spea*), and of polyphenism for wing development in some populations of a water strider (*Aquarius remigis*), as well as the transition from random bilateral asymmetry to consistent "left-handed" or "right-handed" phenotypes. Except for asymmetry, the ancestral polyphenisms are thought to represent adaptive plasticity, presumably the product of natural selection, and the evolutionary change illustrates my "scenario 1," narrowing of an adaptive reaction norm.

Considerably, more research, on a range of taxa and characters, will be needed before we can judge whether or not plasticity often "leads" adaptation. Most of the literature to date represents a biased sample of characters, viz., those in which a role for plasticity might be suspected from the outset. It will be illuminating to determine whether plasticity has any detectable role at all for sets of randomly selected differences among closely related species. I expect that among those cases in which a history of plasticity can be shown, the great majority will be interpretable as modifications of an adaptive reaction norm that had evolved in an ancestor by the action of natural selection on genetic variation (my scenarios 1 and 2). They will be interesting and important to document, but they will not represent a significant departure from the Synthetic Theory.

4.3 Nongenetic Inheritance

Most of evolutionary theory, during and since the Evolutionary Synthesis, has been framed in terms of inheritance based on variation that (as known since 1953) resides in DNA sequence. It has long been known, however, that there exist other forms of inheritance (Bonduriansky 2012). The cortical structure in ciliates, for example, is transmitted in cell division and grows by building onto the template provided by inherited cortex. Cultural characteristics such as language and wealth are nongenetically inherited. Cultural inheritance can be viewed as an example of inheritance of environmentally caused variation. Jablonka and Lamb (2010, p. 137) use Mayr's (1982) term "soft inheritance" to include several processes by which "variations that are the result of environmental effects are transmitted to the next generation." Some authors (e.g., Koonin and Wolf 2009) have enthusiastically welcomed certain of these processes as a return of Lamarckism; the most enthusiastic and prolific such advocacy has been by Jablonka and Lamb (e.g., 1995, 2005). I will comment on only one of these processes, transgenerational epigenetic inheritance.

Waddington (1957) introduced the term "epigenetic" to refer to the developmental processes by which genotypes become expressed as phenotypes. Today, it usually refers to "a mitotically and/or meiotically heritable change in gene function that cannot be explained by changes in DNA sequence" (Gilbert 2006, p. 118). Some of the huge body of research on epigenesis concerns transgenerational inheritance, via meiosis. The most frequently cited molecular mechanism of genomic "imprinting" is methylation of certain cytosine residues, which generally silences the gene. The methylated state persists and is replicated in mitosis; it is usually erased in the germ line or during embryogenesis, but not always—in which case, there is transgenerational inheritance. Methylation and other epigenetic "marking" of genes is often induced by specific environmental stimuli, and often enhances fitness within that environmental context (Bossdorf et al. 2008). For example, the production of some chemical defenses in plants, which are induced by damage by herbivores or pathogens, may be epigenetically inherited (Holeski et al. 2012). It is this inheritance of a potentially adaptive phenotype by a process other than mutation of DNA sequence that stimulates Lamarckian interpretations.

Clearly, epigenesis and epigenetic inheritance are important biological phenomena that have evolutionary implications. But it is necessary to ask whether or not transgenerational epigenetic inheritance fits into or departs from the Synthetic

Theory, whether it represents a true vindication of Lamarckism, and whether, and in what ways, it may be important in evolution.

Many phenomena that were not explicitly considered during the formation and early elaboration of the ST (referring here to both mathematical population genetic theory and the verbal theory that extended from population genetics to macroevolution) subsequently found a place in it quite comfortably. For example, maternal effects based on offspring effects of maternal genotype were modeled by fairly simple elaborations of traditional population genetic theory (e.g., Wade and Beeman 1994; Wolf et al. 1999), as were the evolutionary dynamics of transposable elements (e.g., Charlesworth and Langley 1989). The Synthetic Theory, formulated before Watson and Crick published on DNA, did not specify the nature of mutations. Thus, the population dynamics of epigenetic mutations ("epimutations") can be described in the same terms as sequence mutations (Haig 2007; Slatkin 2009). Population genetic models of epigenetic inheritance and its interaction with genetic inheritance have shown some of its most interesting theoretical effects (Day and Bonduriansky 2011). As befits a hitherto unknown biological process, some potential effects were not envisioned by Fisher, Wright, or Haldane, but neither were many other evolutionary dynamics described by population geneticists since then.

The big question is whether transgenerational epigenetic inheritance is Lamarckian. The key feature of Lamarckism is the production, from within the organism (in response to some stimulus), of inherited variation that is biased, directed, toward an adaptive end. In an incisive analysis, Haig (2007), a leading researcher on evolutionary effects of epigenetics, argues that transgenerational epigenetic inheritance is not Lamarckian, even when the phenotypes expressed enhanced fitness in the environmental context that induces them. Epigenetic inheritance characterizes few genes. Therefore, some feature of the marked gene must distinguish it from others, and make it susceptible to an epigenetic mark that resists erasure in the germ line. Moreover, there is considerable evidence of *genetic variation in the propensity of a gene to be methylated or otherwise marked* (Dickins and Rahman 2012). Genetic variants that act as maladaptive developmental switches will be eliminated by purifying natural selection, whereas variants that enhance fitness will be perpetuated by selection. The simplest interpretation, then, of environmental induction of fitness-enhancing inherited epigenetic switches is that they are adaptations honed by the action of natural selection on genetic variation, just like adaptive, phenotypically plastic reaction norms. As Dickins and Rahman (2012) remark in their critique of the evolutionary role of soft inheritance, epigenetic systems are phenotypes, subject to the standard evolutionary processes of mutation, natural selection, and genetic drift. Haig notes that adaptive directedness, or "intentionality," cannot be intrinsic to the epigenetic process: It must arise by some other process, and the only known candidate process is the "neo-Darwinian" action of natural selection on adaptively undirected variation that is the centerpiece of the Synthetic Theory. What we should like to have, then, is data on phenotypic effects of a large sample of *novel* epigenetic mutations, similar to the extensive data on de novo genetic mutations, that respond to environmental stimuli in species that have not experienced those or similar environments in their evolutionary history. The prediction is that they will show no overall tendency to be directed toward fitness-enhancing phenotypes.

Is epigenetic inheritance important in evolution? Almost surely it is, but importance can mean many things. In their population genetic models, Day and Bonduriansky (2011, also Bonduriansky and Day 2009) find a variety of ways in which epigenetic inheritance can affect the dynamics of gene frequency change; for instance, it can change the adaptive landscape, resulting in evolution toward a different genetic equilibrium. What is far from certain is that inherited epigenetic variation is the source of long-lasting adaptive phenotypes. Inheritance of epigenetic effects is frequently observed to persist for two or three generations; the highest figure I have encountered (in my limited reading) is nine generations. One of the most famous examples of an epigenetic phenotype is the "peloria" form of *Linaria vulgaris*, in which the normally bilaterally symmetrical flower is radially symmetrical (the phylogenetically ancestral condition) instead. This form was named by Linnaeus, and it can be found today, but there is no evidence at all that there has been unbroken descent from the mid-eighteenth century to the present time (a point that Jablonka and Lamb 2010 do not make in describing this example). The marked state of a gene is generally highly unstable, so the low fidelity of transmission will reduce the precision of adaptation (Haig 2007) and make it unlikely that an epigenetic phenotype will be fixed in a population and persist for any appreciable period of evolutionary time.

Instances of fitness-enhancing inherited epigenotypes appear to represent adaptations, not the source of adaptations. (As Dickins and Rahman 2012 remark, Jablonka and Lamb conflate proximate and ultimate causes of phenotypes.) But the adaptive epigenetic phenotype seems seldom to be stable enough to characterize an entire population. Future research might reveal, but so far I know of no evidence, that epigenetic differences distinguish different species or different populations of a single species. Despite the paucity or lack of even modest examples of epigenesis as a source of adaptation, Jablonka and Lamb (e.g., 2010) speculate at length about how this "Lamarckian" mechanism will account for adaptation ("genetic change is not necessary"), how it may accelerate adaptive evolution by enhancing the effectiveness of genetic assimilation, how incompatible chromatin marks may lower the fitness of hybrids and contribute to reproductive isolation, and how it may "play a key role in many macroevolutionary changes," especially if hybridization and polyploidization are accompanied by bursts of epigenomic variation. The claim about adaptation is, I believe, flatly wrong. The other speculations are interesting and enjoyable to read, but it would be good to bear in mind that they are so speculative, so removed from evidence, and so lacking in any compelling, rigorous theoretical foundation that they are wildly premature.

4.4 Evolutionary Developmental Biology and Evolutionary Theory

Many evolutionary biologists react with skepticism, or outright dismissal, to great speculative leaps about the likelihood that developmental mechanisms will replace traditional explanations of macroevolution. Probably most evolutionary biologists strongly disagree with the aversion to genetics some evolutionary developmental biologists evince, and especially with their tendency to proclaim that internal powers of organisms steer their evolutionary fate—a seeming echo of the decades of widespread, deep, almost emotional aversion to Darwin's theory of natural selection on undirected variation. Some evolutionary biologists, especially population geneticists, are inclined to dismiss EDB altogether. But that would be a great mistake, I believe, for the argument that evolutionary theory lacks but needs a theory of the origin of phenotypic variation is convincing—even obvious. As I indicated in the historical background with which this essay begins, most biologists since Darwin, including the architects of the Evolutionary Synthesis, recognized that not all conceivable variations are possible, and that taxon-specific biases or constraints must affect the likely paths of evolution. Subsequently, many population geneticists and other evolutionary biologists came closer to Gould (2002) portrayal of the Synthetic Theory: that it assumed that variation is always small in extent of change, copious in amount, and isotropic in direction. Given the evidence that new variation is limited rather than isotropic, evolutionary biology will clearly be enriched by a theory, founded in mechanistic molecular, cell, and developmental biology, of variation and how it can be shaped by natural selection into diverse, sometimes novel phenotypes.

Such a theory is under construction, with firm foundations in mechanistic biology, population genetic theory, and perhaps systems theory. Much of it stems from the discovery of phylogenetically conserved genes, chiefly regulatory genes, such as the *Distalless* gene, which initiates development of evaginations that form legs and other appendages in a wide range of animal phyla. These genes have often been recruited or co-opted to govern other pathways. For example, the anterior–posterior axis of all bilaterian animals is patterned by *Hox* genes that were recruited, much later, to pattern the proximal–distal pattern of tetrapod limbs. Animal phyla share a "genetic toolkit" of such deeply conserved genes and pathways, as Carroll and collaborators (2005), True and Carroll (2002) have called it. The remodeling of ancestral features and the origin of new ones may therefore be easier than traditionally thought, if existing genetic and developmental pathways can be expressed at different times or in conjunction with other such pathways.

A similar theme has been advanced by Kirschner and Gerhart (2005, 2010) in their theory of "facilitated variation," expressed more in terms of cellular and developmental processes than of genes. Phylogenetically "conserved core processes," such as the formation of the actin-based cytoskeleton, are "the basic machinery" of multicellular organisms that can be expressed, by virtue of gene regulation, in diverse contexts and in various combinations. They can "deconstrain" evolution and increase "evolvability" (the "capacity to generate heritable, selectable phenotypic variation") partly because they consist of a set of elements that are expressed as a functional unit (that was assembled by past genetic variation and selection) and need not be separately evolved anew. Other features that enhance evolvability include compartmentation (expression only in certain parts of the developing organism) and exploration. For instance, an evolutionary change in the length of a femur entrains changes in muscles, nerves, and blood vessels, all of which grow and proliferate in diverse directions, but persist and differentiate only in proper relation to the bone; evolving a longer leg does not require independent genetic change in all these components. These developmental and cellular processes may well be adaptations, formed by an ancient history of genetic variation and selection, but they make subsequent phenotypic evolution easier than it might otherwise be. The roles of developmental processes that Kirschner and Gerhart propose do not go beyond the empirical evidence; as they note, advances in understanding the mechanisms by which phenotypes are formed "have not undermined the previous achievements of evolutionary theory" (p. 276).

Evolvability has also been explored by Riedl (1978) and by Günter Wagner and colleagues, who approach the topic via population genetic and quantitative genetic models and data (e.g., Wagner 2010; Wagner et al. 2007; Pavlicev and Wagner 2012). They have aimed at developing a theory of the evolution of the mapping between genotype and phenotype via development. For example, pleiotropy will tend to reduce evolvability (the potential of a population to evolve under natural selection) if it affects functionally unrelated characters, for a mutation that improves the function of one character is likely to damage the function of another, leading to antagonistic effects on fitness. On the other hand, pleiotropic effects on functionally related features may be more likely to have advantageously correlated effects. Population genetic models show that evolvability can evolve, in that patterns of pleiotropy can be shaped by natural selection. For example, modifier mutations can be selected that reduce the harmful effect of another antagonistically pleiotropic locus on one of the affected characters, and effectively reduce or eliminate the pleiotropic correlation between the characters, thus changing the genotype–phenotype map (Pavlicev and Wagner 2012). Consequently, modularity, similar in concept to Kirschner and Gerhart's compartmentation, can be expected to evolve: Pleiotropy will be more frequent among functionally related characters or measurements than among unrelated ones. Pleiotropy is a major cause of genetic correlations among characters, which are estimated by the methods of quantitative genetics (Pavlicev and Wagner 2012). Studies of both genetic and phenotypic correlations in a variety of species have supported the theoretical expectation. For example, genetic correlations among various measurements of the mandible of mice decompose the structure into two modules, corresponding to the tooth-bearing part and the ascending ramus to which muscles attach. Correlations between corresponding bones of the forelimb and hindlimb are lower in humans (in which very different functions evolved relatively recently) than in other apes (in which some similarity of function, i.e., climbing, is retained). In this and many other instances, quantitative genetics can sketch the developmental map; identifying the genes and the developmental processes to which they contribute can follow.

Wagner's theoretical approach, then, explores the far reaches of the relationship between development and evolution. Not only can we look forward to learning the developmental basis of the evolution of modified and novel characters, and how developmental processes can facilitate, bias, or constrain evolution; we may look forward to understanding how natural selection has shaped the structure of the developmental processes themselves.

5 Accounting for Diversity

I will use "diversity" to mean number of taxa (often species) and "disparity" to mean some measure of the variety of different phenotypes among the members of a clade. A large literature is concerned with accounting for differences in diversity (and with disparity to a much lesser extent) among geological time periods, among geographic regions, and among clades. Numbers of species change by speciation (by which I mean the evolution of reproductive isolation between populations) and extinction. Changes in species diversity are often analogized with population growth, so differences in diversity may be attributed to differences in available time (e.g., since a region became habitable, or since the origin of a clade), in rates of increase (speciation rate minus extinction rate), or in limiting or damping factors (e.g., interspecific competition). Rates need not be constant, of course: A mass extinction caused by a bolide impact is a great increase in extinction rate.

The field of evolutionary ecology includes extensive theory and evidence bearing on topics, such as the evolution of interactions among species, that bear on the processes that influence diversity. However, much of the theory and other discourse on diversity dynamics and differences use species as units of evolution, and does not explicitly include evolutionary (or ecological) processes within species. This includes most of ecological theory, which addresses conditions for coexistence at the level of regional assemblages, taking into account competition and interactions between trophic levels. To the traditional equilibrium theory of ecological diversity, which emphasized the importance of resource partitioning, indirect competition, and predator–prey dynamics, has been added a neutral theory, based on rates of speciation and extinction of ecologically equivalent species (Hubbell 2001). Such ecological models, in which species are the units, have been paralleled in paleobiology, in which changes in diversity have been compared with "random clades" (Raup et al. 1973) and have been explored with species-level analogs of competition between species (Sepkoski 1996). Some paleobiologists have reported that rates of change in the diversity of fossilized taxa are negatively related to diversity (Foote 2010), mirroring long-standing observations of rapid evolutionary radiations on islands and rapid increases in diversity after mass extinction events: circumstances in which competition is thought to have been alleviated. An important contribution of paleobiologists in the 1970s (Eldredge and Gould 1972; Stanley 1975) was to draw attention to *species selection* or *clade selection* (Jablonski 2008), selection above the level of the individual or the local population, which may be detected as nonrandom differences in diversification rate among clades, and may sometimes be attributed to certain characters (see below). Models of species selection can account for some evolutionary trends, especially in characters that affect speciation or extinction rates. This hierarchical approach was important in distinguishing "active" trends (a shift in the entire distribution of character states among species in a clade) from "passive" trends (in which the variance expands from a boundary, carrying with it a change in the mean) (Gould 1988; McShea 1994). A hierarchical perspective, recognizing that selection can act at multiple levels, has been invaluable for understanding macroevolutionary patterns.

Such theory, however, takes speciation rates, extinction rates, and the properties of species as given; it does not include microevolution, i.e., the evolutionary processes within species that might account for speciation and extinction. Williams (1992, p. 31), who perhaps more than anyone else is associated with the defense of individual selection and criticism of group selection, wrote that "the microevolutionary process that adequately describes evolution in a population is an utterly inadequate account of the evolution of the Earth's biota. It is inadequate because the evolution of the biota is more than the mutational origin and subsequent survival or extinction of genes in gene pools. Biotic evolution is also the cladogenetic origin and subsequent survival and extinction of gene pools in the biota." However, speciation is based on genetic changes within populations; extinction occurs when genetic changes (if they occur) are insufficient to enable survival of any of the organisms that make up a population or species. Ideally, a microevolutionary theory of these changes could be scaled up to describe a theory of rates of speciation, extinction, and diversification. A combination of theory and data can account for some examples of speciation and of population extinction, but we are very far from having the empirical information that would be necessary to apply such a theory on the scale of entire clades.

The possible role of species selection in shaping diversity and macroevolutionary trends is viewed by some as an extension of and challenge to the ST (e.g., Erwin 2010). However, advocates of species selection differ in whether the process is based only on features that are "emergent" at the species level (such as breadth of geographic range) or an any "aggregate" feature of the organisms that constitute the species. Few cases of species selection based on emergent properties have been identified, but many features of organisms have been identified that affect diversification rate. Such cases seem to fit squarely within the Synthetic Theory. For example, Mitter et al. (1988) introduced the method of "replicated sister-group comparisons," in which the species diversity in lineages that possess a feature hypothesized to increase diversification, and that has evolved repeatedly, is compared with their sister groups that lack the feature. A causal role in diversification is inferred, based on the assumption that other diversity-enhancing features are randomized among the various lineages. Determining whether a difference in diversification rate resides in the rate of speciation, extinction, or both is difficult, although extinction rate may sometimes be estimated from the fossil record or perhaps from the shape of a phylogeny (a controversial procedure; see Rabosky 2010). Mitter et al. (1988) found that herbivorous lineages of insects usually have more species than their nonherbivorous sister groups. It is not yet known whether herbivorous insects are more diverse because adapting to different host plants causes rapid speciation ("ecological speciation;" Nosil 2012), because specializing on different plants reduces competition and the likelihood of extinction, or both. From such comparisons, diversification rate has been associated with many features (Coyne and Orr 2004), such as resin canals, nectar spurs and the herbaceous growth form in plants, sexual dichromatism and feather ornamentation in birds, and viviparity in fishes. (More powerful phylogenetic methods have since been developed to infer the impact of characters on rates of diversification FitzJohn et al. 2009.)

Identifying features that affect diversification rate may provide a qualitative relationship between evolutionary processes within species (microevolutionary processes) and the rate or extent of diversification, but it falls short of a functional model that would predict diversity differences in different times or places. Population genetic models and data of speciation are extensive (Gavrilets 2004; Nosil 2013), but only in the last few years have there been efforts to scale the models up to the macroevolutionary level. Using individual-based computer models of parapatric populations that adapt to a variety of multidimensional ecological niches, Gavrilets and Vose (2005) simulated adaptive radiation, and obtained results that matched empirical patterns, especially a rate of diversification that is initially high but later declines (cf. also Gavrilets and Losos 2009). In another such model, Aguilee et al. (2013) found that landscape dynamics affect diversification: In a mosaic of several habitat types, the number of ecologically divergent species is greatest if geographic barriers between habitats are alternately stronger (permitting divergent adaptation) and weaker (enabling populations to meet and evolve reinforced reproductive isolation).

Possibly, the theoretically least developed component of macroevolution is extinction. Populations that are small, for any reason, are susceptible to extinction by random fluctuation of population size, an effect that is exacerbated by accumulation of deleterious mutations. However, extinction of entire species is usually attributed to failure to adapt fast enough to a changing environment. This statement finds its theoretical expression in models of a single quantitative (polygenic) trait, in which the rate of population growth declines, and may become negative, as the difference between the trait mean and the new optimum increases. The models assume either a sudden change in environment of a specified magnitude (i.e., different between initial trait mean and trait optimum) (Gomulkiewicz and Holt 1995) or a steadily changing optimum that is tracked, with a lag, by a changing trait mean (Chevin et al. 2010). In the latter case, the rate of trait evolution after initial standing genetic variation has been depleted depends on the rate at which new genetic variance arises by mutation. Because more mutations occur in larger populations, the chance that a population survives is affected by its size. The more a population dwindles in size, the more likely it is to dwindle further.

As is true of many models, these are undoubtedly sufficient to predict survival versus extinction, when conditions meet the assumptions. They could certainly be modified for different assumptions, such as dependence of fitness on more than one character, in which case the genetic variance–covariance matrix (**G**) and its analog for de novo mutation (**M**), would be substituted for the additive genetic variance of the single character. The problem with predicting extinction of any particular population or species, or accounting for variation in extinction rate, is an empirical one, comparable to predicting the weather in New York two years from today, or accounting for the difference in the mean July temperature in two successive years: We are not remotely capable, at this time, of obtaining (or, probably, of processing) all the necessary data. If we ask, for example, what the likelihood is that the American population of the monarch butterfly (*Danaus plexippus*) will survive the next century of climate change, we should need to know the predicted extent and pattern of changes in temperature both in its breeding areas of North America and its overwintering areas in southern Mexico, in relation to the temperature tolerances of the relevant life history stages of the butterfly; and we need to know the magnitude of genetic variation in and genetic correlations among these several physiological measurements, as well as the rate at which these genetic statistics are changed by input of new mutations. That would require a staggering amount of research, but it would by no means be enough. The butterfly will experience other ecological changes than temperature alone: There are now and will be temperature-related changes in precipitation that can affect abundance and quality of its food plants (species of *Asclepias*, milkweeds) and probably the coniferous trees in the Mexican mountains where it overwinters; there will be changes in land use and in the communities of predators, parasites, and competing species. Whether or not the monarch's current host plants can adapt to the climate change, or be replaced by northward-moving alternative species of *Asclepias*, and whether or not the butterfly populations have genetic variation in traits that mediate their ecological interaction with other species are unknowns that might be critical determinants of the species' future. That is, we do not know what ecological factors are likely to require adaptation, much less the butterfly's "evolvability" with respect to those factors. In this area, as with many aspects of evolutionary biology, we have a theory that explains extinction but has very restricted predictive value—perhaps like physics, which explains climate but is unlikely to yield precise predictions of daily weather in the long term.

6 Conclusions

Maynard Smith (1966), surely one of the most open-minded of the great evolutionary biologists, wrote "It is in the nature of science that once a proposition becomes orthodox it should be subjected to criticism…It does not follow that, because a proposition is orthodox, it is wrong." More recently, Wagner (2010), acknowledging criticism of his ideas on the evolution of evolvability, wrote, "But, critics are good because only with relentless rational criticism will any scientific idea mature and serve the scientific community or society at large." I think there is value in all the challenges to the ST that I have discussed in this essay, for at the very least they have forced biologists to examine and defend orthodoxy, and in almost all cases, there has been at least some supportable and valuable content in the new idea. At the same time, I have tried to be critical of these challenges, for two major reasons. First, although science depends on new ideas and challenges to orthodoxy, blind enthusiasm for new ideas can be immensely counterproductive if it is misguided, for it may consume resources, time, and at worst careers, and so the challenges themselves need to be challenged. (And not all challengers are unsung Barbara McClintocks and Alfred Wegeners; some are Velikovskys.) Second, orthodox propositions usually have staying power for good reason. The evolutionary principles articulated in the Evolutionary Synthesis displaced and vanquished anti-Darwinian ideas by force of rigorous theory and multiple lines of evidence consistent with (and in some cases rigorously testing) that theory. The claims embodied in the Evolutionary Synthesis were well founded, and hold up today to an extraordinary extent. It is, of course, inconceivable that they should be complete and sufficient in the face of the vast increase of biological knowledge, especially of molecular, developmental, and physiological processes, but they were well founded enough not to be abandoned lightly. Having considered several challenges to the explanation of macroevolution developed during the Evolutionary Synthesis, I conclude that the ST remains fairly intact, but that the challengers have advanced our understanding or at least introduced considerations worth pursuing. My specific conclusions follow.

Higher taxa, with pronounced morphological differences from related taxa, do not arise saltationally, by single "macromutations," or reorganization of the genome. But there is no strong evidence that *all* character changes proceed by very slight steps, by the substitution of alleles of small effect at multiple loci. Some mutations (and genomic changes such as polyploidy) of fairly large effect are now known to contribute to evolution. It is possible that mutations of critical regulatory genes that switch on certain developmental pathways have caused large evolutionary changes, but as far as I know, this is still an open question.

The pattern of stasis punctuated by rapid evolutionary changes was wrongly interpreted to mean that natural selection cannot readily alter characteristics except via massive genetic change in small populations during speciation. However, stasis, which had been neglected before Eldredge and Gould brought it to the fore, requires explanation and is plausibly explained by fluctuating and geographically variable selection. The possibility that rapid episodes of character evolution do represent speciation, and that speciation facilitates departure from stasis, remains to be tested, but is consistent with data.

Critiques of adaptation have some validity, but have probably been overemphatic and more skeptical than warranted. Probably no evolutionary biologist has ever subscribed to the caricature of the Evolutionary Synthesis in which variation was supposed to be copious and "isotropic," i.e., equally available for all possible modifications. Nevertheless, many evolutionary biologists have supposed that genetic or developmental constraints have been so loose as to be negligible in practice. Identification and characterization of such constraints is now a major area of interest, thanks in large part to critiques of the "adaptationist program," and it is clear that constraints can be very important in biasing the direction of evolution or preventing adaptation altogether. Still, it remains heuristically valuable to ask what kind of selection might have impelled such evolution as has occurred, and in many (perhaps most) cases, it is likely that selection of some form has played a role. There is little reason to doubt a role for selection in the evolution of features that clearly have a close and important bearing on fitness.

The reunion of evolutionary and developmental biology, long overdue, is beginning to fill a major gap in evolutionary theory, the nature of evolutionary changes in the mapping between genotype and phenotype and the origin of phenotypic variation. Before and since the Evolutionary Synthesis, however, some developmental biologists have sought to minimize the significance of natural selection, and even of genetics, in evolution and development, by viewing the physical processes of development, and of biomolecules and cell structures, as the locus of explanation. But these are proximal explanations of form, necessary but not sufficient for explaining evolution. Proximal physical processes can constrain form and are clearly involved in the production of new forms, which cannot exist other than by physical events. But these events cannot explain the fixation of the new forms in species populations, nor the further honing of such features into more precise, effective adaptations. All proteins and cell structures produce effects by physical processes, but genetically based alteration of the proteins and structures alters the processes. Explanation by gene frequency change and explanation by changes in the material, mechanistic properties of organisms are complementary; one need not diminish the significance of the other. Natural selection on genetic variation remains the ultimate explanation of all adaptive evolution.

A reawakening of a major role of phenotypic plasticity in evolution is being presented as another challenge to orthodox theory. Most of the phenotypically plastic traits under discussion appear to be adaptations to environmental heterogeneity that have been shaped by natural selection among genetically variable reaction norms. In some cases, part of such a reaction norm (the phenotype evoked by and adapted to one of the environmental states) has been genetically consolidated or assimilated. In other cases, a more extreme phenotype, developed as a simple extension of the ancestral reaction norm, develops in response to a more extreme state of the environment. Both of these events, viewed only in the immediate context, appear to illustrate "genes as followers" of developmental phenotypic change, but in a longer historical perspective are seen to emerge as a by-product of a history of selection on genetic variation. Perhaps plasticity could be viewed as the leader, and genes as followers, when a plastically produced phenotype is fortuitously "preadapted" to a qualitatively novel environment. I suspect this occurs rarely, but it remains to be seen.

Many or most epigenetic alterations of phenotype can often be viewed as a form of phenotypic plasticity. The developmental switch is usually adaptive; it is often genetically variable, and so it presumably evolved by natural selection. Epigenetic changes that are inherited across generations can be modeled as ordinary mutations, the long-term evolutionary effect of which depends on their stability (or, conversely, on the rate of "back-mutation") and frequency of occurrence. Their stability seems seldom to extend beyond a dozen generations or so, and no cases have yet been described in which epigenetic differences are fixed between different populations. They clearly can affect fitness and may affect immediate, local adaptation, but any macroevolutionary role has yet to be established. There is no evidence, to my knowledge, of a Lamarckian spontaneous origin of adaptively directed "epimutation" arising de novo.

In agreement with some other authors (e.g., Sterelny 2000; Minelli 2010), I conclude that the developmental phenomena described to date can readily be encompassed by the broad principles of the Evolutionary Synthesis.

Variation in rates of diversification stems from dynamics of speciation and extinction, both of which are explicable in microevolutionary terms. Indeed, the theory of speciation is far advanced, even if still controversial. However, attempts to build a theory of diversification from speciation theory have only started. The fairly minimal existing theory of extinction is surely valid, but obtaining the information necessary to predict extinction or to explain differences in extinction rates will be very difficult.

Finally, can microevolution explain macroevolution? It depends on what "explain" means. Existing theory can provide a plausible account of the history and causes of most or all evolutionary phenomena. In many but not all cases, it will be possible to derive some support or counterevidence from data. The degree of detail of the account will satisfy some, but not others: For example, there may be evidence of selection on the genes underlying a phenotype, and of the source and strength of selection, but the developmental events between gene and phenotype may be unknown. Opinion will vary on whether or not the explanation is complete or sufficient in that case. Likewise, if "explanation" requires that evolution be predictable for more than a few generations, the theory and data of microevolution will provide no more satisfying "explanation" than does physics if it is required to make long-term predictions of weather. I do not know of any macroevolutionary phenomena that are inconsistent with existing evolutionary theory, any phenomena that would require us to reject one of its principles as simply false. Nonetheless, the relative importance of many of the factors of evolution is debatable, and I assume that every part of our explanatory theory is incomplete. Of course, the Evolutionary Synthesis will be extended, molded, and modified. But there will not be a Kuhnian "paradigm shift." Science really does accomplish something.

Acknowledgments I am grateful to Michael Bell for calling my attention to a passage in *The Origin of Species* in which Darwin seems to foreshadow punctuated equilibria, to Alan Love for enlightening discussion of the relationship of developmental biology to the Evolutionary Synthesis, and to two anonymous reviewers for suggestions and queries. I wish also to thank members of the several audiences to whom I have presented some of these thoughts, whose questions and criticisms have impelled me to learn more and (I hope) to think more carefully.

References

Adams MB (1980) Sergei Chetverikov, the Kol'tsov Institute, and the evolutionary synthesis. In: Mayr E, Provine WB (eds) The evolutionary synthesis: perspectives on the unification of biology. Harvard University Press, Cambridge, Massachusetts, pp 242–278

Adams DC, Berns CM, Kozak KH, Wiens JJ (2009) Are rates of species diversification correlated with rates of morphological evolution? Proc R Soc B 276:2729–2738

Aguilee R, Claessen D, Lambert A (2013) Adaptive radiation driven by the interplay of eco-evolutionary and landscape dynamics. Evolution 67:1291–1306

Allen CE, Beldade P, Zwaan BJ, Brakefield PM (2008) Differences in the selection response of serially repeated color pattern characters: standing variation, development, and evolution. BMC Evol Biol 8:94

Amundson R (2005) The changing role of the embryo in evolutionary thought: Roots of Evo-Devo. Cambridge University Press, New York

Anderson EA (1949) Introgressive hybridization. Wiley, New York

Arendt J, Reznick D (2008) Convergence and parallelism reconsidered: what have we learned about the genetics of adaptation? Trends Ecol Evol 23:26–32

Aubret F, Shine R, Bonnet X (2004) Adaptive developmental plasticity in snakes. Nature 431:261

Aubret F, Bonnet X, Shine R (2007) The role of adaptive plasticity in a major evolutionary transition: early aquatic experience affects locomotor performance of terrestrial snakes. Func Ecol 21:1154–1161

Avise JC (1977) Is evolution gradual or rectangular? Evidence from living fishes. Proc Nat Acad Sci USA 74:5083–5087

Barker JSF, Thomas RH (1987) A quantitative genetic perspective on adaptive evolution. In: Loeschcke V (ed) Genetic constraints on adaptive evolution. Springer, Berlin, pp 3–23

Barrett RDH, Schluter D (2008) Adaptation from standing genetic variation. Trends Ecol Evol 23:38–44

Barton NH, Charlesworth B (1984) Genetic revolutions, founder effects, and speciation. Annu Rev Ecol Syst 15:133–164

Barton N, Partridge L (2000) Limits to natural selection. BioEssays 22:1075–1084

Barton NH, Briggs DEG, Eisen JA, Goldstein DB, Patel NH (2007) Evolution. Cold Spring Harbor Laboratory Press, Cold Spring Harbor

Baum DA, Donoghue MJ (2010) Transference of function, heterotopy and the evolution of plant development. In: Cronk QCB, Bateman RM, Hawkins JA (eds) Developmental genetics and plant evolution. Taylor & Francis, London, pp 52–69

Beldade P, Koops K, Brakefield PM (2002) Developmental constraints versus flexibility in morphological evolution. Nature 416:844–847

Bokma F (2010) Time, species, and separating their effects on trait variance. Syst Biol 59:602–607

Bonduriansky R (2012) Rethinking heredity, again. Trends Ecol Evol 27(6):330–336

Bonduriansky R, Day T (2009) Nongenetic inheritance and its evolutionary implications. Annu Rev Ecol Evol Syst 40:103–125

Bossdorf O, Richards CL, Pigliucci M (2008) Epigenetics for ecologists. Ecol Lett 11:106–115

Bowler PJ (1983) The eclipse of Darwinism. Johns Hopkins University Press, Baltimore, Maryland

Bradshaw AD (1991) Genostasis and the limits to evolution. Phil Trans R/Soc Lond B 333:289–305

Bradshaw HD, Wilbert SM, Otto KG, Schemske DW (1998) Quantitative trait loci affecting differences in floral morphology between two species of monkeyflower (*Mimulus*). Genetics 149:367–382

Britten RJ, Davidson EH (1971) Repetitive and non-repetitive DNA and a speculation on the origin of evolutionary novelty. Quart Rev Biol 46:111–133

Carroll SB, Grenier JK, Weatherbee SD (2005) From DNA to diversity: molecular genetics and the evolution of animal design, 2nd edn. Blackwell, Malden, Massachusetts

Carson HL, Templeton AR (1984) Genetic revolutions in relation to speciation phenomena: the founding of new populations. Annu Rev Ecol Syst 15:97–131
Charlesworth B (1980) The evolutionary genetics of adaptation. In: Nitecki MH (ed) Evolutionary innovations. University of Chicago Press, Chicago, pp 47–98
Charlesworth B (1982) Hopeful monsters cannot fly. Paleobiology 8:469–474
Charlesworth B (1984) Some quantitative methods for studying evolutionary patterns in single characters. Paleobiology 10:308–318
Charlesworth B, Langley CH (1989) The population genetics of *Drosophila* transposable elements. Annu Rev Genet 23:251–287
Charlesworth B, Rouhani S (1988) The probability of peak shifts in a founder population. II. An additive polygenic trait. Evolution 42:1129–1145
Charlesworth B, Lande R, Slatkin M (1982) A neo-Darwinian commentary on macroevolution. Evolution 36:474–498
Chevin L-M, Lande R (2013) Evolution of discrete phenotypes from continuous norms of reaction. Am Nat 182:13–27
Chevin L-M, Lande R, Mace GM (2010) Adaptation, plasticity, and extinction in a changing environment: towards a predictive theory. PLoS Biology 8(4):e1000357
Clausen J, Keck DD, Hiesey WM (1948) Experimental studies on the nature of species. III. Environmental responses of climatic races of *Achillea*. Carnegie Inst Washington Publ 581:129
Conner JK (2012) Quantitative genetic approaches to evolutionary constraint: how useful? Evolution 66:3313–3320
Coyne JA, Orr HA (2004) Speciation. Sinauer, Sunderland, Massachusetts
Coyne JA, Barton NH, Turelli M (1997) A critique of Sewall Wright's shifting balance theory of evolution. Evolution 51:306–317
Darlington CD (1939) The evolution of genetic systems. Basic Books, New York
Darwin C (1859) On the origin of species by means of natural selection, or preservation of favoured races in the struggle for life. Murray, London
Davidson EH (2011) Evolutionary bioscience as regulatory systems biology. Developmental Biology 357:35–40
Dawkins R (1976) The selfish gene. Oxford University Press, Oxford
Day T, Bonduriansky R (2011) A unified approach to the evolutionary consequences of genetic and nongenetic inheritance. Am Nat 178:E18–E36
de Beer G (ed) (1938) Evolution: essays on aspects of evolutionary biology. Oxford University Press, Oxford
de Beer G (1940) Embryos and ancestors. Clarendon Press, Oxford
Depew DJ, Weber BH (2013) Challenging Darwinism: expanding, extending, replacing. In: Ruse M (ed) The Cambridge encyclopedia of Darwin and evolutionary thought. Cambridge University Press, Cambridge, pp 405–411
DeWitt TJ, Sih A, Wilson DS (1998) Costs and limits of phenotypic plasticity. Trends Ecol Evol 13:77–81
Dickerson GE (1955) Genetic slippage in response to selection for multiple objectives. Cold Spring Harbor Symp Quant Biol 20:213–224
Dickins TE, Rahman Q (2012) The extended evolutionary synthesis and the role of soft inheritance in evolution. Proc R Soc B 279:2913–2921
Dobzhansky Th (1937) Genetics and the origin of species. Columbia University Press, New York
Dobzhansky Th (1951) Genetics and the origin of species, 3rd edn. Columbia University Press, New York
Dobzhansky Th (1956) What is an adaptive trait? Am Nat 40:337–347
Dobzhansky Th (1970) Genetics of the evolutionary process. Columbia University Press, New York
Dobzhansky Th, Ayala FJ, Stebbins GL, Valentine JW (1977) Evolution. Freeman, San Francisco
Duchene D, Bromham L (2013) Rates of molecular evolution and diversification in plants: chloroplast substitution rates correlate with species richness in the Proteaceae. BMC Evol Biol 13(65). doi:10.1186/1471-2148-13-65

Eldredge N (2001) The sloshing bucket: how the physical realm controls evolution. In: Crutchfield J, Schuster P (eds) Evolutionary dynamics: exploring the interplay of selection, neutrality, accident, and function. Oxford University Press, New York, pp 3–32

Eldredge N, Gould SJ (1972) Punctuated equilibria: an alternative to phyletic gradualism. In: Schopf TJM (ed) Models in paleobiology. Freeman, Cooper, San Francisco, pp 82–115

Eldredge N, Thomson JN, Brakefield PM et al (2005) The dynamics of evolutionary stasis. Paleobiology 31(suppl 5):133–145

Endler JA (1986) Natural selection in the wild. Princeton University Press, Princeton

Eroukhmanoff F, Svensson EI (2008) Phenotypic integration and conserved covariance structure in calopterygid damselflies. J Evol Biol 21:514–526

Erwin DH (2010) Microevolution and macroevolution are not governed by the same evolutionary processes. In: Ayala FJ, Arp R (eds) Contemporary debates in philosophy of biology. Wiley, Chichester, pp 180–193

Etterson JR, Shaw RG (2001) Constraint to adaptive evolution in response to global warming. Science 294:151–154

Fisher RA (1930) The genetical theory of natural selection. Clarendon Press, Oxford

FitzJohn RG, Maddison WP, Otto SP (2009) Estimating trait-dependent speciation and extinction rates from incompletely resolved phylogenies. Syst. Biol. 58:595–611

Foote M (2010) The geological history of biodiversity. In: Bell MA, Futuyma DJ, Eanes WF, Levinton JS (eds) Evolution since Darwin: the first 150 years, pp 479–510. Sinauer Associates, Inc., Sunderland, Massachusetts

Ford EB, Huxley JS (1929) Genetic rate factors in *Gammarus*. Ark Entw Mech Org 117:67

Freeman S, Herron JC (2001) Evolutionary analysis, 2d edn. Prentice Hall, Upper Saddle River

Futuyma DJ (1979) Evolutionary biology. Sinauer, Sunderland, Massachusetts

Futuyma DJ (1984) Neo-Darwinism under opprobrium. Science 226:532–533 (Review of M.-W. Ho and P.T. Saunders, *Beyond Neo-Darwinism.*)

Futuyma DJ (1987) On the role of species in anagenesis. Am Nat 130:465–473

Futuyma DJ (2010) Evolutionary constraint and ecological consequences. Evolution 64:1865–1884

Futuyma DJ, Keese MC, Funk DJ (1995) Genetic constraints on macroevolution: the evolution of host affiliation in the leaf beetles genus *Ophraella*. Evolution 49:797–809

Gavrilets S (2004) Fitness landscapes and the origin of species. Princeton University Press, Princeton

Gavrilets S, Losos JB (2009) Adaptive radiation: contrasting theory with data. Science 23:732–737

Gavrilets S, Vose A (2005) Dynamic patterns of adaptive radiation. Proc Nat Acad Sci USA 102:18040–18045

Ghalambor CK, McKay JK, Carroll SP, Reznick DN (2007) Adaptive versus non-adaptive phenotypic plasticity and the potential for contemporary adaptation in new environments. Func Ecol 21:394–407

Gilbert SF (2006) Developmental biology, 8th edn. Sinauer, Sunderland, Massachusetts

Gilbert SF, Epel D (2009) Ecological developmental biology: integrating epigenetics, medicine, and evolution. Sinauer, Sunderland, Massachusetts

Gingerich PD (1983) Rates of evolution: effects of time and temporal scaling. Science 222:159–161

Goldie X, Lanfear R, Bromham L (2011) Diversification and the rate of molecular evolution: no evidence of a link in mammals. BMC Evol Biol 11(286). doi:10.1186/1471-2148-11-286

Goldschmidt R (1940) The material basis of evolution. Yale University Press, New Haven, Connecticut

Gomulkiewicz R, Holt RD (1995) When does evolution by natural selection prevent extinction? Evolution 49:201–209

Goodwin BC (1984) A relational or field theory of reproduction and its evolutionary implications. In: Ho M-W, Saunders PT (eds) Beyond Neo-Darwinism: an Introduction to the new evolutionary paradigm, pp 219–241. Academic Press, London

Gottlieb LD (1984) Genetics and morphological evolution in plants. Am Nat 123:681–709
Gould SJ (1977) Ontogeny and phylogeny. Harvard University Press, Cambridge, Massachusetts
Gould SJ (1980) Is a new and general theory of evolution emerging? Paleobiology 6:119–130
Gould SJ (1983) The hardening of the modern synthesis. In: Grene M (ed) Dimensions of Darwinism. Cambridge University Press, Cambridge
Gould SJ (1988) Trends as changes in variance: a new slant on progress and directionality in evolution. J Paleont 62:319–329
Gould SJ (2002) The structure of evolutionary theory. Harvard University Press, Cambridge, Massachusetts
Gould SJ, Eldredge N (1993) Punctuated equilibrium comes of age. Nature 366:223–227
Gould SJ, Lewontin RC (1979) The spandrels of San Marco and the Panglossian paradigm: a critique of the adaptationist programme. Proc R Soc Lond B 205:581–598
Goulson D, McGuire K, Munro EE et al (2009) Functional significance of the dark central floret of *Daucus carota* (Apiaceae) L.: is it an insect mimic? Plant Species Biol 24:77–82
Grigg JW, Buckley LB (2013) Conservation of lizard thermal tolerances and body temperatures across evolutionary history and geography. Biol Lett 9:20121056
Haig D (2007) Weismann rules! OK? Epigenetics and the Lamarckian temptation. Biol Phil 22:415–428
Haldane JBS (1932a) The causes of evolution. Longmans, Green, New York
Haldane JBS (1932b) The time of action of genes and its bearing on some evolutionary problems. Am Nat 66:5–24
Hamburger V (1980) Evolutionary theory in Germany: a comment. In: Mayr E, Provine WB (eds) The evolutionary synthesis: perspectives on the unification of biology. Harvard University Press, Cambridge, Massachusetts, pp 303–308
Hartl DL, Taubes CH (2008) Towards a theory of evolutionary adaptation. Genetica 102(103):525–533
Heberer E (ed) (1943) Die Evolution der Organismen. Ergebnisse und Probleme der Abstammungsgeschichte, Gustav Fischer, Jena
Holeski LM, Jander G, Agrawal AA (2012) Transgenerational defense induction and epigenetic inheritance in plants. Trends Ecol Evol 27:618–626
Holt RD, Gaines MS (1992) Analysis of adaptation in heterogeneous landscapes—implications for the evolution of fundamental niches. Evol Ecol 6:433–447
Houle D (1998) How should we explain the variation in the genetic variance of traits? Genetica 102(103):241–253
Hubbell SP (2001) The unified neutral theory of biodiversity and biogeography. Princeton University Press, Princeton
Hunt G (2010) Evolution in fossil lineages: paleontology and the origin of species. Am Nat 176(SupplS1):S61–S76
Huxley J (ed) (1940) The new systematics. Clarendon Press, Oxford
Huxley JS (1942) Evolution, the modern synthesis. Allen and Unwin, London
Innocenti P, Morrow EH, Dowling DK (2011) Experimental evidence supports a sex-specific selective sieve in mitochondrial genome evolution. Science 332:845–848
Jablonka E, Lamb MJ (1995) Epigenetic inheritance and evolution: the Lamarckian dimension. Oxford University Press, Oxford
Jablonka E, Lamb MJ (2005) Evolution in four dimensions. MIT Press, Cambridge, Masachusetts
Jablonka E, Lamb MJ (2010) Transgenerational epigenetic inheritance. In: Pigliucci M, Müller G (eds) Evolution: the extended synthesis. MIT Press, Cambridge, Massachusetts, pp 137–174
Jablonski D (2008) Species selection: theory and data. Annu Rev Ecol Evol Syst 39:501–524
Jacob F, Monod J (1961) Genetic regulatory mechanisms in the synthesis of proteins. J Mol Biol 3:318
Kellermann VM, van Heerwaarden B, Hoffmann AA, Sgró CM (2006) Very low additive genetic variance and evolutionary potential in multiple populations of two rainforest *Drosophila* species. Evolution 60:1104–1108

Kimura M (1968) Evolutionary rate at the molecular level. Nature 217:624–626
Kimura M (1983) The neutral theory of molecular evolution. Cambridge University Press, Cambridge
King JL, Jukes TH (1969) Non-darwinian evolution. Science 164:788–798
King M-C, Wilson AC (1975) Evolution at two levels: molecular similarities and biological differences between humans and chimpanzees. Science 188:107–116
Kingsolver JG, Hoekstra HE, Hoekstra JM et al (2001) The strength of phenotypic selection in natural populations. Am Nat 157:245–261
Kirkpatrick M (2009) Patterns of quantitative genetic variation in multiple dimensions. Genetica 136:271–284
Kirkpatrick M, Barton NH (1997) Evolution of a species' range. Am Nat 150:1–23
Kirschner MW, Gerhart JC (2005) The plausibility of life: resolving Darwin's dilemma. Yale University Press, New Haven, Connecticut
Kirschner MW, Gerhart JC (2010) Facilitated variation. In: Pigliucci M, Müller G (eds) Evolution: the extended synthesis. MIT Press, Cambridge, Massachusetts, pp 253–280
Koonin EV, Wolf YI (2009) Is evolution Darwinian and/or Lamarckian? Biol Direct 4:42
Kutschera U, Niklas KJ (2004) The modern theory of biological evolution: an expanded synthesis. Naturwissenschaften 91:255–276
Lande R (2009) Adaptation to an extraordinary environment by evolution of phenotypic plasticity and genetic assimilation. J Evol Biol 22:1435–1446
Ledón-Rettig CC, Pfennig DW, Crespi EJ (2010) Diet and hormonal manipulation reveal cryptic genetic variation: implications for the evolution of novel feeding strategies. Proc R Soc B 277:3569–3578
Levin DA (1983) Polyploidy and novelty in flowering plants. Am Nat 122:1–25
Levinton JS (2001) Genetics, paleontology, and macroevolution, 2d edn. Cambridge University Press, Cambridge
Levinton JS, Futuyma DJ (1982) Macroevolution: pattern and process introduction and background. Evolution 36:425–473
Lewontin RC (1974) The genetic basis of evolutionary change. Columbia University Press, New York
Lewontin RC (1977) Caricature of Darwinism. Nature 266:283–284 (Book review of R. Dawkins, *The selfish gene.*)
Lewontin RC (1979) Sociobiology as an adaptationist program. Behav Sci 24:5–14
Lieberman BS, Dudgeon (1995) An evaluation of stabilizing selection as an explanation for stasis. Palaeogeogr, Palaeoclimatol Palaeoecol 127:229–238
Love AC (2003) Evolutionary morphology, innovation, and the synthesis of evolutionary and developmental biology. Biol Phil 18:309–345
Love AC (2009) Marine invertebrates, model organisms and the modern synthesis: epistemic values, evo-devo, and exclusion. Theor Biosci 128:19–42
Love AC, Raff RA (2003) Knowing your ancestors: themes in the history of evo-devo. Evol Devel 5:327–330
Lynch M, Walsh B (1998) Genetics and analysis of quantitative traits. Sinauer, Sunderland, Massachusetts
Macnair M (1997) The evolution of plants in metal-contaminated environments. In: Bijlsma R, Loeschcke V (eds) Environmental stress, adaptation and evolution. Birkhäuser, Basel, pp 3–24
Magnuson-Ford K, Otto SP (2012) Linking the investigations of character evolution and species diversification. Am Nat 180:225–245
Malmgren BA, Berggren WA, Lohmann GP (1983) Evidence for punctuated gradualism in the Late Neogene *Globorotalia tumida* lineage of planktonic Foraminifera. Paleobiology 9:377–389
Marroig G, Cheverud JM (2005) Size as a line of least resistance: diet and adaptive morphological radiation in New World monkeys. Evolution 59:1128–1142

Martin M, Orgogozo V (2013) The loci of repeated evolution: a catalog of genetic hotspots of phenotypic variation. Evolution 67:1235–1250
Masel J, King OD, Maughan H (2007) The loss of adaptive plasticity during long periods of environmental stasis. Am Nat 169:38–46
Mather K (1955) Response to selection: synthesis. Cold Spring Harbor Symp Quant Biol 20:1128–1142
Mattila TM, Bokma F (2008) Extant mammal body masses suggest punctuated equilibrium. Proc R Soc B 275:2195–2199
Maynard Smith J (1966) Sympatric speciation. Am Nat 100:637–650
Maynard Smith J (1978) Optimization theory in evolution. Annu Rev Ecol Syst 9:31–56
Maynard Smith J, Burian R, Kaufman S et al (1985) Developmental constraints and evolution. Quart Rev Biol 60:265–287
Mayr E (1942) Systematics and the origin of species. Columbia University Press, New York
Mayr E (1954) Change of genetic environment and evolution. In: Huxley J, Hardy AC, Ford EB (eds) Evolution as a process, pp 157–180. Allen and Unwin, London. (Reprinted in Mayr E, Evolution and the diversity of life, pp 188–210, Harvard University Press, 1976.)
Mayr E (1960) The emergence of evolutionary novelties. In: Tax S (ed) The evolution of life. University of Chicago Press, Chicago, pp 157–180
Mayr E (1961) Cause and effect in biology. Science 134:1501–1506
Mayr E (1963) Animal species and evolution. Harvard University Press, Cambridge, Massachusetts
Mayr E (1976) Evolution and the diversity of life: selected essays. Harvard University Press, Cambridge, Massachusetts
Mayr E (1982) The growth of biological thought. Harvard University Press, Cambridge, Massachusetts
Mayr E (1993) What was the evolutionary synthesis? Trends Ecol Evol 8:31–34
Mayr E, Provine WB (eds) (1980) The evolutionary synthesis: perspectives on the unification of biology. Harvard University Press, Cambridge, Massachusetts
Mc Guigan K, Nishimura N, Currey M et al (2010) Cryptic genetic variation and body size evolution in threespine stickleback. Evolution 65:1203–1211
McCauley DE (1993) Evolution in metapopulations with frequent local extinction and recolonization. Oxford Surv Evol Biol 9:109–134. Oxford University Press, Oxford
McShea DW (1994) Mechanisms of large-scale evolutionary trends. Evolution 48:1747–1763
Minelli A (2010) Evolutionary developmental biology does not offer a significant challenge to the neo-Darwinian paradigm. In: Ayala FJ, Arp R (eds) Contemporary debates in philosophy of biology. Wiley, Chichester, pp 213–226
Mitter C, Farrell BD, Wiegmann B (1988) The phylogenetic study of adaptive zones: has phytophagy promoted insect diversification? Am Nat 132:107–128
Moczek AP (2012) The nature of nurture and the future of evodevo: toward a theory of developmental evolution. Integr Comp Biol 52:108–119
Moczek AP, Sultan S, Foster S et al (2011) The role of developmental plasticity in evolutionary innovation. Proc R Soc B 278:2705–2713
Monroe MJ, Bokma F (2009) Do speciation rates drive rates of body size evolution in mammals? Am Nat 174:912–918
Moran NA (1992) The evolutionary maintenance of alternative phenotypes. Am Nat 139:971–989
Müller G (2007) Evo-devo: extending the evolutionary synthesis. Nature Rev Genet 8:943–949
Müller G (2010) Epigenetic innovation. In: Pigliucci M, Müller G (eds) Evolution: the extended synthesis. MIT Press, Cambridge, Massachusetts, pp 307–332
Murray JD (1981) A pre-pattern formation mechanism for animal coat markings. J Theor Biol 88:161–199
Newman SA (2010) Dynamic patterning modules. In: Pigliucci M, Müller G (eds) Evolution: the extended synthesis. MIT Press, Cambridge, Massachusetts, pp 281–306
Nosil P (2012) Ecological speciation. Oxford University Press, Oxford

Orr HA (1999) An evolutionary dead end? (Review of *Phenotypic evolution*, by CD Schlichting and M Pigliucci). Science 285:343–344
Orr HA, Coyne JA (1992) The genetics of adaptation revisited. Am Nat 140:725–774
Oster G, Odell G, Alberch P (1980) Mechanics, morphogenesis, and evolution. Lect Math Life Sci 13:165–255
Pavlicev M, Wagner GP (2012) A model of developmental evolution: selection, pleiotropy and compensation. Trends Ecol Evol 27:316–322
Pfennig DW, Wund MA, Snell-Rood EC et al (2010) Phenotypic plasticity's impacts on diversification and speciation. Trends Ecol Evol 25:459–467
Pigliucci M (2001) Phenotypic plasticity: beyond nature and nurture. Johns Hopkins University Press, Baltimore
Pigliucci M, Müller G (eds) (2010) Evolution: the extended synthesis. MIT Press, Cambridge, Massachusetts
Price TD, Qvarnström A, Irwin DE (2003) The role of phenotypic plasticity in driving genetic evolution. Proc R Soc Lond B 270:1433–1440
Provine WB (1986) Sewall Wright and evolutionary biology. University of Chicago Press, Chicago
Provine WB (2001) The origins of theoretical population genetics, 2nd edn. University of Chicago Press, Chicago
Rabosky D (2010) Extinction rates should not be estimated from molecular phylogenies. Evolution 64:1816–1824
Rabosky DL (2012) Positive correlation between diversification rates and phenotypic evolvability can mimic punctuated equilibrium on molecular phylogenies. Evolution 66:2622–2627
Ramsey J (2011) Polyploidy and ecological adaptation in wild yarrow. Proc Nat Acad Sci USA 108:7096–7101
Raup DM, Gould SJ, Schopf TJM et al (1973) Stochastic models of phylogeny and the evolution of diversity. J Geol 81:525–542
Reif W-E, Junker T, Hoßfeld U (2000) The synthetic theory of evolution: general problems and the German contribution to the synthesis. Theor Biosci 119:41–91
Rensch B (1939) Typen der Artbildung. Biol Rev 14:180–222
Rensch B (1943) Die paläontologischen Evolutionsregeln in zoologischer Betrachtung. Biologia Generalis 17:1–55
Rensch B (1947) Neuere Probleme der Abstammungslehre. Enke, Stuttgart
Rensch B (1959) Evolution above the species level. Wiley, New York
Ricklefs RE (1980) Phyletic gradualism versus punctuated equilibrium: applicability of neontological data. Paleobiology 6:271–275
Riedl R (1978) Order in living organisms: a systems analysis of evolution. Wiley, New York
Robson GC, Richards OW (1936) The variation of animals in nature. Longmans, Green, London
Rovito SM (2010) Lineage divergence and speciation in the web-toed salamanders (Plethodontidae: *Hydromantes*) of the Sierra Nevada. California Mol Ecol 19:4554–4571
Schindewolf OH (1950) Grundfragen der Paläontologie. Schweitzerbart, Stuttgart
Schlichting CD, Pigliucci M (1998) Phenotypic evolution: a reaction norm perspective. Sinauer, Sunderland, Massachusetts
Schluter D (1996) Adaptive radiation along genetic lines of least resistance. Evolution 50:1766–1774
Schmalhausen II (1949) Factors of evolution. Blakiston, Philadelphia
Schwander T, Leimar O (2011) Genes as leaders and followers in evolution. Trends Ecol Evol 26:143–151
Schwenk K, Wagner GP (2004) The relativism of constraints on phenotypic evolution. In: Pigliucci M, Preston K (eds) Phenotypic integration: studying the ecology and evolution of complex phenotypes. Oxford University Press, Oxford, pp 390–408
Scoville AG, Pfrender ME (2010) Phenotypic plasticity facilitates recurrent rapid adaptation to introduced predators. Proc Nat Acad Sci USA 107:4260–4263

Sepkoski JJ Jr (1996) Competition in macroevolution: the double wedge revisited. In: Jablonski D, Erwin DH, Lipps J (eds) Evolutionary paleobiology. University of Chicago Press, Chicago, pp 211–255

Sheppard PM, Turner JRG, Brown KS et al (1985) Genetics and the evolution of Müllerian mimicry in *Heliconius* butterflies. Phil Trans R Soc Lond B 308:433–613

Simpson GG (1944) Tempo and mode in evolution. Columbia University Press, New York

Simpson GG (1949) Essay-review of recent works on evolutionary theory by Rensch, Zimmermann, and Schindewolf. Evolution 3:178–184

Simpson GG (1953a) The Baldwin effect. Evolution 7:110–117

Simpson GG (1953b) The major features of evolution. Columbia University Press, New York

Slatkin M (1977) Gene flow and genetic drift in a species subject to frequent local extinctions. Theor Pop Biol 12:253–262

Slatkin M (1996) In defense of founder-flush theories of speciation. Am Nat 147:493–505

Slatkin M (2009) Epigenetic inheritance and the missing heritability problem. Genetics 182:845–850

Smocovitis VB (1996) Unifying biology: the evolutionary synthesis and evolutionary biology. Princeton University Press, Princeton

Smocovitis VB (2010) "Where are we?" Historical reflections on evolutionary biology in the twentieth century. In: Bell MA, Futuyma DJ, Eanes WF, Levinton JS (eds) Evolution since Darwin: the first 150 years. Sinauer, Sunderland, Massachusetts, pp 49–58

Snell-Rood EC, Van Dyken JD, Cruikshank T et al (2010) Toward a population genetic framework of developmental evolution: the costs, limits, and consequences of phenotypic plasticity. BioEssays 32:71–81

Stanley SM (1973) An explanation for Cope's rule. Evolution 27:1–26

Stanley SM (1975) A theory of evolution above the species level. Proc Nat Acad Sci USA 72:646–650

Stebbins GL (1950) Variation and evolution in plants. Columbia University Press, New York

Stebbins GL (1974) Flowering plants: evolution above the species level. Harvard University Press, Cambridge, Massachusetts

Stebbins GL, Ayala FJ (1981) Is a new evolutionary synthesis necessary? Science 213:967–971

Sterelny K (2000) Development, evolution, and adaptation. Phil Sci (Proc) 67:S369–S387

Stern DL (2011) Evolution, development, and the predictable genome. Roberts, Greenwood Village, Colorado

Stolfus A (2006) Mutationism and the dual causation of evolutionary change. Evol Develop 8:304–317

Storz JF, Scott GR, Cheviron ZA (2010) Phenotypic plasticity and genetic adaptation to high-altitude hypoxia in vertebrates. J Exp Biol 213:4125–4136

Strotz LC, Allen AP (2013) Assessing the role of cladogenesis in macroevolution by integrating fossil and molecular evidence. Proc Nat Acad Sci USA 110:2904–2909

Suzuki Y, Nijhout HF (2006) Evolution of a polyphenism by genetic accommodation. Science 311:650–652

Templeton A (1982) Why read Goldschmidt? Paleobiology 8:474–481

Thompson DW (1917) On growth and form. Cambridge University Press, Cambridge

Thompson JN (2005) The geographic mosaic of coevolution. University of Chicago Press, Chicago

Thompson JN (2013) Relentless evolution. University of Chicago Press, Chicago

True JR, Carroll SB (2002) Gene co-option in physiological and morphological evolution. Annu Rev Cell Devel Biol 18:53–80

Turing AM (1952) The chemical basis of morphogenesis. Phil Trans R Soc Lond B 237:37–72

Uyeda JC, Hansen TF, Arnold SJ, Pienaar J (2011) The million-year wait for evolutionary bursts. Proc Nat Acad Sci USA 108:15908–15913

Venditti C, Pagel M (2010) Speciation as an active force in promoting genetic evolution. Trends Ecol Evol 25:14–20

Voss SR, Smith JJ (2005) Evolution of salamander life cycles: a major-effect quantitative trait locus contributes to discrete and continuous variation for metamorphic timing. Genetics 170:275–281
Waddington CH (1953) Genetic assimilation of an acquired character. Evolution 7:118–126
Waddington CH (1957) The strategy of the genes. Allen and Unwin, London
Wade MJ, Beeman RW (1994) The population dynamics of maternal-effect and selfish genes. Genetics 138:1309–1314
Wade MJ, Goodnight CJ (2000) The ongoing synthesis: a reply to Coyne et al. Evolution 54:317–324
Wagner GP (2010) Evolvability: the missing piece in the neo-Darwinian synthesis. In: Bell MA, Futuyma DJ, Eanes WF, Levinton JS (eds) Evolution since Darwin: the first 150 years, pp 197–213. Sinauer, Sunderland, Massachusetts
Wagner GP, Pavlicev M, Cheverud JM (2007) The road to modularity. Nature Rev Genet 8:921–931
Wake DB (2009) What salamanders have taught us about evolution. Annu Rev Ecol Evol Syst 40:333–352
Walsh B, Blows MW (2009) Abundant genetic variation + strong selection = multivariate genetic constraints: a geometric view of adaptation. Annu Rev Ecol Evol Syst 40:41–59
West-Eberhard MJ (2003) Developmental plasticity and evolution. Oxford University Press, Oxford
Wiens JJ, Graham CH (2005) Niche conservatism: integrating evolution, ecology, and conservation biology. Annu Rev Ecol Evol Syst 36:519–539
Williams GC (1992) Natural selection: domains, levels, and challenges. Oxford University Press, New York
Winsor MP (2006) The creation of the essentialism story: an exercise in metahistory. Hist Phil Life Sci 28:149–174
Wolf JB, Brodie ED Jr, Moore AJ (1999) The role of maternal and paternal effects in the evolution of parental quality by sexual selection. J Evol Biol 12:1157–1167
Wood TE, Burke JM, Rieseberg LH (2005) Parallel genotypic adaptation: when evolution repeats itself. Genetica 123:157–170
Wright S (1932) The roles of mutation, inbreeding, crossbreeding, and selection in evolution. Proc 6th Int Cong Genet 1:356–366
Wund MA (2012) Assessing the impact of phenotypic plasticity on evolution. Integr Comp Biol 52:5–18
Wund MA, Valeney S, Wood S, Baker JA (2012) Ancestral plasticity and allometry in threespine stickleback reveal phenotypes associated with derived, freshwater ecotypes. Biol J Linn Soc 105:573–583
Yeung CKL, Tsai PW, Chesser RT et al (2011) Testing founder effect speciation: divergence population genetics of the spoonbills *Platalea regia* and *Pl. minor* (Threskiornithidae: Aves). Mol Biol Evol 28:473–482

Evolution as a Largely Autonomous Process

Folmer Bokma

Abstract Evolutionary theory has proven generally successful in predicting phenotypic changes over one or a few generations of natural or artificial selection, but fails to predict evolutionary dynamics over longer periods of time, which is a major shortcoming: At longer timescales, existing theory is largely concerned with a posteriori explanations and cannot even predict whether a population or species will adapt to environmental change, or go extinct. Based on a review of key literature from before Darwin to today, I argue that the reason for this shortcoming is that in the Modern Synthesis fitness is regarded as determined exclusively by how well traits are suited to the biotic and abiotic environment. I argue that much can be gained by explicitly considering that fitness has a significant intrinsic component, determined by how well different traits are adapted to each other. Due to adaptation of traits to each other, those traits that are important for the functioning of many other traits can vary only within narrow tolerance limits. Short-term selection experiments and year-to-year fluctuations in natural populations taking place within these tolerance limits give the appearance of rapid evolution. Yet the tolerance limits will prevent changes to accumulate over time, and hence, these traits evolve in a million years no more than they do in a decade. Only traits like coloration that have little influence on other traits can evolve freely, but that will rarely be sufficient to prevent extinction. Significant evolutionary departures require a reshuffling of the interactions between traits and will often coincide with speciation. Emerging from a complex system of interacting traits, the magnitude and direction of these changes will be largely independent of the factors that triggered them, rendering macroevolution a largely autonomous process.

Keywords Extinction · Macroevolution · Microevolution · Natural selection · Speciation

F. Bokma (✉)
IceLab and Department of Ecology and Environmental Science,
Umeå University, Umeå, Sweden
e-mail: folmer.bokma@emg.umu.se

E. Serrelli and N. Gontier (eds.), *Macroevolution*,
Interdisciplinary Evolution Research 2, DOI 10.1007/978-3-319-15045-1_3

1 Historical Examples

In the study of biological evolution, a distinction is often made between microevolution and macroevolution. Microevolution refers to the process of change over time in a species or population, for example genetic or phenotypic change, or changes in its geographic distribution, or in its interactions with other species. Macroevolution on the other hand refers to processes that typically involve a timescale spanning the origination and extinction of species. Speciation and extinction dynamics, but also long-term genetic and phenotypic evolution are typical subjects of macroevolutionary study. Thus, micro- and macroevolution concern partly overlapping processes, but at timescales that typically differ by at least three orders of magnitude.

A long-standing issue in evolutionary biology, which has been debated ever since Darwin and Wallace presented their theory of evolution in 1859, is whether micro- and macroevolution are one and the same process observed at different timescales, or essentially distinct processes. Because microevolution can be readily studied experimentally, our mechanistic understanding of microevolutionary processes is much better than our understanding of macroevolution, which is more restricted to patterns, rather than processes. There is abundant evidence that microevolution is due to natural (and sometimes sexual) selection, as Darwin and Wallace originally hypothesized. Consequently, the debate whether micro- and macroevolution are equivalent processes has focused on the question whether or not macroevolution is an adaptive process governed by survival of the fittest individuals in a population.

Here, I will illustrate the history of this debate using examples from various fields of study that have bearing on the issue. I will show that fossil records as well as present-day species indicate that extrapolating findings from microevolutionary studies to a macroevolutionary timescale leads to predictions that differ dramatically from observed patterns. I will then review findings from developmental and population genetics during the later part of the past century that could provide an explanation for this discrepancy. Based on these insights, I will finally address the relation between micro- and macroevolution and the critical importance of speciation in biological evolution.

1.1 The Predicted Hawk Moth

Despite its beautiful, bright flowers, the Madagascan orchid *Angraecum sesquipedale* is seldom grown by amateur breeders. Nevertheless, to many biologists, the species is well known as a key player in one of the early success stories of evolutionary biology. When Charles Darwin was sent some specimens of this orchid by James Bateman, he must have recognized immediately the characteristic feature of the flowers: an unusually long nectar spur. Darwin described these extremely long spurs in his 1862 "On the various contrivances by which British and foreign orchids are fertilized by insects, and on the good effects of intercrossing." He argued that although the flowers' up to 40 cm nectar spurs were unusually long, there presumably existed a pollinator with a suitably long proboscis to obtain the nectar and pollinate the plant. By itself, this prediction was perhaps not as daring as some authors have later suggested: Darwin was aware of the existence of species with flowers with very long spurs (albeit not as long as those of *A. sesquipedale*), and he also knew that these species were typically pollinated by hawk moths. Therefore, he predicted not only that given the existence of a flowering plant with 40 cm nectar spurs, there presumably had to exist a pollinator with an equally long proboscis, but also that this pollinator presumably was a hawk moth. What makes this prediction remarkable is that Darwin proposed a mechanism for the process that led to the unusual length of the nectar spurs, as well as the proboscis of the (as yet hypothetical) pollinator: evolution by natural selection.

Darwin (1862) hypothesized that pollinators with longer tongues would obtain nectar from an orchid without pollinating it, causing selection for longer spurs. Longer spurs, in turn, would exert selection on pollinators for longer tongues. This evolutionary arms race may sound familiar nowadays, but in Darwin's time, it was a contentious issue. Indeed, George Campbell, the 8th duke of Argyll wrote in his 1867 "The Reign of Law" that the intricate complexity of the flowers of *A. sesquipedale* reflected divine creation. As a reaction in the same year, Alfred Russel Wallace wrote "Creation by Law," in which he defended Darwin's coevolutionary model. A pollinator for *A. sesquipedale* was not found during Darwin's life. Only in 1903, Lionel Walter Rothschild and Karl Jordan found a moth with a suitably long tongue on Madagascar. It was named *Xanthopan morganii praedicta*—because Darwin had predicted its existence. (It was later found that the specimen did not represent a subspecies of the continental *X. morganii*, but the story of its discovery is far better known than the species itself, so it remains widely known by its subspecific epithet.)

The arms race that Darwin and Wallace hypothesized for the evolution of long spurs and probosces is of course not unique to orchids and hawk moths. In 1973, Van Valen famously called this type of mutually antagonistic coevolution for the Red Queen, a character in Lewis Carroll's 1871 "Through the Looking-Glass, and What Alice Found There," who explained to Alice about Looking-Glass Land: "Now, here, you see, it takes all the running you can do, to keep in the same place." The analogy with an evolutionary arms race is evident: If orchids evolve longer nectar spurs, hawk moths have to evolve longer tongues to obtain the same amount of nectar, and vice versa, if the moths evolve longer tongues, the orchids have to keep up with longer spurs to attain the same pollination success. Van Valen used the Red Queen as a hypothetical explanation for the observation that the extinction risk of genera is constant through time. Older genera are no more likely to go extinct than young genera, which could be due to species continuously coevolving.

Presumably, because of the contrast between the duke of Argyll's rigid religious views on the natural world and Darwin's idea of a highly dynamic arms race between evolving species, the discovery of *X. morganii praedicta* has often been heralded as a prime example of the explanatory power of evolution. However, the coevolutionary arms race suspected by Wallace (1867) and Darwin (1862) suggests that the Red Queen keeps on running: It would lead to ever longer spurs and tongues. Yet she appears to have halted, as there is no evidence that the nectar spurs of *A. sesquipedale* are any longer now than they were in Darwin's time. What's more, most orchid species have shorter spurs, and most pollinators have shorter tongues. Indeed, among all of the thousands of species of flowering plants, *A. sesquipedale* is virtually unique in having such long nectar spurs. By far, most species have very short spurs compared to the few species that are pollinated by hawk moths. As Bradshaw pointed out in his 1991 Croonian lecture to the Royal Society of London, "most species are very stable, and in situations where evolution is observed in one species often none is found in others despite equivalent opportunity."

As recently as 2007, Whittall and Hodges studied the coevolution of nectar spurs and the probosces of their pollinators, not in *Angraecum* but in a clade of columbine flowers, using modern phylogenetic methods. They found that there are several different sizes of spurs facilitating several different pollinators, from bumblebees with relatively short tongues to hummingbirds and hawk moths with long tongues. They found evidence that—as Darwin (1862) and Wallace (1867) had predicted—spur length increases rather than decreases, but only seldom, in rapid bursts instead of continuously. Therefore they hypothesized that when a new pollinator with a longer tongue enters the range of a flowering plant species, its spurs catch up rapidly, while pollinators with shorter tongues continue to effectively facilitate pollen transfer in the species with shorter spurs.

This finding is interesting for two reasons: It illustrates the possibilities that modern molecular techniques, in particular molecular phylogenetic analysis of species' phenotypes, offer to study macroevolution. More importantly, it illustrates that evolutionary biologists' interpretation of the history of *A. sesquipedale* and *X. morganii praedicta* as an example of the predictive power of evolution is rather optimistic. Darwin (1862) and Wallace's (1867) theory predicts that plants and their pollinators engage in an evolutionary arms race to explain unusual phenotypes in a minority of species, but fails to predict why the vast majority of species do not engage in the same arms race, despite equivalent opportunity. In addition, the theory explains that flowers and tongues get longer, not whether and when that process stops. Many evolutionary biologists take this for granted as just another aspect of nature's great and dazzling variety.

1.2 Fossil Elephants

The observation that evolution is not quite as dynamic as Darwin and Wallace perhaps first envisioned, had been pointed out to Darwin already soon after the first edition of "The Origin" was published (1859). One of the first to read Darwin's book was Scottish paleontologist Hugh Falconer. In India, where he was head of the Saharánpur botanical gardens, Falconer made detailed studies of the fossils pertaining to species of mastodon, rhinoceros, and elephant. Forced by deteriorating health, he returned to Europe and began summarizing his work on fossil remains of elephants when he received a copy of Darwin's book. In a letter dated

September 1862, Falconer wrote to Darwin: “I am bringing out a heavy memoir on Elephants—an *Omnium gatherum* affair—with observations on the fossil and recent species. One section is devoted to the persistence in time of the specific characters of the mammoth. I trace him, from before the Glacial period, through it, and after it, unchangeable and unchanged, so far as the organs of digestion (teeth) and locomotion are concerned. Now the glacial period was no joke—it would have made ducks and drakes of your dear pigeons and doves.” Darwin was quick to reply, and his reply to Falconer is particularly interesting. He wrote: “You speak of these animals as having been exposed to vast range of climatal changes from before to after the Glacial period; I should have thought from analogy of sea-shells, that by migration (or local extinction when migration not possible) these animals might and would have kept under nearly the same climate.”

From this letter, it is clear that Darwin was aware that “seashells” responded to climate change by tracking their preferred habitat rather than by adapting, because he suggested that the same could apply to the European mammoth. Darwin was right in his suggestion: Migration of whole associations of species during Quaternary climatic fluctuations has now been described in detail from, for example, pollen records (Bradshaw 1991; Thomas et al. 2004; Davis and Shaw 2001; see also Eldredge et al. 2005). Further support comes from extant species that experience climate change and follow their preferred environmental conditions in space (Kelly and Goulden 2008; Tingley et al. 2009). As Bradshaw pointed out in his Croonian lecture: “It is customary to take these migrations for granted as a reflection of the ecological preferences of the species concerned. Yet is this acceptable? There is a valid alternative scenario that the species concerned … could have remained in situ and coped with the environmental improvement by evolutionary change. There is no sign of this; the stability of the ecological preferences of the species in the face of such major environmental alterations is impressive.” Thus, although Darwin was correct that mammoths could probably, like most other species, have responded to climate change by tracking their niche in space, his reply failed to answer Falconer’s question. Falconer did not ask “how would mammoths have responded to climate change,” but why they did not show evolutionary change despite great climatic vicissitudes. Generations of evolutionary biologists have followed this example, ascribing change to natural selection when it occurs, but failing to account for the frequent cases where no evolution is observed. Hence, as recently as 2010, Futuyma wrote that “we are led to confront an overridingly important question that highlights the immaturity of our science: how can we predict whether a population (or species) will adapt or become extinct in the face of environmental change?”

2 Punctuated Equilibria

From Futuyma’s remarks, it may be clear that the development of evolutionary theory has anything but come to an end with the Modern Synthesis: There are too many instances where evolution is not observed while it would be expected, and evolutionary theory fails to predict when and where and to what extent evolution will take place, and when no evolution should be expected. To better understand what revisions of the Modern Synthesis are needed, let us return to Falconer's critical 1862 letter to Darwin. During the following years, they exchanged several letters, and Falconer pointed out that in the fossil records he studied, species often appear suddenly differentiated, to remain largely unchanged until going extinct. In other words, judging from the fossil record, evolutionary change is largely constrained to incipient species. Eventually, Darwin acknowledged in the third edition of "The Origin" that "It is a more important consideration, clearly leading to the same result, as lately insisted on by Dr. Falconer, namely, that the periods during which species have been undergoing modification, though very long as measured by years, have probably been short in comparison with the periods during which these same species remained without undergoing any change." Thus, Darwin acknowledged Falconer's observations, but was not worried because they were "clearly leading to the same result." It is difficult to judge exactly what "result" Darwin refers to. It is possible that he meant that species are not divinely created units, but emerge from other species. In that case, Darwin was correct that irrespective of whether change is rapid or slow, the species are historically related.

It is also conceivable, however, that Darwin meant that species diverge due to natural selection, and that it does not matter whether this divergence happens rapidly in incipient species or gradually over time. We may never know precisely what Darwin meant, but it is clear that later students of evolution disagreed on this issue; whether rapid change in incipient species leads to the same result, as Darwin claimed. The periods of rapid change were themselves not contested. Rather, discussion concerned the forces that were responsible. In his seminal 1944 "Tempo and Mode in Evolution," George Gaylord Simpson indicated the great relevance of this question when he wrote that the most important distinction between students of evolution is between "those who believe that discontinuity arises by intensification or combination of the differentiating processes already effective within a potentially or really continuous population and those who maintain that some essentially different factors are involved. This is related to the old, but still vital problem of microevolution as opposed to macroevolution."

It is interesting that Simpson called the relation between microevolution and macroevolution an "old" question already in the first half of the twentieth century, because most present-day evolutionary biologists presumably associate the issue not with Simpson's 1944 "Tempo and Mode" or still older works, but with Eldredge and Gould's 1972 theory of punctuated equilibria. In a seminal contribution to the paleobiological revolution (Maynard Smith 1984), they claimed that evolution is not due to the gradual accumulation of adaptive differences in established species from one generation to the other, as Darwin envisioned. They claimed that the fossil record provides abundant evidence that species change rapidly when they originate, to subsequently remain static until going extinct (Eldredge and Gould 1972). Subsequent studies of fossil populations confirmed this notion (e.g., Lieberman et al. 1995), so that the frequent occurrence of stasis in the fossil record now has good statistical support (Hunt 2007; see also

Uyeda et al. 2011). As may be evident from the sections above, Eldredge and Gould (1972) were not the first to make these points. Falconer had pointed out to Darwin that change is often absent in established species despite dramatic environmental change and questioned whether this was compatible with Darwin's theory of gradual transformation of species. Several students of evolution had expressed doubts whether evolution by natural selection could, even accumulated over geological periods of time, account for more substantial differences between higher taxa, and hence, already Simpson (1944) referred to this as an old issue. However, between "Tempo and Mode in Evolution" and Eldredge and Gould's theory of punctuated equilibira, the Modern Synthesis had "hardened" (Gould 2002) to the extent that most evolutionary biologists not only rejected punctuated equilibrium theory, but even claimed that stasis interspersed with rapid change is not an accurate description of biological evolution. In other words, Simpson still considered these "discontinuities" in the fossil record as a generally accepted phenomenon, discussion focusing on the underlying mechanism, but 25 years later many evolutionary biologists dismissed even the phenomenon as an artifact of incomplete fossil records. For example, Stebbins and Ayala (1981) argued that fossil species can be distinguished *only if* morphologically distinct, and others claimed that discontinuities were due to incomplete fossil records. Gould (2002) replied to such claims with "stasis is data": The observation that species persist "unchangeable and unchanged," as Falconer put it in his 1862 letter to Darwin, cannot be due to missing data. Some students of evolution expressed support for punctuated equilibria. Stanley, for example, (in)famously claimed (1975) that "if most evolutionary change occurs during speciation and if speciation events are largely random, natural selection, long viewed as the process guiding evolutionary change, cannot play a significant role in determining the overall course of evolution. Macroevolution is decoupled from microevolution…".

2.1 Phylogenetic Evidence for Punctuated Equilibria

In Darwin and Falconer's time, evidence for stasis had to come from the fossil record. At present, there are more than just fossil data pertaining to the frequency of stasis. Soon after Eldredge and Gould had published their theory of punctuated equilibria in 1972, it was realized that molecular phylogenies can complement the time frame provided by the fossil record (Avise and Ayala 1975). Avise (1977) compared phenotypic variation between clades of fishes: In a clade with few species, few speciation events would have contributed to morphological variation, whereas in a species-rich clade, many speciation events would have contributed. If evolution is restricted to incipient species, then the species-rich clade would show greater phenotypic variation than the species-poor clade. This approach circumvents the problem that fossil species may not represent biological species, or that they would be distinguishable only if substantially different (Stebbins and Ayala 1981), and has the additional advantage that one can study traits that cannot be studied from the fossil record. Later methods to distinguish punctuated equilibria from gradual change using present-day species are based on the same principle as Avise and Ayala's pioneering attempts (Ricklefs 1980; Douglas and Avise 1982; Mooers and Schluter 1998, 1999; Mooers et al. 1999; Bokma 2002, 2008). Details of methods differ, but their principle is the same: If evolutionary divergence takes place gradually over time, then recent sister species cannot be significantly diverged, because time has been insufficient for differences to accumulate. By contrast, if evolution is dramatically accelerated in incipient species, then even recent sister species can be substantially diverged. After the initial phase, differentiation accumulates, but under gradual evolution this is due to greater time since divergence, whereas under speciational evolution it is due to a growing number of speciation events separating species. Molecular phylogenies provide an estimate of the time and number of speciation events separating any two extant species, and hence, it is possible to distinguish the contributions of gradual and speciational change to variation between species.

Phylogenetic methods have now been applied to data from present-day species in a number of studies, and evidence accumulates suggesting that Falconer was right when he wrote to Darwin that species change rapidly initially, to remain virtually unchanged until going extinct. A particularly interesting example was already mentioned above and concerns spur length in columbine flowers: Whittall and Hodges (2007) found that there are several different sizes of nectar spurs facilitating an array of pollinators, from bumblebees with relatively short tongues to hummingbirds and hawk moths with long tongues. Evolution of spur length seems to be almost exclusively toward longer spurs. However, this does not seem to be due to an arms race between flowers and pollinators, as Darwin (1862) and Wallace (1867) initially suggested. Instead, increases in spur length appear rapid, following the punctuated equilibrium pattern, and are often accompanied by a change in pollination syndrome. Whittall and Hodges therefore suggested that when a new pollinator with a longer tongue enters the range of a columbine population, the spurs evolve rapidly, but the pollinators with shorter tongues keep pollinating the species with shorter spurs. Similarly, stasis in floral morphology in Malpighiaceae appears to be due to the relation between plant and its specialist New World oil-bee pollinators (Davis et al. 2014). One of the best studied groups of species are the mammals, and therefore we used mammals to test whether body size evolution shows signs of rapid change in incipient species. We found that presumably around 60 % of variation in body mass between present-day mammalian species is due to short periods of rapid change in incipient species. For primates and carnivores exceptionally good phylogenies are available, as well as good measurements of species' average body masses. Interestingly, we found that primates appear to diverge about three times faster if evolution is assumed to be purely gradual, but when rapid evolution during speciation is taken into account, both clades appear to have identical rates of evolution (Mattila and Bokma 2008). We also showed that variation in the rate of body size diversification between mammalian clades could be explained by differences in speciation and extinction rates (Monroe and Bokma 2009). These findings do not only provide evidence for a punctuated equilibrium model of evolution, they also question the role of natural selection, because if body masses were locally adaptive, we would not expect great differences in rates of diversification between clades that share the same environment. In other words, if speciation occurs more often in one clade, one would expect the body size changes during speciation to be smaller in this clade. The observation that the magnitude of rapid change during evolution is independent of how frequently such changes occur (Mattila and Bokma 2008; Monroe and Bokma 2009) suggests that the environment has little effect on this process.

Because phylogenetic analyses are based entirely on present-day species, they allow analyses of characters that are not preserved in the fossil record. A good example of the use of this feature is studies by Pagel and colleagues, who tested for acceleration of nucleotide sequence evolution during speciation (Webster et al. 2003; Pagel et al. 2006; Venditti and Pagel 2010). They used phylogenies of a wide variety of taxa and found that speciation has a significant accelerating effect on evolution also at the level of DNA.

3 The Role of Natural Selection

Given the accumulating evidence that species change rapidly when they originate to subsequently remain static, the question becomes whether that pattern can be reconciled with the Modern Synthesis' view of macroevolution as the long-term consequence of microevolution driven by natural selection. Rapid change per se can hardly be argued to challenge the role of natural selection. Wallace already noted in 1889 that "Mr Darwin was rather inclined to exaggerate the necessary slowness of the action of natural selection … but with the knowledge we now possess of the great amount and range of individual variation, there seems no difficulty in an amount of change, quite equivalent to that which usually distinguishes allied species, sometimes taking place in less than a century, should any rapid change in conditions necessitate an equally rapid adaptation." His statement was confirmed mathematically about a century later by theoretical work suggesting that selection as well as genetic drift could lead to rapid evolution especially in small populations (Lande 1976, 1980; Lynch and Hill 1986). Thus, the mere observation that evolution is sometimes rapid does not warrant conclusions about whether this is caused by intensification of selection, or "some essentially different factors" (Simpson 1944).

Assuming that evolution is due to natural selection on heritable genetic variation, evolutionary biologists reasoned that the other component of punctuated equilibrium, stasis, resulted either from lack of genetic variation, a lack of selection, or from stabilizing selection. There are examples of traits that lack the genetic variation needed to adapt to changing environmental conditions (e.g., Hoffmann et al. 2003; Kellermann et al. 2009). However, it is exceedingly unlikely that low rates of evolution in established species are generally due to a lack of genetic variation. For example, bill dimension in *Geospiza fortis* has been observed to respond to selection due to fluctuations between wet and dry years, with selection differentials of up to one standard deviation units in 1977 and 2004 (Grant 1986; Grant and Grant 2011). Calculated over the entire 30-year period of observation, however, selection differentials are at least an order of magnitude lower, and the rate of long-term evolution estimated from differences between species is yet much lower. That indicates that the changes do not add up from year to year, but rather oscillate around a constant long-term beak size. This observation, which originally led to the use of mean-reverting stochastic processes such as Ornstein-Uhlenbeck to model evolution, is not unique for Darwin's finches (e.g., Lieberman et al. 1994). Gingerich (1983) already documented various examples of this phenomenon from a variety of species: Evolutionary change is independent of time, so that if calculated over short periods of time evolution seems rapid, while if calculated over long periods of time it seems slow (Gingerich 2009; Eldredge et al. 2005; Bell 2010; Uyeda et al. 2011). Lande (1976; see also Roff 2000) pointed out that "A simple formula for estimating the minimum selective mortality per generation necessary to explain observed rates of phenotypic evolution is derived (assuming that genetic drift was not involved). The minimum mortality rates needed to explain observed rates of evolution in tooth characters of Tertiary mammals are very small, typically about one selective death per million individuals per generation." In addition, artificial selection (domestication) has resulted in rates of change orders of magnitude higher than observed in nature. Clearly, many traits are long-term (but not short-term) static despite having more than enough genetic variation to evolve.

If low rates of evolution are not generally due to a lack of genetic variation, could it be that they are then due to stabilizing selection due to constancy of environment? Perhaps, for example, Darwin's finches (Grant and Grant 2011) do not experience greater climatic fluctuations in a millennium than they do between years. Stabilizing selection has regularly been proposed as an explanation for low rates of evolution. For example, Lynch (1990) concluded that "a comparison of the evolutionary rates of cranial morphology in mammals with the neutral expectation suggests that stabilizing selection is a predominant evolutionary force keeping the long-term diversification of lineages well below its potential." Also other groups of species exhibit low rates of long-term evolution (e.g., Lieberman et al. 1994). In fact, mammals are known to exhibit exceptionally "rapid" evolution compared to other groups (Van Valen 1985; Lynch 1990). The hypothesis that stasis is due to stabilizing selection is problematic, however. Species remain static even though both their biotic and abiotic environments change markedly (Lieberman and Dudgeon 1996). Indeed, Falconer had already pointed out that if the European mammoth, like many other species, was driven to extinction by climate change, then very likely its changing environment exerted a selective pressure, and we may safely assume that this pressure exceeded one selective death in a million (cf. Lande 1976). In other words, while observed rates of evolution suggest pervasive stabilizing selection, environmental changes that simultaneously take place rather suggest strong directional selection.

Here, it should be emphasized that the mechanisms behind stasis are relatively rarely addressed (but see e.g., Bradshaw 1991; Lieberman and Dudgeon 1996;

Merilä et al. 2001; Hansen and Houle 2004; Bell 2010; Futuyma 2010), as the majority of evolutionary biologists remain concerned with microevolution, ostensibly because macroevolution is not amenable to experiments. Unfortunately, this has led many to believe that the pattern of stasis and rapid evolution in incipient species is fully in line with the Modern Synthesis: Stasis could be due to stabilizing selection and rapid change due to directional selection. This was summarized succinctly by Maynard Smith (1989): "The relation between population genetics and paleontology is unsatisfactory. It is not uncommon for paleontologists to assert that population genetics cannot account for the fossil record, whereas population geneticists hold that there is nothing in the record that they cannot explain." Although 25 years have passed since Maynard Smith made that statement, it applies quite well still today. Of course, stasis as well as rapid change can be explained by population genetic theory, just like evolutionary theory could explain the long nectar spurs of *A. sesquipedale* as the result of an arms race with its pollinator, or stasis in mammoths as a consequence of their migrations. The problem is that theory does not predict why spurs became elongated in *A. sesquipedale* (and not in most other species) and why they became about 30 cm long (and not 10 or 100 cm), or why *A. sesquipedale* evolved, but the mammoth tracked its niche. It does not explain which factors exert stabilizing selection in established species, and why this would suddenly be different when a new species emerges. Put bluntly, when it comes to long-term evolution, current theory provides little else than ad hoc explanations.

3.1 The Genetic Basis of Evolution

In order to understand why stasis and punctuated equilibria pose a challenge for current evolutionary theory, we will once more return to Darwin, who derived the most famous of his many scientific contributions, "*On the Origin of Species by Means of Natural Selection, or the Preservation of Favoured Races in the Struggle for Life*" (1859) long before he sent it to the Linnaean Society for scrutiny when Wallace (1855) had independently come to the same conclusions. An important reason for this delay was that Darwin wanted to build a stronger scientific basis for his argument so as to avoid the scorn attending the publication of *Vestiges of Creation* (Chambers 1844), not to mention concern about the church's reaction. A subsidiary reason for the delay may have been that Darwin was well aware that an important piece was missing from his theory: a theory of inheritance. Darwin was aware that offspring tend to resemble their parents, but did not know what caused this resemblance. Mendel had already discovered the mechanism behind inheritance from flower color and shape of peas in the abbey of St. Thomas monastery garden, but did not present his ideas publicly until 1865 (see Mendel 1866). Hence, Darwin considered some form of blending inheritance to provide his theory of evolution with a mechanistic backbone, something which was criticized and may have been the reason why the concept of natural selection remained controversial (Larson 2004), while the idea of evolution garnered much support among the readers of "The Origin" (with the exception of many church representatives).

Even though Darwin, unaware of Mendel's (1866) progress, had to base his theory on an imperfect understanding of inheritance, the fundamental principle underlying evolutionary theory has remained unchanged (albeit amended) until present: (i) Organisms have traits that are transmitted from parent to offspring, (ii) organisms tend to produce more offspring than required to replace themselves, which causes competition for resources so that (iii) those among the offspring whose traits make them better suited to their local environment have the best chances to survive and reproduce. Evolution by natural selection "naturally follows" from these three principles (Reznick and Ricklefs 2009), and therefore, some biologists claim (as a counterargument to religious movements still questioning the validity of evolutionary theory) that adaptive evolution is inevitable. However, if that is the case, then at the stage where niche tracking was no longer possible (if not already much earlier), natural selection should have acted on mammoths and other species alike so that they adapted to the changing climate. Apparently, that hardly happened or was not possible at all, and they went extinct. Hence, extinction, the fate of most if not all species, is perhaps the best and most ubiquitous evidence of species' failure to adapt (Maynard Smith 1989) rather than of competitive exclusion due to microevolution (Reznick and Ricklefs 2009). This leads us to the core of the "unsatisfactory relation" between genetics and paleontology: Why, if evolution is an obvious, perhaps even inevitable process, do species frequently fail to adapt to environmental change?

Mendel happened to work on a system in which the laws of inheritance could be shown relatively easily. Of the traits that he investigated in *Pisum sativum*, the majority contained just two alleles. If repeated on other traits or species (as he attempted with bees and mice), these experiments would have yielded results much harder to interpret. In fact, such results were available already: As a member of the nineteenth-century British upper class, Darwin was well acquainted with animal breeding. He was particularly fond of doves and pigeons (hence, Falconer's joke about ducks and drakes), but general interest focused on horse and cattle breeding, and when two lines of cattle with high milk yield were crossed to obtain even higher yield, the offspring could show average yield, lower than each of the parental lines. Such observations frustrated not only the breeders aiming for improved economy, but also scientific investigations looking for a simple mechanism of inheritance.

Mendel's laws of inheritance were rediscovered by De Vries (1900) and Correns (1900), and a decade later, Johannsen (1905) explained that both heritable and non-heritable factors are responsible for differences in the weight of bean seeds (introducing the words gene, genotype, and phenotype). Nilsson-Ehle (1909) suggested that multiple hereditary factors were responsible for determining the phenotype of grain color in wheat. Breeding experiments like Castle's (1916) with guinea pigs demonstrated that selection could shift the trait value outside of the range of values previously observed and eventually Fisher (1918, 1930) hypothesized that the additive action of multiple genes can explain the inheritance of quantitative characters. These developments convinced many that natural selection could cause changes in the average trait value from one generation to the next, as Darwin had hypothesized. Studies in natural populations confirmed this potential. The most famous of these early studies was the work by Kettlewell (1956) on industrial melanism in Lepidoptera, most famously the peppered moth *Biston betularia.* Before the English industrial revolution, the lighter phenotype was favored as it was hard to find for birds against a background of light-colored trees and lichens. During and after the industrial revolution, when pollution killed the lichens and soot darkened the trees, the darker phenotype became favored.

As experiments and theory suggested that phenotypic changes in laboratory and natural populations could be explained as the result of natural selection, the "Modern Synthesis" emerged, in which natural selection, a century earlier regarded with skepticism (Larson 2004), became firmly established as the driving force behind evolution. The belief in the potential of natural selection was so great that many geneticists "moved straight from the gene level to that of macroevolution" and "accepted as an article of faith … that all macroevolutionary phenomena were consistent with the laws of genetics" (Mayr 1982). More pluralistic views about differentiation of species that had existed in the early days of the Modern Synthesis were replaced more and more by an undisputed role of natural selection as the driving force behind both micro- and macroevolution. This development is what Gould (2002) called the "hardening of the synthesis". Hilarious examples of "adaptive storytelling" can be found in Gould and Lewontin's famous 1979 paper on the spandrels of San Marco, in which they dubbed this "Panglossian paradigm", after a character in Voltaire's 1759 satire "Candide"—Dr. Pangloss—who is convinced that "all is for the best in the best of all possible worlds". This Panglossian paradigm was hardly challenged until the early 1970s, when a group of young paleontologists started a "paleobiological revolution" (Sepkoski and Ruse 2009) that eventually led paleontologists back to the "high table" of evolutionary biology (Maynard Smith 1984).

In more recent years, molecular genetic techniques have provided a picture of the genetic basis of traits that early geneticists could only have dreamed of. Despite these technological advances, theory remains heavily influenced by very early findings, especially the concept of a "gene." Flower color in peas as studied by Mendel (1866) is a trait with a relatively simple genetic basis: It can be described assuming two factors (nowadays known as alleles). Such factors were dubbed "genes" by De Vries (1900), and nowadays, we can precisely determine the genomic location of genes controlling simple traits like flower color—with greater effort even the functional nucleotide polymorphisms. When Fisher (1930) explained that quantitative characters could be thought of as the additive effect of a number of Mendelian "genes," the word gene still did not mean much more than "heriditary factor," because the existence of DNA as carrier of the hereditary information still had to be discovered. However, when during the 1980s and 1990s most known Mendelian characters were successfully related to genes (factors) with a specific location on the genome, it became widely assumed that the "genes" underlying more complex traits were also real, existing entities that could be located on the genome, based on a simple interpretation of Fisher's influential explanation. Only very recently has it been found (with the help of high-throughput whole-genome genotyping) that heritable genetic variation can often only be partially explained by allelic variants: The rest has become known as "missing heritability." Thus, the simplifying assumption that most traits have a largely additive genetic basis is based on theory rather than empirical observation and is only slowly being replaced by a view of heritability as an emerging property of a complex nonlinear genotype-to-phenotype map.

Selection led to rapid change in coloration in peppered moths, and would Mendel (1866) have selected for purple flowers in peas, that would presumably have led to fixation of purple flowers in his study population equally rapidly. Such rapid evolution is facilitated by the simple genetic basis of these traits, a linear genotype-to-phenotype map (Van't Hof et al. 2011). Examples of evolution from such simple traits are not necessarily representative for evolution of complex traits or entire organisms. Polygenic traits may respond to selection in a similar fashion as Mendelian characters, if their genetic basis is additive (e.g., Lynch and Walsh 1997), but this seemingly simple observation leads to a paradox: It is widely assumed that natural selection acts most strongly on traits that are most important for fitness. Hence, selection prunes deleterious alleles especially in traits important for fitness, which should lead to a reduction in genetic variation. This would lead to the paradox that evolution of traits crucial for fitness should be slow, because variation remaining in such traits is mainly of environmental origin (Merilä and Sheldon 1999). The solution to this paradox lies in genetic architecture: There is genetic variation in fitness-related traits (Houle 1992), even more than in morphological traits, but the variation is largely nonadditive (Merilä and Sheldon 1999). This is an important observation, because it is far from obvious whether and how selection can act on traits governed by complex nonlinear genotype-to-phenotype maps. Cattle breeders experienced this already in Darwin's time, and it may have been an important reason why the principles of inheritance remained elusive until the twentieth century.

Why would especially fitness-related traits have complex genetic backgrounds that, paradoxically, potentially impair the efficiency of natural selection? A growing body of evidence suggests that complex genetic backgrounds aid stable development (Nijhout 2002; Levine and Davidson 2005; Davidson and Levine 2008; Higgins et al. 2010). The environment of species fluctuates perpetually (Bell 2010), as exemplified by Darwin's finches experiencing a virtually random sequence of wet and dry years during their lifetime (Grant and Grant 2011). To survive in unpredictable environments, many traits are canalized (Waddington 1942; see also Meiklejohn and Hartl 2002). Perhaps the best known example of canalization is phenotypic plasticity, where the trait value achieved depends in a systematic (non-random) fashion on some particular aspects of the environment (Nijhout 2002; West-Eberhard 2005; Pigliucci and Müller 2010; see also Hallson and Björklund 2012). Plasticity can be achieved through regulatory feedback in developmental pathways, which are well known to exist not only in various regulatory gene complexes (Nijhout 2002; Davidson and Levine 2008) and operons, but also above the level of DNA as activator proteins. Also on the inside of an organism, the "environment" fluctuates, as mutations accumulate during lifetime, concentrations of enzymes fluctuate, and genes are copied and lost. Developmental pathways involving feedback loops may buffer against such variations (Bergman and Siegal 2003). This idea, that complex genetic backgrounds provide a platform for stabilized development in unpredictably fluctuating environments, is the central idea behind Lerner's (1954) "genetic homeostasis," Waddington's (1942) canalization (see also Schmalhausen 1949), and goes back to Bernard (1878) who wrote that "all vital mechanisms, varied as they are, have only one objective, that of preserving constant the conditions of life in the internal environment."

4 Organisms' Independence of Their Environment

Bernard's remark alludes to a simple principle: The fitness of an organism is determined not only by its external environment, but also, or even predominantly, by how well it functions internally (Schwenk and Wagner 2001). Consequently, traits will evolve to form coadapted complexes (Seaborg 1999). This principle is largely neglected in modern evolutionary biology, which focuses almost exclusively on natural selection from the outside, but it is certainly not new. Whyte (1965; see also Frazetta 1975; Riedl 1977) expressed his surprise that biologists had not incorporated it in their theories and encountered few similar ideas in contemporary literature, even though the importance of the internal environment and coadaptation of parts was understood already long before Waddington (1942), Lerner (1954), and especially Schmalhausen (1949) emphasized it. Indeed, it can be traced back to Aristotle's (350 b. Chr.) "On the Parts of Animals" and is largely equivalent to Cuvier's principle of correlation of parts (1798 in Rudwick 1997), which influenced Bernard (1878) to emphasize that the "milieu intérieur" renders an organism relatively independent of its environment. Chetverikov (1926) referred to this as the "genotypic milieu," and Mayr (1963) as "cohesion of the genotype." Mayr also strongly advocated more holistic views of the genotype, explaining that the contribution to fitness of any gene depends on all other genes. Indeed, Cuvier—who studied fossil and living elephants before Falconer did—opposed the idea of gradual evolution (50 years before Darwin) for two reasons: First because his studies of fossils led him to conclude that one form does not generally evolve into another, and second because his studies of comparative anatomy made him regard "organisms as integrated wholes, in which each part's form and function were integrated into the entire body. No part could be modified without impairing this functional integration" (1798, in Rudwick 1997). More than 150 years later, substantial evidence had accumulated supporting Cuvier's view: Lerner (1954) pointed out that artificial selection for extreme phenotypes tends to reduce fitness, as Dawson (1964) concisely summarized: "Genetic analysis of a variety of traits in populations of several species has revealed a similar pattern for many traits which are important components of fitness. In general such characters have intermediate optima, and prolonged artificial selection for extreme expression leads to a reduction in fitness." The principle that it is advantageous when the different parts of an organism function well together (Schwenk and Wagner 2001) is the ultimate explanation for canalization and developmental plasticity, for if it were unimportant how well different traits function together, there would be little benefit from canalizing their development!

Cuvier (1798) explained his *conditions d'existence* as follows: "if an animal's teeth are such as they must be in order for it to nourish itself with flesh, we can be sure without further examination that the whole system of its digestive organs is appropriate for that kind of food, and that its whole skeleton and locomotive organs, and even its sense organs, are arranged in such a way as to make it skillful at pursuing and catching its prey. For these relations are the necessary conditions of existence of the animal; if things were not so, it would not be able to subsist." Thus, evolutionary changes in an animal's teeth are useless unless accompanied by simultaneous changes in digestive, locomotor, and sensory organs. Cuvier's holistic view of an organism was reflected in the concept of a *bauplan* which was common in continental Europe in the early twentieth century. This view, however, became more and more replaced by the atomist Anglo-American view of the organism as a collection of separate traits (Eldredge et al. 2005; see also Schwenk and Wagner 2001; Minelli and Fusco 2012), perhaps because in the laboratory "[t]here appears to be no character … that cannot be selected in *Drosophila*" (Lewontin 1974). The view that traits form coadapted complexes, however, implies that fitness of any trait, and hence of the organism as a whole, depends on the traits themselves and their mutual relations, rather than on the relation of individual traits with the outside environment as the Modern Synthesis assumes.

4.1 The Cost of Selection

These different views of what makes an organism have more than just philosophical implications. Haldane attempted already in 1957 "to make quantitative the fairly obvious statement that natural selection cannot occur with great intensity for a number of characters at once unless they happen to be controlled by the same genes." To illustrate this, Haldane used the well-known example of the peppered moth *B. betularia*. There are two color morphs of this moth, light and dark, the latter of which is much easier to spot on a background of pale lichens and white birch bark. The dark morph increased in numbers following the industrial revolution as soot darkened the trees. Kettlewell (1956) had shown that differential predation by birds was so intense that the frequency of the conspicuous morph could be halved in a single day. Haldane wrote that "if the change of environment had been so radical that ten other independently inherited characters had been subject to selection of the same intensity as that for colour, only 0.5^{10}, or one in 1,024, of the original genotype would have survived. The species would presumably have become extinct."

As Van Valen (1963) put it, Haldane's "dilemma" was that "in the process of the evolutionary substitution of one allele for another, at any intensity of selection and no matter how slight the importance of the locus, a substantial number of individuals would usually be lost because they did not already possess the new allele. For most organisms, rapid turnover in a few genes precludes rapid turnover in the others. A corollary of this is that, if an environmental change occurs that necessitates the rather rapid replacement of several genes if a population is to survive, the population becomes extinct." Apparently, most evolutionary biologists did not understand this "fairly obvious statement" because Kimura remarked in 1995 that "it seems to be widely accepted among biologists that Darwinian selection can act almost without limit as to the number of loci or sites that are simultaneously substituting advantageous alleles."

A recent study suggests that the "substitutional load" (Kimura 1995) may be smaller than Haldane initially calculated (Nunney 2003), but that applies only to the "independently inherited characters" that Haldane considered. "Haldane's dilemma" (as Van Valen dubbed it) becomes much more severe when the traits form a coadapted complex, where change in any trait is beneficial only if all the other traits change simultaneously, by the right amount, and in the right direction, as Cuvier (1798) implies: Evolutionary changes in an animal's teeth are useless unless accompanied by simultaneous changes in digestive, locomotor and sensory organs. For complexes of many traits, the chance to find even a single individual in a very large population who unites these beneficial mutations is remote (Haldane 1957; Van Valen 1963; Kimura 1995). Hence, might a change of environment necessitate a change in any component of such a complex, it could quite rapidly drive even a large population to extinction.

Not all traits are important for the proper functioning of other traits. In mammals, for example, body temperature is crucially important (and virtually static over tens of millions of years) even though it is highly unlikely that for placental mammals all over the planet, the optimum body temperature is 37 °C in all their different *environments* (Hansen and Houle 2004). Therefore, it seems highly implausible that the body temperature of 37 °C is locally adaptive. However, it is quite possible that 37 °C is the temperature to which most processes are adapted inside all these different *species*. By contrast, a change in color or thickness of the fur causes far less internal disruption than a change in body temperature would, and therefore, polar bears evolved a thick white fur to compensate for a body temperature that is quite extravagant where they live. Darwin's finches may have various beak shapes, yet they are all finch-sized homeotherm oviparous birds, and "most species displaying geographic variation vary clinally only in ecotypic characters affecting primarily size, proportions, and coloration, but show no significant evolutionary departures" (Mayr 1992). Indeed, neglecting traits that do not show variation, evolutionary biologists have documented a splendid variety of supposedly adaptive variation in traits amenable to change like flower color in peas, wing color in *B. betularia*, fur thickness in mammals, and bill dimension in birds.

4.2 Epistasis

In a recent essay on the same topic as this chapter, Futuyma (2010) considered "genetic integration" an important but not sufficient explanation for species' failure to adapt, because population geneticists did not find evidence for extensive gene interactions, while stasis is observed in all the traits of an organism (Futuyma 2010 p. 1873). However, if some aspects of an organism resist adaptive change because they are part of coadapted complexes, then species will in nature track their niche, even though other aspects of their phenotype might be amenable to change. If species track their niche, also traits with the potential to evolve will tend to remain unchanged (Björklund and Merilä 1993; Seaborg 1999). This could explain organism-wide stasis, although mosaic evolution (Simpson 1944) may still be the norm. More importantly, widespread gene interactions (if not undetected for methodological reasons) are not necessary for traits to exert stabilizing selection on each other.

Genes may interact in various ways. Pleiotropy means that a gene affects more than a single trait (Stearns 2011; Wagner and Zhang 2011). Conversely, however, traits do not have to be controlled by the same genes to influence each other. For example, body temperature does not have to be genetically correlated with all the processes it affects to incur stabilizing selection. Thus, the internal component of fitness does not require pleiotropy. Genes may also act epistatically. Mayr promoted the importance of the internal cohesion of organisms for a long time (Mayr 1954, 1963, 1982, 1992) and often used the term epistasis to describe it. It seems that to him epistasis referred to the fitness effect of a gene, which "is always in the context with other genes, and the interaction with those other genes makes a particular gene either more favorable or less favorable." He criticized "atomist" geneticists from R.A. Fisher to B. Charlesworth and R. Lande for neglecting this (Mayr 1992). To most geneticists, however, epistasis implies that the effect of a gene on the phenotype, not fitness, depends on the genetic background. Coadapted traits may well affect each other's contribution to fitness [epistasis sensu Mayr (1992)] without any epistatic gene action. Therefore, lack of evidence for widespread epistatic interactions is no evidence against an intrinsic fitness component.

More recently, it has been suggested that absence of response to selection in the wild could be due to genetic correlations, referred to as "the" G-matrix. It is possible that in special situations genetic correlations change, cancel, or even reverse response to selection (Lande and Arnold 1983; Schluter 1996). However, the G-matrix is a locally valid statistical simplification of the genotype–phenotype map and not a persistent property of an individual or population. Therefore, G-matrices are expected to change as, e.g., allelic frequencies evolve and hence cannot meaningfully be argued to constrain evolution (Pigliucci 2006). Of course, the actual developmental processes may constrain evolution (Minelli and Fusco 2012), but through internal selection (Schwenk and Wagner 2001) and not simply by redirecting response to selection (Schluter 1996), because traits that are long-term static often show considerable short-term responses (Gingerich 1983; Bell 2010;

Uyeda et al. 2011). In summary, stabilizing selection can result from the functional interaction between traits (Schwenk and Wagner 2001) and does not directly require or predict genetic correlations, genetic linkage, pleiotropy, or epistasis.

4.3 The Strength of Internal Factors

Could coadaptation of characters exert stabilizing selection strong enough to let incipient species rapidly evolve into "clearly defined and well marked species" (Wallace 1905) that subsequently become static as their body parts become so coadapted that they hardly respond to environmental change? Yoo (1980) selected *Drosophila melanogaster* for increased abdominal bristle numbers. In 90 generations bristle numbers increased from approximately 10 initially to reach a plateau at about 40. When selection was suspended, bristle numbers dropped equally rapidly with about 10 bristles disappearing in just 10 generations. Such reversals and selection plateaus have traditionally been ascribed to (unintended) selection on the various other traits that are pleiotropically controlled by the same or linked genes (Mackay and Lyman 2005). However, Teotonio et al. (2009) studied reversed selection in *D. melanogaster* and found "that despite the complete convergence to ancestral levels of adaptation, allele frequencies only show partial return," which would be a remarkable coincidence if the reversal were due to pleiotropy. It seems therefore more plausible that reversals are instead due to a decrease in fitness due to strong selection (Lerner 1954; Dawson 1964; Meiklejohn and Hartl 2002) disrupting coadapted character complexes. The speed of reversal in this and other, similar studies suggests that the strength of stabilizing selection is comparable to that of the typically strong directional selection that preceded it.

5 A Largely Autonomous Process

When Darwin acknowledged Falconer's comments that species may change rapidly when they emerge, to subsequently remain largely static, he remarked that this was "clearly leading to the same result." As discussed above, it remains unclear what Darwin meant with that remark. It is clear, however, that rapid change in incipient species became a long-standing issue in evolutionary biology. Simpson (1944) referred to the discussion on micro- versus macroevolution as "old," but he also wrote that the most important distinction between students of evolutionary biology is whether they believe that bursts of rapid evolution represent a fundamentally different process or not. When Eldredge and Gould (1972) and Stanley (1975) revived the discussion, intense debate followed once more and continues (Pennell et al. 2014; Lieberman and Eldredge 2014).

In the early part of the twentieth century, several students of evolution developed theories to explain phenotypic differences between groups of species, which they deemed too substantial to be explained by a gradual process of evolution. These theories, often aimed at explaining hypothetical processes leading to novel genera or taxa of still higher rank, have invariably caused intense debate. I have therefore chosen not to address these theories here, but to focus on the processes that govern evolutionary change within established species, and in incipient species—which would generally be considered congeners. That is because we can be sure that these processes exist (although we do not know their exact nature) and because I believe that one does not need to involve the origin of higher taxa to address the relation between micro- and macroevolution.

In the simplest relation between micro- and macroevolution, the latter is nothing else than the long-term consequence of the former. Above I have presented various lines of evidence suggesting this is not the case. In the first place, estimates of rates of evolution based on phenotypic differences between species are orders of magnitude lower than those obtained from short-term studies (Lande 1976; Lynch 1990; Lynch and Hill 1986). More precisely, estimated rates are inversely related to the time period over which they are calculated (Gingerich 1983, 2009). In addition, both fossil records and phylogenetic studies of present-day species suggest that evolutionary change is concentrated in short periods of time, often associated with speciation (Webster et al. 2003; Pagel et al. 2006; Hunt 2007; Whittall and Hodges 2007; Mattila and Bokma 2008; Monroe and Bokma 2009). These lines of evidence all demonstrate that macroevolution is not microevolution writ large.

This raises two questions: What prevents short-term, ostensibly adaptive evolutionary changes to accumulate over time, and what happens during speciation. To address the first question first: Long-term stasis has been observed despite substantial changes in species' biotic and abiotic environments, and even in species that apparently suffer mortality due to these environmental changes. This indicates that long-term stasis is not due to stabilizing selection exerted by the environment (Lieberman and Dudgeon 1996), which implies that it instead originates from within the organism, from traits exerting selection on each other. The idea that traits are adapted to each other can be traced back to Cuvier and even Aristotle (see also Chetverikov 1926; Schmalhausen 1949; Whyte 1965; Frazetta 1975; Riedl 1977; Schwenk and Wagner 2001), but I am not aware of direct experimental evidence. [Experiments showing reversed evolution when directional selection is suspended (e.g., Yoo 1980) are in line with this idea (Meiklejohn and Hartl 2002), but do not rule out alternative explanations]. However, theory predicts that selection cannot act effectively on many traits at the same time (Haldane 1957; Van Valen 1963; Kimura 1995), especially not if those traits form a coadapted complex. Theory also predicts that systems of multiple interacting factors will have stable equilibria (May 1972) to which they return after small deviations. Moreover, traits that intuitively appear less important for the proper functioning of other traits, for example coloration, appear to be far less constrained than more central traits like body temperature in homeotherms, which is what one would expect if traits exert selection on each other. Therefore, even without direct evidence, I think it is very likely that long-term stasis is caused by traits exerting selection on each other.

This leads us to an answer on the second question; what happens during speciation. It is hard to imagine that when a system of interacting traits moves from one stable equilibrium to another, it would remain compatible with the previous system. Thus, a transition from one equilibrium to another will very generally be associated with speciation. (That does not, of course, mean that every single case of speciation will involve a shift to a new equilibrium.) This idea was proposed by Simpson as "quantum evolution," by Mayr (1954, 1982) as "genetic revolution" (see also Barton and Charlesworth 1984; Seaborg 1999; Provine 2004), and by Eldredge and Gould as "punctuated equilibria" (1972; Gould and Eldredge 1986, 1993). This is supported in the first place by studies indicating that reproductive isolation is driven not by accumulation of genetic mutations in general, but by regulatory changes (Prager and Wilson 1975). Further support comes from theoretical analysis of complex systems of interacting factors (May 1972): Such systems are not expected to evolve gradually, but to shift from one equilibrium either back to the same equilibrium, to chaos (extinction), or to a new equilibrium (speciation), as Simpson already anticipated in 1944. Finally, it is supported by empirical evidence that evolutionary change is concentrated in short bursts, often associated with speciation (Eldredge and Gould 1972; Webster et al. 2003; Pagel et al. 2006; Hunt 2007; Whittall and Hodges 2007; Mattila and Bokma 2008; Monroe and Bokma 2009). Thus, traits that do not influence the functioning of other traits may evolve gradually, but more substantial evolutionary change that involves persistent changes of coadapted trait complexes represents shifts between stable equilibria of a complex system.

There has been some discussion in literature on what factors could cause a population to shift away from a stable equilibrium. Simpson (1944) referred to such factors as "superthreshold," as opposed to the subthreshold factors causing only microevolutionary deviations from which the system returns to the old equilibrium. Simpson apparently considered the possibility that intense natural selection could be a superthreshold force. Mayr (1954) and also Eldredge and Gould (1972) seemed more inclined to think that small population size plays a crucial role. Perhaps because Mayr never provided a specific definition of a "genetic revolution," the concept became associated with small population size. When later researchers of speciation found no evidence of widespread bottlenecking, many took this as evidence against genetic revolutions (Coyne and Orr 2004), and current speciation research largely disregards functionally entangled evolutionary factors (Minelli and Fusco 2012). However, recent findings suggest that superthreshold changes may be achieved in more ways than through intense selection or extreme bottlenecking. One powerful but rare cause could be whole-genome duplications (De Bodt et al. 2005). A far more common potential trigger is hybridization (McCarthy 2008). Studies of gene expression have shown that hybridization of species (as is common in plants) or isolated populations [which is harder to detect in retrospect, but may have occurred in our own species (Patterson et al. 2006)] is often accompanied by substantial regulatory changes that may result in novel ontogenies (Nielsen et al. 2000). Thus, rather than through very small numbers of individuals suffering inbreeding depression (Mayr 1954, 1982), genetic revolutions may occur in large numbers of outbred individuals experiencing hybrid vigor.

Rapid change in incipient species experiencing a novel environment may also be achieved through phenotypic plasticity (Pigliucci and Müller 2010) followed by acquisition of the necessary adaptations through a decrease in plasticity or "genetic assimilation" (Waddington 1942).

If a shift from one equilibrium to another drives rapid change in incipient species, its direction and magnitude will be determined by the selective forces exerted by traits on each other and not by the environment—even if environmental change initially triggered the shift. If that is the case, then rapid change in incipient species is largely independent of the environment and independent of microevolutionary changes. Our analyses of body mass evolution in mammals suggest that this is indeed the case: The magnitude of rapid body mass change in incipient species appears to be independent of how frequently such changes occur (Mattila and Bokma 2008; Monroe and Bokma 2009), which suggests that the environment has little effect on this process. Many groups of ecologically similar species show substantially different rates of evolution. These differences are so common that they are often taken for granted, but I argue that they indicate that the tempo of macroevolution is set not by the pace of (a)biotic environmental change, but largely by interactions between traits, rendering macroevolution a largely autonomous process. [Of course, species may still appear well adapted to their environment due to exaptation, pre-adaptation, and ecological fitting (Bock 1959; Gould and Vrba 1982; Janzen 1985)].

As Anderson (1972) pointed out in his famous essay "More is Different," the elementary entities of solid state physics obey the principles of particle physics, yet solid state physics is not just applied particle physics. Likewise, cells consist of molecules, but cell biology is not applied molecular biology, and psychology is not applied physiology. At each hierarchical stage, entirely new concepts, principles, and generalizations are necessary. It is in this fashion that macroevolution is related to microevolution. The elementary entities of macroevolution follow the principles of microevolution, but we cannot make macroevolutionary predictions from mere knowledge of microevolutionary processes. Entirely new concepts and principles are necessary to describe a largely autonomous process.

References

Anderson PW (1972) More is different. Science 177:393–396

Avise JC (1977) Is evolution gradual or rectangular? Evidence from living fishes. Proc Nat Acad Sci USA 74:5083–5087

Avise JC, Ayala FJ (1975) Genetic change and rates of cladogenesis. Genetics 81:757–773

Barton NH, Charlesworth B (1984) Genetic revolutions, founder effects, and speciation. Annu Rev Ecol Syst 15:133–164

Bell G (2010) Fluctuating selection: the perpetual renewal of adaptation in variable environments. Philos Trans R Soc B Biol Sci 365:87–97

Bergman A, Siegal ML (2003) Evolutionary capacitance as a general feature of complex gene networks. Nature 424:549–552

Bernard C (1878) Leçons sur les phénomènes de la vie communs aux animaux et aux végétaux. J.-B. Baillière et fils, Paris

Björklund M, Merilä J (1993) Morphological differentiation in Carduelis finches: adaptive versus constraint models. J Evol Biol 6:359–374
Bock WJ (1959) Preadaptation and multiple evolutionary pathways. Evolution 13:194–211
Bokma F (2002) Detection of punctuated equilibrium from molecular phylogenies. J Evol Biol 15:1048–1055
Bokma F (2008) Detection of "punctuated equilibrium" by Bayesian estimation of speciation and extinction rates, ancestral character states, and rates of anagenetic and cladogenetic evolution on a molecular phylogeny. Evolution 62:2718–2726
Bradshaw AD (1991) Genostasis and the limits to evolution. Philos Trans R Soc B Biol Sci 333:289–305
Campbell GJD (1867) The reign of law. Strahan, London
Castle WE (1916) Size inheritance in Guinea-Pig crosses. Proc Nat Acad Sci USA 2:252–264
Chambers R (1844) Vestiges of the natural history of creation. John Churchill, London
Chetverikov SS (1926) On certain aspects of the evolutionary process from the standpoint of modern genetics. Philos Trans R Soc B Biol Sci 105:167–195 (trans: Barker M, Lerner IM 1961)
Correns C (1900) G. Mendel's Regel über das Verhalten der Nachkommenschaft der Rassenbastarde. Berichte der Deutschen botanischen Gesellschaft 18:158–168
Coyne JA, Orr HA (2004) Speciation. Sinauer, Sunderland
Cuvier JLNF (1798) Unpublished manuscript MS628, Bibliothèque Centrale, Museum National d'Histoire Naturelle, Paris. Translated in Rudwick MJ (1997)
Darwin C (1859) On the origin of species by means of natural selection, or the preservation of favoured races in the struggle for life. John Murray, London
Darwin C (1862) On the various contrivances by which British and foreign orchids are fertilized by insects, and on the good effects of intercrossing. John Murray, London
Davidson EH, Levine MS (2008) Properties of developmental gene regulatory networks. Proc Nat Acad Sci USA 105:20063–20066
Davis MB, Shaw RG (2001) Range shifts and adaptive responses to quaternary climate change. Science 292:673–679
Davis CC, Schaefer H, Xia Z, Baum DA, Donoghue MJ, Harmon LJ (2014) Long-term morphological stasis maintained by a plant–pollinator mutualism. Proc Nat Acad Sci USA 111:5914–5919
Dawson PS (1964) Genetic homeostasis and developmental rate in Tribolium. Genetics 31:873–885
De Bodt S, Maere S, Van de Peer Y (2005) Genome duplication and the origin of angiosperms. Trends Ecol Evol 20:591–597
De Vries H (1900) Sur la loi de disjonction des hybrides. Comptes Rendus de l'Academie des Sciences 130:845–847
Douglas ME, Avise JC (1982) Speciation rates and morphological divergence in fishes, tests of gradual versus rectangular modes of evolutionary change. Evolution 36:224–232
Eldredge N, Gould SJ (1972) Punctuated equilibrium: an alternative to phyletic gradualism. In: Schopf TJM (ed) Models in paleobiology. Freeman Cooper & Co, San Francisco, pp 82–115
Eldredge N, Thompson JN, Brakefield PM, Gavrilets S, Jablonski D, Jackson JBC, Lenski RE, Lieberman BS, McPeek MA, Miller W (2005) The dynamics of evolutionary stasis. Paleobiology 31:133–145
Fisher RA (1918) The correlation between relatives on the supposition of Mendelian inheritance. Proc R Soc Edinb 52:399–433
Fisher RA (1930) The genetical theory of natural selection. Clarendon Press, Oxford
Frazetta TH (1975) Complex adaptations in evolving populations. Sinauer, Sunderland
Futuyma DJ (2010). Evolutionary constraint and ecological consequences. Evolution XX:1865–1884
Gingerich PD (1983) Rates of evolution: effects of time and temporal scaling. Science 222:159–161
Gingerich PD (2009) Rates of evolution. Annu Rev Ecol Evol Syst 40:657–675
Gould SJ (2002) The structure of evolutionary theory. Harvard University Press, Cambridge
Gould SJ, Eldredge N (1986) Punctuated Equilibrium at the third stage. Syst Zool 35:143–148
Gould SJ, Eldredge N (1993) Punctuated equilibrium comes of age. Nature 366:223–227
Gould SJ, Lewontin RC (1979) The spandrels of San Marco and the Panglossian paradigm: a Critique of the Adaptationist Programme. Proc R Soc B Biol Sci 205:581–598

Gould SJ, Vrba E (1982) Exaptation—a missing term in the science of form. Paleobiology 8:4–15
Grant PR (1986) Ecology and evolution of Darwin's Finches. Princeton University Press, Princeton
Grant PR, Grant BR (2011) How and why species multiply—the radiation of Darwin's Finches. Princeton University Press, Princeton
Haldane JBS (1957) The cost of natural selection. J Gentics 55:524–551
Hallson LR, Björklund M (2012) Selection in a fluctuating environment leads to decreased genetic variation and facilitates the evolution of phenotypic plasticity. J Evol Biol 25:1275–1290
Hansen TF, Houle D (2004) Evolvability, stabilizing selection, and the problem of stasis. In: Pigliucci M, Preston K (eds) The evolutionary biology of complex phenotypes. Oxford University Press, Oxford, pp 130–150
Higgins JA, Bailey PC, Laurie DA (2010) Comparative genomics of flowering time pathways using Brachypodium distachyon as a model for the temperate grasses. PLoS ONE 5:e10065
Hoffmann AA, Hallas RJ, Dean JA, Schiffer M (2003) Low potential for climatic stress adaptation in a rainforest Drosophila species. Science 301:100–102
Houle D (1992) Comparing evolvability of quantitative traits. Genetics 130:195–204
Hunt G (2007) The relative importance of directional change, random walks, and stasis in the evolution of fossil lineages. Proc Nat Acad Sci USA 104:18404–18408
Janzen DH (1985) On ecological fitting. Oikos 45:308–310
Johannsen WL (1905) Arvelighedslærens elementer. Gyldendal, Copenhagen
Kellermann V, van Heerwaarden B, Sgro CM, Hoffmann AA (2009) Fundamental evolutionary limits in ecological traits drive Drosophila species distributions. Science 325:1244–1246
Kelly AE, Goulden ML (2008) Rapid shifts in plant distribution with recent climate change. Proc Nat Acad Sci 105:11823–11826
Kettlewell HBD (1956) A résumé of investigations on the evolution of melanism in the Lepidoptera. Proc R Soc London B 145:297–303
Kimura M (1995) Limitations of darwinian selection in a finite population. Proc Nat Acad Sci USA 92:2343–2344
Lande R (1976) Natural selection and random genetic drift in phenotypic evolution. Evolution 30:314–334
Lande R (1980) Genetic variation and phenotypic evolution during allopatric speciation. Am Nat 116:463–479
Lande R, Arnold SJ (1983) The measurement of selection on correlated characters. Evolution 37:1210–1226
Larson EJ (2004) Evolution: the remarkable history of a scientific theory. Random House, New York
Lerner IM (1954) Genetic homeostasis. Wiley, New York
Levine M, Davidson EH (2005) Gene regulatory networks for development. Proc Nat Acad Sci USA 102:4936–4942
Lewontin RC (1974) The genetic basis of evolutionary change. Columbia University Press, New York
Lieberman BS, Dudgeon S (1996) An evaluation of stabilizing selection as a mechanism for stasis. Palaeogeogr Palaeoclimatol Palaeoecol 127:229–238
Lieberman BS, Eldredge N (2014) What is punctuated equilibrium? What is macroevolution? A response to Pennell et al. Trends Ecol Evol 29:185–186
Lieberman BS, Brett CE, Eldredge N (1994) Patterns and processes of stasis and change in brachiopod species lineages. American Museum of Natural History Novitates No. 3114, 23 pp
Lieberman BS, Brett CE, Eldredge N (1995) Patterns and processes of stasis in two species lineages from the Middle Devonian of New York State. Paleobiology 21:15–27
Lynch M (1990) The rate of morphological evolution in mammals from the standpoint of the neutral expectation. Am Nat 136:727–741
Lynch M, Hill WG (1986) Phenotypic evolution by neutral mutation. Evolution 40:915–935
Lynch M, Walsh B (1997) Genetics and analysis of quantitative traits. Sinauer, MA
Mackay TF, Lyman RF (2005) Drosophila bristles and the nature of quantitative genetic variation. Philos Trans R Soc B Biol Sci 360:1513–1527

Mattila TM, Bokma F (2008) Extant mammal body masses suggest punctuated equilibrium. Proc R Soc London B 275:2195–2199
May RM (1972) Will a large complex system be stable? Nature 238:413–414
Maynard Smith J (1984) Palaeontology at the high table. Nature 309:401–402
Maynard Smith J (1989) The causes of extinction. Philos Trans R Soc B Biol Sci 325:241–252
Mayr E (1954) Change of genetic environment and evolution. In: Huxley J, Hardy A, Ford E (eds) Evolution as a process. Allen & Unwin, London, pp 157–180
Mayr E (1963) Animal species and evolution. Harvard University Press, Cambridge
Mayr E (1982) Speciation and macroevolution. Evolution 36:1119–1132
Mayr E (1992) Speciational evolution or punctuated equilibria. In: Somit A, Peterson S (eds) The dynamics of evolution. Cornell Univ. Press, New York, pp 21–48
McCarthy EM (2008) On the origins of new forms of life: a new theory. http://www.macroevolution.net/support-files/forms_of_life.pdf
Meiklejohn CD, Hartl DL (2002) A single mode of canalization. Trends Ecol Evol 17:468–473
Mendel JG (1866) Versuche über Pflanzenhybriden. Verhandlungen des naturforschenden Vereines in Brünn 4:3–47
Merilä J, Sheldon BC (1999) Genetic architecture of fitness and nonfitness traits: empirical patterns and development of ideas. Heredity 83:103–109
Merilä J, Kruuk LE, Sheldon BC (2001) Explaining stasis: microevolutionary studies in natural populations. Genetica 112(113):199–222
Minelli A, Fusco G (2012) On the evolutionary developmental biology of speciation. Evol Biol 39:242–254
Monroe MJ, Bokma F (2009) Do speciation rates drive rates of body size evolution in mammals? Am Nat 174:912–918
Mooers AO, Schluter D (1998) Fitting macroevolutionary models to phylogenies: an example using vertebrate body sizes. Contrib Zool 68:3–18
Mooers AO, Schluter D (1999) Reconstructing ancestor states with maximum likelihood: support for one- and two-rate models. Syst Biol 48:623–633
Mooers AO, Vamosi SM, Schluter D (1999) Using phylogenies to test macroevolutionary hypotheses of trait Evolution in cranes (Gruinae). Am Nat 154:249–259
Nielsen MG, Wilson KA, Raff EC, Raff RA (2000) Novel gene expression patterns in hybrid embryos between species with different modes of development. Evol Dev 2:133–144
Nijhout HF (2002) The nature of robustness in development. BioEssays 24:553–563
Nilsson-Ehle H (1909) Kreuzungsuntersuchungen an Hafer und Weizen. Academic dissertarion, Lund University, Sweden
Nunney L (2003) The cost of natural selection revisited. Ann Zool Fenn 40:185–194
Pagel M, Venditti C, Meade A (2006) Large punctuational contribution of speciation to evolutionary divergence at the molecular level. Science 314:119–121
Patterson N, Richter DJ, Gnerre S, Lander ES, Reich D (2006) Genetic evidence for complex speciation of humans and chimpanzees. Nature 441:1103–1108
Pennell MW, Harmon LJ, Uyeda JC (2014) Is there room for punctuated equilibrium in macroevolution? Trends Ecol Evol 29:23–32
Pigliucci M (2006) Genetic variance–covariance matrices: a critique of the evolutionary quantitative genetics research program. Biol Philos 21:1–23
Pigliucci M, Müller GB (2010) Evolution—the extended synthesis. MIT Press, Cambridge
Prager EM, Wilson AC (1975) Slow evolutionary loss of the potential for interspecific hybridization in birds: a manifestation of slow regulatory evolution. Proc Nat Acad Sci USA 72:200–204
Provine WB (2004) Ernst Mayr: genetics and speciation. Genetics 167:1041–1046
Reznick DN, Ricklefs RE (2009) Darwin's bridge between microevolution and macroevolution. Nature 457:837–842
Ricklefs RE (1980) Phyletic gradualism vs. punctuated equilibirum, applicability of neotological data. Paleobiology 6:271–275
Riedl R (1977) A systems-analytical approach to macro-evolutionary phenomena. Q Rev Biol 52:351–370

Roff D (2000) The evolution of the G matrix: selection or drift? Heredity 84:135–142
Rudwick MJS (1997) Georges Cuvier, fossil bones, and geological catastrophes. University of Chicago Press, Chicago
Schluter D (1996) Adaptive radiation along genetic lines of least resistance. Evolution 50:1766–1774
Schmalhausen II (1949) Factors of evolution, the theory of stabilizing selection (trans: Dordick I). Blakiston, Philadelphia, and Toronto
Schwenk K, Wagner GP (2001) Function and the evolution of phenotypic stability: connecting pattern to process. Am Zool 41:552–563
Seaborg DM (1999) Evolutionary feedback: a new mechanism for stasis and punctuated evolutionary change based on integration of the organism. J Theor Biol 198:1–26
Sepkoski D, Ruse M (eds) (2009) The paleobiological revolution: essays on the growth of modern paleontology. University of Chicago Press, Chicago
Simpson GG (1944) Tempo and mode in evolution. Columbia University Press, New York
Stanley SM (1975) A theory of evolution above the species level. Proc Nat Acad Sci USA 72:646–650
Stearns FW (2011) One hundred years of pleiotropy: a retrospective. Genetics 186:767–773
Stebbins GL, Ayala FJ (1981) Is a new evolutionary synthesis necessary? Science 213:967–971
Teotonio H, Chelo IM, Bradic M, Rose MR, Long AD (2009) Experimental evolution reveals natural selection on standing genetic variation. Nat Genet 41:251–257
Thomas CD, Cameron A, Green RE, Bakkenes M, Beaumont LJ, Collingham YC, Erasmus BF, De Siqueira MF, Grainger A, Hannah L, Hughes L, Huntley B, Van Jaarsveld AS, Midgley GF, Miles L, Ortega-Huerta MA, Peterson AT, Phillips OL, Williams SE (2004) Extinction risk from climate change. Nature 427:145–148
Tingley MW, Monahan WB, Beissinger SR, Moritz C (2009) Birds track their Grinnellian niche through a century of climate change. Proc Nat Acad Sci USA 106(Suppl 2):19637–19643
Uyeda JC, Hansen TF, Arnold SJ, Pienaar J (2011) The million-year wait for macroevolutionary bursts. Proc Nat Acad Sci USA 108:15908–15913
Van Valen L (1963) Haldane's Dilemma, evolutionary rates, and heterosis. Am Nat 47:185–190
Van Valen L (1973) A new evolutionary law. Evol Theory 1:1–30
Van Valen L (1985) Why and how do mammals evolve unusually rapidly? Evol Theory 7:127–132
Van't Hof AE, Edmonds N, Dalikova M, Marec F, Saccheri IJ (2011) Industrial melanism in British peppered moths has a singular and recent mutational origin. Science 332:958–960
Venditti C, Pagel M (2010) Speciation as an active force in promoting genetic evolution. Trends Ecol Evol 25:14–20
Waddington CH (1942) Canalization of development and the inheritance of acquired characters. Nature 150:563–565
Wagner GP, Zhang J (2011) The pleiotropic structure of the genotype-phenotype map: the evolvability of complex organisms. Nat Rev Genet 12:204–213
Wallace AR (1855) On the law which has regulated the introduction of new species. Ann Mag Nat Hist 26:184–196
Wallace AR (1867) Creation by law. Q J Sci 4:471–488
Wallace AR (1889) Darwinism: an exposition of the theory of natural selection, with some of its applications. MacMillan, London
Wallace AR (1905) My life: a record of events and opinions. Chapman & Hall, London
Webster AJ, Payne RJH, Pagel M (2003) Molecular phylogenies link rates of evolution and speciation. Science 301:478
West-Eberhard MJ (2005) Developmental plasticity and the origin of species differences. Proc Nat Acad Sci USA 102:6543–6549
Whittall JB, Hodges SA (2007) Pollinator shifts drive increasingly long nectar spurs in columbine flowers. Nature 447:706–709
Whyte LL (1965) Internal factors in evolution. George Braziller, New York
Yoo BH (1980) Long-term selection for a quantitative character in large replicate populations of Drosophila melanogaster. II. Lethals and visible mutants with large effects. Genet Res 35:19–31

Visualizing Macroevolution: From Adaptive Landscapes to Compositions of Multiple Spaces

Emanuele Serrelli

Abstract The adaptive landscape is an important diagrammatic concept that was conceived in population genetics. During the Modern Synthesis, in the first half of the twentieth century, the landscape imagery was used to represent evolution on a large scale, aiding in the construction of a common language for a new evolutionary biology. Not only historic adaptive landscapes by Dobzhansky, Simpson, and others are a record of how macroevolution was thought of in those decades; they stimulate reflection on "combination spaces" that underlie them. In fact, any landscape diagram is the three-dimensional transposition of a multidimensional space of combinations of genes, morphological traits, or other kinds of variables. This is an important and enduring general point of awareness: The diagram displays some aspects of the considered space while hiding others, exposing the author and the user to incomplete understanding and to conflating different spaces. Today, macroevolution is studied as a multifarious exploration of spaces of possibilities of all different sorts, interconnected in complex ways: genotype spaces, molecular spaces, morphospaces, geographical spaces, ecological spaces, and genealogical spaces. Actual macroevolutionary stories and outcomes are a subset of the universes of possible combinations—of genes, nucleotides, morphological traits, and environmental variables. Visualizations of macroevolution are a challenge of showing both distinction and correlation between spaces of possibilities.

Keywords Adaptation · Speciation · Macroevolution · Visualization

The "adaptive landscape" was one of the earliest ways of representing evolution graphically. It was devised in the early 1930s in the context of population genetics, but it was soon reused to represent global biodiversity, speciation, and

E. Serrelli (✉)
"Riccardo Massa" Department of Human Sciences,
University of Milano Bicocca, Milan, Italy
e-mail: emanuele.serrelli@epistemologia.eu

E. Serrelli and N. Gontier (eds.), *Macroevolution*,
Interdisciplinary Evolution Research 2, DOI 10.1007/978-3-319-15045-1_4

adaptive radiation. Due to its flexible semantic scope, this pictorial concept played a major role in the Modern Synthesis, contributing in particular to the establishment of a common language between experimental geneticists, mathematicians, paleontologists, naturalists, and all biologists in general. In addition to facilitating communication among scientists, it helped shoring up the idea that micro- and macroevolution were different degrees of magnification of the same, few, long-reaching fundamental evolutionary mechanisms. Historic adaptive landscapes also reflected their authors' idea of macroevolution as fundamentally connected with adaptation to environmental conditions and, sometimes, to a perceived global trend of progressive optimization and complexification.

Today, macroevolution is no longer viewed as driven exclusively by adaptation to local conditions, least of all by optimization, and evolution is studied at many scales of observation, each with its peculiar mechanisms and entities. The diminished use of macroevolutionary adaptive landscapes reflects this changed epistemology. Landscape diagrams are limited to a much narrower scope of applications, while other forms of visualization (e.g., diversity diagrams, distribution maps, and phylogenetic trees) are more prominent in illustrating macroevolutionary phenomena. Yet, most graphical representations of evolution share at least one characteristic with adaptive landscapes: They show *spaces of possibilities* and their changing occupancy through time. Macroevolutionary studies today approach *multiple* spaces of possibilities—for example, in phylogeny, morphology, and geography—combining them and showing their relationships. Not only is analyzing historic adaptive landscapes useful to understand the view of macroevolution that they contributed to consolidate, and helpful to appreciate a fundamental logic of visual representations of evolution. Over time, historic landscapes generated overinterpretations and misinterpretations, because they were based on some conflations and ambiguities that were deeply related to their unifying goals. Their story is therefore also instructive about some enduring risks created by powerful and flexible graphical representations of macroevolution.

1 Adaptive Landscapes in the Modern Synthesis

The earliest use of landscape images with peaks and valleys representing evolution is usually attributed to Sewall Wright, who used these diagrams for a presentation at the 6th Congress of Genetics in Ithaca, New York, in 1932 (Wright 1932, see Figs. 1 and 2). Theodosius Dobzhansky, who was present at the talk, immediately adopted Wright's diagram for the genetics course he taught at the California Institute of Technology, and incorporated it in his *Genetics and the Origin of Species* (1937) and all its subsequent editions (Provine 1986). Dobzhansky's immensely influential book became the principal direct source of the landscape imagery across the vast literature on evolution (Ruse 1990). Among the most significant treatises on evolutionary theory that included references to the adaptive landscape were George Gaylord Simpson's *Tempo and Mode in Evolution* (1944, see Figs. 3, 4 and 5) and

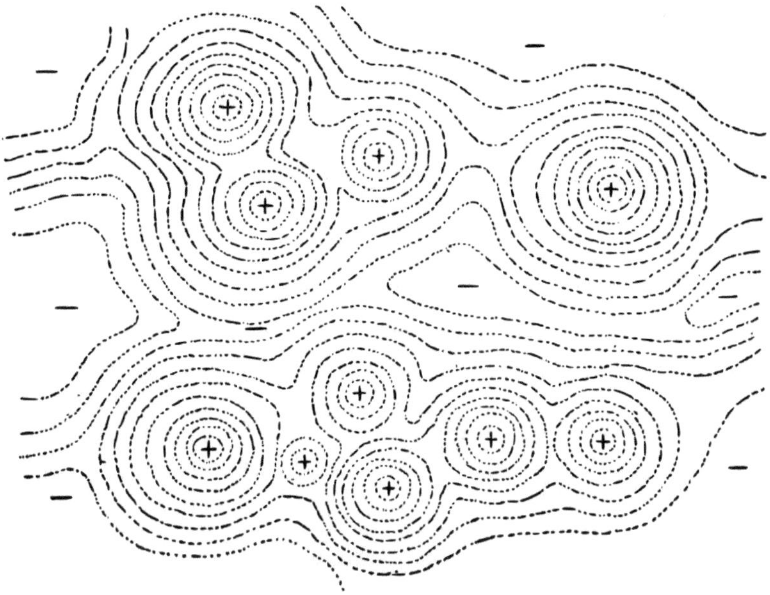

Fig. 1 Dozhansky's symbolic landscape picture of the relations between the organism and the environment. The picture is a topographic map of all the possible genetic combinations, in which the "contours" follow combinations with equal adaptive values. Adaptive peaks and valleys are marked, respectively, by plus and minus signs [*Source* Dobzhansky (1937), 3rd ed. 1951, p. 8]

Huxley's *Evolution: The Modern Synthesis* (1942). The presence of the landscape in the formative years of the Modern Synthesis raises the question of what a role the landscape imagery may have played in advancing its principal arguments.

The Modern Synthesis—henceforth, MS—is often regarded as "the general acceptance of two conclusions: gradual evolution can be explained in terms of small genetic changes ("mutations") and recombination, and the ordering of this genetic variation by natural selection; and the observed evolutionary phenomena, particularly macroevolutionary processes and speciation, can be explained in a manner that is consistent with the known genetic mechanisms" (Mayr 1980: 1). The focus on "acceptance" and "conclusions" determines considering the MS as a "product." To elucidate the possible roles of the adaptive landscape in the MS, however, a view of the MS as a *process* is more effective than a view of the MS as the product of some process.

1.1 The Modern Synthesis as a Process

The book *The Evolutionary Synthesis* (Mayr and Provine 1980), based on a conference organized by Ernst Mayr at the end of the 1970s, documented evidence that the MS had occurred simultaneously and at different rates in different fields

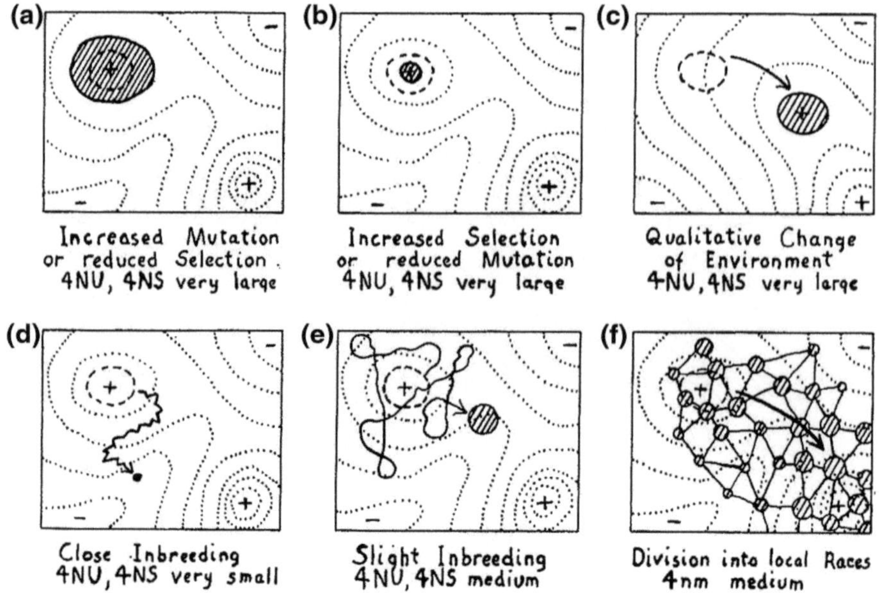

Fig. 2 Possible courses on Wright's adaptive surface. The *dotted circle* is the initial location of the population inside the range of all possible genetic combinations. The subsequent dynamics depends on the population's demography (number of individuals, spatial subdivision) and genetics (number of loci, number of alleles, topography of the landscape), as well as on selection pressures, on the environment influencing the fitness values of genotypes, and on the (extremely low) rate of viable mutations that modify the topography [*Source* Wright's original (1932) reproposed by Dobzhansky (1937)]

(e.g., genetics, cytology, or botany) dealing with many different levels of biological organization (from chromosomes to higher taxa) and in different countries, "or even in the same fields in different divisions of the same university" (Provine 1980a: 405). As pointed out by Provine in the same book, "The evolutionary synthesis was a very complex process; its historical development cannot be encompassed accurately by any simple thesis […] No simple historical thesis, however, brilliant, can describe all essential elements of the evolutionary synthesis" (1980a: 405). There are indeed a few accounts of the MS as a process. They even disagree about the time span in which they locate the MS.[1] Yet, most authors recognize two main phases over the years from the early 1910s to the early 1940s.

The "Darwinian–Mendelian synthesis" of the 1910s and 1920s reconciled the Darwinian theory of natural selection with the Mendelian theory of inheritance.

[1] When did the MS begin? Provine, finding no trace of origin in any of the major works of the MS, cited Thomas Kuhn saying that "the actual origins of a scientific field generally will not be found in the major books that embody the fundamental beliefs of the field" (Provine 1980a: 400–401). He even expressed the feeling of evolutionary synthesis having "been a part of biology for a long time, almost since Darwin" (Ivi: 400).

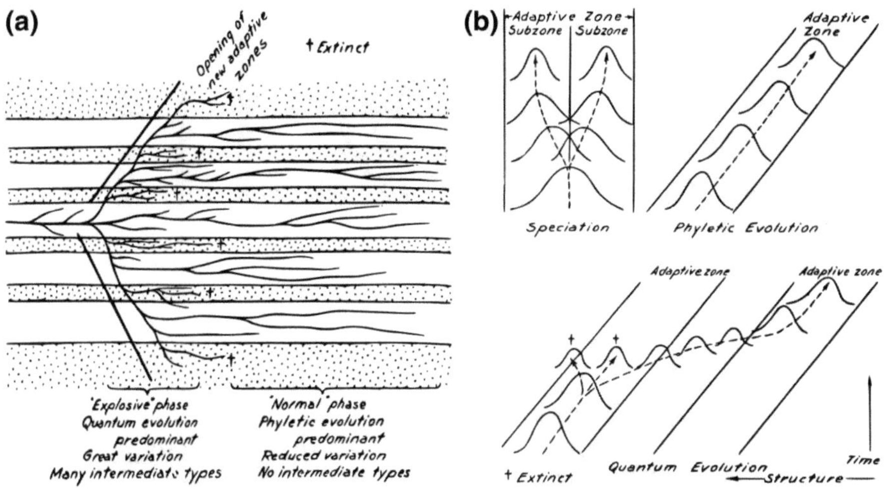

Fig. 3 **a**, **b** Two pictures by George Gaylord Simpson exemplifying his use of the adaptive landscape in a unified picture of evolution. **a** "Diagram of explosive speciation" belonging to a traditional kind of tree diagrams drawn in paleontology. **b** Landscape pictures of the three major modes of evolution—speciation, phyletic evolution, and quantum evolution—showing an interpretation of macroevolutionary branching explicitly linked to population genetics [*Source* Simpson (1944), Figs. 31 and 35]

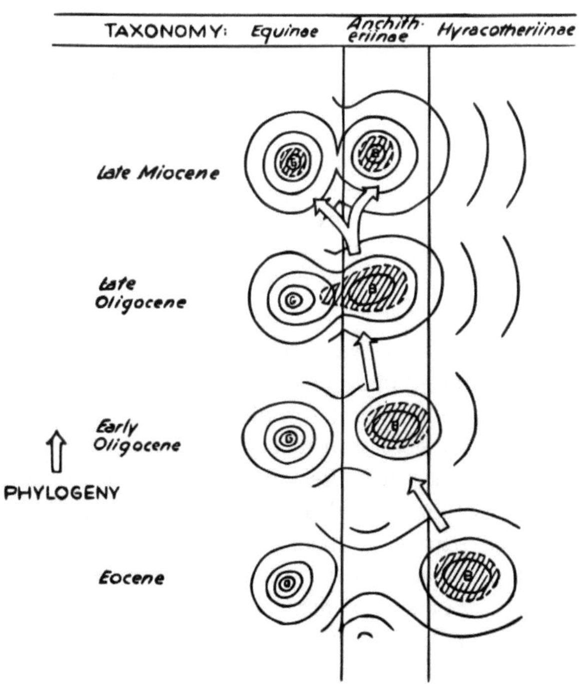

Fig. 4 Simpson's landscape of the evolution of horses from browsers (*right*) to browsers and grazers (*left*). Eocene: browsing and grazing are two well-separated peaks; only the browsing peak is occupied. Oligocene: the browsing peak moves toward the grazing peak due to climate change and size increase. Late Oligocene and Early Miocene: The two peaks are close enough and asymmetrical variation of the family causes some animals to be on the saddle. Miocene: A segment of the population breaks away under selection pressure and climbs the grazing peak [*Source* Simpson (1944), Fig. 13]

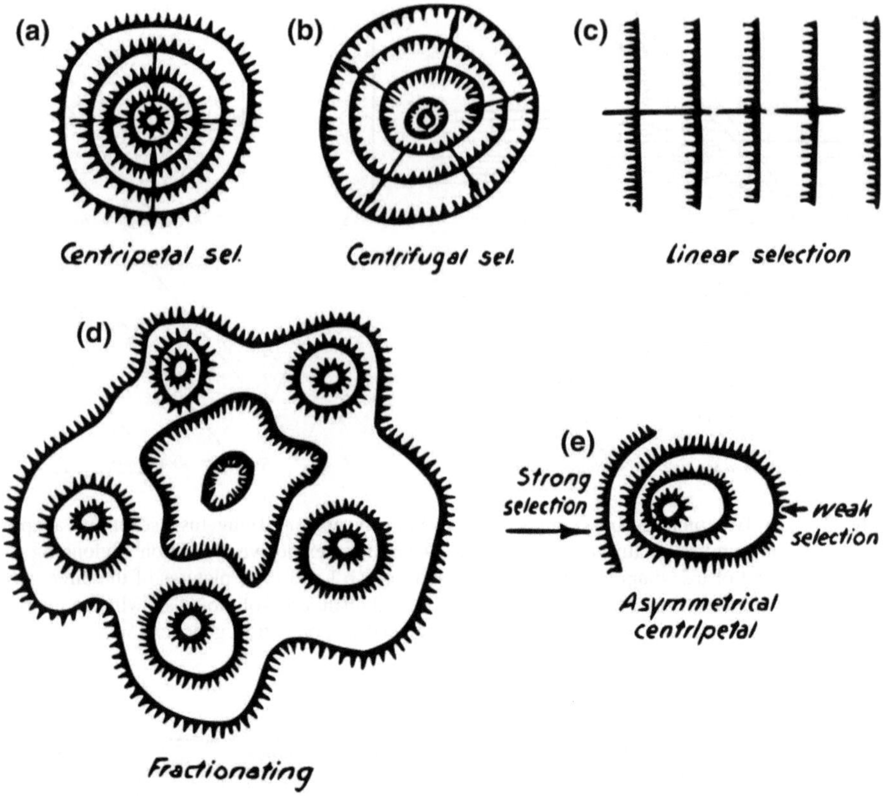

Fig. 5 Simpson's graphical portrait of five different selection regimes. "Contours are analogous to topographic maps, with hachures placed on downhill side. Direction of selection is uphill, and intensity proportional to slope": **a** the adaptive optimum is the "typical" character, selection concentrates or stabilizes the population; **b** the "typical" character is ill-adapted, selection favors divergence; **c** selection drives to an optimum which is out of the current adaptive range of the population; **d** centrifugal toward five different optima; **e** asymmetrical–centripetal, employed also in Late Oligocene of Fig. 4 [*Source* modified from Simpson (1944), Fig. 11]

In those years, the Mendelian theory, increasingly supported by experimental evidence featuring discrete variation ("mutations"), was considered to contradict the conditions for natural selection to occur. Population genetics was essentially a powerful mathematical theory of multiple factors (e.g., selection, mutation, drift, population size, and allele frequency), a theory that guided laboratory experiments and some agricultural applications (Provine 1980b). Population genetics introduced a "particulate theory of inheritance" (cf. Mayr 1980), whereby continuous variation began to be considered as the product of a huge array of discrete variants of discrete genetic elements. In the "gene pool" of a population, small fitness differences could have major impacts and evolution by natural selection could occur. Sewall Wright, the inventor of the adaptive landscape, was undisputedly one of the main figures in this phase. For Lewontin (1980), Wright's "synthesis," based on the importance of gene interactions, was particularly important as a step to the subsequent phase of synthesis, because it considered a number of factors (e.g., random processes, demography, and geographical subdivision), their balance, and their configurations in relation to outcomes like speciation or extinction. For Lewontin, the synthetic work of Wright provided hints "of the way in which both speciation and extinction can flow mechanically from the processes of modulation of variation" (Lewontin 1980: 61).

The "second synthesis" of the 1930s and 1940s was characterized by the engagement into evolutionary theory of previously separated fields, such as systematics and taxonomy, zoology and botany, paleontology and morphology, and cytology and embryology. In the early 1930s, all these fields had been dominated by theories that were incompatible with, or irrelevant to, the theory of evolution by natural selection, or any theory of evolution at all. By 1947, according to Ernst Mayr's memories, there was "an essential agreement among [scientists of all fields] on the gradual mode of evolution, with natural selection as the basic mechanism and the only direction-giving force" (Mayr 1980: 42). A synthesis *had* occurred. Between 1930 and 1947, books such as Dobzhansky's (1937), Simpson's (1944) and Huxley's (1942) had been published. A renewed evolutionary theory had united geneticists on the one hand, and those called by Mayr "the naturalists" on the other. Mayr described the MS as a process engaging separate fields of research by virtue of common language, communication, reciprocal awareness, and familiarity. The MS would consist in the spread of a very general way of thinking, with resolution of oppositions into complementarities, yielding a sense of a single theory potentially explaining in a consistent way a unified set of phenomena. In this process, liable to different views, we will focus on the role of adaptive landscapes.

1.2 Mayr's Communication View Versus Lewontin's Export View of the Modern Synthesis

Under the assumption that the MS was a complex, two-step historical process that took place between the 1910s and the 1940s, the nature of steps and the relationship between them can still be viewed in different ways. We will compare two views, one held by Ernst Mayr, the other by Richard Lewontin. For Mayr, the second step of the MS was a phase of enhanced communication. For Lewontin, it consisted in the diffusion and application of mathematical models. It is easy to guess that the adaptive landscape's role in the MS depends on the view of the MS we embrace. Adopting Mayr's view, we will consider Sewall Wright's visualization as a migrant metaphor, generating new common pictorial (rather than mathematical) language between disparate fields.

The main reason why Ernst Mayr had organized the 1980 Evolutionary Synthesis conference was to revise a dominant view of the MS held by geneticists since the 1950s. In Provine's words:

> ...most geneticists in the 1950s [...] believed that the evolutionary synthesis was a function or product of advances within the field of genetics and that these advances were applied to other fields like systematics, paleontology, embryology, cytology, and morphology [..., that] the real advances took place in genetics and were *exported to* other fields of evolutionary biology, thus creating the evolutionary theory (Provine 1980a: 402–403, emphasis added).

At one level, Mayr's discontent concerned the relative contributions of different professionals to the second synthesis: He was opposing the idea that geneticists had been the main contributors to the synthesis, with a minor or even passive involvement of those whom Mayr called the "naturalists," e.g., systematists and paleontologists (cf. Mayr 1959, 1973). At the same time, the target of Mayr's campaign was *the view of synthesis as an export* (from genetics). As Provine recognized at the end of the 1980 conference, the evolutionary synthesis was "more than a simple application of new concepts in genetics to other facets of evolutionary biology, as earlier accounts have suggested" (1980a: 405).

Interestingly, the previously standard view was maintained by at least one participant of the 1980 conference: population geneticist Richard Lewontin. At the conference, Lewontin presented his own version of the "export view" of the synthesis, as an increasingly extensive use of population genetics mathematical models as a guide for empirical research and hypothesis testing in all biological fields. He stuck to the view that the "second synthesis" was an extension, a true export from genetics to other fields, insisting on the importance of mathematical theory and on the potentially tight relationship between theory and observations. For Lewontin, all the mathematical theory produced in the "Darwinian–Mendelian synthesis" by Sewall Wright and the other "architects" would have been be able to guide the naturalists' work by providing detailed predictions and means to subtly distinguish among competing hypotheses for explaining observations; the obstacle to this was the scarce mathematical knowledge and understanding of most biologists: "...a tremendous amount of understanding and synthesis of evolutionary ideas could have been derived chiefly from the theoretical work of Fisher and Wright. These insights were explicit in the writings of Fisher and Wright, but simply unavailable to most biologists for reasons of literacy" (Lewontin 1980: 58). Fields like systematics and paleontology were making use of too few of the models that mathematical geneticists had worked out. The genetical theory was indeed not "correctly incorporated" in the work of other biologists, so, if theory had to say something, it was that "we have not yet found the observations sufficient to distinguish among the [evolutionary] hypotheses" (Ivi: 66). These limitations deprived evolutionary hypotheses of sufficient support and, ultimately, rendered the synthesis incomplete.

Mayr's view radically differed from that of Lewontin's. The former asked to the latter: "Wasn't the evolutionary synthesis possible on the basis of a very minimal agreement that there are small genetic changes and that the phenotypes produced by even every small genetic changes may differ and probably usually differ in their selective values? […] perhaps we should not complain about the failure of application of the sophisticated and advanced theoretical analyses…" (Q&A section in Lewontin 1980: 67). Lewontin, of course, replied: "I really disagree." His point was that much of mathematical theory was destined to be "a guide for perplexed experimentalists" that was not exploited yet (Lewontin 1980: 65).

A core aspect of synthesis in Mayr's view was the construction of a common language among specialists in fields that had grown apart. Terminological inconsistencies were in need of full resolution, as in the case of "mutation," used "to describe aspects of the phenotype by naturalists and to describe aspects of the genotype by geneticists" (1980: 14). The construction of a new, shared vocabulary was then, for Mayr, a mainstay of the MS: "The creation of a new evolutionary terminology greatly contributed to the eventual synthesis. At least some of the misunderstandings resulted from the lack of an appropriate and precise terminology for certain evolutionary phenomena. Nearly all the architects of the new synthesis contributed terminological innovations" (Ivi: 29). The MS had to fill in communication gaps among specialists throughout the world who might not have been aware of each others' work and advancements: Before synthesis, for example, much progress made by experimentalists had "percolated only very incompletely" (Mayr 1980: 28) to the naturalists, and vice versa:

> The conceptual advances made by either [experimental geneticists and naturalists] were not perceived by their opponents; in fact, they were usually unknown to them. As a result, the construction of a unified and comprehensive theory of evolution during the first three decades of the century was impossible. The naturalists had wrong ideas on the nature of inheritance and variation; the experimental geneticists were dominated by typological thinking that resulted in their distrust of natural selection and a belief in the importance of pure lines and mutation pressure. Like the naturalists, the geneticists had many misconceptions about the nature of variation. They had at least as great an ignorance of the excellent taxonomic literature on species and speciation as the naturalists had of the genetic literature (Ivi: 13).

Distinctions were clarified, resolving long-standing oppositions into complementarities, and allowing for untroubled division of scientific research programs. Important, for example, was the clarification that "the naturalists and the experimental geneticists were concerned with different levels" (Mayr 1980b: 11) of the "biological hierarchy," whereas in earlier years, much confusion was brought about by "discussions that made no distinction between phyletic evolution (temporal genetic changes in populations) and the multiplication of species—that is, the splitting of a phyletic line into two or more reproductively isolated lines" (Ivi: 35). *Consistency* was a key word in Mayr's view of synthesis: Authors such as Haldane, Simpson, Huxley, Mayr, and Dobzhansky had been showing consistency between experiments and field work, genetic and phenotypic data in the wild, theoretical and field genetics, and microevolution and macroevolution. Notice that for Lewontin, "consistency" was a pejorative term compared to "entailment" or "necessity": "As an evolutionary geneticist, I do not see how the origin of higher taxa are the necessary consequence of neo-Darwinism. They are sufficiently explained, but they are not necessary consequences" (Lewontin 1980: 60).

The MS *was*, in Mayr's view, the described process of concrete communication across disciplines, which, in turn, incubated some convictions. For example, different fields of biology came to be viewed as focusing on different aspects of the same unified Darwinian theory, rather than advancing multiple competing theories, as was believed in the first decades of the century. Experimental and mathematical evidence of the agency and efficacy of natural selection was acknowledged across the board, along with what Mayr called a neo-Darwinian or population thinking, characterized by the centrality of variation and by the importance of ecological factors and population structure. In this social dynamics of science, Mayr vindicated the importance of naturalists who developed a new, evolutionary systematics, and a population understanding of species. The adaptive landscape imagery, devised in one particular field, became a *metaphor*, i.e., a way of building reciprocal confidence and common language, as well as serving as a vehicle for carrying over scientific content across different disciplines.[2]

1.3 Adaptive Landscapes in the Communication View of the Modern Synthesis

In Mayr's account, an ingredient of the MS was the increased confidence of scientists toward each other's methods. In the MS, Mayr emphasized the role of "bridge builders," i.e., individual scientists active in overcoming the hostility that developed among many fields as a result of the lack of communication between them. Sewall Wright, as a mathematician and, to a limited extent, former laboratory scientist, was a bridge builder between geneticists and naturalists.[3] For example, his 'shifting balance theory' and the observations he drew from the work of naturalists were explicitly mutually reinforcing, "an early direct link" to speciation theory (Pigliucci 2008: 596). Another "astoundingly early recognition" was the compatibility of population genetics with the fossil record. Wright's landscape metaphor was yet another communication device, purposefully tailored to an audience of naturalists.

We have seen in the beginning that the adaptive landscape had been proposed by Wright to summarize the general dynamics of a Mendelian population in a form understandable to other biologists. Wright consciously aimed at showing

---

[2] A dictionary definition of metaphor states that "a metaphor is defined as a figure of speech concisely comparing two things, saying that one is the other," while its etymology contains the idea of transfer. The English "metaphor" derives from the sixteenth century's old French *métaphore*, from the Latin *metaphora* "carrying over." In Greek metaphorá (μεταφορά) "transfer," from metaphero (μεταφέρω) means "to carry over," "to transfer": from meta (μετά) "between" + phero (φέρω), "to bear," "to carry." With this etymology, metaphor seems just the right word to qualify the adaptive landscape in the MS as seen *à la* Ernst Mayr.

[3] Ironically, for personal and arbitrary reasons, Wright was absent from Mayr's list of bridge builders (Provine 1986).

the potential relevance of population genetics to other fields of biology where mathematical models were largely dismissed as abstract and obscure. His effective visualization allowed any biologist to intuitively grasp the meaning of particular concepts employed in mathematical models, such as genetic combinations, mutation pressure, selection pressure, and drift, and to relate them to his or her own research field. The diffusion of the adaptive landscape was thus not the direct transfer of mathematical concepts expected in "export" views of the MS, but a way of engaging colleagues into new, open-ended, seemingly promising research questions (Ruse 1990). On the other hand, the mathematical foundation of the visualization was the source of its authority, even though the confidence in mathematical models was a rather recent acquisition:

> [for many years experimental scientists had] insisted that the solution to the problems of evolution would have to be found by asking entirely different questions and by using entirely different methods. They insisted that the hypothetico-deductive approach was pure speculation and that the looked-for laws could be found only by induction from experiments (Mayr 1980: 27).

Was the great success of the adaptive landscape *due to* the biologists' increased trust toward mathematical models? Or did the diagram itself *stimulate* confidence in mathematical models by showing that they could be consistent with—and relevant to—empirical observations in many fields?

In its visual form, the landscape metaphor became particularly popular among paleontologists and morphologists, used to visual representations of macroevolution as we shall see in Simpson's work (see Fig. 3a). In this pictorial convergence, the sense of a shared language was amplified. After the MS period, the landscape diagram had a long and rich history in evolutionary biology. In the 1990s, it was used as the principal iconic image for *Climbing Mount Improbable* (1996) by Dawkins (Fig. 6). Even though the latter might not be of particular theoretical significance, it clearly illustrates the efficacy of the landscape imagery as a compelling visual aid in communicating scientific ideas for general audience in the context of popular scientific writing.

In today's evolutionary literature, terms such as "peaks," "valleys," "climb," "rugged," or, more recently, "ridges" (Gavrilets 1997) are still common, even in cases where an adaptive landscape is not represented graphically. This demonstrates that, during the MS, adaptive landscapes opened up an enduring semantic area for the verbal description of evolutionary phenomena. The adaptive landscape was thus also a linguistic creation, a figure of speech important in the development of a new, shared evolutionary language.

1.4 Dobzhansky's Landscape

Let us now begin an analysis of the landscape pictures proposed during the MS. In general, a landscape image, despite what is first evoked by its name, does not directly represent the environment. Rather, the landscape is an image of something

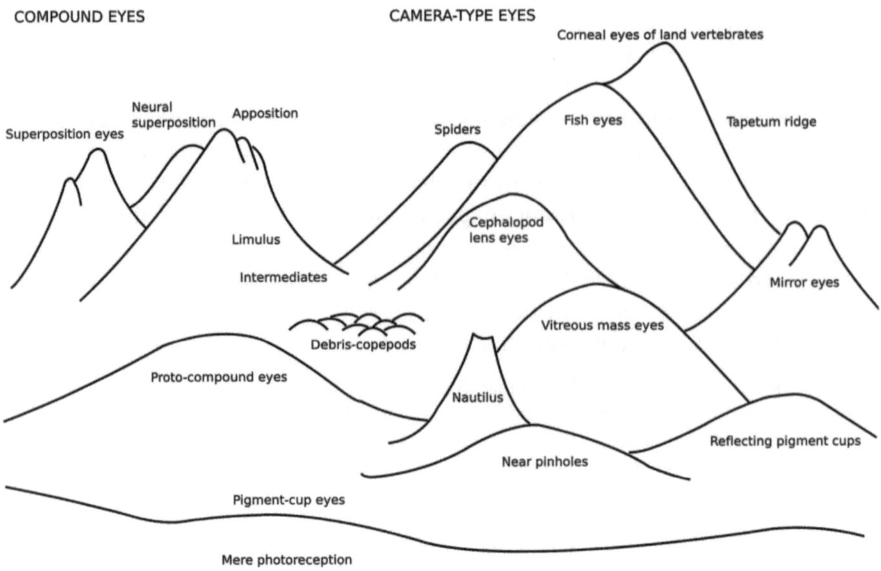

Fig. 6 According to ethologist and evolution writer, Dawkins (1996), a landscape like this depicts "the eye region of Mount Improbable" (p. 177) [*Source* redrawn after Dawkins (1996), Fig. 5.30]

that lives in an environment, e.g., a population or a species. The landscape surface in fact represents the entire virtual *space of possibilities* of the living entity in question. The vertical dimension of the evolutionary landscape represents "adaptedness," i.e., fitness.[4] Therefore, the physical environment, albeit not depicted directly, is extremely important because it influences the whole shape of the surface by codetermining the elevation of each and every location of the surface.

Wright's diagrams were designed to represent a population. They would have their scope extended in Dobzhansky's and Simpson's work, but a verbal description with the original scope is found in Huxley's influential book (1942), specifically in chapter "Mendelism and evolution." Huxley drew upon the Wrightian idea of a "peak shift" to describe the theoretical and empirical pattern of a population that transitions from a phase of relaxation of selective pressure and excess of variability to a phase when the pressure is re-established and the population turns into a "stable type" slightly modified from the starting "type":

> Wright (1932), in discussing such problems in more general terms, concludes that there must be available to most species a number of gene-combinations all of about the same survival value, he compares them to peaks separated by "valleys" of intermediate

[4] In some versions of the landscape, including Wright's (1932), the vertical dimension also depends on absolute, intrinsic values, such as "harmony" of the genetic combination (see also Bokma, *this volume*). As we see below, in Dawkins's landscape, the vertical elevations of the surface are largely determined by some intrinsic and absolute measure such as "complexity" or "perfection."

> combinations which are less favourable. Normally it is difficult or impossible for selection to shift the type from an established peak to another, although this might be equally satisfactory if reached: but when the intensity of selection is reduced (or when low size of population promotes the accidental survival of genes and gene-combinations), many "valley" combinations are realized, the species can cross dryshod to other peaks, and it will be a matter of chance on which Ararat the type eventually remains perched when the rising tide of selection again floods out the valleys (Huxley 1942: 113).

Huxley's "landscape of gene combinations" is close to Wright's original idea: Its "evolving entity" is a population or, at most, a species. Each point of the surface represents one possible individual in the population or species.

A landscape picture (Fig. 1) appeared in the first pages of Dobzhansky's book *Genetics and the Origin of Species* as "a symbolic picture of the relations between the organism and the environment." The context was an appreciation of life's diversity and of the strong correlation between diversity and adaptedness (Dobzhansky 1937: 3, 6). The "evolving entity" on Dobzhansky's landscape was, again, a population of individuals, and each point of the landscape corresponding to an individual, actual or potential, seen as a particular genetic combination. The "adaptive value"[5] of each genetic individual, represented by elevation, was also a measure of its genetic contribution to the following generation. The population's peak position would thus be maintained because, by definition, the next generation was mostly produced by genotypes corresponding to the peaks of the landscape, re-proposing parental genetic combinations. In other words, peaks represented groups of related genetic combinations favored by natural selection.

Figure 2, reproduced by Dobzhansky from Wright (1932) without modifications in the last chapter of *Genetics and the Origin of Species* entitled "Patterns of evolution," showed that natural selection was not only a mechanism for stabilization of peak positions, but also a mechanism for evolutionary change in the context of a "balance of factors." The figure showed different cases of evolutionary dynamics on the landscape. At any particular time, the population was found in a small area of the landscape, and several factors, beside natural selection, such as mutation, population size, and population subdivision, could determine its movement on the landscape. In certain cases, the population chased a peak that was escaping due to environmental change. But the figure made it apparent that the population's exploration of the landscape, determined by many factors, was not an inflexible and close-ended movement of peak climbing.

Up to this point explored, the present discussion was concerned with similarities between the landscapes of Dobzhansky's and those of Wright and Huxley. There is, however, one significant difference in the landscape use by these authors pertaining to macroevolution. Dobzhansky's landscape is definitely a macroscopic one: It illustrates the idea that living beings are "...a great array of families of

[5] "The relative capacity of a given genotype to transmit their genes to the gene pool of the following generations constitutes the adaptive value, or the Darwinian fitness, of that genotype. The adaptive value is, then, a statistical concept which epitomizes the reproductive efficiency of a genotype in a certain environment" (chapter "Toward a Natural Philosophy of Macroevolution", "Selection": 78).

related gene combinations, which are clustered on a large but finite number of adaptive peaks" (Ivi: 10), and comprehends not only all existing species, but also nonexisting ones:

> Every organism may be conceived as possessing a certain combination of organs or traits, and of genes which condition the development of these traits. Different organisms possess some genes in common with others and some genes which are different. The number of conceivable combinations of genes present in different organisms is, of course, immense. The actually existing combinations amount to only an infinitesimal fraction of the potentially possible, or at least conceivable, ones. All these combinations may be thought of as forming a multidimensional space within which *every existing or possible organism* may be said to have its place (Ivi: 8, emphasis added).

For Dobzhansky, the scope of population genetics is evolution itself: Since species are made of populations, the same rules that work within a population also govern "the genetic constitution of species" (Dobzhansky 1937: 16). Stated very explicitly, "Evolution is a change in the genetic composition of populations. The study of mechanisms of evolution falls within the province of population genetics" (Ibidem). Change in the genetic composition populations, in accordance with environments, creates diversity and discontinuities at all taxonomical levels, up to "phylogenetic changes leading to the origin of new classes of organisms" (Ibidem).

Physical environments are an invisible, fundamental, and macroscopic shaping factor. The environmental scope of the landscape logically matches the *complete* set of *all* "the environments that exist in the world" (Ivi: 8). On Dobzhansky's landscape, existing organisms are on peaks because they are *well* adapted to environments. Valleys "symbolize the gene combinations the adaptive values of which are low in the existing environments" (Ivi: 277); therefore, they are "deserted and empty," and they host species and individuals that do not actually exist.

Dobzhansky's peaks are arranged in ranges, reflecting the hierarchically nested structure of taxonomy:

> …the adaptive peaks and valleys are not interspersed at random. "Adjacent" adaptive peaks are arranged in groups, which may be likened to mountain ranges in which the separate pinnacles are divided by relatively shallow notches. Thus, the ecological niche occupied by the species "lion" is relatively much closer to those occupied by tiger, puma, and leopard than to those occupied by wolf, coyote, and jackal. The feline adaptive peaks form a group different from the group of the canine "peaks." But the feline, canine, ursine, musteline, and certain other groups of peaks form together the adaptive "range" of carnivores, which is separated by deep adaptive valleys from the "ranges" of rodents, bats, ungulates, primates, and others (Ivi: 10).

Lion, tiger, puma, and leopard are reciprocally distinct, while altogether they belong to a group (felines) distinct from the one containing wolf, coyote, and jackal (canines). Two individuals on the same peak grossly belong to the same population or species. Two individuals on the same "range" belong to a same higher taxon, e.g., genus or family. In Dobzhansky's argument, the hierarchical structure of diversity mirrors, through adaptedness, the discrete variation of habitats: "The enormous diversity of organisms may be envisaged as correlated with the immense variety of environments and of ecological niches which exist on earth. But the variety of ecological niches is not only immense, it is also discontinuous" (Ivi: 9).

Let us now summarize some of the messages concerning macroevolution conveyed by Dobzhansky's landscape. A pictorial representation devised for population genetics could be extended in scope to include the entire global biota. This possibility reinforced the idea that evolution at large fell "within the province" of population genetics: Dobzhansky suggested global diversity patterns to be explained as simply a global cumulative pattern summed over all the contemporaneous populations. Organic diversity and diversity of habitats were tightly linked due to the process of adaptation which produced a condition of adaptedness and distributed biodiversity on adaptive peaks. Groups differed because they were adapted to different habitats and, vice versa, environmental variation in time and space corresponded to organized variation in organisms: "It is a natural surmise as well as a profitable working hypothesis, that the diversity and discontinuity on one hand, and the adaptation to the environment on the other, are causally related" (Ivi: 8). The constant production of diversity at macroscopic level was basically due to progressive adaptation, since "[t]he evolutionary changes not only enable life to endure the shocks emanating from the environment; they permit life to conquer new habitats, and to establish progressively firmer control of the older ones" (Ivi: 4).

1.5 Simpson's Landscapes

Simpson's book *Tempo and Mode in Evolution* (1944) set out "to determine how populations became genetically and morphologically differentiated, to see how they passed from one way of living to another or failed to do so, and to examine the figurative outline of the stream of life and the circumstances surrounding each characteristic element in that pattern" (Ivi: xxx). The "figurative outline of the stream of life" well describes the tradition of graphical representations of paleontologists and morphologists, and in Fig. 3, we see that the adaptive landscape—a typical representation of population genetics—resonated, in Simpson, with those images. In fact, tempo and mode in evolution were not explained, for Simpson, *despite* population genetics models but *by means of* them, expanding, where necessary, with models specifically designed to answer macroevolutionary questions. So, after the first chapter on paleontological data and models of rates of evolution, Simpson devoted a long chapter to the "Determinants of evolution" studied by population geneticists, i.e., factors like variability, mutation rate, character of mutations, length of generations, size of population, and selection, and tried to use these factors in explanatory hypotheses of paleontological patterns. Simpson's whole book was an attempt to synthesize paleontology and genetics through the illustration of "the basic evolutionary phenomena" and of "the grand pattern and great processes of life" (Ivi: xxvii) that risked to be overlooked due to the increasing disciplinary fragmentation and specialization. Unified ways of depicting "the pattern of evolution" (Eldredge 1999) suggested a unified account of evolution.

Simpson focused in particular on two topics on which "the paleontologist enjoys special advantages" (Ivi: xxix): The first was suggested by the word "tempo" of evolution, and it had "to do with evolutionary rates under natural conditions, the measurement and interpretation of rates, their acceleration and deceleration, the conditions of exceptionally slow or rapid evolutions, and phenomena suggestive of inertia and momentum" (Ivi: xxix). The second topic was labeled "mode" of evolution, i.e., the possible ways in which new species and new groups are born (Fig. 3b).

Whereas Dobzhansky's landscape was about genes, Simpson's landscapes were about morphology. In this phenotypic context, Simpson observed that "in any phyletic series various different characters are changing over the same period of time" (Simpson 1944: 4). While some characters depended in different ways on the external environment, being under incomplete genetic control, other characters appeared to be genetically related and to change in a correlated way, although this relation could break down after a period of evolutionary stability. This justified Simpson's approach to landscapes that concerned with just a few characters or a *character complex*, chosen and weighted case by case. Consideration of a subset of characters, perceived to be evolutionarily significant, also had a pragmatic advantage: It alleviated an epistemological limit of considering of too many characters simultaneously. In his first chapter on "Rates of evolution," Simpson affirmed that although "[considering] whole organisms, as opposed to selected characters of organisms, would be of the greatest value for the study of evolution […] it cannot be said that the problem is quite insoluble, but certainly it is so complex and requires so much knowledge not now at hand that no solution is in sight at present" (Simpson 1944: 15–16). We see that while Dobzhansky acknowledged adaptation as a result of *all* of the organism's traits, Simpson considered the "evolving entity" as the ensemble of all variations *of* the chosen character (or character complex) in a population. In this new framework, Simpson's landscapes still represented all variations that were possible but unobserved. It was this virtuality that enabled the landscape to represent change through time, as a realization of previously virtual possibilities.

Simpson's account of horse evolution in North America exemplified how character complexes and changing physical environment interact in shaping the evolutionary landscape. Figure 4 depicted the evolutionary phases by which a family of browser species, Hyracotheriinae, split into browsers and grazers under the effect of environmental change combined with characters correlation. For Simpson, in the Eocene (56–33.9 million years ago), browsing and grazing in the Equidae represented two well-separated peaks, but only the browsing peak was occupied by members of the family. Grazing was possible but not realized: It was an empty peak. Later, in the Oligocene, the environment played an important role: While the local climate became drier, vast forests started to shrink, grasses evolved, and animals became larger.[6] Anchitherinae evolved from Hyracotheriinae. Due to charac-

[6] See Stigall, *this volume*, for updated evidence on the nonadaptive nature of the matching between climate and speciation in this evolutionary radiation.

ter correlation, larger size drew the transformation of other characters in the direction of the grazing adaptation. For example, teeth crowns became higher (Simpson 1944: 93), and, incidentally, high crowns were in the direction of the grazing adaptation. The increased reachability of a possible adaptative condition is shown as a kind of change of the shape of the phenotypic landscape: The browsing peak "moved toward the grazing peak." In the Late Oligocene, the two peaks were close enough that there was a saddle between them: Variation around the browsers peak was asymmetrical, specifically biased toward the grazing peak which was still unoccupied. Some of the animals were therefore on the saddle between the two peaks. Being relatively ill-adapted, they were subject to centrifugal, fractionating selection in two directions: back to the browsers peak and forward toward the grazing peak. With relative ease, by the Late Miocene selection shaped a branch of grazers, because the saddle was asymmetrical too: "the slope leading to grazing […] is steeper than those of the browsing peak, and the grazing peak is higher (involves greater and more specific, less easily reversible or branching specialization to a particular mode of life)" (Ivi: 93).

Another figure of Simpson's book (Fig. 5) showed a repertory of general selection patterns. Simpson commented on the figure as follows:

> The field of possible structural variation is pictured as a landscape with hills and valleys, and the extent and directions of variation in a population can be represented by outlining an area and a shape on the field. Each elevation represents some particular adaptive optimum for the characters and groups under consideration, sharper and higher or broader and lower, according as the adaptation is more or less specific. The direction of positive selection is uphill, of negative selection downhill, and its intensity is proportional to the gradient. The surface may be represented in two dimensions by using contour lines as in topographic maps (Ivi: 89).

A peak in the center of the population (Fig. 5a) indicated that the "typical character" was favored, and selection was "centripetal," concentrating the population around its mean. Conversely, when a valley was in the center of the population (Fig. 5b), the typical character was disadvantaged, and selection was "centrifugal" driving the members of the species away from the typical but maladaptive condition. In cases where there were no peaks in the population, but the population was on a slope (i.e., there was no optimum in its immediately possible characters but there was an adaptive tendency in a particular direction), there would be linear selection, probably driving the population toward a new, previously unexplored peak (Fig. 5c). When a valley was surrounded by peaks (Fig. 5d), selection would not only be generically centrifugal but would also drive different parts of the population on distinct adaptive peaks (fractionating selection). Sometimes the slopes surrounding a peak could be nonuniformly steep (Fig. 5e), i.e., variations of an optimal character in a direction could bring to maladaptive situation faster than in other directions; in such cases, centripetal selection would be asymmetrical: more severe on some variations and more permissive on others.

Two of these selection patterns are the ones employed by Simpson in the horses case: weak asymmetrical centrifugal selection toward grazing (Fig. 5e) in the Late Oligocene, when variation in the direction of grazing was more strongly favored by selection; and fractionating selection (Fig. 5d) during the Miocene, when a segment of the population broke away under selection pressure and climbed the grazing peak with relative rapidity, consolidating into a new taxon, Equidae.

Simpson's adaptive landscapes had, as usual, adaptation as their central organizing principle, and they represented adaptive dynamics, but Simpson was much less radical than Dobzhansky in seeing adaptation as a central organizing principle *of evolution*. The adaptive landscape of the evolution of horses in Oligocene and Miocene was, in fact, *adaptive* in that it considered the evolution of food habit, "an essential element in the progress of the Equidae" (Simpson 1944: 90). It considered the characters that directly or indirectly relate to food habits, to begin with the height of teeth and including many others. It was a dynamic model, articulated in four phases, and portraying natural selection as a vector on the local landscape which, in turn, changed dynamically. But natural selection, determined by the changing local environment and constrained by correlations within the considered character complex, was not the only mechanism of differentiation. Simpson wrote:

> No theorist, however radically non-Darwinian, has denied the fact that natural selection has some effect on evolution. An organism must be viable in an available environment in order to reproduce, and selection inevitably eliminates at least the most grossly inadaptive types of structure. Aside from this obvious fact, theories as to the role and importance of selection range from belief that it has only this broadly limiting effect to belief that it is the only really essential factor in evolution [...]. The last word will never be said, but all these disagreements can be reconciled, and the major discrepancies can be explained. In the present synthesis adaptation, preadaptation, and nonadaptation all are involved, and all can be assigned immediate, if not ultimate, causes (Simpson 1944: 74, 77).

In fact, adaptive events were time-limited patterns in Simpson's view of evolution. We have seen how Dobzhansky's landscape linked the micro- and the macroscales of evolution. Simpson's adaptive landscapes did the same with a different strategy. They represented those patterns that were "fundamental in nature and broad in scope." They did not represent the whole organic world, nor they dealt with all the characters of organisms and populations. However, they constituted a flexible tool for describing a vast array of evolutionary events from the morphological point of view, with specific reference to connecting micro- and macroevolution.

1.6 The Lasting Communication Efficacy of Adaptive Landscapes: Richard Dawkins's Mount Improbable

In a communication perspective, the MS consisted in the development of interactions between once disparate fields. The improvement of a common language was both a consequence and an enabling condition. In this context, the adaptive landscape summarized the researches of mathematical population genetics in a widely understandable form that, furthermore, looked consistent with observations of "naturalists," opening up tentative reconceptualizations and research possibilities. If a metaphor is an object flexible and open enough to enable generative contaminations between contexts, then the adaptive landscape was a metaphor for evolutionary biology. We can see the enthusiastic acceptance of the metaphor as both a consequence of the credit gained among biologists by mathematical models, and a contribution to such credit made possible just by its communicational effectiveness.

The communicational virtues of the landscape, such as comprehensiveness, conciseness, and context-independence, were again exploited many years later by science writer Richard Dawkins, notorious for his ability to convey his "big picture" of evolution by using eloquent examples and simple and powerful metaphors (see Elsdon-Baker 2009). In the popular book *Climbing Mount Improbable* (1996), Dawkins proposed a vision of evolution as "mountain climbing." Taking for example the fig tree as a particularly complex and "perfect" organism, Dawkins wrote: "...the fig tree stands atop one of the highest peaks on the massif of Mount Improbable. But peaks as high as the fig's are best conquered at the end of the expedition" (Dawkins 1996: 1–2). Dawkins's main point was that the most perfected forms, seen from the standpoint of simple forms, were only apparently unreachable. Like in mountain hiking, there were gradual paths to the highest peaks:

> ...the story of the fig is, at the deepest level, the same story as for every other living creature on this planet [...]. We shall reconstruct the slow, gradual evolution of wings and of elephant trunks [...]. We shall program computers to assist our imagination in moving easily through a gigantic museum of all the countless creatures that have ever lived and died, and their even more numerous imaginary cousins, who have never been born. We shall wander the paths of Mount Improbable, admiring its vertical precipices from afar, but always restlessly seeking the gently graded slopes on the other side (Dawkins 1996: 2).

The task of evolutionary biology was to relentlessly seek graded slopes on hidden sides of the mountain, because evolutionary paths are linear trajectories of exploration of forms: They are series of gradual modifications, where each form is a slight improvement of the previous. The moving force on Dawkins's landscape was natural selection, described as a cumulative process of *finding* (also called "exploration," "discovery," and "expedition"). Accidental improvements were cumulated because of heredity. The resulting vector on Mount Improbable always pushed upwards. Hence, mountain climbing, once started, had to keep going, and when a peak was reached, there was no possible improvement nor "jump" from a kind of structure to another. Dawkins was very clear on this point: Natural selection cannot accept downhill movements. Valleys cannot be crossed. Rather, through small steps, simpler and more probable structures turned into complex and improbable ones. Every step along the Mount's slopes consisted in the discovery of a solution (shape, mechanism, and function) which was slightly but significantly more complex, perfect, improbable, and "designoid" than the previous steps. Dawkins explained the idea of designoid as follows:

> Accidental objects are simply found. Designed objects aren't found at all, they are shaped, moulded, kneaded, assembled, put together, carved: in one way or another the individual object is pushed into shape. Designoid objects are cumulatively found, either by humans as in the case of domestic dogs and cabbages, or by nature in the case of, say, sharks.

> The fact of heredity sees to it that the accidental improvements found in each generation are accumulated over many generations. At the end of many generations of cumulative finding, a designoid object is produced which may make us gasp with admiration at the perfection of its apparent design. But it is not real design, because it has been arrived at by a completely different process (Ivi: 22–23).

One of *Mount Improbable*'s pictorial representations, reproduced here in Fig. 6, was based on a map of the alternative forms of the eye structure found in different living and extinct species.[7] At the bottom of the drawing, there were the simplest forms of photoreception, those performed by a slight photosensitive spot on the body surface. At the top, instead, were the most perfect forms, divided into two comprehensive groups: "composite" and "photographic" eye. The vertebrate eye was the most perfect form of photographic eye: (1) It had specialized cells with many strata of pigments that effectively captured a huge percentage of photons; (2) single cells were sensitive to the direction of light, and they were arranged as a concave surface (in humans, the retina) with a small, pinpoint opening or "pinhole" that projected a single image on the surface itself; (3) it had a transparent lens which reduced diffraction and improved the image through an appropriate refraction index; and (4) it was provided with muscular apparatuses capable of modifying focus, regulating pupil size and moving the eye, as well as with neural systems necessary for controlling the eyes and elaborating information from them. Between the top and bottom boundaries of the landscape, many other eye forms of varying complexity—at least forty—were possible, and many of them were actually found in different species. For example, depicted at the center of the drawing

[7] By framing the concept of "design" in an evolutionary view, and by taking just the eye evolution example, Dawkins harked back to a historical case in the development of Darwin's theory of natural selection: The eye had been a classical example by which natural theologists built the "argument from design," advocating the need for an intelligent designer to account for the most "adaptively complex" structures. William Paley, one of Darwin's intellectual interlocutors, was the oft-cited champion of this school of thought. He had defended the argument from design in his influential book, *Natural Theology*: "…there is precisely the same proof that the eye was made for vision, as there is that the telescope was made for assisting it. They are made upon the same principles; both being adjusted to the laws by which the transmission and refraction of rays of light are regulated" (Paley 1828: 17). Darwin had looked at the eye with some worry while he was developing the theory of natural selection, and perceiving that the theory would require the demonstration of gradual implementation of complex structures. In his *Notebook C* (1838), he had written: "We never may be able to trace the steps by which the organization of the eye, passed from simpler stage to more perfect, preserving its relations. The wonderful power of adaptation given to organization. This really perhaps greatest difficulty to whole theory." And again, in the *Origin of Species*, Darwin had declared: "To suppose that the eye with all its inimitable contrivances for adjusting the focus to different distances, for admitting different amounts of light, and for the correction of spherical and chromatic aberration, could have been formed by natural selection, seems, I freely confess, absurd in the highest degree […]. If it could be demonstrated that any complex organ existed, which could not possibly have been formed by numerous, successive, slight modifications, my theory would absolutely break down" (Darwin 1859: 143, 146). The main concern of Dawkins's book was, in fact, a new clarification of the classic "problem of design." Mount Improbable answered the problem by giving a visual shape to the classical argument of "chance *and* necessity" (cf. Monod 1970). Darwinism was defended, for Dawkins, by demonstrating the existence and viability of several intermediate forms (Elsdon-Baker 2009).

as a vulcan, there was *Nautilus*, a marine cephalopod, so ancient to be considered a living fossil, had a particularly perfected pinhole eye. Of the above list of vertebrate eye characteristics, *Nautilus* had only (1) and (2), but its structure and its rough images suited well the functional needs of the species.[8]

With the vertical dimension of Fig. 6, Dawkins attempted to represent several properties: perfection, complexity, improbability, and the time dimension. The eyes at the bottom of the figure were supposed to be the early evolutionary precursors of all forms. An initial path of improvement could split into different alternatives, leading to peaks in different directions, like the "composite" and the "photographic" mountain ranges. Each range contained different alternatives that were legitimate and viable forms, exhibited by living species that were adapted to their respective environments. Dawkins's landscape of eye evolution focused on a character "complex" like Simpson's, but without the same taxonomical cohesiveness: It had an extensive taxonomical scope, spanning all animals, but taxonomy was not the core criterion of the landscape; groups and "ranges" were assembled according to the eye form.

2 Thinking Deeper: Adaptive Landscapes and Spaces of Possibilities

In recent years, there has been much debate on adaptive landscapes (reviewed in Serrelli 2011). Here, we focus on a single major theme that has not been considered in depth elsewhere, i.e., the observation that any three-dimensional landscape is *always* inextricably linked to a space of potential and explored possibilities (*space of virtual combinations*) the number of dimensions of which typically exceeds three. The link between the landscape and space is sometimes implicit. The subsequent sections examine the virtual spaces that underlied the various historic landscapes discussed above. Up to this point, we have focused on their macroevolutionary scope and on strong connection to adaptation. We now turn our attention to their role as representations of spaces of virtual combinations in attempt to draw general insights about macroevolutionary discourse.

2.1 Genetic Spaces: Wright and Dobzhansky

In population genetics, any realized population is a small subset of the virtual space of possible combinations of its alleles. As the population evolves, it moves through the space of genetic possibilities, exploring new combinations. Many analyses have observed that, in a space of genetic combinations, any particular combination has a huge numbers of *neighbors*, where neighbors are combinations

[8] For detailed information on the eye of *Nautilus* and for updates with respect to Dawkins' notions, see Saunders and Landman (1987), Muntz (1999), Warrant (1999), Colicchia (2006).

that can be reached by small modifications of the chosen combination. Each and every combination in the space has a fixed number of neighbors, a number that is called *dimensionality* of the space. Wright's original landscape (Figs. 1 and 2) was thus the three-dimensional "packing" of a multidimensional space of combinations: While fitness was represented on the vertical dimension, neighborhood on the surface was thought to reflect neighborhood in the space of genetic combinations. Some recent criticisms of the adaptive landscape have focused on the impossibility to preserve the same neighborhood relationships on a two-dimensional surface, but the point is not very relevant here.[9] Rather, we shall consider how a diagrammatic picture can blanket different combination spaces and movement mechanisms, incubating conceptual misunderstandings: We will look at the discrepancies between Dobzhansky's and Wright's virtual spaces that, nonetheless, were mapped on aesthetically identical landscape pictures.

One difference between Dobzhansky's and Wright's genetic spaces concerned the exploration mechanism. Dobzhansky saw *mutation* as a major mechanism. Sewall Wright, on the other hand, regarded mutation to be either a frequent reversible event, or a rare and mainly lethal occurrence. In 1932, in presenting the dynamics on the landscape, Wright had *excluded* mutation as a major evolutionary mechanism. Commenting on the peak equilibrium situation, (Fig. 2 square a) Wright reflected:

> …further evolution can occur only by the appearance of wholly new (instead of recurrent) mutations, and ones which happen to be favorable from the first. Such *mutations would change the character of the field* itself, increasing the elevation of the peak occupied by the species. Evolutionary progress through this mechanism *is excessively slow* since the chance of occurrence of such mutations is very small and, after occurrence, the time required for attainment of sufficient frequency to be subject to selection to an appreciable extent is enormous (Wright 1932: 360, emphasis added).

A clarification of the expression "recurrent mutations" will be useful to understand Wright's mutation notion and its role in his landscape.

Wright's population genetics (e.g., 1931) considered the possibility of nonrecurrent, "wholly new" mutations, but they were rare, almost exceptional. These

---

[9] Kaplan (2008), for example, explains: "The problem […] is that this compression misrepresents the distances between most of the genotypes—accurate representations of distance cannot survive the packing of many dimensions into a few" (630). Pigliucci (2008) writes: "there is no metric that allows one to 'pack' genotypes side-by-side" (593). Sewall Wright was actually in agreement with this 'impossible packing' objection: "The two dimensions […] are a very inadequate representation of such a field" (Wright 1932: 356–357), and a surface picture "cannot accurately represent relations that are multidimensional" (Wright 1988: 116). But, while he would probably have considered it a major objection for a geometrical model, he shielded the surface picture by declaring the latter "useless for mathematical purposes" (Ibidem). The surface is not to be considered a point-by-point map of the combination space: It captures and displays some general features of the space. It is, indeed, a metaphor. Kaplan (2008), like others, accepts Wright's defense of metaphor in general, but then, he criticizes this particular metaphor as poor and misleading with respect to the complexity of the space it represents. Other authors (e.g., Gavrilets 1997, 2004) accept Wright's idea of metaphors *and* the clarity of this particular metaphor, but think that its messages are *wrong*, i.e., that the genotype space is not like Wright thought. For further discussion of what I call the "impossible packing objection," see Serrelli (2011).

mutations *introduced novel alleles* into the combination space, adding a number of new possible combinations. They modified the space and, consequently, the surface. The population moved as a secondary effect of the transformed space. Far more important were recurrent mutations, namely mutations that, in the offspring, changed one parental allele into another one which was always *already present* in the population's genotype space. They were formalized as two opposite mutation pressures acting on the frequency of one allele. Mutation pressure *toward* the considered allele was the chance for other alleles to turn into the considered allele. Mutation pressure *away from* the considered allele was the chance for the considered allele to turn into other alleles. Recurrent mutations did influence the population movement, but slightly, because the mutation rate was always very low: Wright admitted that some mechanisms (e.g., cosmic rays, as suggested by some authors) could increase mutation rates, but they did "not appear adequate to explain evolution to an important extent" (Wright 1932: 361). The most important movement mechanism in Wright's Mendelian space was recombination[10] among a constant set of alleles, moderately influenced by recurrent mutations. Therefore, Wright's space included most of the possible evolutionary change, because such a change consisted in the realization of different combinations in the same space.

In Dobzhansky's account, the Wrightian "trial and error" mechanism for the population to explore the landscape space was "provided primarily by mutation and sexual reproduction, which are able to generate a practically limitless variety of genotypes" (Dobzhansky 1937: 278).[11] For example, when a peak moved due to environmental change, mutations were the possible source of new adaptive genotypes (Ivi: 277). But Dobzhansky's mutation clearly meant the origin of novel alleles, whereas in Wright's population genetics, mutation was totally conceptualized in terms of existing alleles. Dobzhansky's landscape *introduced* the mutation concept as a movement mechanism *into* Wright's landscape.[12] But mutation can

[10] Neo-Darwinian population genetics strongly held recombination as the major genetic evolutionary mechanism. For Ernst Mayr (e.g., 1980), the underemphasis of mutation by Wright and others in the first decades of the twentieth century could be straightforwardly explained by the fact that macromutations were studied by authors such as Hugo DeVries, who considered it as the non-Darwinian mechanism for the origin of species. The redefinition of mutation was, for Mayr, achieved later by the second phase of the MS.

[11] Dobzhansky recognizes that mutation works together with recombination, so that even the right mutations may fail to prevent extinction "if the requisite constellations of these elements do not appear in time" (Dobzhansky 1937: 277).

[12] In the second phase of the MS, mutation was reintegrated into population genetics and began to be seen as a major mechanism for the origin of evolutionary novelties. Those were the years in which Dobzhansky worked and wrote. But the absence (almost "eclipse"?) of recombination in those later works was, for Ernst Mayr, an "astonishing" fact: "It would be decidedly whiggish to suppress the fact that even [many biologists] slighted recombination in the 1930s. Dobzhansky, who later did much to establish the evolutionary importance of recombination, hardly referred to it in the first edition of his book (1937) […]. Recombination was, of course, well known since 1900 as one of the basic Mendelian processes and described in every genetics textbook. Remarkably, only a few evolutionary geneticists used it as a source of material for selection" (Mayr 1980: 23).

be hardly seen as a mechanism for movement in the completely defined allele space envisaged by Wright. In other words, while Dobzhansky still visualized the evolution of a population as movement in a defined space, by changing the idea of mutation, he unnoticeably violated the assumptions that made that visualization possible.

More puzzles were created by the macroevolutionary scope of Dobzhansky's genetic space, and they were concealed by the landscape picture. One concerns the wide environmental scope, another has to do with neighborhood criteria.

Wright's landscape had maintained very clearly that the "niche"[13] inhabited by the population was supposed to shape the *whole* landscape by influencing the fitness of *each and every* possible combination in the space. The whole idea of an adaptive landscape was based on all possible and realized organisms sharing the same gene pool. Dobzhansky's landscape, instead, assumed that multiple niches could simultaneously be represented on the same landscape. Beside the fact that this bold extension of the taxonomical scope would hinder the measurement of distances among the genotypes in the space, the real problem seems to lie in postulating a single global gene pool on which all selection pressures on earth would act (although, as we have seen, Richard Dawkins might like this idea).

An even more serious source of inconsistency for Dobzhansky's space is the conflation of different and independent neighborhood criteria. The Wright–Dobzhansky version of the landscape has no graded axes: The surface is formed "by aggregation" of related combinations.[14] A peak or a valley is a group of related combinations. *Relatedness* is thus the organizational criterion of the surface, rooted in the combination space and to be preserved in the space-to-surface transition. Distance on the surface is proportional to relatedness, but Dobzhansky talked about peaks and ranges in terms ecological niches, conflating similar ecology with close genealogical relationship. Why should closely related taxa also occupy the same ecological niche?

There were different criteria in conflict in Dobzhansky's landscape, and the attempt to connect different spaces of possibilities results in their conflation. Genealogical and ecological groupings do not coincide. We shall see that the fascinating challenge of representing macroevolution is, indeed, the task of disentangling distinct though correlated spaces. Geographic, genealogical, morphological, and ecological groupings do not coincide. Today we are aware that they should be represented on different landscapes: geographical landscapes (based on spatial distribution), genealogical landscapes (based on relatedness), morphological landscapes (based on shared combinations characters), and ecological landscapes (based on the combinations of environmental variables that are suitable for the considered organisms).

---

[13] Attributing the term "niche" to Wright is slightly anachronistic. I use niche in the "evolutionary" sense (sensu Odling-Smee et al. 2003), as the set of selective pressures acting on a population. See below for further analysis of ecological niches.

[14] This answers, in passing, an often expressed doubt on Wright's landscape (e.g., Provine 1986).

2.2 *Phenetic Spaces: Simpson*

Simpson's landscapes hinged upon phenotypic combination spaces. How does a phenotypic combination space differ from a genotypic one? How relevant are such differences? Sewall Wright, reviewing Simpson's landscapes, wrote that

> Phenotypes are, no doubt, more appropriate units for dealing with selection, whether between individual or groups, but genotypes seem more appropriate for mutation or random drift. The choice, however, is practically irrelevant in connection with pictorial representation of changes that occur in populations (Wright 1988: 120).

Pigliucci (2008) considered the transition from a genotypic to a phenotypic space as "a bold and questionable move" and criticized Wright for oversimplification: "Wright acknowledges that Simpson took the idea of adaptive landscapes—defined in genotypic space—and translated it with little fanfare at the phenotypic level" (Pigliucci 2008: 597). Despite Pigliucci's argument, however, a phenotypic landscape actually refers to a peculiar combination space that can be studied with great autonomy from any "underlying" genotypic space, a space made up by all the possible combinations of a limited number of distinct phenotypic characters.

An important difference between phenotypic and genetic spaces is, rather, that phenotypic characters vary in a *continuous* fashion, whereas genetic characters (loci) have discrete variants (the alleles). In a phenotypic space, a trait—e.g., body size—can in principle assume any value in a continuous range. The concept of neighborhood is not applicable in the same way as in a space of discrete characters and so is not dimensionality. Different, continuous measures of phenotypic distance should be devised in Simpson's space. As for the movement mechanism, the character complex in an individual could still be seen as a recombination of the parents' traits, and Simpson's landscape might supposedly integrate reproduction and inheritance rules, but they would be more complicated than the relatively simple Mendelian recombination between parents in a genotypic space.

Simpson's landscapes, differently from Dobzhansky's and more similarly to Wright's, are local in their environmental and taxonomical scope: They explain singular evolutionary events. In these events, local ecological conditions determine selection pressures, causing different morphological structures to fall "under the influence" of a new peak, and local populations to adaptively split.

Simpson's landscapes are limited also in scope of considered characters. Whereas Wright and Dobzhansky had defined the individual organism as an assembly of *all* its (genetic) traits, Simpson's concept of character complex strongly reduces the number of considered traits. This decreases the dimensionality of the space of combinations. The concept of a character complex also implies constraints: The various characters in a character complex cannot come in any combination of values, but a change in one character causes a consequent change in the other. In horse evolution, for example, body size influences teeth structure. Constraints among characters are not considered universal, nor stable over long periods of time: Simpson specifies that such constraints themselves evolve. But *within* the scope of the considered evolutionary event, the combination space is completely defined.

We see that Wright's, Dobzhansky's, and Simpson's landscapes are three-dimensional pictures that refer to distinct combination spaces. Their spaces share important features while differing by other, equally important, ones.

2.3 Evolution Imagined on a Universal Landscape: Dawkins

Dawkins's Mount Improbable expressed some of the central pillars of his view of evolution: the universal scope of a few simple powerful mechanisms, the primacy of function as a major organizing principle of the living world, and the idea of a measure of absolute complexity and perfection. That being said, Richard Dawkins's landscape of eye evolution (Fig. 6) inherited some characteristics from both Simpson and Dobzhansky.

Like Simpson's, Dawkins' landscape was phenotypic and dealt with a particular character complex: all and only the taxa that did have any kind of eyes appeared on the landscape. In some of his most cautious works, Dawkins qualified himself as an "epistemological adaptationist," limiting his explanatory framework to some features that are adaptively complex, because adaptively complex characters are those characters needed an evolutionary explanation the most : "I shall be an adaptionist in the [...] sense that I shall only be *concerned* with those aspects of the morphology, physiology, and behaviour of organisms that are undisputedly adaptive solutions to problems" (Dawkins 1983: 17). Consequently, the sole movement mechanism on Dawkins's landscape was natural selection, a force that constantly cumulated small improvements toward adaptive complexity. In *Climbing Mount Improbable,* there was little mention of the possible mechanisms by which a species could climb *down* a peak (*contra* Wright, see Fig. 2). In "Universal Darwinism," we find Dawkins's interpretation of Wright's view of evolution:

> The phrase 'random genetic drift' is often associated with Sewall Wright, but [...] he clearly sees selection as the driving force of adaptive evolution. Random drift may make it easier for selection to do its job by assisting the escape from local optima, but it is still selection that is determining the rise of adaptive complexity (Dawkins 1983: 31).

Unlike Simpson's landscape, Dawkins's had a universal taxonomical and environmental scope: Like Dobzhansky's, it included species from all over the natural world. In this extension with respect to Simpson, the morphological criterion overrode the taxonomical criterion, or, better, the two were assumed as coincident for the organization of the landscape surface.

An elusive aspect of Dawkins's phenetic landscape is its lack of consideration of phenotypic evolution and constraints. Natural selection was to be intended, in Dawkins's view, as competition among genes or "replicators." Dawkins's famous argument for this was that units of selection need to be stable in time, and genetic replicators are the only candidates for such stability. At the immense scale of Mount Improbable, competition among replicators proceeded, for Dawkins, in some universal genetic space, where there was also little room for a Wrightian "balance of factors" such as migration rate, population subdivision and size with consequent random fixation, plurality of equally fit combinations, and epistasis (implying that alleles can never be thought as actually having individual fitness values). Phenotypic evolution is the *result* of the exploration of the universal genetic space.

A radical novelty of Dawkins's landscape concerned the measure of adaptiveness: Some peaks and ranges were higher than others, giving the landscape a "staircase" shape. Organisms with "mere photoreception" or "pigment cup-eye" were at valley bottoms, regardless their being adapted and well fit to their own environment and life conditions. Fitness was in fact less relative to a particular environment: It had an absolute component called, by Dawkins, "perfection" or "adaptive complexity." The latter concept had been explained, for example, in another essay titled "Universal Darwinism" (Dawkins 1983): "A complex thing is a statistically improbable thing [...]. Living things are not just statistically improbable in the trivial sense of hindsight: their statistical improbability is limited by the a priori constraints of design. They are *adaptively* complex" (Ivi: 16–17).

The solution to the eye "dilemma" in contemporary evolutionary biology (e.g., Lamb et al. 2007, 2009; Gregory 2008b; Lamb 2011) lies in the integration of phylogenetic and adaptive approaches, where adaptation is seen as a complex dialectic between structures and functions, with the chance—already considered by Darwin—of co-option of preexisting structures to contribute to novel functions in different ecological contexts. None of this can be shown in Dawkins's landscape, where structures evolved with little trace of the complex and taxon-specific characters correlations that were hinted to in Simpson's landscapes. Dawkins's absolute measure of "adaptive complexity" hid, in fact, the configuration of selective pressures that explained specialization and speciation in Simpson's evolutionary episodes, such as the Equidae evolution is the Oligocene and Miocene. Dawkins's metaphor, though effective, was thus an oversimplified picture, and we find many inconsistencies when we try to derive the hypothetical virtual space that would underlie this three-dimensional picture.

2.4 Lessons Learned About Graphical Representations of Evolution

Our analysis of historic macroevolutionary landscapes has brought to light several avenues of reflection on representing macroevolution graphically.

Graphical representations may work as a bridge between fields that are mutually independent, sometimes suspicious of each other, and locked. Adaptive landscapes, in the 1930s and 1940s, were an important means for shaping the emerging common language of evolution.

Graphical representations are not only means of communication, but also ways of thinking. In the MS, adaptive landscapes expressed fundamental ideas about the evolutionary process. Notably, they embodied the idea of a unity between micro-evolution and macroevolution (considering them just different focalizations or magnifications of evolution) and assumed the association between macroevolution and adaptation.

A visual representation of evolution frequently displays virtual possibilities for an evolving entity, encoded as combinations that lie in a particular structured space. We have seen Wright's and Dobzhansky's genetic spaces, and how genetic spaces can slightly differ, for example, by their exploration mechanism. We have appreciated that a landscape illustration can somehow hide those differences. With Simpson, we have seen spaces of combinations of morphological characters, and the idea of a character complex, i.e., a small subset of correlated characters that evolve together. We have also appreciated the difference between spaces of combinations of continuous versus discrete traits.

When we employ a visual depiction of evolution, a good practice therefore will be to ask ourselves which space of virtual combinations underlies the image, and, consequently, what dimensionality, which movement rules, and which constraints do hold. An important feature of a combination space is its scope. If an illustration privileges ecological aspects, a larger scope will imply the consideration of larger areas, more ecosystems, more avatars, considering a larger part of a compositional hierarchy (see Tëmkin and Eldredge, *this volume*). A large morphological scope will mean more morphological possibilities. A broader genealogical scope will mean distantly related taxa, in a completely different logic. The possibilities and effects of changing the scope of a landscape depend, then, upon the aspects of evolution that are being visualized.

In the next section, we shall see the challenge of current macroevolutionary studies, that is, representing the exploration of multiple interrelated spaces by means of compositions of multiple graphical representations, including—why not?—landscapes, where appropriate.

3 Visualizing Macroevolution Today

Some tasks that were urgent in the MS are definitely less pressing today. Evolutionary biologists do not shoehorn macroevolution into microevolutionary terms—for example, they do not force macroevolution into population genetics models. Rather, they study macroevolution as an autonomous domain with a variety of methods. Macroevolution is also peacefully decoupled from adaptation, although, of course, the two are still related in complex ways (see Bokma, *this volume*). Macroevolutionary phenomena such as speciation, diversity, and disparity, with their peculiar patterns, do not necessarily constitute adaptive "peak climbing," although, for instance, patterns of adaptation such as specialization or niche breadth are integral part of macroevolutionary explanations, as Stigall explains in *this volume*. This epistemological milieu may count as a reason why adaptive landscapes are less and less recruited as visual aids in macroevolutionary studies.

Dawkins's view of macroevolution, embodied in his Mount Improbable metaphor and in his adaptive landscape of eye evolution, took evolution to be a giant enterprise of adaptive search performed by the living world at large, through the sorting of genes in some supposed global pool. Other macroevolutionary landscapes, such as Dobzhansky's and Simpson's, were more circumstantial and devoted to a sense of adaptation more relative and local. They all shared the focus on adaptation as a driver of macroevolution.

Compared to the years of the MS, contemporary evolutionary biology approaches macroevolution with a much greater variety of notions and tools, some of which will be sketched out in what follows. Examples could be taken from literally everywhere in contemporary evolutionary publications. For reasons of space, however, the examples will be mainly restricted to a single area, i.e., studies of the Ediacara biota. We shall also see landscapes surviving in some interesting forms, two of which are particularly relevant to macroevolution: One is related to low-dimensional spaces, the other to high-dimensional molecular spaces.

3.1 Phylogenetic Diagrams

A very important way of visualizing macroevolution is "evolutionary trees" or "phylogenetic trees" or "dendrograms" (Fig. 7). Trees allow for mapping characters, distinguishing between ancestral and derived character states, and formulating hypotheses about multiple origins and evolutionary mechanisms of particular characters. In the tree imagery, the spatial disposition of taxa reflects common descent and some sort of divergence. Such use of spatial distance is reminiscent of the macroevolutionary landscapes by Dobzhansky and Dawkins. Relatedness, although mixed with ecology, was the foundational principle of Dobzhansky's landscape. Individuals and groups sharing recent common ancestors were located

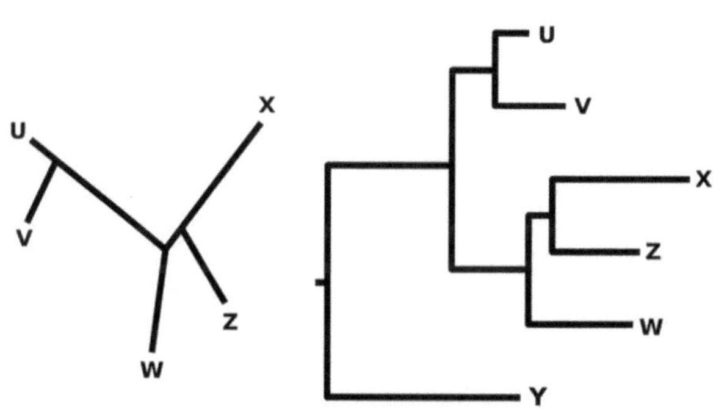

Fig. 7 A toy dendrogram in two versions: unrooted (*left*) and rooted (*right*). The two trees differ by one taxon, Y, hypothetically added as an "outgroup" to root the tree. Branch spans might reflect elapsed time (if the tree is a chronogram) or amount of change (if the tree is a phylogram). Trees are a fundamental way of visualizing and studying macroevolution, but they are prone to misinterpretations that can be avoided by correct "tree thinking" [*Source* modified from Gregory (2008a)]

close to each other. Recent ancestry was evoked also by their gathering in ranges. Dawkins's landscape lined up existing species in a map of branching evolutionary trails, from the most ancient to the most recent. Unlike landscapes, phylogenetic trees are not constrained to displaying the virtual possibilities that are available at a given time: They are devoted to history. Unlike the movement on adaptive landscapes, especially on Wright's, the divergence represented by evolutionary trees does not happen in a predefined space. Divergence is, in principle, indefinite; a space can be constructed to measure distance and reconstruct divergence, but only in hindsight, and edges represent relationships, not precise trajectories whose points are specific individuals or groups. Also, with respect to landscapes, phylogenetic trees lose the adaptive dimension: They become pure genealogy.[15]

Reading a tree, just like understanding the meaning of an adaptive landscape, requires an appropriate way of thinking. Many studies have revealed the cognitive properties and difficulties of tree thinking.[16] Tests on people's tree-reading abilities have suggested that typical tree misreadings are expression of cognitive tendencies and that some tree-drawing habits can trigger and shore up the misreadings. The habit of laying out taxa left to right according to prejudices about ancestry and primitiveness, or species-poor to species-rich (even performed by the automatic "ladderize" function in phylogeny computer programs!), can reinforce the habit of reading trees from left to right combined with a persistent progressionist view of evolution. Certain present-day species are described as "primitive" and considered "ancestral" to other extant species, even though evolution has not stopped in any lineage (Omland et al. 2008: 855).

Looking at the phylogram in Fig. 7, many people think they can infer, for example, that Y is older than W and that W is, in turn, older than Z and X, because of how far "back in time" their origins are shown. This is a misinterpretation because the nodes do not represent the origin of Y or Z or X, but rather the last common ancestor shared between Y and all the others, and between Z and X. The inference of which taxon is the oldest is an overinterpretation: It depends on a tendency to *project the tip back to the node*.

---

[15] An indirect link between trees and landscapes is the following. Phylogenetics today is carried out with the aid of computer programs that, given a dataset of genetic sequences and other characters, explore the space of possible trees and determine *which* tree is the most likely under the given data, as well as *how much* we can be confident in it being the solution (see Wiley and Lieberman 2011). The process of discovering the tree is often imagined as a series of blind "robot *walks*" *on the landscape of all possible trees*, with the most likely trees on top of likelihood peaks, surrounded by similar trees that are good, but not quite!

[16] Some examples of this flourishing literature are O'Hara (1992, 1998), Baum (2005), Gregory (2008a), Omland et al. (2008), Thanukos (2009, 2010), McLennan (2010), Meisel (2010), Halverson (2011), Torrens and Barahona (2012). Perhaps similar discourses could be made on network thinking, concerning for example network drawing choices (that are also available as options in computer programs), cognitive tendencies in reading them, biases in recognizing modules and hierarchical levels in networks (Papin et al. 2004). But, unlike the tree-thinking "story," this one has not been written yet.

An analog mistake consists in considering a species as ancestral to the others: The *reading along the tips* habit. Extinct ancestor species *could* be included among the tips, but the tree says nothing about descent of tips from one another. As a rule, ancestors should not be looked for on tips. Chimps are not ancestors to humans, while the two groups share ancestors.

The same species that seems ancestral is often considered as globally primitive with respect to the others. This is *the primitive taxon fallacy*. In Fig. 7, Y would appear primitive, but basal with respect to the others. Omland et al. (2008) provide the example of some animals, like the platypus or the marsupial opossum, that are often labeled "primitive mammals" even in the scientific literature. The authors show that this idea can be suggested by cladograms, wherein the opossums are not the "focal species." Every tree indeed does have a focus, it is always incomplete, and it is enriched according to the focus. In Fig. 7, Y is unlikely to be the focal species. Y is probably introduced as an *outgroup* to study the other groups. The side of Y would be more populated with relatives in a cladogram where Y is the focal species; therefore, Y would not appear ancestral or primitive.

In fact, tree-drawing habits can probably be traced back to the same cognitive tendencies we see in tree reading. To contrast these misrepresentation loops, tree-thinking authors propose specific training and constant effort toward tree-drawing choices that are less common. The order of taxa is a free choice since a tree can be rotated around any node, like a hanging mobile. The tree reader can be put into a cognitive tension by means of "zigzag" rotation or random rotation. In Fig. 8, bilaterian animals are far right, indulging our idea that they are somehow advanced with respect to the others; yet, we can imagine to invert sponges and bilaterian animals by rotating the section of the tree around the node of their common ancestor. How advanced would we look then?

A more fundamental misconception consists in interpreting the tree as a tree for making inferences about species, as opposed to a tree for making inferences about characters. This is the *tree of species fallacy*. To summarize these and other conceptual issues, Omland et al. write:

> Which of the species is the oldest? Which is the youngest? Which is most ancestral? Most derived? Most primitive? Most simple? Most complex? The answer is that a phylogeny provides NO INFORMATION about any of these questions! (Omland et al. 2008: 856).

Omland et al. emphasize that any taxon is a mix of ancestral and derived characters. Trees provide a structure for inferring which character states are ancestral and which are derivative. In fact, the addition of taxa is instrumental to getting a suitable nesting pattern to resolve the evolution of the character at hand. The branching pattern in Fig. 7 may be sufficient to make inferences about some of the characters of U, V, X, Z, and W. Here, Y plays the role of an outgroup, added because there is evidence of it exhibiting ancestral states of those characters. This tree, where Y is alone in the species-poor area, is not the right context to infer or test such ancestral state: That would be another instance of the primitive taxon fallacy. In order to study the evolution of a character, taxa should be added until a suitable nested pattern is obtained: Y should be nested into a sufficient and meaningful phylogenetic context.

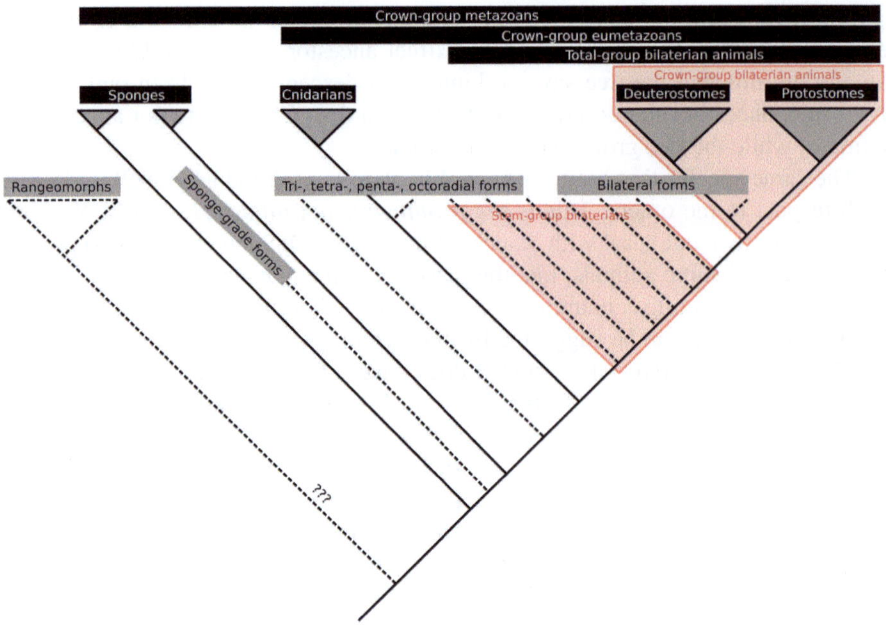

Fig. 8 Cladogram of Ediacaran groups (*dashed branches*, all extinct) and living groups (*solid branches*) by Xiao and Laflamme (2009), an example of how morphology and descent are decoupled. All sponge-like taxa together form a "grade," i.e., a morphological grouping with no immediate phylogenetic implication. Ediacaran sponge-like forms are a clade, distinct from the two clades of modern sponges. Ediacaran radial forms belong partly to extinct clades, partly to clades that survive today (Cnidarians, e.g., jellyfishes). Ediacaran bilateral forms belong to various clades; some of them are Ddeuterostomes and their clade flourishes today in the form of, e.g., worms, sea urchins, and vertebrates. "Crown group," a technical term of cladistics, means a collection of taxa composed by the living representatives of the collection together with all their ancestors back to their most recent common ancestor, as well as all of that ancestor's descendants. Bilaterian animals are a crown group of all living bilaterian animals with their last common ancestor and all its descendants, therefore also some Ediacaran bilaterians [*Source* redrawn from Xiao and Laflamme (2009)]

3.2 Stratigraphies and Geographies

Adaptive landscapes did not visualize the physical environment or the geographical space. Interpreting adaptive landscapes as geographical landscapes was possible—for example by identifying genetic separation with geographical separation—but wrong, being it based on a conflation between distinct and incommensurable spaces (Serrelli 2011). In fact, geography and ecology are invaluable sources of data and patterns for macroevolution. They are recorded and exposed visually by stratigraphy, geography, and paleogeography.

Stratigraphy is an ancient way of studying the accumulation of sediments and rocks over geological time. The relationship between stratigraphy and living beings is very intimate. Stratigraphic diagrams usually imitate the spatial disposition of layers: oldest at the bottom and most recent at the top. Same age formations get

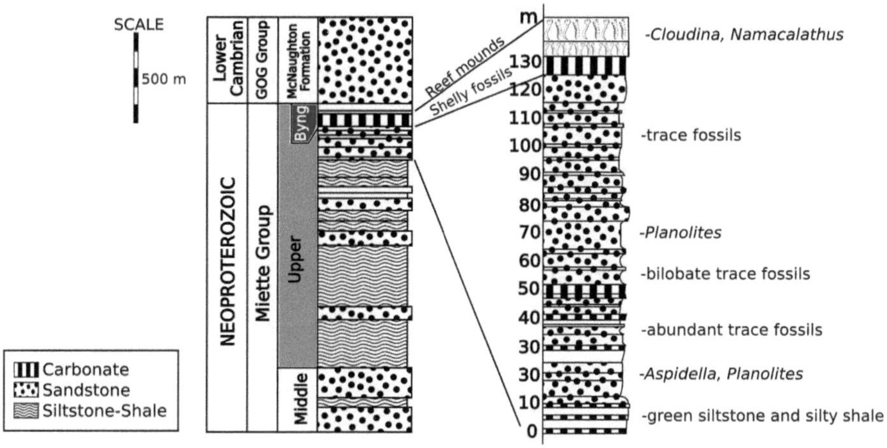

Fig. 9 Stratigraphic section of the *Namacalathus–Cloudina* assemblage in the Miette Group, Salient Mountain area, British Columbia. Geologic depositions are graphically transformed into horizontal layers to show formation sequence. *Right* part of the figure: magnification of the Byng formation layers, with respective elevation in meters. Filling patterns inform about the composition of each layer (carbonate or sandstone). Labels on the far right mostly indicate fossil findings, where present. *Cloudina* and *Namacalathus* (also drawn inside Byng formation layers) cannot be found in sandstone, where carbon has been eliminated by dolomitization. *Left* part of the figure: larger geological context. Layers are flanked by labels: *right* to *left*, the subformations, formations, and periods [*Source* simplified redrawing from Hofmann and Mountjoy (2001)]

aggregated into "floors." However, such vertical disposition is largely the result of theoretical transformations. In actual rocks, depth and age are not proportional. A layer in a stratigraphy is rarely homogeneous: Intrusions of atypical areas with atypical processes or organisms are a rule. Sediments can also be found in different states according to their history after formation. An example of a stratigraphic diagram is Fig. 9, from Hofmann and Mountjoy's (2001). Figure 10 shows the current geographical disposition of the formations.[17]

A challenge for stratigraphy is to integrate knowledge about local geology into a global picture of the world's geological evolution. As we go back in time, localities

[17] Hofmann and Mountjoy studied a geologic formation called the "Miette Group," in British Columbia, Canada. Within it, they focused on the Byng carbonate platform, a recent part of the Upper Miette Group. The Byng formation within the Upper Miette context is a case of spatial and geographical heterogeneity. Furthermore, parts of the Byng Formation are "dolomitized." Dolomitization is a chemical process by which calcium is replaced by magnesium, obliterating fossil shells. Other parts of the Byng formation are predominantly limestone. In the beds of limestone made by accumulations of biogenic carbonate—that is, shells—millimeter-scale tiny animals such as *Cloudina* and *Namacalathus* are found. These fossil shelly organisms are from the Neoproterozoic, a very ancient period—from 1,000 to 541 million years ago—that preceded the so-called Cambrian explosion. The Cambrian explosion is the relatively rapid and abundant appearance of animal phyla, including all those surviving today (except Bryozoa, which appeared later). See "Cambrian fauna" in Fig. 7a.

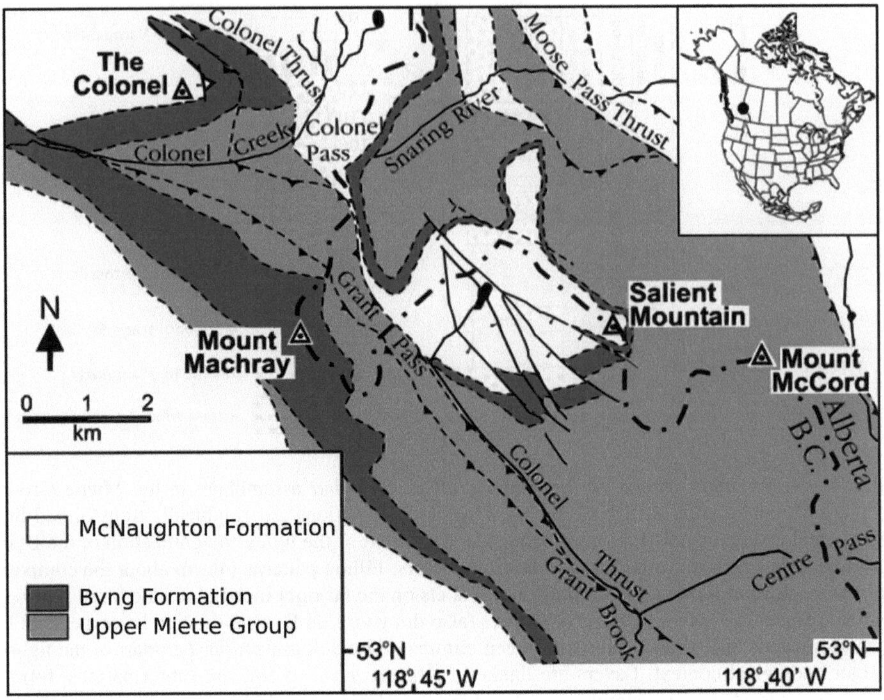

Fig. 10 Geography of the Miette Group, Salient Mountain area, British Columbia. *Dark gray* is the Byng formation, wrinkled and tilted in "sheets" due to tectonic movements (see the "Colonel Thrust"), while shown as a horizontal layer in stratigraphy (Fig. 9) [*Source* modified from Hofmann and Mountjoy (2001)]

move, subside, and become closer or slide apart (Fig. 11).[18] The geological history of the planet appears as an exploration of different geographies. Knowledge of the rules of this exploration allows for inferring past geographies and for interpreting fossils and other traces of macroevolution in the evolving geographical space. Global and local data must be used to reconstruct the correct paleogeography and paleoecology, posing interesting challenges to the visual representation of macroevolution and to the scientific imagination of both evolutionary biologists and

[18] *Cloudina* and *Namacalathus* are part of the worldwide Ediacaran biota, the earliest known complex of multicellular organisms. Most Ediacaran life forms are known by indirect traces ("trace fossils"), but *Cloudina* and *Namacalathus* are particular for their skeleton which is liable to fossilization. *Cloudina* has global distribution, while *Namacalathus* is rare. The *Cloudina–Namacalathus* association studied by Hofmann and Mountjoy (2001) in British Columbia was first found in Namibia, in the "Nama sequence." Today, Canada and Namibia are antipodal, but, back to 1 million years ago when rocks formed, their latitude was very different, and so, was the global configuration of landmasses as we see in Fig. 11. Hofman and Mountjoy argue that finding a new instance of the Nama sequence constitutes a widening of the known geographic range of such assemblage, making it "cosmopolitan" and enhancing expectations that the assemblage will be found at least in other areas that, in the Neoproterozoic, were between current Canada and current Namibia.

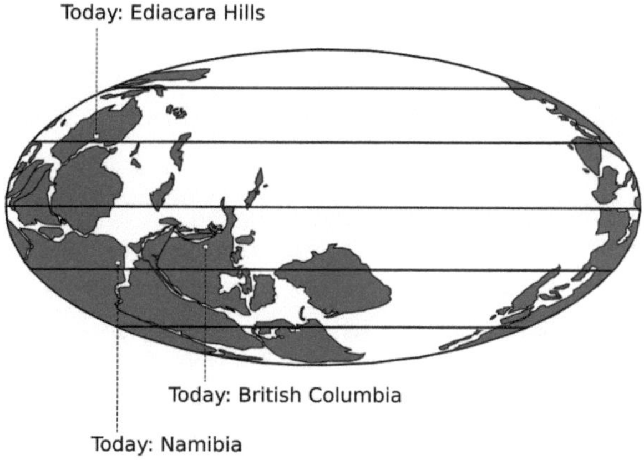

Fig. 11 Reconstruction of continent positions in the early Vendian (the Vendian is 650–543 million years ago), prior to the Ediacaran biota. Some paleocoordinates of Ediacaran assemblages are shown [*Source* redrawn and simplified from Waggoner (2003)]

laypeople. There are many sources of data to do that, from the most classical paleomagnetism (rocks containing polarized metals that "lock" the north–south orientation at the time of solidification) to contemporary computer GIS techniques.[19]

Centuries of geology have deposited different inconciliable naming systems. Strata are often named after modern reference places, which, furthermore, reflect the history of geological findings more than any meaningful aspect of geological history. Rhode (2005) identifies the necessity to "sort out the mess that man has made of the geological timescale" in introducing an interesting database of geological information called *GeoWhen*. The same goal is pursued by official organizations such as the International Commission on Stratigraphy.[20] The Commission has to "precisely define global units (systems, series, and stages) of the International Chronostratigraphic Chart that, in turn, are the basis for the units (periods, epochs, and age) of the International Geologic Time Scale; thus setting global standards for the fundamental scale for expressing the history of the Earth."

3.3 Morphospaced Stratigraphies

A very classical diagram of macroevolution marks, on a timeline, the oldest and the most recent finding of fossil taxa. The oldest finding is considered as a proxy for the origin (e.g., speciation), and the most recent is considered close to the extinction date. The arrangement of several fossil taxa with their life spans

[19] See the *Paleomap Project* by Chrisptopher R. Scotese (http://www.scotese.com/). Some interesting works with GIS in paleontology have been carried out by Lieberman (Rode and Lieberman 2004; Hendricks et al. 2008; Abe and Lieberman 2009, 2012; Lieberman 2012; Myers and Lieberman 2011) and Jablonski (e.g., 2008). GIS is routinely used to study current biodiversity. GIS techniques are also used the context of "geophylogenies," i.e., integrations between geographical and genealogical data (Kidd and Liu 2008; Kidd 2010) that we shall see below.

[20] http://www.stratigraphy.org/.

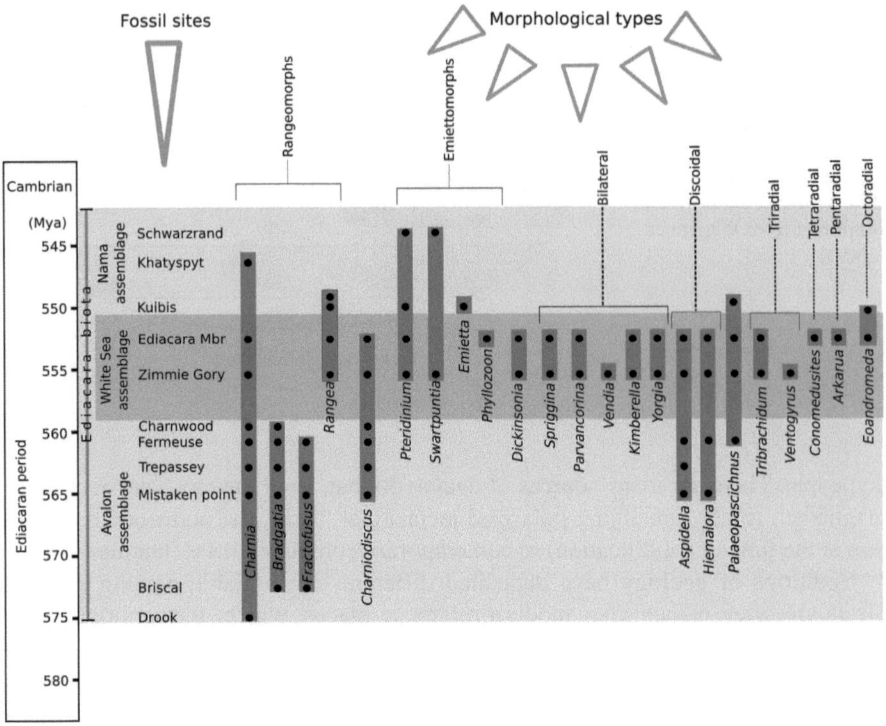

Fig. 12 Morphospaced stratigraphy of the Ediacara biota. Vertical dimension is time. Avalon, White Sea, and Nama assemblages are considered as noncontemporaneous stages composed, in turn, of fossils from different sites of various ages within their time interval. Each vertical column represents a taxon with its life span and is labeled with its name. The horizontal arrangement is meant to show clusters of morphologically similar taxa. Morphospaced stratigraphies have no immediate phylogenetic implication, and they are made possible by evolutionary stasis (taxa remain stable along their lifespan) [*Source* drawn after Xiao and Laflamme (2009), modified from Brasier and Antcliffe (2004)]

shows important macroevolutionary patterns. An example of this way of representing is Fig. 12, from a Xiao and Laflamme (2009). The subject of the figure is the Ediacaran biota, a distinct group of complex macroscopic organisms that flourished just before the Cambrian radiation of animals. We see that fossil species remain relatively stable spanning different periods (see Bokma, *this volume*, Tëmkin and Eldredge, *this volume*). In fact, this kind of diagram is made possible by the paleontological fact of stasis. In the early years of evolutionary biology after the Modern Synthesis, stasis was downplayed as an evolutionary pattern. Accordingly, divergence representations like Simpson's Fig. 3a prevailed on the kind of graphical depiction we are examining.

The vertical arrangement of Fig. 12 is stratigraphic, and in this case, taxa span multiple strata. But taxa are included according to a particular criterion, unrelated to the identification of layers or to geography: in this case, a hybrid ecological–phylogenetic

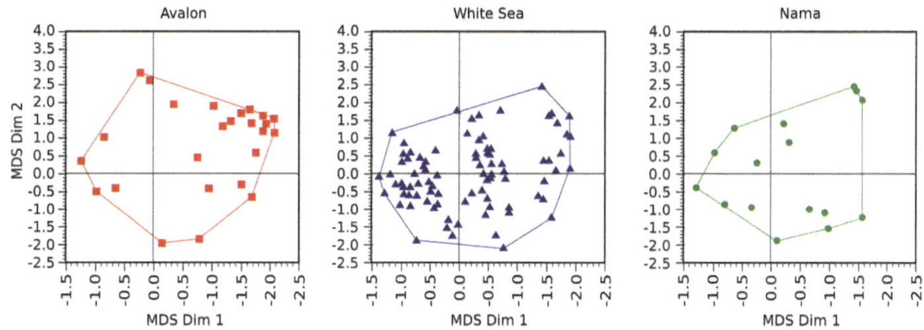

Fig. 13 Distribution of the fossils of the three Ediacara assemblages in a morphospace of 50 characters, mathematically transformed into two dimensions by means of "nonparametric multi-dimensional scaling" (MDS). The structure of diversity within each assemblage differs, although the global range of the realized morphologies remains remarkably stable. The relative stability of morphospace occupancy shows, for Shen et al. (2008), that shifts in paleoenvironments are not the controlling factor of morphological evolution in this case [*Source* modified from Shen et al. (2008)]

criterion. Horizontally, taxa are ordered according to morphological criteria.[21] Avalon, White Sea, and Nama are three major Ediacaran assemblages. The Ediacara biota is named after Australia's Ediacara Hills; the Avalon assemblage gets the name from a dreamy peninsula in Newfoundland, Canada; the Nama assemblage of Namibia was the subject of Figs. 9, 10 and 11; and White Sea is in Russia. Notice in Fig. 12 that Xiao and Laflamme (2009) accept the hypothesis that the three assemblages represent successive, although partly intersecting, evolutionary stages, rather than contemporaneous biogeographic provinces or environmental/ecological associations (for an analysis, see Shen et al. 2008). Besides that, the diagram makes it clear that, at each point in time, the various species represent explorations of different morphologies and lifestyles. The horizontal arrangement is a heuristic disposition aiming to show the existence of clusters of species sharing similar anatomies, i.e., grades. In this way, some forms, like *Dickinsonia*, are shown as "transitional," not because they are unstable and maladapted, but rather because they are located at crossroads of morphological clusters or large-scale splitting trends.

Shen et al. (2008) represent, in another way, the morphological features of the organisms in the three Ediacara assemblages. Figure 13 plots the Ediacara fossil findings considering 50 morphological characters—e.g., overall shape, first-order

[21] From left, *Rangeomorphs* are "fractally quilted with frondlets arranged to form a repetitive, self-similar pattern;" *Erniettomorphs* "have biserially quilted tubes that are alternately arranged along a midline," and they are not bilaterally symmetric; *Bilateral forms* are characterized by "anterior–posterior differentiation with a differentiated 'head' region;" *Discoidal* forms are characterized by "concentric and sometimes radial features;" taxa such as *Charniodiscus* and *Palaeopascichnus* are morphologically unique intermediates between more abundant morphologies; *Triradial forms* are characterized by triradial symmetry or consist of three spiral arms; *Tetraradial forms* are rare, *Pentaradial forms* are characterized by a five-fold symmetry, *Octoradial forms* include "one species which consists of eight spiral arms tightly wrapped into a disk".

symmetry, central–peripheral differentiation, and the presence of peculiar features—and shows how each assemblage occupies the morphological space so constructed. Each point is a fossil, and the realized morphospace of each assemblage is visualized as a convex "hull." The two coordinates in the "scatterplot" (Dim1 and Dim2) are worked out from the 50 characters by means of "nonparametric multidimensional scaling" (MDS). The figure shows that throughout the three Ediacaran assemblages, the overall size and position of the realized morphologies have remained markedly static, while taxonomic diversity *within* each biota, i.e., the number of taxa and their distances, has changed. Shen et al. talk about "the decoupling of taxonomic and morphological evolution," likening, under this respect, the "Avalon explosion" to the Cambrian explosion.

While macroevolution is undoubtedly a simultaneous exploration of multiple configurations of characters in morphospaces, it also consists in geographical exploration, as well as exploration of ecological possibilities and of genealogical outgrowths. These explorations, while related, are not necessarily coupled. Shen et al., for example, are careful in decoupling morphology from ecology and genealogy. They deal with morphology and "make no inference on the phylogenetic homology and functional biology of the coded characters." On the other hand, they try to test the claim that Ediacaran morphology was primarily controlled by paleoenvironments. Avalon-type biotas occur in deep marine habitats, whereas Nama-type biotas occur in distributary-mouth bar shoals, yet, according to Shen et al.'s analysis, morphologically the "groups are indistinguishable or strongly overlapping," implying that "paleoenvironments were not a major factor controlling the extent of Ediacara morphospace." Morphological proliferation is thus not tightly coupled with ecological exploration.

Genealogical relationships in the Ediacaran biota are not shown in Figs. 12 and 13. Figure 8, modified from Xiao and Laflamme (2009), is a tree showing hypothetical broad-scale phylogenetic relationships among some major Ediacaran fossil groups in the context of later, still living groups. The tree shows how morphological groups or "grades" can be either monophyletic like *Rangeomorphs*, nonmonophyletic like *Bilateral* forms, and surely polyphyletic like *Discoidal* forms, scattered across different lineages. The example shows that descent relationships, even when known with reasonable confidence, may not be fully reflected in morphology, generating visualization conflicts.

3.4 Diversity Curves

The dynamics of biodiversity variables over geological time is a crucial aspect of macroevolution. Graphical representations of those dynamics are essential in macroevolution studies. Figure 14 shows two versions of a familiar diagram of extinctions. In charts of this kind, the horizontal axis (left to right) represents time. The vertical axis reports a variable of biodiversity. The first version (a) traces the number of families over the whole history of multicellular life, marking the "Big Five"

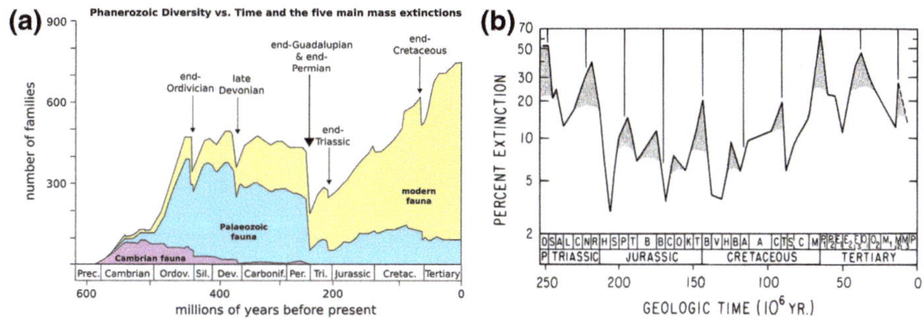

Fig. 14 Profiling large-scale biodiversity variables is crucial to macroevolutionary studies, but requires the construction of huge datasets by the integration of many different studies **a** number of families since the dawn of multicellular life, **b** extinction rate (in percentage of extinct taxa per unit of time) since the Late Permian [*Source* **a** Redrawn and modified from Ian Metcalfe's research page (http://www-personal.une.edu.au/~imetcal2/web-data/Research/Mass-Extint/Extinct.html), accessed on June 13th, 2014; **b** from Raup and Sepkoski (1984)]

mass extinctions. A taxonomical or phylogenetic criterion is used to split out the rise and survivorship of different groups (Cambrian, Paleozoic, and modern faunas). The second chart (b) is from the original study in which Raup and Sepkoski (1984) demonstrated the existence of the Big Five. Their chart shows not an absolute value, but a frequency: percent extinctions. The two paleontologists were working in the context of the "birth of paleobiology" (Sepkoski 2012), when computational methods were run on large datasets to expose important macroevolutionary patterns, legitimating macroevolution as a proper field of study.[22] Temporal biodiversity graphs, whether local or global in scale, always require the assembly of many heterogeneous data and their elaboration through statistical and computational procedures.

Temporal charts may visualize dynamic values. Origination/speciation *rates* and extinction *rates* are a fundamental aspect of macroevolution. Rates are dynamic values, although they can also be expressed as frequencies (a speciation rate, for example, can be well seen as number of speciation events per unit of time). Extinction and speciation rates are not constant over time, and species abundance or scarcity results from the balance between these two processes.

Other temporal charts may focus on biometrical measures. These are continuously varying traits that are measurable in living organisms or fossils, and that can be mapped over groups and over time. An increase or decrease of the average value of a particular character through time may constitute a macroevolutionary trend. *Niche breadth* is a particularly interesting example of a biometrical trait measured and plotted over macroevolutionary time. There are several concepts of

[22] The importance of this pioneering study can hardly be overestimated. Later, many "mass extinctions" beyond the Big Five have been revealed and studied in the fossil record, and knowledge in this field is constantly evolving.

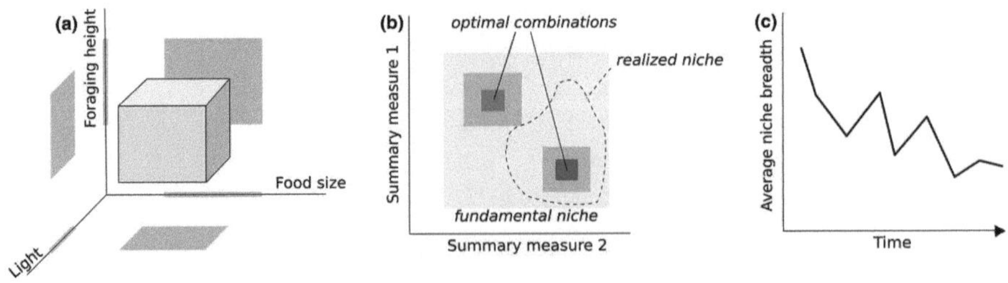

Fig. 15 **a** Niche of an organism considering three environmental variables. For each variable, the niche includes the range in which the organism/species can survive and reproduce indefinitely. **b** Realistic niches have many more than three dimensions. Summary measures can be extrapolated to go back to two dimensions. Within the fundamental niche, there are "peaks" of optimal combinations of values; the realized niche, constrained by community organization, is a subset of the fundamental niche. **c** A hypothetical macroevolutionary trend toward specialization (i.e., niche breadth reduction) (original drawing)

ecological niche (Peterson 2003). The Hutchinsonian niche (Hutchinson 1957, 1965), foundational to modern ecology, is an n-dimensional hypervolume whose dimensions are environmental variables that define the requirements of an individual or a species, e.g., elevation, temperature, precipitations, soil dimensions such as nitrogen, and mean solar insolation. Along each of these dimensions, there is a range where the considered organism can live. The resulting hyperspace is the organism's (Hutchinsonian) niche—see Fig. 15a.[23] The niche hypervolume comprises all combinations of the environmental conditions which do permit an individual of the species to survive and reproduce indefinitely. Hutchinson defined the niche *breadth* for an organism as the habitable range along each particular environmental variable (Fig. 15b). Niche breadth may widen or shrink in macroevolution, as organisms become, respectively, more generalist or more specialist (Fig. 15c). Something will be said below on the correlation between niche breadth and geographic range, in relation to connecting multiple spaces of possibilities.

3.5 Legacy of Macroevolutionary Landscapes

The genotypic landscapes of Wright and Dobzhansky predate by several decades the advent of molecular biology and DNA sequencing. Since then, molecular biology has shown lower level *sequence spaces* (Smith 1962, 1970). These are very similar to Wright's allele spaces, but in sequence spaces, individuals are not sets alleles, they are sequences of nucleotides; the number of loci is replaced by the number of nucleotides in the sequence; and the number of possible states

[23] Hutchinson distinguished the *fundamental* niche—the maximum inhabitable hypervolume in the absence of biotic constraints—from the *realized* niche, a smaller hypervolume occupied by the species when under competition, predation, parasitism, and other constraints.

for each nucleotide is four (A, C, T, G). The conception of mutation has also changed. Mutation can still be seen from the alleles point of view, as the origin of brand new elements, and as a "different rules" mechanism altering the combination space (as it was in Wright and, less clearly, in Dobzhansky). On the other hand, mutation (at least point mutations) can now be treated as movement in the sequence space, i.e., not as the origin of a brand new factor, but rather as the realization of a possibility which was already specified in the combination space.

The relevance of sequence spaces for macroevolution does not derive, of course, from the illusion of constructing mega-landscapes of nucleotides (as Dawkins had imagined with alleles).[24] Rather, sequence spaces are guides to mapping what macroevolution has been realizing, as a subset of a whole lot of possible alternatives. Sequence spaces are source of evidence to explain, for example, why some combinations were not realized in macroevolution, opening up new research questions.

Relevant to macroevolution studies are, for example, low-dimensional sequence spaces of short genetic regions, for which a meaningful fitness value can be calculated. RNA models (Schuster et al. 1994) match nucleotide sequences with the determined RNA's secondary structure, which in turn can be assigned a fitness value. This modeling tradition has shown that many sequences in the sequence space produce the same RNA structure, or structures that are equally fit with respect to some biochemical function, implying that a lot of nearly neutral movement can happen in the sequence space. Low-dimensional sequence spaces can also be used to study proteins. While the entire sequence space of an average protein is intractable owing to the "hyperastronomical" number of variants (Voigt et al. 2000), a significant region of the protein can be correlated to some measure of function or fitness (Rowe et al. 2010). Low-dimensional biochemical spaces have become amenable to experimental investigation: Knowledge of neighborhood and fitness rules is applied to guide chemical synthesis; engineered proteins (e.g., Stadler and Stadler 2002) can be put to work in organisms, and their function measured; and empirical fitness landscapes can then be compared to theoretical ones (Carneiro and Hartl 2010; Lunzer et al. 2005; Poelwijk et al. 2007). Another model, the block model (Perelson and Macken 1995), reduces dimensionality by envisioning genes or proteins as combinations of distinct functional domains and can be studied with suitable mathematical methods (Orr 2006).

Given the interest of low-dimensional spaces, many modeling works use biologically unrealistic exploration mechanisms to study their general properties. In some models, for example, a DNA sequence recurrently mutates to a finite number of alternative sequences (Smith 1962, 1970; Gillespie 1983, 1984, 1991). General space properties such as the relation among correlation, ruggedness, and expected outcome of adaptive walks were found by Kauffman (1993) (see also Kauffman

[24] A hypothetical sequence space accommodating whole genomes would be absurdly high dimensional (a human sequence space, for instance, would have 2.9 billion dimensions); not to mention the inconceivable fitness effect of every nucleotide in the sequence and of any single nucleotide substitution.

and Levin 1987). His *NK* models, with *N* genes and *K* fitness interactions between genes, were tunably "rugged." Correlated landscapes have great *K*, i.e., many interactions among genes. They are smoother than completely random landscapes (i.e., they feature fewer local optima, Weinberger 1990), and this affects probabilistically the dynamics of adaptive walks therein. These studies are more a way of exploring the properties of combination spaces than a way of simulating macroevolution, but more and more empirical studies (cf. Poelwijk et al. 2007) take advantage of metrics and methods developed for these artificial fitness landscapes.

So far, I have provided examples of low-dimensional character spaces that are relevant to macroevolution and partly liable to two-dimensional representation in the form of landscapes. The genetics of macroevolution, however, also requires a completely different kind of spaces, now that high-throughput sequencing is readily available. To go beyond molecular phylogenies and explore the mechanisms of genomes evolution, scientists need spaces with huge dimensionality that imitate whole gene pools, in which each combination is a whole genome. But low-dimensional and high-dimensional spaces are separated by an epistemological "phase transition": High-dimensional spaces need another kind of mathematics, basically holistic, probabilistic, with a strong role of statistics (Hansen 2012; Doebeli 2012). Even here, however, "undrawn" landscapes are an indispensable aid. Terms like "ruggedness" have precise mathematical, statistical meaning. They are used in mathematical theories even in the frequent cases where no landscape picture can be usefully drawn.

Simpson's phenotypic landscapes also maintain their service to macroevolutionary studies. They had been formalized and researched at first by Lande (1979a, b) and then by others (see Arnold et al. 2001, 2008) who introduced the name "Simpson–Lande landscapes." Methods such as "principal component analysis" extract a series of bidimensional maps that together show the "shape of variability."[25] Simpson–Lande landscapes are essentially phenotypic, but, expanding the biometrical tradition of population genetics (Provine 1971), and they track not only the movement of populations in phenotypic space, but also its genotypic effects, e.g., changes in heritability and emergence of correlations. The task is performed by mathematical tools such as the "G-matrix" that constitute the field of "quantitative genetics." The G-matrix is a multivariate version of classic heritability, *h2*, that measures the additive genetic variance and covariance of traits within a population. The evolution of the population can be theoretically studied by looking at how the G-matrix reacts to an adaptive landscape. Rice (2004, 2012) also developed mathematical formalisms to introduce a phenotype landscape that explicitly maps genetic and developmental traits onto the phenotypic traits that are acted upon by selection. The old adaptive landscape with a few characters *à la* Simpson is thus not abandoned; on the contrary, it explodes into a series of useful representations.

---

[25] For another example of a two-dimensional measure extracted from the variability in a multidimensional space, see Fig. 13.

Simpson–Lande and similar landscapes address the complex *dynamics* of phenotypic and genotypic combinations. They refer not only to spaces of phenotypic characters, but to true dynamical systems that produce their exploration vectors.[26] *Morphospaces* are phenotypic spaces of a different kind, devoided of immediate dynamic aspects. We have seen in Fig. 13 an analysis of how Ediacaran taxa occupy the "regions" of a 50-dimensional morphospace. A space with so many dimensions cannot be fully represented in two or three dimensions, but several kinds of distances can be calculated, for example, with the principal components method, and plotted onto those morphological "topographies." The distances and neighborhood in another space, for example, the genealogical space (Fig. 8), might be added, but this would require different visualization strategies. In a genealogical space—for example, Wright's landscape (Fig. 1) cleared up from Dobzhansky's conflations—closely related taxa are close to each other, regardless their morphology.

Ever since Wright, Dobzhansky, and Simpson, landscapes are two-dimensional transformations of multidimensional spaces of possibilities. Coupled with spaces, landscapes are used to explore what is possible, what is realized, what is observed, and what is unobserved, in the biological domains. Sometimes, drawing a landscape is not possible or useful, for example, when dimensionality of the space is too high. Even here, however, landscape thinking is a fundamental tool to approach combination spaces. Genotypic spaces, sequence spaces, morphospaces, ecological spaces, and others are essential in studying macroevolution. As we will see in the next section, the combinations observed in these spaces are mapped onto each other—e.g., phylogenetic, geographical—and vice versa, to yield a complex understanding of macroevolution.

3.6 Composing Multiple Spaces

Creative ways of connecting different spaces of possibilities and their exploration have become the challenge of macroevolutionary studies and visual representations. A clear example of the interaction among different spaces of possibilities in macroevolution, with the consequent necessity of appropriate representations and connecting bridges, is provided by the simultaneous evolution of morphological features, niche breadth, and geographic range.

---

[26] A notion of landscapes tightly related to dynamical systems is extremely interesting (see Fusco et al. 2014). Here, a dynamic vector is associated with each and every combination of the considered variables, so that the space—so to speak—fully includes the rules of its own exploration. The idea diverges somehow from all the presentation made in this chapter, which relates landscapes to static combination spaces whose exploration mechanisms exceed the combinations. As a matter of fact, macroevolution studies often employ these kinds of spaces (e.g., morphospaces, ecological niches) and connect them in different ways. They are rarely able to define dynamical systems.

As we have seen with Simpson and other evolutionists, individuals can be seen as realizations in a space of possible combinations of characters. Possible combinations of morphological characters make up a morphospace (Fig. 13), whereas possible genomes make up a genotypic space (Fig. 1). Evolutionary change of morphology can be seen as movement inside morphospaces, although structural renovation of these spaces is possible as well, especially in macroevolution, as noted in Simpson's discussion of the transience of character complexes. The same applies to genetic evolution. In turn, movement in genetic and morphological spaces affects movement in ecological spaces, as the modified characters affect the ecological conditions that are inhabitable and their "fitness" with respect to the organisms.

As we have seen, a niche is a hypervolume in the space of combinations of environmental variables that are relevant to an organism (Fig. 15). Being a multidimensional space of combinations, the niche can be conveniently imagined as a landscape and, in certain cases, visualized that way (Fig. 15b). All the possible combinations of environmental variables will be aggregated into a surface. For a particular species, each combination will be valued for its suitability. Different species, perhaps related by common descent, will have different landscapes, affecting their possible coexistence (Holt 2009). Through macroevolutionary time, species or their descendants will move to different and new ecological niches. A species niche is related, on the one hand, to the particular combination of characters of its members, and, on the other hand, to geographical range.

The ecological niche of a species concurs in determining its geographical distribution, so much so indeed that niche theory allows for inferring the ecological niche *from* geographical distribution: The niche for particular organisms is determined by statistically examining the combinations of environmental conditions that occur at the geographical locations currently inhabited by that organism. Geographic range is a sovraindividual measure, at least in the sense that it is not appreciated by measuring some trait in the individual fossil, rather it is appreciated by looking at the collective property geographical range.[27] The inferred niche hypervolume can then be used to project changes in geographic range under possible scenarios of environmental change (Peterson and Vieglais 2001; Peterson et al. 2011; see also Stigall, *this volume*).

In sum, we have a plethora of interconnected spaces of possibilities that are explored interdependently in macroevolution. Scientists are becoming more and more sensitive to integrating genealogy, morphology, genetics, ecology, and geography, without conflating them as the early, pioneering neo-Darwinists did. For

[27] "Range size heritability" is a strong concept that has been also tied to the issue of species selection. In an interesting study on carnivores, Machac et al. (2011) clarify that range size heritability *patterns* are accounted for by phylogeny only in part. The patterns actually emerge as a consequence of the interplay between evolutionary *and geographic* constraints: Geographic constraints (e.g., temperatures, precipitations) are the most proximate factor shaping species' ranges, but related species are sensitive to shared sets of geographic constraints.

example, a considerable effort is put into relating phylogenies to ecological events. Composite pictures are created to aid this task. Any evolutionary tree is, of course, a very incomplete representation of macroevolution, because it does not show why and where in the world the branches originate and expand or contract. It does not show biogeographical information. But evolution does not happen in a vacuum, for a sort of internal tendency to change toward a direction. If macroevolution happens by the formation, persistence, and disruption of ecological equilibria (Tëmkin and Eldredge, *this volume*), then phylogenetic trees must be combined with information on the geographical distribution of taxa that belong to the phylogeny, and with ecological representations. On the other hand, patterns of geographical distribution reconstructed from the fossil record may be mapped onto a phylogenetic tree, and used as a criterion to discriminate between different allopatric speciation processes. This is possible if different ecological modes of speciation are developed by the researcher into contrasting phylogenetic predictions.[28]

Macroevolution is a multispatial process: It happens in several intertwined "spaces," not only geographical spaces, but also genealogical spaces, ecological spaces, and genetic and morphological spaces. In these spaces, macroevolution develops as a bundle of stories that are interrelated, but not coincident. Landscapes can be a useful tool for thinking to some of these spaces. Although microevolutionary concepts—one for all: fitness—are not scalable to macroevolution, integration can be attempted. Pictures can play an important bridging function here, as they did in the MS (Sidlauskas et al. 2009).

4 Conclusion

Looking at graphical representations of macroevolution is a way of appreciating that macroevolution is a bundle of explorations of different and interconnected spaces of possibilities. Life on earth explores morphospaces, genetic spaces, geographic and ecological spaces, and genealogical spaces. Of course, macroevolution includes the outgrowth of new spaces of possibilities. "Major transitions" such as the origin of the eukaryotic cell by endosymbiosis or the inception of multicellularity, for example, opened unprecedented spaces of possibilities. Evolutionary biologists avail themselves of more and more advanced methods to map these changing spaces of possibilities and to discover the exploration mechanisms at work inside them. Macroevolutionary adaptive landscapes were an important

[28] Stigall, *this volume* exemplifies the methodology of integrating biogeography and phylogeny at the macroevolutionary scale. When a daughter species occupies an area different or additional to the ancestral distribution, the speciation mode of that branching is classified as *dispersal*: The new species must have originated as a migrating subpopulation. When a daughter species occupies a subset of the ancestral range, speciation mode is considered *vicariance*: The ancestral species must have been passively fragmented by ecological barriers.

device of communication during the Modern Synthesis. They enforced reciprocal knowledge, confidence, and trust among separate disciplinary communities, and they helped in constructing a shared language to talk about evolution. They were a real scientific metaphor, if by metaphor, we mean a medium connecting different places. Macroevolutionary landscapes were meant to present evolution as a unitary process, from micro to macro, from genetics to paleontology, spanning all life on earth. While landscapes showed their flexibility in representing all this breadth, they also operated conflations between genealogical, ecological, geographic, and morphological spaces of possibilities. The rich history of evolutionary landscapes is useful to reflect on how scientific visualizations are imbued with deep ways of thinking, and to analyze the visualizations that appear in contemporary macroevolutionary studies. Here, we have seen evolutionary trees, stratigraphies and geographies, morphospaces, ecological spaces, diversity curves, and again landscapes of different kinds of spaces. We have also begun to appreciate the challenge of connecting knowledge of macroevolution, intrinsically fragmented over different spaces of possibilities, by means of composite pictures and, more importantly, composite and interdisciplinary studies.

References

Abe FR, Lieberman BS (2012) Quantifying morphological change during an evolutionary radiation of Devonian trilobites. Paleobiol 38(2):292–307

Abe FR, Lieberman BS (2009) The nature of evolutionary radiations: a case study involving Devonian trilobites. Evol Biol 36(2):225–234

Arnold SJ, Pfrender ME, Jones AG (2001) The adaptive landscape as a conceptual bridge between micro- and macroevolution. Genetica 112–113:9–32

Arnold S, Bürger R, Hohenlohe O, Ajie B, Jones A (2008) Understanding the evolution and stability of the G-matrix. Evolution 62:2451–2461

Baum DA, DeWitt Smith S, Donovan SSS (2005) The tree-thinking challenge. Science 310:979–980

Brasier M, Antcliffe J (2004) Decoding the Ediacaran enigma. Science 305:1115–1117

Carneiro M, Hartl DL (2010) Adaptive landscapes and protein evolution. Proc Nat Acad Sci USA 107(Suppl):1747–1751

Colicchia G (2006) Ancient cephalopod scavenges successfully with its pinhole eye. Phys Educ 41(1):15–17

Darwin CR (1838) Notebook C: transmutation of species 2-7. Online on http://darwin-online.org.uk cod. CUL-DAR122

Darwin C (1859) On the origin of species, 1st edn. John Murray, London (quoted sixth edition, 1872)

Dawkins R (1983) Universal Darwinism. In: Bendall DS (ed) Evolution from molecules to man. Cambridge University Press, Cambridge. Reprinted in Hull DL, Ruse M (eds) The philosophy of biology. Oxford University Press, Oxford, pp 15–37

Dawkins R (1996) Climbing mount improbable. W.W. Norton & Company, New York

Dobzhansky T (1937) Genetics and the origin of species, 3rd edn. Columbia University Press, New York 1951

Doebeli M (2012) Adaptive dynamics: a framework for modeling the long-term evolutionary dynamics of quantitative traits. In: Svensson EI, Calsbeek R (eds) The adaptive landscape in evolutionary biology. Oxford University Press, Oxford, pp 227–242

Eldredge N (1999) The pattern of evolution. Freeman & Co, New York

Elsdon-Baker F (2009) The selfish genius. How Richard Dawins rewrote Darwin's legacy. Icon Books, London
Fusco G, Carrer R, Serrelli E (2014) The landscape metaphor in development. In: Minelli A, Pradeu T (eds) Towards a theory of development. Oxford University Press, Oxford, pp 114–128
Gavrilets S (1997) Evolution and speciation on holey adaptive landscapes. Trends Ecol Evol 12:307–312
Gavrilets S (2004) Fitness landscapes and the origin of species. Princeton University Press, Princeton
Gillespie JH (1983) A simple stochastic gene substitution model. Theor Popul Biol 23:202–215
Gillespie JH (1984) Molecular evolution over the mutational landscape. Evolution 38:1116–1129
Gillespie JH (1991) The causes of molecular evolution. Oxford University Press, Oxford
Gregory TR (2008a) Understanding evolutionary trees. Evol Edu Outreach 1(2):121–137
Gregory TR (2008b) The evolution of eyes. Evol Edu Outreach Spec Issue 1(4):351–516
Halverson KL (2011) Improving tree-thinking one learnable skill at a time. Evol Edu Outreach 4(1):95–106
Hansen TF (2012) Adaptive landscapes and macroevolutionary dynamics. In: Svensson EI, Calsbeek R (eds) The adaptive landscape in evolutionary biology. Oxford University Press, Oxford, pp 205–226
Hendricks JR, Lieberman BS, Stigall AL (2008) Using GIS to study palaeobiogeographic and macroevolutionary patterns in soft-bodied Cambrian arthropods. Palaeogeogr Palaeoclimatol Palaeoecol 264(1–2):163–175
Hofmann HJ, Mountjoy EW (2001) Namacalathus-Cloudina assemblage in Neoproterozoic Miette Group (Byng Formation), British Columbia: Canada's oldest shelly fossils. Geology 29(12):1091–1094
Holt RD (2009) Bringing the Hutchinsonian niche into the 21st century: ecological and evolutionary perspectives. Proc Natl Acad Sci USA 106(Suppl 2):19659–19665
Hutchinson GE (1957) Concluding remarks. Cold Spring Harbor Symp Quant Biol 22:415–427
Hutchinson GE (1965) The niche: an abstractly inhabited hypervolume. The ecological theatre and the evolutionary play. Yale University Press, New Haven, pp 26–78
Huxley JS (1942) Evolution: the modern synthesis. Allen & Unwin, London
Jablonski D (2008) Species selection: theory and data. Ann Rev Ecol Evol Syst 39(1):501–524
Kaplan JM (2008) The end of the adaptive landscape metaphor? Biol Philos 23(5):625–638. doi:10.1007/s10539-008-9116-z
Kauffman SA (1993) The origins of order: self-organization and selection in evolution. Oxford University Press, New York
Kauffman SA, Levin S (1987) Towards a general theory of adaptive walks on rugged landscapes. J Theor Biol 128:11–45
Kidd DM (2010) Geophylogenies and the map of life. Syst Biol 59(6):741–752
Kidd DM, Liu X (2008) Geophylobuilder 1.0: an arcgis extension for creating "geophylogenies". Molec. Ecol Res 8(1):88–91
Lamb TD (2011) Evolution of the eye. Sci Am 305(1):64–69
Lamb TD, Arendt D, Collin SP (2009) The evolution of phototransduction and eyes. Philos Trans R Soc B 364(1531):2791–2793
Lamb TD, Collin SP, Pugh EN (2007) Evolution of the vertebrate eye: opsins, photoreceptors, retina and eye cup. Nat Rev Neurosci 8:960–975
Lande R (1979a) Effective deme size during long-term evolution estimated from rates of chromosomal rearrangements. Evolution 33:234–251
Lande R (1979b) Quantitative genetic analysis of multivariate evolution, applied to brain: body size allometry. Evolution 33:402–416
Lewontin RC (1980) Theoretical population genetics in the evolutionary synthesis. In: Mayr E, Provine WB (eds) (1980) The evolutionary synthesis: perspectives on the unification of biology. Harvard University Press, Cambridge, pp 58–68
Lieberman BS (2012) Adaptive radiations in the context of macroevolutionary theory: a paleontological perspective. Evo Biol 39(2):181–191

Lunzer M, Miller SP, Felsheim R, Dean AM (2005) The biochemical architecture of an ancient adaptive landscape. Science 310:499–501
Machac A, Zrzavý J, Storch D (2011) Range size heritability in Carnivora is driven by geographic constraints. Am Nat 177(6):767–779
Maynard Smith J (1970) Natural selection and the concept of a protein space. Nature 225:563–564
Mayr E (1959) Where are we? Cold Spring Harb Symp Quant Biol 24:1–14
Mayr E (1973) The recent historiography of genetics. J Hist Biol 6:125–154
Mayr E (1980) Prologue: some thoughts on the history of the evolutionary synthesis. In: Mayr E, Provine WB (eds) (1980) The evolutionary synthesis: perspectives on the unification of biology. Harvard University Press, Cambridge, pp 1–48
Mayr E, Provine WB (eds) (1980) The evolutionary synthesis: perspectives on the unification of biology. Harvard University Press, Cambridge
McLennan DA (2010) How to read a phylogenetic tree. Evol Edu Outreach 3(4):506–519
Meisel RP (2010) Teaching tree-thinking to undergraduate biology students. Evol Edu Outreach 3(4):621–628
Monod J (1970) Le hasard et la nécessité. Le Seuil, Paris. Tr. Chance and necessity: an essay on the natural philosophy of modern biology. Alfred A. Knopf, New York, 1971
Muntz WR (1999) Visual system, behaviour, and environment in cephalopods. In: Archer SN et al (eds) Adaptative mechanisms in the ecology of vision. Kluwer, London, pp 467–483
Myers CE, Lieberman BS (2011) Sharks that pass in the night: using geographical information systems to investigate competition in the cretaceous western interior seaway. Proc R Soc B 278(1706):681–689
Odling-Smee J, Laland K, Feldman MW (2003) Niche construction. Princeton University Press, Princeton
O'Hara RJ (1992) Telling the tree: narrative representation and the study of evolutionary history. Biol Philos 7:135–160
O'Hara RJ (1998) Population thinking and tree thinking in systematics. Zool Scripta 26(4):323–329
Omland KE, Cook LG, Crisp MD (2008) Tree thinking for all biology: the problem with reading phylogenies as ladders of progress. BioEssays 30:854–867
Orr H (2006) The population genetics of adaptation on correlated fitness landscapes: the block model. Evolution 60:1113–1124
Paley W (1828) Natural theology, 2nd edn. J. Vincent, Oxford
Papin A, Reed JL, Palsson BO (2004) Hierarchical thinking in network biology: the unbiased modularization of biochemical networks. Trends Biochem Sci 29(12):641–647
Perelson AS, Macken CA (1995) Protein evolution on partially correlated landscapes. Proc Natl Acad Sci USA 92:9657–9661
Peterson AT (2003) Predicting the geography of species' invasions via ecological niche modeling. Q Rev Biol 78(4):419–433
Peterson AT, Soberón J, Pearson RG, Anderson RP, Martínez-Meyer E, Nakamura M, Araújo MB (2011) Ecological niches and geographic distributions. Princeton University Press, Princeton
Peterson AT, Vieglais DA (2001) Predicting Species invasions using ecological niche modeling: new approaches from bioinformatics attack a pressing problem. Bioscience 51(5):363
Pigliucci M (2008) Sewall Wright's adaptive landscapes: 1932 vs. 1988. Biol Philos 23:591–603
Poelwijk FJ, Kiviet DJ, Weinreich DM, Tans SJ (2007) Empirical fitness landscapes reveal accessible evolutionary paths. Nature 445:383–386
Provine WB (1971) Origins of theoretical population genetics. University of Chicago Press, Chicago
Provine WB (1980a) Epilogue. In: Mayr E, Provine WB (eds) (1980) The evolutionary synthesis: perspectives on the unification of biology. Harvard University Press, Cambridge, pp 399–411

Provine WB (1980b) Genetics. In: Mayr E, Provine WB (eds) (1980) The evolutionary synthesis: perspectives on the unification of biology. Harvard University Press, Cambridge, pp 51–58
Provine WB (1986) Sewall Wright and evolutionary biology. University of Chicago Press, Chicago
Raup DM, Sepkoski JJ Jr (1984) Periodicity of extinctions in the geologic past. Proc Natl Acad Sci USA 81:802
Rhode RA (2005) Introduction. GeoWhen Database. http://www.stratigraphy.org/upload/bak/geowhen/index.html. Accessed June 2014
Rice SH (2004) Evolutionary theory: Mathematical and conceptual foundations. Sinauerassociates, Sunderland
Rice SH (2012) Phenotype landscapes, adaptive landscapes, and the evolution of development. In: Svensson EI, Calsbeek R (eds) The adaptive landscape in evolutionary biology. Oxford University Press, Oxford, pp 283–295
Rode AL, Lieberman BS (2004) Using GIS to unlock the interactions between biogeography, environment, and evolution in Middle and Late Devonian brachiopods and bivalves. Palaeogeog Palaeoclim Palaeoecol 211(3–4):345–359
Rowe W, Platt M, Wedge DC, Day PJ, Kell DB, Knowles J et al (2010) Analysis of a complete DNA-protein affinity landscape. J R Soc Interface 7(44):397–408
Ruse M (1990) Are pictures really necessary? The case of Sewell Wright's "adaptive landscapes". In: PSA: proceedings of the biennial meeting of the philosophy of science association vol 2, pp 63–77
Saunders WB, Landman NH (1987) Nautilus: the biology and paleobiology of a living fossil. Plenum, New York
Schuster P, Fontana W, Stadler PF, Hofacker IL (1994) From sequences to shapes and back: a case study in RNA secondary structures. Proc R Soc B 255:279–284
Sepkoski D (2012) Rereading the fossil record. The growth of paleobiology as an evolutionary discipline. The University of Chicago Press, Chicago
Serrelli E (2011) Adaptive landscapes: a case study of metaphors, models, and synthesis in evolutionary biology. PhD thesis in Education and Communication Sciences, University of Milano Bicocca. http://hdl.handle.net/10281/19338
Serrelli E (under review) The Extended Evolutionary Synthesis: limits and promises of a metascientific view of evolutionary biology. Stud Hist Philos Biol Biomed Sci (under review)
Shen B et al (2008) The Avalon explosion: evolution of Ediacara morphospace. Science 319(5859):81–84
Sidlauskas B et al (2009) Linking big: the continuing promise of evolutionary synthesis. Evolution 64(4):871–880
Simpson GG (1944) Tempo and mode in evolution. Columbia University Press, New York
Smith JM (1962) The limitations of molecular evolution. In: Good IJ (ed) The scientist speculates: an anthology of partly baked ideas. Basic Books, New York, pp 252–256
Stadler BMR, Stadler PF (2002) Generalized topological spaces in evolutionary theory and combinatorial chemistry. J Chem Inf Comput Sci 42:577–585
Thanukos A (2009) A name by any other tree. Evol Edu Outreach 2(2):303–309
Thanukos A (2010) Evolutionary trees from the tabloids and beyond. Evol Edu Outreach 3(4):563–572
Torrens E, Barahona A (2012) Why are some evolutionary trees in natural history museums prone to being misinterpreted? Evol Edu Outreach 5(1):76–100
Voigt CA, Kauffman S, Wang ZG (2000) Rational evolutionary design: the theory of in vitro protein evolution. Adv Protein Chem 55:79–160
Waggoner B (2003) The Ediacaran biotas in space and time. Integr Comp Biol 43(1):104–113
Warrant EJ (1999) The eye of deep-sea fishes and the changing nature of visual scenes with depth. Phil Trans R Soc Lond B 355:1155–1159
Weinberger E (1990) Correlated and uncorrelated fitness landscapes and how to tell the difference. Biol Cybern 63:325–336

Wiley EO, Lieberman BS (2011) Phylogenetics: theory and practice of phylogenetics systematics. Wiley-Blackwell, Hoboken
Wright S (1931) Evolution in Mendelian populations. Genetics 16(97):97–159
Wright S (1932) The roles of mutation, inbreeding, crossbreeding and selection in evolution. In: Proceedings of the 6th international congress of genetics vol 1, pp 356–366
Wright S (1988) Surfaces of selective value revisited. Am Nat 131(1):115–123
Xiao S, Laflamme M (2009) On the eve of animal radiation: phylogeny, ecology and evolution of the Ediacara biota. Trends Ecol Evol 24(1):31–40

Toward a Natural Philosophy of Macroevolution

Stanley N. Salthe

Abstract This paper understands macroevolution from a general perspective focused upon energy and thermodynamics. Its biological perspective is ecological, more particularly regarding energy flows. The basic image is the spontaneous dispersion of energy gradients, which, while microscopic, entrains and enables the hierarchical organization of material systems, including the living. The paper will deal with the philosophy of development (involving final cause), the dissipative structure concept, the maximum entropy production, and maximum power principles. The origin of life was the origin of detailed informational control of energy flows. Key processes in organic evolution relating to energy flows were the tendency to generate a plenitude of ecological niches, as well as the evolution of endothermy, involving increases in both the size and complexity of organisms. Organisms serve the universe by serving as exemplary channels speeding up the dissipation of energy gradients. In my perspective, the course of human evolution can be understood, not as being a goal of organic evolution as such, but as entrained by a universal development toward thermodynamic equilibration.

Keywords Development · Dissipative structure · Diversity · Ecological niche · Energy flow · Entropy production · Hierarchies-subsumptive · Compositional

1 Introduction

Natural philosophy is an attempt to construct a scientifically based 'big picture' understanding of the world. It attempts to synthesize understandings gathered from any and all of the specialized sciences. Its epitome was FWJ Schelling (1775–1854), with his global developmental model (Esposito 1977). Natural philosophy projects—both directly and indirectly—a culture's vision of the universe,

S.N. Salthe (✉)
Brooklyn College, City University of New York, New York City, USA
e-mail: ssalthe@binghamton.edu

E. Serrelli and N. Gontier (eds.), *Macroevolution*,
Interdisciplinary Evolution Research 2, DOI 10.1007/978-3-319-15045-1_5

and suggests a role for humanity in that picture. Using the ethnographic definition of 'myth' as generative stories that are believed, we can see that natural philosophy elaborates a myth (Salthe 1992, 2004a), in effect directly challenging religious myths.

Natural philosophy lost its place as an avowed discipline around the end of the nineteenth century, when science began to become largely entrained into supporting engineering, medicine, and technological advance. Individuals, especially some philosophers (e.g., Peirce, Whitehead, Buchler, Bunge), continued to contribute to natural philosophy, while some scientists, capping off their careers, contributed as well (Duhem, Einstein, Elsasser, Monod). My own approach emphasizes thermodynamics as a unifying science in a world that is capable, as well, of being modeled using hierarchical structures (Salthe 2007, 2012). My objective with regard to natural philosophy is to display biological evolution as an aspect of cosmic evolution.

2 The Energy Flow Perspective

In a joint paper with Eldredge and Salthe (1984), I opted, as a materialist, to focus upon the energy connections in the world, as contrasted with the more idealistic information perspective that has dominated biology since the middle of the twentieth century. This necessarily means that my outlook was, and is, 'ecological' inasmuch as ecology is at base the study of energy flow relations taking place on earth between the influx of solar radiation and its reradiation into space (Lotka 1922; Kleidon 2010). This perspective was most fully presented by the Odum brothers, between Odum (1971) and Odum and Odum (2000). Of course, it is information that channels most energy flows in the world, but my focus is upon the flows rather than the informational constraints.

Information regulates the flows of energy everywhere that matter exists, either as spontaneous configurations (crystals, river drainage systems) or as organizations (dissipative structures including the living and machines) (Bejan and Lorente 2006, 2013; Bejan et al. 2008). Even when not guided by forms, energy will flow, as in diffusion and wave front spreading, and so it will have been taking place even before gravitating matter appeared in the universe. Energy flow is the change that is the source of all other changes, including biological evolution, as energy disperses away from regions of higher concentration (Annila 2010; Annila and Salthe 2010). Energy dispersion is everywhere spontaneous, and everything of interest in the world emerges from tapping into that dispersion, temporarily diverting some of it into forms of every kind, as demonstrated, for example, in the records of macroevolution. I note that energy dissipation is logically prior to its capture in events and by processes—that is, it is spontaneous, entraining everything that occurs.

What is energy? It is that which, when gathered together, emerges as matter and which scatters about when matter dissipates, which it tends to do. Everything emerges from energy flow by tapping into that flow and extracting a small portion for work (the exergy), leaving the rest (heat energy) to disperse as entropy. Thus

$$\text{available energy} \rightarrow \text{exergy} + \text{entropy.}$$

Entropy can be visualized as undirected particle motion, causing particles to gradually disperse and slow down as their local situation cools. Heat is the rapid motion of locally concentrated particles. So energy is basically a microscopic concept, but its flow can result in organizing meso- and macroscopic forms at many scales, including living systems and their constructions (Gladyshev 2006; Salthe 2007).

3 The Larger Framework

What is needed here are a few lines to orient us in the bigger picture—which is required because energy is integral to that picture and because the big picture is naturally the focus of natural philosophy. Before there was gravitating matter in the universe, it is hypothesized in the big bang theory to have been energy in a universal quark-gluon 'soup,' followed by an expanding matter-radiation 'soup,' followed by the formation of stars and galaxies, pulled together by the gravitational attraction that was emergent with baryonic matter. In one galaxy, the sun and its planets formed. We can use a subsumptive, or specification, hierarchy (Salthe 2012) to make a general summary, modeling successive originations of modes of being, or integrative levels, as subclasses, as follows:

$$\{\text{physical realm} \rightarrow \{\text{chemical or material realm} \rightarrow \{\text{biological realm } \{\text{etc.}\}\}\}\}$$

displaying what one might call the really 'mega' aspects of macroevolution. From each realm that is more generally present in the universe (the outer brackets), more highly specified realms will have originated in particular locales by way of the addition of new informational constraints. The more primitive realms remain in place, making possible and supporting the more derived ones wherever these occur.

The biological realm may have emerged in only one locale in the universe. As a general scientific perspective, that seems rather unsatisfactory, and the discipline of the origin of life studies takes, the view that life is funded upon just another phase of matter, and so we should eventually be able to duplicate its origin in vitro. However, implicitly mitigating this view is our understanding of the elaborate mechanisms of the genetic system. It is this system that makes life possible. It is so complicated that no single spontaneous 'origin' as such seems plausible, but rather—given that this *was* a 'spontaneous' event—a long concatenation of happenstances in special, perhaps unique, environments (Salthe 2009). That is to say, here we are dealing with history. The origin of the genetic system is basically unknown, likely unknowable. It was, in effect, the origin of informational control, of machinery, and of digitality as well, and it requires a semiotic perspective to understand it (Pattee 2007; Barbieri 2011). Prior to that origin, all energy utilization resulted only in mass action. After it, by way of the genetic apparatus, mass action supports a piecemeal linking of molecules in an orderly manner, resulting in the generation of highly specific catalysts (enzymes). This difference is physically enormous. Life adds to the list of constraints that could characterize a locale, informing activities that may take place there. These constraints include generally various boundary conditions—temperature, roughness, etc., as well as the chemical affordances. Genetics adds technique!

4 Subsumptive Logic and the Philosophy of Development

Subsumptive logic can be succinctly shown in the following hierarchy (Salthe 2012):

{physical world {chemical world {biological world {social world}}}}

In words, this might be summed up (paraphrasing G.G. Simpson) as 'physics is the science that applies to everything, while biology is a realm to which all the sciences (to the left of it in the hierarchy) apply.' Given a particular format like this, our ideas become entrained in certain directions. The logic of a phylogeny requires, if it is taken to provide a model of biological evolution, that refinement is the basic process at work through time. Materially, this means that new kinds are formed by way of the addition of new, modifying, informational constraints and new information (which may cause loss of form as well as its modification). Thus, reptiles are, logically, more highly specified amphibians and birds more highly specified reptiles—and so birds are also more highly specified amphibians. There can be nothing radically, *totally* new. Everything has a precursor; nothing can come from nothing. This is a materialist as well as a developmental principle. (This 'developmental' point is a challenge for evolutionary biology, which has denied any developmental tendencies in biological evolution.)

I have argued (Salthe 1993) that the basic logic of development *as a kind of change* is refinement and that evolution, as a kind of change (not as the label for that particular example, organic evolution), is without logic. By this, I mean that evolution is just the willy-nilly accumulation of marks from the effects of fortuitous encounters with unfamiliar situations. Darwinian adaptation is a purely opportunistic 'grasping at straws' and could never be planned logically. Note that any example of development will also 'evolve'—that is, a developing system will *individuate* by accumulating accidental marks. In biology, these can sometimes be registered in the genetic system (Jorgensen 2011; Shapiro 2011) and may then be passed along as part of the, now changed, genetic heritage of a lineage.

This idea that development is fundamentally a process of restriction (or refinement) is not a claim that there can be nothing new in biological evolution (assuming here that evolution is basically a developmental process); any modification would trivially register something new. Locally, within an evolving clade, a changed form could present the possibility of opening up new opportunities for resource exploitation. But each opening necessarily closes off other possibilities that might have been explored instead. This is better appreciated by stepping back and examining a larger chunk of a phylogeny. Thus, bird evolution has precluded the possibility of a fossorial life—despite some digging out of nests and the kakapo's burrow. This adaptive zone is being explored in most regions by other vertebrate clades, such as amphibians (caecilians) and mammals (moles, mole rates, etc.). Thus, while the more detailed picture shows openings up into new Hutchinsonian niches (the niche of a given population, Salthe 2001), the larger picture shows a gradual closing off of lifeway possibilities (Eltonian niches—general life ways, like, say, the cat way of life) for each lineage. Then, local evolutionary openings are into increasingly more restricted (more highly specialized) possibilities (Fig. 1). As local diversity increases, global disparity increase stalls. This general pattern, like any, will have exceptions and would, of course, be nullified by catastrophic extinction events, which tend to 'reset' an evolving system. We humans appear to be one of these exceptions.

By mapping numbers of species to information theoretic concepts, Brooks and Wiley (1986) produced a general interpretation of biological evolution that is concordant with the present view. In their view—and the record shows that—the variety (informational or Shannon entropy) of biological kinds increased rapidly early in evolution, with a subsequent gradual decline in the origination of new kinds continuing onto an asymptotically declining future (Fig. 2; see also, e.g., Vermeij 1987, Fig. 13.1). Viewing the biosphere as a pool of informational variety, or

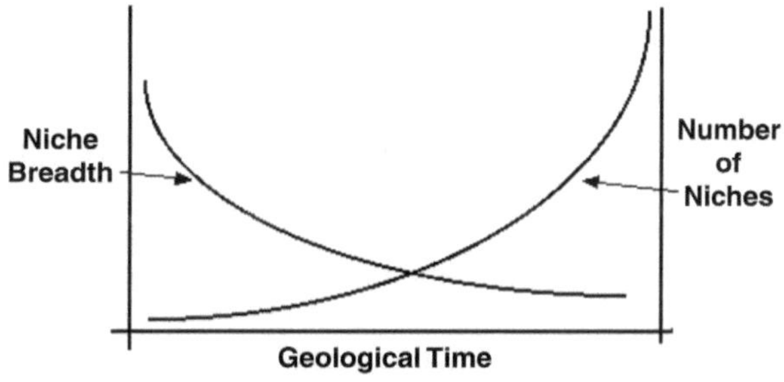

Fig. 1 General view of Hutchinsonian ecological niches through geological time, based on developmental logic

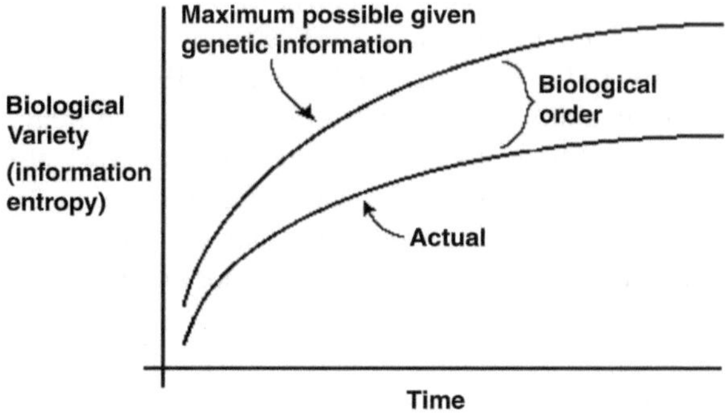

Fig. 2 The evolution of biological information, based on Brooks and Wiley (1986)

'disorder,' Brooks and Wiley represented generic and specific biological forms as embodying a degree of order compared to the informational disorder that would be the case if there were no individuated biological *kinds* and instead only masses of individuals of every possible form. That is, the order in biological kinds represents less disorder/variety than might be maximally possible given the coding possibilities of the genetic information held in living systems. Maximal disorder here is being conceived as the complete expression of all genetic information, while order would be represented by the larger than molecular bodies making up the various specific kinds of actual living things as based on the species-specific portions of genomes. Brooks and Wiley note that, overall, the variety of definable biological kinds has increased over time, expressing informational entropy increase.

From the ecological perspective, it is easily understood that variety of biological forms also represents a variety of Hutchinsonian ecological niches (Fig. 1). It also therefore reflects the variety of sources and modes of utilization of energy gradients. Biological variety can continue to increase over time in part because biological kinds themselves generate new ecological affordances. The 'actual' curve in Fig. 2 increases in a diminishing returns pattern. This rate of increase diminishes over time because, even though the universe appears to be expanding, the earth is finite. As well, the energy efficiency of the capture of incident solar energy by photosynthesis is quite low—around six percent—representing a kind of 'rock-bottom' constraint, and there is no reason to suppose that it could increase over time. As well, the more kinds that exist, the more will tend to go extinct at any time.

5 Dissipative Structures

The world in which evolution occurs is not at equilibrium in any aspect. This means that energy flows are maintained everywhere all the time. This might refer to the slow, steady expansion of a high-pressure region of the atmosphere driven by solar insolation or, more dynamically, the flow of ocean currents called for by the tendency for energy (here temperature differences) to even out. But, at smaller scales, everything dynamic, from thunderstorms to dust devils, from schools of fishes to individual organisms, are in more or less constant motion—in some cases with organisms, even (internally) while 'at rest.' These individuated active systems, from a flame to a hurricane, are examples of 'dissipative structures' (Prigogine and Nicolis 1977; Prigogine and Stengers 1984). Nothing that moves is not either a dissipative structure, or part of one, and often both, with one dissipative structure nested within another. Simple spontaneous types at meso- and macroscopic scales are waves, vortices, and branching tree forms like lightening. At the chemical scale, dissipative structures would be continually maintained exothermic pathways from substrates to end products, while endothermic reactions would be nestled within exothermic ones.

Dissipative structures appear when energy gradients become too steep to be significantly degraded by gradual heat conduction. At that point, they emerge as convective forms of various kinds as allowed by local conditions (Schneider and Kay 1994; Salthe 2007). The processes of life's origination—that is, the origin of the genetic system—will necessarily have occurred within dissipative structures. Some of these dissipative structures will have developed later at several scales into living cells and organisms, being captured by 'genetic takeover' (Cairns-Smith 1993). The viewpoint of this paper is that living things as we know them are internally informed dissipative structures (Wicken 1987). This means that they are especially stable and can 'skate across' gaps in energy supply that would lead to the dispersion of abiotic dissipative structures. The energy supply for this stability is internally stored ATP, which would, of course, need to be regenerated. Living systems have internalized their primary energy source as well as the (genetic) source of their actual forms, which relate appropriately (adapt) to particular environments.

Current discussions of dissipative structures have been centering on the maximum entropy production principle (MEPP; e.g., Annila 2010; Kleidon 2010; Martyushev 2013). As I put it in 2010, 'an energy dissipative system that can assume several to many conformations will tend to take up one, or frequently return to one, that maximizes the entropy production from the energy gradients it is dissipating—to a degree consistent with the system's survival.' The latter qualification is required because literal maximization—an explosion!—would destroy the system. MEPP is a universal, global principle, on the reasonable supposition that our universe is a thermodynamically isolated system, thereby being subject to the second law of thermodynamics. How this cashes out locally is by way of the maximum energy dispersion principle (MEDP). Local energy dispersion will eventually result in the production of heat energy (entropy), but much of it immediately would be in the form of various waste products (imagine the feeding of a shark). In living systems, energy dispersion is produced by work; the faster/harder the work, the more of the supporting energy gradient gets dispersed as heat energy. Living systems are always, with intermittent rest periods, striving in one way or another; and while at rest, energy continues being used for healing. MEDP, then, refers to energy gradient dispersion by non-equilibrium systems such as organisms located within an out-of-equilibrium, thermodynamically isolated system (the universe) which elicits MEPP as an expression of its tendency toward thermodynamic equilibrium known as the second law of thermodynamics. Local MEDP serves global MEPP.

In the ultimate macroevolutionary perspective, Chaisson (2001, 2008, 2012) has shown that evolution (cosmic, biological, and technological) has produced systems that depend upon, and produce, greater and greater intensities of energy flows through them (Fig. 3). Energy throughput per unit mass is known as power. In dissipative structures, this power is the source of the energy (the exergy) that organizes and maintains them. In social systems, we are used to referring to this portion and that expended in various projects, as supporting 'work,' and I think we can safely refer to the portion used by living systems to maintain and reproduce themselves as work as well. I would further extend that usage to all dissipative structures (Salthe 2010a)—thus, tornadoes work hard at destroying houses.

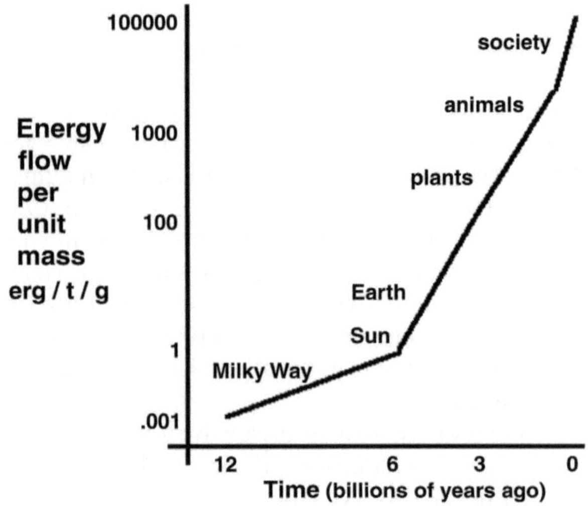

Fig. 3 Energy flow intensity in selected entities through time. Redrawn from Chaisson (2012). erg/t/g is ergs per unit time per gram weight

An extremely curious fact about our world on earth is the constitutively poor energy efficiency of any effective work, which is never better, with significant loadings, than 50 % (Odum 1983). There has been no explanation of this known to me (Salthe 2003). However, if we consider ourselves, and all dissipative structures, to be part of the universe, we could see that all the multiple works manifesting this poor efficiency are activities *of the universe*, working through us, in its project of universal thermodynamic equilibration. This perspective (see also Martyushev 2013) is uncongenial to our fundamental cultural outlook of human individualism. On the template of the following subsumptive hierarchy (Salthe 2013), which would be in effect anywhere on earth:

{physical world {chemical world {biological world}}}

showing how some realms of nature, modeled as classes and subclasses, relate to each other logically, we have

{entropy production {free energy dissipation {work}}}

That is, work is a mode of free energy usage, which in turn is a mode of entropy production.

The line of reasoning in the previous paragraph is not 'scientific.' That is, it has no relation to the motivations for most scientific investigations or to the world view of the subsequent reports. It is concerned with the possible *meanings* of scientific knowledge, which is, in fact, the province of natural philosophy and of this paper.

We may ask: are species dissipative structures? No. What about populations? As parts of local ecosystems, possibly. Species cannot be said to be active agents in the ecological world. My current view is that species are essences, meaning that, as such, they have no dynamic ecological role, which is carried out instead, and not always identically, by local populations. Insofar as they are causal entities, species are coded scripts in DNA within cells. The essence of this would be that small portion of the script carried by every member of the collection of organisms making up what we discern as a species *and* by no member of any other species. It could be just a few dozen kilobases, and it could be located in only one organism or free living cell of a species on the brink of extinction. Practically, it could never actually be identified. In short, species are constructs of the informational discipline, Systematic Biology. Populations, however, occupying actual space in time, and working Hutchinsonian niches, would be functioning parts of local ecosystems, where they transform energy in their position in a food chain and food web. But this sort of function—e.g., the ecological efficiency—is actually only the sum of the activities of the included organisms, which are the actual dissipative structures involved. However, it seems possible that a population's relations to other local populations would be what allocates the characteristic number of individuals in that population. This would be a greater-than-organism function that effects the total energy transformations of that population, giving the population an actual ecological function of apportioning the energy throughput of a local ecosystem.

6 Senescence and Reproduction

All dissipative structures, if they survive assaults, develop through to senescence, which sets the stage for recycling. I have proposed (Salthe 1993) that development can be characterized by three general stages, as shown in Table 1.

Table 1 General stages of development

Immature stage
Relatively high energy density (per unit mass) flow rate
Relatively small size and/or gross matter and energy throughputs
Rate of increase of informational constraints relatively high, as part of high growth rate
Internal stability relatively low (it is changing fast), but dynamical stability (persistence) is high
Homeorhetic stability to same-scale perturbations relatively high
Mature stage (only in relatively very stable systems)
Declining energy density flow rate still sufficient for recovery from perturbations
Size and gross energy and matter throughput typical for the kind of system
Form is definitive for the kind of system
Internal stability adequate for system persistence
Homeostatic stability to same-scale perturbations adequate for recovery from insults
Senescent stage
Energy density flow rate gradually dropping below functional requirements
Gross matter and energy throughput high but its increase is decelerating
Form increasingly accumulates deforming marks as a result of encounters, as part of individuation
Internal stability of system becoming high to the point of inflexibility
Homeostatic stability to same-scale perturbations declining

Modified from Salthe (2010b), based on thermodynamic (Zotin 1972; Polenttini 2013) and information theoretic criteria

Senescence is most often treated as a human disease, even though it is the natural culmination of all dissipative structures. Basically it results, in my perspective, from the inevitable taking on of new information (any modifications) after the definitive form of a kind of system has been achieved. Living systems have not been able to escape it. And so senescence is not a particular by-product of natural selection. It has had an enormous consequence. After the evolution—or was it the development? or might it even have been the incorporation?—of the genetic apparatus, the division of a (now living) cell entailed the multiplication of a particular identity. Protobiotic systems likely fragmented and spread over an abiotic ecosystem. Once the genetic system had made its appearance, fragments could take with them information that made their predecessor successful enough to grow enough under bearing conditions to fragment. This allowed them to split more readily when that was promoted by instability due to increasing size. Identity and its reproduction had been born. Before that, every piece of matter was different from all others; after that, we have definite kinds, enforced by genetic identity. Which living systems, after the origin of life, would inherit the earth? Those that could maintain a presence in the face of inevitable senescent decline and, as well, those that could utilize their energy sources faster than other competing kinds (Matsuno and Swenson 1999). This latter involved being able to respond successfully with an increased rate of energy acquisition when presented with larger energy sources to some extent even during senescent decline (Polenttini 2013). Abiotic dissipative systems would do this spontaneously, but the more complicated living dissipative structures could be inhibited by their own delicate form to different degrees, giving rise to competition for moderate energy gradients. We now had particular kinds, in effect competing with other particular kinds, for energy. This situation placed a premium upon the ability to switch to new energy sources as those being utilized became depleted locally, driving the possibility of 'speciation.'

7 Ecosystems as Energy Flow Pathways

Energy flows are everywhere in the universe and would have been happening on, and within, the primitive earth as well. The energy viewpoint is useful in allowing us to realize that there always have been ecosystems on the earth after it formed, perhaps at first only a single global one, with weather phenomena, orogenic activity, erosions, and so on (Kleidon 2010). Logically, we need to see protoliving systems emerging within, and getting intercalated into, preexisting liquid water flow systems. This would have been contextualized by the precipitation/evaporation cycle, where the emerging living systems likely were fostered within moderate flow eddies created and maintained within faster flowing hydrological systems that could deliver substrates. All of this will have been contextualized by local diurnal cycles produced by the spinning of the earth, allowing temporary survival (at night) of delicate forms at different scales, promoting further epigenesis. So the origin of life would have occurred within some abiotic dissipative structure(s) in an abiotic ecosystem where the energy flows were moderate, as well as intermittent (Branscom and Russell 2013; Pross and Pascal 2013). In this sort of setting, we know pretty well how phospholipid membranes, microspheres, protobionts, and metabolism could have been produced spontaneously within increasingly lively 'living ponds.' With life being characterized as a system having propagation and interpretation (Pattee 2007), we can fairly well imagine crude forms of propagation going on, but biosemioticians would insist that interpretation became associated with the genetic apparatus, yet we have little idea about how this may have originated. (With this fact, I could assert that the energy perspective is actually on a sounder scientific basis than the alternative information perspective! Of course, both perspectives are required to understand living systems.)

The earth spins in a stream of electromagnetic radiation dissipated from the sun's fusion dynamics. On earth, some of this energy (the sun's entropy!) is further dissipated by the evaporation of water and the production of weather phenomena. This leads in turn to geophysical dissipation in the weathering of rocks and the production of dissipative structures such as drainage systems and ocean currents. I trace this particular dissipative pathway because it is solar radiation that is tapped also by most known current living systems, by way of photosynthesis. Living systems dissipate this energy gradually through food webs and beyond that by contributing to the mass wasting of rocks and the production of soils, as well as by plant transpiration, producing the humidity required by storms.

A given biotic storage of energy could be dissipated in many ways—action, predation, fire, floods, and ultimately by way of detrivory. Quoting my 2010 paper again, 'an energy dissipative system that can assume several to many conformations will tend to take up one, or frequently return to one, that maximizes the entropy production from the energy gradients it is dissipating—to a degree consistent with the system's survival.' This is, again, MEPP, as carried out locally by way MEDP. So, at any locale, we can expect 'things to happen' that rapidly dissipate whatever energy gradients are present. This might at any moment call for a lightning strike, a predator's strike, decay processes, and photosynthesis. On earth, biological systems have been entrained into this consequence of the second law of thermodynamics in a big way. It can be asserted that biological entities are almost always 'striving.' A major source of evidence for this viewpoint has been produced by Adrian Bejan as reported in numerous papers, in order to illustrate his 'constructal theory.' He has convincingly shown that all flow systems on earth, natural or manufactured, are formed so as to 'develop the flow architecture that maximizes flow access under the constraints posed to the flow' (Reis and Bejan 2006). This is exactly what would be predicted by the MEDP perspective. It has been noted that MEDP is often not a testable phenomenon in particular cases. Constructal form has however been amply demonstrated in many material systems, allowing plausible conjecture. I note here in passing that many adaptive scenarios in neo-Darwinian discourse are similarly hobbled by non-testability, with much less indirect evidence to back them up! In my 1975 paper (Salthe 1975) I cited references showing that natural selection functions more intensely during times of strenuous activity—which would, of course, be times of greatest entropy production.

The MEDP label has been objected to by noting that, for example, living systems do not literally maximize the energy flows through them, which would kill them. Instead, they optimize the energy flows, as discussed by Lotka (1922) and Odum and Pinkerton (1955), who dubbed this principle the 'maximum power principle' (MPP) (Salthe 2010a). Literal maximization anywhere would be explosive. In fact, dissipative structures generally intervene as a natural energy source increases before it can become explosive (Schneider and Kay 1994; Salthe 2007). Thus, while the expanding universe demands MEPP, this tends to be realized locally as MEDP, and this is further modulated in the more delicate living systems to MPP. The trade-off for the universe in the latter case is an increasing diversity of energy flows, producing some that might not occur at all abiotically. The predominant perspective on organismic energy use focuses upon energy efficiency, taking energy savings to be a major desideratum (e.g., Shepard et al. 2013). This is derived from the engineering mind-set, where energy supply is an important limitation. But organisms generally have reliable sources of energy within their niche spaces, as well as internal sources of energy such as ATP, both of which facts allow effective actions to trump energy savings as immediate goals. Excessive exertions might lead to death, but survivors are successful at—precisely—exertions. This means that energy efficiency would evolve within the overriding context of striving (Annila and Salthe 2010). The MEDP perspective is that energy efficiency would evolve mostly in order to promote striving. This would mean that, while effectiveness is the true focus of selection, efficiency increase during extreme activity could perhaps contribute to that effectiveness, but it would hardly count as a major focus of selection, except as weeding out really inefficient abnormalities.

8 Biological Evolution

As Darwin pointed out, evolution can be visualized as producing a tree of biological forms (e.g., Rieppel), which can be modeled by a subsumptive hierarchy, {{class {order {genus { }}}}}. Each form comes out working its own ecological niche. From this, evolution can be viewed as an exploration of ways to degrade increasingly more particular energy gradients, creating a diversity, and increasing number, of energy sources. The big picture shows life beginning with chemosynthesis and/or photosynthesis, followed, in the animal branch, at length by the evolution of detrivores to utilize an increasing layer of dead biomass. The key idea here is that, given a significant energy source, sooner or later a dissipative system will emerge to disperse it. This led further to more macroscopic forms with mouths and digestive systems utilizing the dead organic mass more rapidly and converting some of it to a form more readily accessed by microorganisms as well. Eventually, some of these detrivores evolved the ability to prey on others of like kind, becoming carnivores. After a while, other detrivores acquired endosymbiont microbes that allowed them to explore herbivory, which led eventually to the establishment of the typical modern food chain with the addition of predation. And then, carnivores and herbivores began exploring increasingly more specialized ways of life.

With regard to MEDP, the evolution to carnivory and herbivory entailed more rapid energy conversions, with top carnivores, for example, tending to speed up local food chain transformations. Viewing populations as dissipative structures, predation dissipates prey energy faster than would slower microorganismic decay of individuals from prey populations. With animals, apex predation can also elicit faster prey reproduction, at earlier ages, since older individuals tend to be preferentially harvested. These effects have been observed in fisheries (e.g., Odum 1983). Herbivory and carnivory also entailed increased organismic striving (in plants for increased growth rates) as a functional requirement, and this entails more energy throughput and heat energy production. This leads us to note Van Valen's (1973) Red Queen's hypothesis, which energized and generalized its forerunner, Gause's (1934) competitive exclusion principle. On the supposition that there are characteristic energy availabilities in a given habitat, if one population begins to reproduce more successfully, it will impact other populations working the same general Eltonian niche (syntopic species of woodland herbs in a forest, for example), requiring these to compensate in order to keep up. The result will also affect others in a food web, delivering a constant jockeying for energy between populations in a habitat.

Eventually, evolution resulted in a hierarchy of biological systems (e.g., Eldredge 1989) that can be modeled as a compositional hierarchy (wholes and parts, Salthe 2012). But how did it accomplish this? As it happens, nature is not intrinsically opposed to this structuring, which would in that case require specific kinds of work to create it. Gladyshev (2006) has shown that such a hierarchy is compatible, as a steady state, with the second law of thermodynamics and so can come about spontaneously (Fig. 4). This suggests that the abiotic ecosystem prior

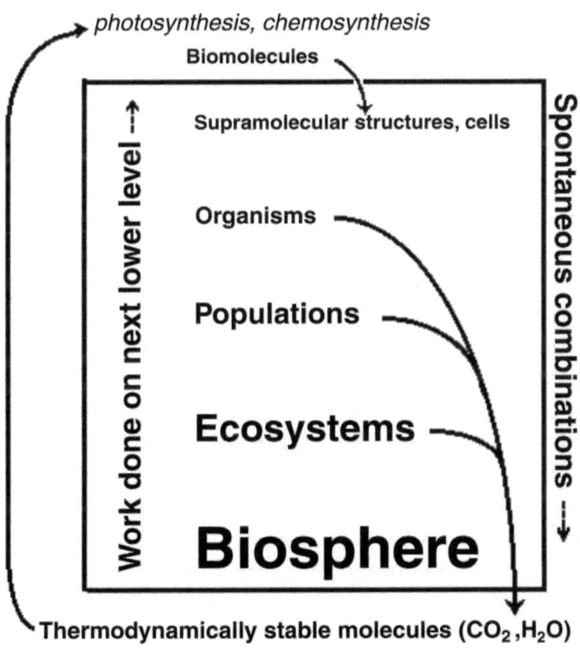

Fig. 4 Steady-state thermodynamic relations among levels in the compositional hierarchy of biological systems on earth. Redrawn from Gladyshev (2006)

to the origin of life will already have self-assembled into such a structure in crude form; biological systems will then have come along and filled in the slots, as it were. Of course work needs to be done to keep any system going, but we need to understand in just what way the work is done to maintain the hierarchical form biologically. Figure 4 shows the biological hierarchy as a spontaneously achieved thermodynamic steady state resulting from self-assembly of the levels. Each level acts in its role as a higher level to the next lower level in a way that fosters the hierarchy as a quasi-stable, steady-state entity. The work that is done—in addition to the primal work of photosynthesis—is done by each level maintaining an environment that keeps eliciting the contributions of its next lower level. That is, it maintains a situation favorable to the activities of the next lower level. For example, the organism creates conditions that promote the activities of its incorporated cells. This work creates a top-down affordance inviting the continued bottom-up contributions of the next lower level that structurally makes possible its next higher level. The work accomplished at one level creates, at the next lower level, conditions that allow it—the higher level—to continue to emerge.

But the Gladyshev model is not complete. It is one thing to understand that a compositional hierarchy could be a thermodynamically secure structure—unlike a tall building, for example—but there is also the question of why it would continue to be thermodynamically favored once achieved. This question moves us from the classical thermodynamics informing Gladyshev to non-equilibrium thermodynamics. In that format, we can bring in MEPP. That is, I would posit that this hierarchical form is the one that elicits the most favorable kinds of flow structures throughput the hierarchy that promote the MPP overall and therefore maximizes as much as possible the system's global rate of entropy production (Salthe 2004b).

So, we have two basic thermodynamic principles working here. One is the classical Gibbs free energy perspective utilized by Gladyshev, stating that a system will spontaneously move to a least free (available, usable) energy configuration. The other is MEPP, stating that this will emerge in a non-equilibrium system as rapidly as possible. In living systems, the solar energy captured by photosynthesis is routed metabolically on its way toward ATP synthesis and the heat energy released from the cytochrome system. Living systems that did not sufficiently moderate the rate the energy dispersion through them burned out and went extinct.

Moving now from the more speculative to more solid evidence in favor of the idea that evolution has increased the rate of earth's energy flows/entropy production, we can recall Bejan's multiple evidences that facilitating energy flows acts as a final cause of any and all dynamic material structures, as well as Chaisson's more particular evidence (Fig. 3) that per unit mass energy flows have increased during evolution. But in my view, the clearest evidence for the influence of MEDP/MEPP in biological evolution is the convergent evolution of homeothermy (endothermy) in mammals and dinosaurs/birds. There are tendencies toward this in other phyla as well, for example, the endothermy of tunas and some other large, fast swimming predatory fishes. Even a plant has achieved it—the flower of the skunk cabbage, which stimulates beetle pollinator activity in cool weather. Behavioral homeothermy is another common tendency, as in lizards. Endothermy requires significant amounts of energy dissipation and is a direct result of physical entropy (heat) production, with no other more elaborate product. It, and behavioral homeothermy, do nothing more than increase the rates of activity of the animals in question. Regardless of particular adaptations that may have been facilitated by homeothermy (such as living in colder climates), its 'universal' achievement is simply to directly increase the entropy production of an animal per second of life. Surely, a frog might be imagined to think, 'What a waste of energy! Look at me—I get along quite well without it (even in Siberia)!'

9 Humanity's Role in the Universe

Evolutionary Biology has not identified any goal of organic evolution. Neither increase in complexity, nor increase in consciousness has been accepted as biological evolution's goal on earth. One may suspect that this may be an effect of the neo-Darwinian domination of evolutionary studies, where all effects are viewed as the results of local caprice, allowing no tendency to evolve out of anything beyond contingeny, or to long survive environmental change if it did emerge. The concept of natural selection allows of no directional tendencies that are not merely local, provisional, and contingent. Even humanity is not viewed as occupying a privileged position at the top of a perceived chain of evolved forms—an image that in any case likely reflects rejected ancient religious opinion.

However, as alluded to above, there is a scientifically based concept of a universal goal of all events and actions—the dispersion of energy gradients toward a universal final state approaching thermodynamic equilibrium, where orderly fluctuations here and there will soon get damped out. This 'heat death of the universe' has had many objections on various grounds, but in the long run, it seems to me that it cannot be avoided given our current knowledge. This knowledge amounts to the empirically derived fact that entropy must increase if it changes in a thermodynamically isolated system. Since it is produced each time any action occurs—and copiously with significant work—we may reasonably infer that the universe, or our section of it, is a thermodynamically isolated system. Given that assumption, the second law of thermodynamics, perhaps second only to the law of gravitation, is the physical law most immediately impacting our lives—disorder threatens everywhere, and we must work harder than seems reasonable to achieve anything (Salthe 2003).

I have argued that the second law is a 'natural' example of final cause (e.g., Salthe 1993; Salthe and Fuhrman 2005). Directional tendencies in nature have been parsed (Mayr 1988) as:

{teleomaty {teleonomy {teleology}}}

or

{natural tendency {function {purpose}}}

on the template:

{physicochemical world {biological world {human socioeconomic world}}}

where each innermost subclass (or integrative level) is a more specific example of the next outermost class. Thus, e.g., purpose is a kind of function. The essential point is that all the labels in the second hierarchy here are end-directed. The second law of thermodynamics is a teleomatic principle, and thus

{entropy production {free energy decline {work {social projects}}}}

on the template:

{physical world {chemical world {biological world {socioeconomic world}}}}

As Lotka pointed out, some solar energy reradiation from earth's surface is delayed while being rerouted through living systems, but this has the 'universal' payoff of slowly undermining earth's gravitated mass by way of activities that stir up soils, undermine rocks, and spray water—slight events with slowly accumulating effects, adding to the much more powerful abiotic disturbances such as earthquakes and storms. Over geological time, events with slight effects gradually accumulate to global significance.

Spontaneous diffusion and mass wasting are insufficient to dissipate all energy gradients formed by gravitation and its sequelae. Dissipative structures such as storms speed things up in the presence of massively steeper energy gradients, but still many gradients will have been very slowly dispersed until life appeared: Thus, lichens speed up the mass wasting of rocks; microorganisms at work even in depths of earth's oceans and mantle produce corrosive chemistry and heat energy; tree roots and burrowers in the soils do so as well and open soils up for further chemical transformations. Microorganisms everywhere transform chemical gradients to simpler forms. (I can't resist an exception that 'proves the point'—bacterial action has resulted in the buildup of small gradients of gold, but subsequent human activities have torn apart many regions of the earth to collect it.) The role of biological diversity in this is clear; it increases the number of energy gradients being dissipated and works on some smaller and more cryptic energy gradients that are not touched significantly by abiotic dissipative structures.

Humanity's role regarding MEPP emerged significantly with industrialization, as visualized in Fig. 3. Human economies have added to all of the dissipative effects on earth's surface and, after the industrial revolution, have magnified some of them many fold, by way of various mining activities, modern travel and warfare. But, as Machu Picchu and China's buried terracotta army show, intense labor was not a consequence of the industrial evolution. Yet, after that direction was taken, hard labor was magnified manyfold and mechanized. Our current throwaway economy may be its crowning achievement. Warfare may have a special role in human entropy production. It is deplored by everyone, yet is indulged continually somewhere on earth's surface—taking a large-scale glimpse, it would be everywhere anytime. Entrainment by the second law can serve as an explanation for the paradoxical role of war as an attractor of human activity regardless of everyone's voiced disapproval, working as a final cause of human activity. The second law would provide a very slight pull at any moment, but is always reliably beckoning so as to tip the scales of decision making in cases of uncertainty. Yet industrial activity may be an even greater, if somewhat more careful, producer of heat energy and debris. The keynote seems to be, Build → Destroy → Rebuild, with heat energy being the only definite result in the long run.

10 Conclusion

Have I, in this paper, reduced all the complexities of biology and sociology to physics? In one sense, yes. On the other hand, my perspective in this paper is that humanity is a part, or aspect, of the universe in addition to being the growing point of a lineage in a cladogram. We no longer can view ourselves as a goal of organic evolution. In the phylogenetic perspective, we are merely one more primate, while in the physical perspective, we have become mighty workers in the service of universal dissipation.

Acknowledgments I thank Malcolm Dean and Richard Sternberg for references and two thoughtful readers who stimulated useful revision and additions.

References

Annila A (2010) The 2nd law of thermodynamics delineates dispersal of energy. Int Rev Phys 4:29–34
Annila A, Salthe SN (2010) Physical foundations of evolutionary theory. J Non Equilibr Thermodyn 35:301–321
Barbieri M (2012) What is information? Biosemiotics 5:147–152
Bejan A, Lorente S (2006) Constructal theory of generation of configuration in nature and engineering. J Appl Phys 100:041301
Bejan A, Lorente S, Lee J (2008) Unifying constructal theory of tree roots, canopies and forests. J Theor Biol 254:529–540
Bejan A, Lorente S (2013) Constructal law of design and evolution: physics, biology, technology and society. J Appl Phys 113:151301
Branscom E, Russell MJ (2013) Turnstiles and bifurcators: the disequilibrium converting engines that put metabolism on the road. Biochim Biophys Acta 1827:62–78
Brooks DR, Wiley EO (1986) Evolution as entropy: toward a unified theory of biology. University of Chicago Press, Chicago
Cairns-Smith AG (1993) Genetic takeover: and the mineral origins of life. Cambridge University Press, Cambridge
Chaisson EJ (2001) Cosmic evolution: the rise of complexity in nature. Harvard University Press, Cambridge
Chaisson EJ (2008) Long-term global heating from energy usage. Eos Trans Am Geophys Union 89:253–255
Chaisson EJ (2012) A singular universe of many singularities: cultural evolution in a cosmic context. In: Eden AH, Moor JH, Sotaker JH, Steinhart E (eds) Singularity hypotheses: a scientific and philosophical assessment. Springer, Berlin

Eldredge N (1989) Macroevolutionary dynamics: species, niches and adaptive Peaks. McGraw-Hill, New York
Eldredge N, Salthe SN (1984) Hierarchy and evolution. Oxf Surv Evol Biol 1:184–208
Esposito JL (1977) Schelling's idealism and philosophy of nature. Bucknell University Press, Lewisburg
Gause GF (1934) The struggle for existence. Williams and Wilkins, Baltimore
Gladyshev GP (2006) The principle of substance stability is applicable to all levels of organization of living matter. Int J Mol Sci 7:98–110
Jorgensen RA (2011) Epigenetics: biology's quantum mechanics. Front Plant Sci 2, Article 10
Kleidon A (2010) Life, hierarchy and the thermodynamic machinery of planet Earth. Phys Life Rev 7:424–460
Lotka AJ (1922) Contribution to the energetics of evolution. Proc Nat Acad Sci 8:147–151
Martyushev LM (2013) Entropy and entropy production: old misconceptions and new breakthroughs. Entropy 2013:1152–1170
Matsuno K, Swenson R (1999) Thermodynamics in the present progressive mode and its role in the context of the origin of life. BioSystems 51:53–61
Mayr E (1988) Toward a new philosophy of biology. Harvard University Press, Cambridge
Odum EP (1971) Fundamentals of ecology. Saunders, New York
Odum HT (1983) Systems ecology: an introduction. Wiley, New York, p 102–116
Odum HT, Odum EC (2000) Modeling for all scales. Academic Press, New York
Odum HT, Pinkerton RC (1955) Time's speed regulator, the optimum efficiency for maximum output in physical and biological systems. Am Sci 43:331–343
Pattee HH (2007) The necessity of biosemiotics: the matter-symbol complementarity. In: Barbieri M (ed) Biosemiotics: information, codes and signs in living systems. Nova Publishers, New York
Polenttini M (2013) Fact-checking Ziegler's maximum entropy production principle beyond the linear regime and towards steady states. Entropy 15:2570–2584
Prigogine I, Nicolis G (1977) Self-organizing systems in non-equilibrium systems. Wiley, New York
Prigogine I, Stengers I (1984) Order out of chaos: man's new dialog with nature. Bantam Books, New York
Pross A, Pascal R (2013) The origin of life: what we know, what we can know and what we will never know. Open Biol 3:120190
Reis AH, Bejan A (2006) Constructal theory of global circulation and climate. Int J Heat Mass Transf 49:1857–1875
Rieppel O (2010) The series, the network, and the tree: changing metaphors of order in nature. Biol Philos 25:475–496
Salthe SN (1975) Problems of macroevolution (molecular evolution, phenotype definition, and canalization) as seen from a hierarchical perspective. Am Zool 15:295–331
Salthe SN (1992) Science as the basis for a new mythological understanding. Uroboros 2:25–45
Salthe SN (1993) Development and evolution: complexity and change in biology. MIT Press, Cambridge
Salthe SN (2001) Theoretical biology as an anticipatory text: the relevance of Uexküll to current issues in evolutionary systems. Semiotica 134:359–380
Salthe SN (2003) Entropy: what does it really mean? Genet Syst Bull 32:5–12
Salthe SN (2004a) To be and then not to be: our myth from science. Kutagubilig J Philos Sci Res 5:179–197 (English version)
Salthe SN (2004b) The spontaneous origin of new levels in a scalar hierarchy. Entropy 2004(6):327–343
Salthe SN (2007) The natural philosophy of work. Entropy 2007(9):83–99
Salthe SN (2009) A review of signature in the cell: DNA and the evidence for intelligent design. In: Meyer SC (ed) Philos pathways (146). Harper One, New York, 19 Aug 2009
Salthe SN (2010a) Maximum power and maximum entropy production: finalities in nature. Cosmos Hist 6:114–121

Salthe SN (2010b) Development (and evolution) of the universe. Found Sci 15:357–367
Salthe SN (2012) Hierarchical structures. Axiomathes 22:355–383
Salthe SN (2013) To naturally compute (something like) biology. Prog Biophys Mol Bio 113:57–60
Salthe SN, Fuhrman G (2005) The cosmic bellows: the big bang and the second law. Cosmos Hist 1:295–318
Schneider ED, Kay JJ (1994) Life as a manifestation of the second law of thermodynamics. Math Comput Model 19:25–48
Shapiro JA (2011) Evolution: a view from the 21st century. FT Press, Upper Saddle River, NJ
Shepard ELC, Wilson RP, Rees WG, Grundy E, Lambertucci SA, Vosper SB (2013) Energy landscapes shape animal movement ecology. Am Nat 182. http://www.jstor.org/stable/10.1086/671257
Van Valen L (1973) A new evolutionary law. Evol Theor 1:1–30
Vermeij GJ (1987) Evolution and escalation: an ecological history of life. Princeton University Press, Princeton
Wicken JS (1987) Evolution, thermodynamics, and information: extending the Darwinian program. Oxford University Press, New York
Zotin AI (1972) Thermodynamic aspects of developmental biology. Karger, Basel

Networks and Hierarchies: Approaching Complexity in Evolutionary Theory

Ilya Tëmkin and Niles Eldredge

Abstract This expansion of the hierarchy theory of evolution provides a new perspective in which biological phenomena are conceptualized. In this work, we (1) attempt to revise the ontology of levels of biological organization and clarify the relationship between the economic and genealogical hierarchies; (2) explore the implications of network theory for evolutionary dynamics in a hierarchical context; and (3) elucidate evolutionary causality by disentangling abiotic drivers from proximal evolutionary processes (the origin and sorting of variation) and their integration across hierarchies. We suggest that a pervasive pattern of stability in living systems across scale results from the architecture of nature's economy itself—biological systems consisting of hierarchically nested, complex networks are extremely robust to extrinsic perturbations. We further argue that instances of evolution are episodic and rapid; they are transient between equilibrial states that ensue when network stability is compromised by sufficiently strong disturbances affecting biological entities at multiple levels of organization. We also claim that environmental abiotic factors are ultimately responsible for these perturbations that, when filtered through the economic hierarchy, shape the patterns of diversity and disparity of life as we know it.

Keywords Causality · Complexity · Disparity · Diversity · Emergence · Patterns · Trends

I. Tëmkin (✉)
Biology Department , Northern Virginia Community College, Annandale, VA, USA
e-mail: TemkinI@si.edu

I. Tëmkin
Department of Invertebrate Zoology, Smithsonian National Museum of Natural History, SmithsonianInstitution, Washington, D.C., USA

N. Eldredge
Division of Paleontology, American Museum of Natural History, New York, NY, USA

E. Serrelli and N. Gontier (eds.), *Macroevolution*,
Interdisciplinary Evolution Research 2, DOI 10.1007/978-3-319-15045-1_6

1 Introduction

The natural world is infinitely complex and hierarchically structured. Hierarchy theory is an approach to understanding the way complex systems work by identifying levels of organization and their relationships in the context of scaling. In a broad sense, a *system* is a network of functionally interdependent and structurally interconnected components comprising an integrated whole. The complexity arises from intricate, nonlinear interactions of a large number of parts that such systems have, where the whole is more than the sum of its parts: That is, given the properties and the laws of interactions of parts, it is not a trivial matter to infer the properties of the whole.

Coining the term "hierarchy" is attributed to Pseudo-Dionysius the Areopagite, a philosopher and a theologian of the late fifth–early sixth century, who used the Greek word *ἱεραρχία* ("rule by priests") in reference to ranks of celestial beings and ecclesiastical power in early Christian church (Hathaway 1969). Despite its narrow original meaning, hierarchical organization can be recognized in almost every system in the world from the structure of the natural world to all the domains of human life. Hierarchies are manifest in the physical composition of natural and artificial objects, engineered mechanisms, genealogical relationships, classification schemes, and socioeconomic organizations. The relative arrangement of levels and the nature of their interactions vary greatly depending upon the specific kind of hierarchy considered.

That hierarchy is a key structural principle of biological systems was first recognized by Woodger (1929), whose work influenced the development of hierarchical approaches across biological disciplines (e.g., Hennig 1950, 1966; Pattee 1973; Whyte 1969). In biology, the concept of hierarchy acquired a plethora of disparate meanings ranging from the descriptions of organization of knowledge to models of functional interactions of living systems to taxonomic classifications. The pervasiveness of hierarchies in nature was amply captured by Francois Jacob's maxim, "every object that biology studies is a system of systems" (Jacob 1974).

Biological evolutionary theory is ontologically committed to the existence of nested hierarchies in nature and attempts to explain natural phenomena as a product of complex dynamics of real hierarchical systems. Consistent with the general tendency of complex systems to attain and remain in equilibrium, living systems display remarkable metastability despite non-equilibrial dynamics at all levels of organization triggered by extrinsic disturbances, suggesting that hierarchical systems have some common properties that are independent of their specific content and can be applied across physical, biological, or social sciences.

2 General Properties of Hierarchical Systems

2.1 Architecture of a Hierarchy

A *hierarchy* is an arrangement of entities in which some are represented as being above, below, or at the same level as other entities. A *level* in a hierarchy refers to a class of entities of the same rank or significance. The entities that comprise a level are referred to as "holons" (Koestler 1967, 1978) in the literature on hierarchy theory. The meaning of levels and the relationship among them depend on what kind of hierarchy is considered: order, inclusion, control, or level hierarchy (Lane 2006). The applications of hierarchies in explaining natural world range from ontologically agnostic theoretical constructs as instrumental in making sense of complex processes (Allen 2008; Allen and Starr 1982) to specific ontological claims regarding the structure of reality (Salthe 1985). While different kinds of hierarchies exist in biological systems, we focus on a particular class of hierarchies—nested compositional hierarchies—as a fundamental structural principle of real biological systems that lies at the heart of evolutionary phenomenology. In subsequent sections, the term "hierarchy" is used in this narrower sense.

A *nested compositional hierarchy* is a pattern of relationship among entities based on the principle of increasing inclusiveness, so that entities at one level are composed of parts at lower levels and are themselves nested within more extensive entities. Levels are classes of such parts and wholes, and represent ordered organization in the context of scale. The rank of levels is assigned according to the scale of the entities that are their members. The term *focal level* designates a level at which a particular phenomenon is observed. *Higher* (or *upper*) and *lower* levels refer to more inclusive and less inclusive levels relative to the focal level, respectively. Mathematically, a nested hierarchy is defined as an ordered set that can be represented graphically as a rooted tree (an acyclical graph) or a Venn diagram (Fig. 1).

Small differences in scale among entities at levels of adjacent ranks can blur clear distinction between levels. The higher levels tend to appear increasingly

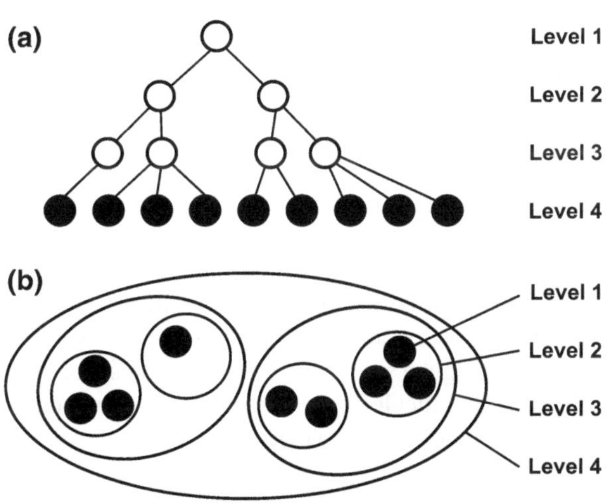

Fig. 1 Diagrammatic views of hierarchical systems. **a** A rooted tree, or acyclical graph; **b** a Venn diagram. Note that the tree structure is a general representation, where links, or edges (*solid lines*), can designate different kinds of relationships among entities (*circles*) at different levels, such as order, inclusion (nesting), or control. The Venn diagram, even though entirely consistent with the graph, visually emphasizes a recursive organization of entities and, therefore, represents a nested compositional hierarchy more specifically

diffuse and less well defined, possibly due to a slower pace dynamics among entities of greater scale (Wimsatt 1976). The perceived discontinuity between levels largely results from differences in the nature of interactions between entities at a given level and entities at different levels. A general compositional asymmetry was also implicated as a general property of nested hierarchies: Systems express much greater variation at lower levels than at higher levels (Weiss 1973).

At any given level, entities have specific attributes or traits that can either be aggregate or emergent. *Aggregate traits* are cumulative properties or combined attributes of entities at the lower level; *emergent traits* are properties that cannot be reduced to or be expressed in terms of properties of entities at the lower level. For example, sex ratio (the ratio of male and female organisms) is an emergent trait of a species or population because it characterizes a collective state of many organisms that cannot be characterized by having this property individually. On the other hand, the ecological specialization of a species is an aggregate trait because it refers to the cumulative range of ecological circumstances occupied by each individual of a species. Approaches for distinguishing between emergent and aggregate traits in biological systems were reviewed by Grantham (2007). At a given level, all entities may either share the same traits or display variation in the traits.

2.2 Hierarchical Dynamics

Two major types of interactions can be distinguished in a hierarchy: within and between levels. Entities at a given level interact directly with each other in the same dynamic process, whereas entities at different levels only interact in an aggregate fashion (Eldredge and Salthe 1984). The differences in the dynamics of processes between versus within levels ultimately arise from scalar differences (frequently of different orders of magnitude) in process rates, yielding a distinction between strong interactions with high-frequency dynamics within levels and weak interactions with low-frequency dynamics among levels (Simon 1962, 1973). Such non-transitivity of direct effects across levels (Salthe 1985) establishes the levels as quasi-independent ("nearly decomposable" sensu Simon 1962) systems allowing for investigating dynamics of individual levels on their own right (Bunge 1979; Levins 1970). Consequently, the details of within-level interactions can be ignored when considering between-level dynamics (Simon 1962). Scalar differences are also responsible for weaker integration of more inclusive units and weakening the strength of interactions at successive, higher levels of a hierarchy, making it more difficult for the investigator to draw boundaries around and characterize these units from the epistemological standpoint.

2.2.1 Interactions Within a Level: Network Dynamics

A commensurate scale of entities and rate of processes within a level allow for representing the intralevel interactions as complex networks in the context of network theory. A *complex network* is a system of interacting entities that is conventionally represented by a graph, a collection of nodes (vertices) connected by links (edges), often having non-trivial topological features (Fig. 2). Links can have different directions, weights, and signs. Typically, directed edges represent the unidirectional flow of information, matter, or energy from a source (starting point) to a target (endpoint); non-directed edges show mutual interactions, where information, matter, or energy are exchanged between a pair of nodes. Networks allow for undirected interactions, such as cyclical relationships and feedback loops, so that the functional dynamics of interactions within a network is temporally restricted. Consequently, entities from different hierarchical levels cannot effectively be members of the same network.

Networks can be characterized by different topological properties, including the degree distribution (the frequency distribution of the number of links per node, or vertex connectivity), clustering coefficient, and average path length (the mean number of nodes along the shortest path connecting two nodes). For comprehensive reviews of general network theory, see Albert and Barabási (2002), Barabási (2002), Dorogovtsev and Mendes (2003), Newman (2003), and Strogatz (2001).

In biology, networks are present at all levels of organization from metabolism and regulation of gene expression to ecological trophic webs and social

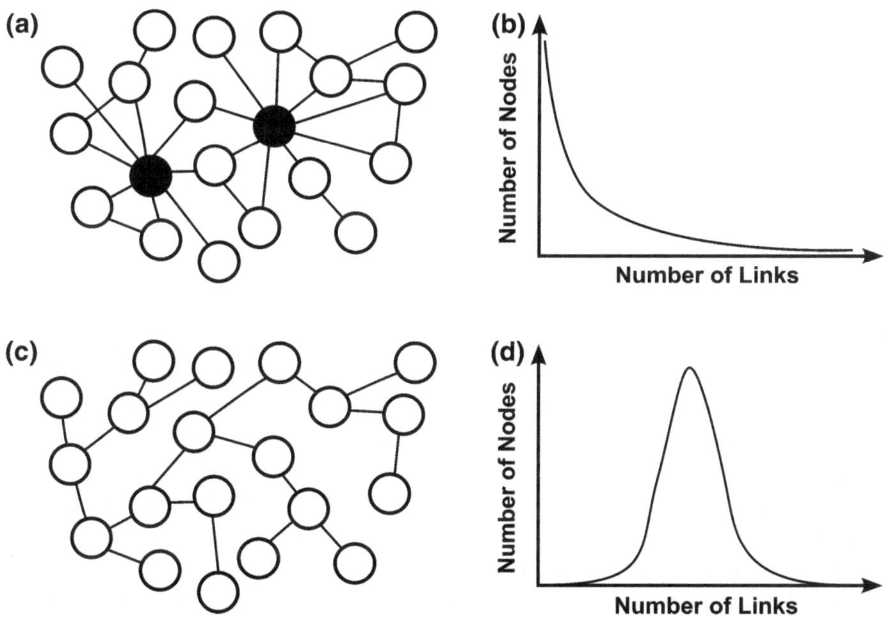

Fig. 2 Diagrammatic views of networks. **a** A scale-free networks with most nodes (*white circles*) having few links (*solid lines*) and few nodes, or hubs (*filled circles*), having a very large number of links. In scale-free networks, the distribution of node connections, the frequency of nodes plotted against the number of links per node (node degree) follows a power law (**b**). **c** A random network with most nodes having approximately the same number of links producing a bell-shaped curve of the degree distribution graph (**d**)

networks within populations. Even though relationships among entities at each level may be defined by unique laws and rules that govern the dynamics within a given level (Pattee 1970), empirical studies of real-world networks reveal that many biological, social, and technological networks share some fundamental architectural principles (Barabási and Albert 1999). Biological complex networks are typically hierarchical and highly modular and have a high clustering coefficient, a heavy tail in the degree distribution, and a short mean path length. The latter characteristic is frequently referred to as "small-world" property (Amaral et al. 2000; Barthélémy and Amaral 1999; Milgram 1967; Watts and Strogatz 1998). The significance of network isomorphism across levels of biological organization lies in the fact that the topology of the network's constituting elements is translated into particular network-wide emergent properties that have profound implications for evolution.

Topology is the principal attribute of any network that places boundaries on how it functions and how it might have formed. Many biological networks have few nodes with a disproportionately large number of connections ("hubs"), while the rest of the nodes have relatively few (Barabási and Oltvai 2004; Proulx et al. 2005; Fig. 2a). Such degree distribution of vertex connectivity that follows a power law function is often referred to as "scale-free" because it lacks a modal hump characteristic of Poisson-distributed mean path length (an average number of nodes along the shortest path connecting two nodes) (Albert and Barabási 2002; Barabási and Albert 1999; Barabási and Bonabeau 2003; Barabasi et al. 1999; Dorogovtsev and Mendes 2002; Simon 1955; Fig. 2b). Scale-free networks have different global features than random networks in which all nodes have relatively the same number of links (Fig. 2c, d).

The most astonishing common emergent property of non-random networks is robustness, an exceptionally high degree of tolerance against random failures and external perturbations (Albert et al. 2000; Dorogovtsev and Mendes 2003; Newman 2003). These networks are relatively insensitive to randomly losing the more highly abundant but less connected nodes but become vulnerable when losing hubs, the rare highly connected nodes. This property is generally attributed to power law distribution of connectivity because hubs are less likely to be affected by random perturbation than other nodes. The adverse effect of this architecture is that perturbations increasing the chance of affecting hubs or resulting in the preferential removal of hubs can cause a severe disruption or collapse of the entire network (Dorogovtsev and Mendes 2002). Thus, the dynamics of the network response in face of different degrees of perturbation are not linear: Until the threshold is reached, the network is unaffected and remains stable, but beyond the threshold, it breaks down. It is important to note that such overarching generalizations greatly oversimplify real functional network dynamics: Scale-free distribution alone is insufficient to account for robustness in actual and simulated molecular networks (Albert et al. 2000; Siegal et al. 2007).

Topology alone is not sufficient for understanding complex networks. The nodes and their interactions within a network may vary greatly and display complex, nonlinear dynamics that can change over time (Strogatz 2001).

2.2.2 Interactions Among Levels: Emergence

Dynamics between any two adjacent levels can either be unidirectional, if only one level exerts an effect on the other, or bidirectional, where both levels mutually affect each other. In bidirectional, or dual control, systems (Polanyi 1968), *upward causation* refers to the effect that the lower level exerts on the upper level; *downward causation* refers to the effect that the upper level exerts on the lower level. One fundamental feature of a hierarchical organization is the asymmetry of inter-level processes (Salthe 1985; Valentine and May 1996) that entail in the asymmetry of the causal effects across hierarchical levels. Consider a minimal hierarchical structure consisting of a focal level and contiguous higher and lower levels ("the triad" of Salthe 1985; Fig. 3). The upward and downward causal effects exerted upon the focal level differ in function. Interactions among entities at a higher level exert constraints, or determine *boundary conditions*. Thus, downward causation establishes the direction of control, affecting simultaneously all subsystems contained within the system (level) where they originate. A consequence of this asymmetry is that dynamics at lower levels might not be manifest at higher levels, although dynamics at higher levels must always propagate downward. Only emergent characters can exert downward causation to processes at the next lower level. The constraints ("non-holonomic constraints" of Pattee 1977) are most effective across contiguous levels, and their importance drops off as the levels involved are increasingly remote (Eldredge and Salthe 1984). Non-contiguous levels may affect the dynamics at the focal level indirectly through cascading upward and downward effects across levels. Interactions among entities at a lower level establish

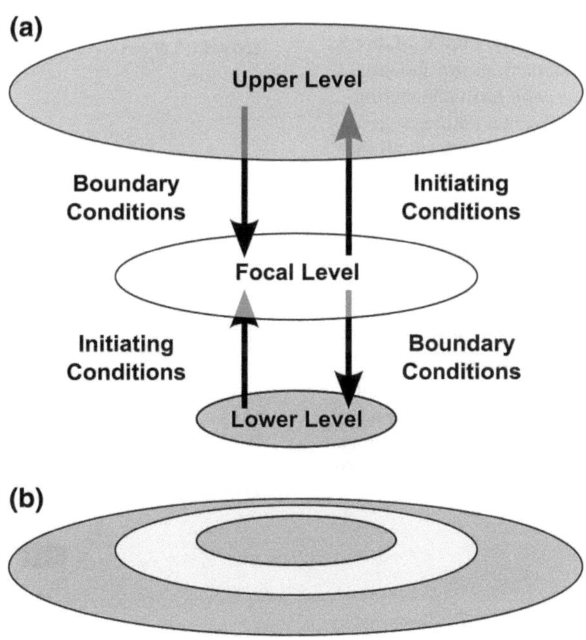

Fig. 3 Dynamics of bidirectional interactions among levels in a nested compositional hierarchy. **a** The *upper* and *lower levels* establish boundary conditions (*downward constraints*) and initiating conditions (*upward constraints*), respectively, for the *focal level*. The *focal level*, in turn, creates initiating conditions for the *upper level* and constrains the processes at the *lower level* through upward and downward causation, respectively. Even though the levels are depicted separately, they are hierarchically nested (**b**)

initiating conditions ("upward causation") (Salthe 1985). Upward causation can be exerted by both aggregate and emergent characters.

Considering the entire dynamics of the triad, the processes occurring at a focal level can simultaneously be initiating conditions for upper, and boundary conditions for lower levels, and, in turn, be affected by boundary and initiating conditions established by these levels, respectively.

2.2.3 Global Dynamics

Complexity of biological evolutionary phenomena stems from the synergetic effect of idiosyncratic processes at different organizational levels and the dynamics of interlevel interactions. A global architecture of a hierarchy has a quasi-fractal quality resulting from nesting of self-similar elements: An entity is composed of a network of its parts and, at the same time, is an element of a network that makes up a higher-level entity (Fig. 4). Such a structure of hierarchically nested networks has a profound effect on evolutionary processes, as will be discussed in later sections. While the appropriate level of mechanistic description for a particular phenomenon rests on the level at which it manifests (Didion 2003), to comprehensively elucidate the causal processes involved one must consider (1) the network dynamics of entities at the focal level, (2) the boundary conditions (constraints), (3) the initiation conditions emerging from network dynamics of entities at adjacent higher and lower levels, respectively, and (4) cascading upward and downward effects from more remote levels.

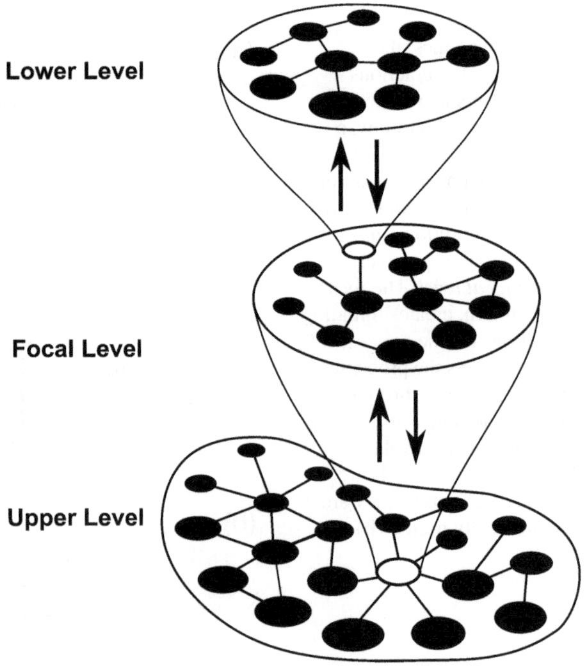

Fig. 4 Diagrammatic view of global dynamics in a hierarchy. Intralevel direct interactions are shown as *solid links* connecting individual entities (*circles*) within networks at all the levels; interlevel indirect interactions representing *upward* and *downward* causation are shown as *up* and *down arrows*, respectively. Note that the global architecture of such compositional hierarchy consisting of nested units of similar organization (i.e., networks) has a fractal dimension (Mandelbrot 1977, 1982)

3 Biological Hierarchies

3.1 The Dual Nature of Life

The two defining processes of life are (1) *interaction*, or active exchange of matter and energy, and (2) *replication*, or transmission of heritable information. It has been previously suggested that the dualism of these two different kinds of phenomena—dynamic and informational—is a general feature of complex systems (Pattee 1970) and, more specifically, they are the two elemental processes of evolving systems (Hull 1980, 1981, 1988). *Replicators* are those entities that pass on their structure (i.e., information) intact through successive generations (Dawkins 1976). *Interactors* are those entities that interact with their environments in such a way as to make replication differential. They are causally related to replicators in such a way that the survival of the former is causally responsible for the differential propagation of the latter (Hull 1980, 1988).

In evolutionary history, the tasks of heritability and dynamic interactions became irreversibly decoupled, producing differentiated and specialized sets of entities at each level of biological organization of greater functional efficiency (Maynard Smith and Szathmáry 1995). In the RNA world at the dawn of history of life, the dynamic and informational kinds of processes—metabolism and heredity—were integrated by ribozymes, molecules capable of catalysis and self-replication (Gilbert 1986). Increasing biological complexity demanded an emergence of highly efficient accurate mechanisms of genetic transmission, so the flow of information became the sole function of highly specialized DNA, whereas extremely versatile proteins specialize in catalysis. At the cellular level, the genetic nucleus in eukaryotes is separated from metabolic cytoplasm and other organelles; at the organismal level, the genetically transmitted germ line is segregated from the mortal soma; and at the population level, as is the case in many eusocial insects, non-reproductive castes are distinct from reproductive individuals.

To represent the entirety of the biotic realm in terms of survival (dynamics of matter and energy exchange) and reproduction (transmission of heritable information), a system of two interconnected hierarchies was advanced that included (1) the economic, or ecological, and (2) the genealogical, or reproductive, hierarchies (Eldredge 1985a; Eldredge and Salthe 1984; Salthe 1985; Fig. 5). This model is capable of representing diachronic (time-extensive) and synchronic (simultaneous) processes. Some levels are fully or partially congruent between the two hierarchies (contain the same or overlapping classes of individuals), whereas others—particularly above the organism—are not. The lack of exact one-to-one correspondence of entities between the hierarchies precludes establishment of a single consistent hierarchical structure (Eldredge and Salthe 1984). In essence, the economic and genealogical hierarchies represent, respectively, the temporal and spatial dimensions of the organic realm. This entails a fundamental dissimilarity in the nature of interactions in the two hierarchies: The time vector in the genealogical hierarchy allows only unidirectional control of information flow, making it time-irreversible, which reflects the historical nature of biological systems and yields a static hierarchy of classification (Grene 1987).

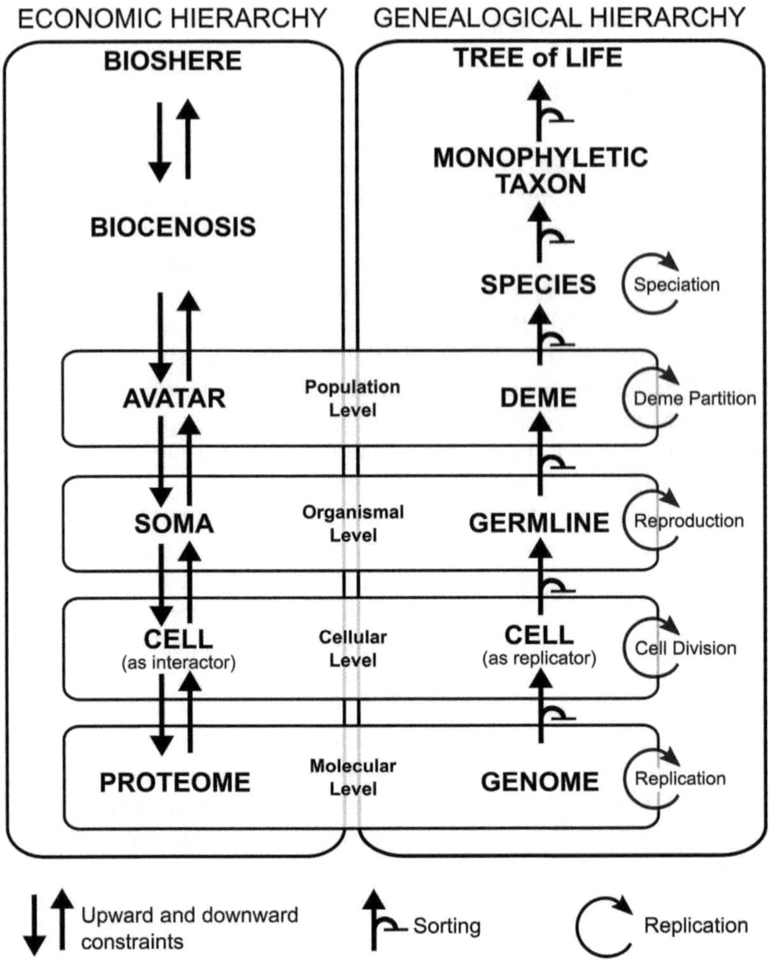

Fig. 5 The dual system of nested compositional biological hierarchies. The *economic*, or *ecological*, hierarchy represents dynamics of matter and energy exchange; the *genealogical*, or *reproductive*, hierarchy describes transmission of heritable information

The economic systems form a hierarchy of control, allowing for dynamic interactions of entities within and across levels through upward and downward causation. The distinction between the two types of hierarchies allows for teasing apart causal relationships between them.

3.2 Biological Entities

The entities that make the economic and genealogical hierarchies are integrated, spatiotemporally constrained systems distributed across the scale from molecular to planetary sizes. Such entities are referred to as *individuals* in a philosophical context

(Ghiselin 1974, 1981; Hull 1976, 1978, 1980, 1981; "mereological sums" sensu Brogaard 2004). It is important to note that the definition of individuals does not require that they replicate. As historical units, individuals differ from classes, or collections of entities whose membership is based on shared properties. A relationship does exist between classes and individuals: Individuals of the same level are particular instances of concrete entities of the same class (Bunge 1979). Levels themselves are not individuals, but classes of individuals of the same rank. Depending on their composition, size, and scale, biological individuals vary greatly in duration and degree of cohesion, which can present difficulties for differentiating them in practice. Taken together, biological hierarchies consist of nested individuals, where individuals at different levels are of different kinds and of different scales (Eldredge and Salthe 1984).

3.3 The Genealogical Hierarchy

The basis of genealogical hierarchy is the flow of information through time. The transmission of information relies on the ability of genealogical entities to reproduce (replicate). The actual physical mechanism of information transfer is DNA replication, but the historical fate of this information depends on replication processes of genealogical individuals at higher levels: Entities can successfully replicate only as parts of the larger whole. Such interdependency of genealogical units across hierarchical levels precludes regarding more inclusive individuals simply as packages of genetic information [or "vehicles" sensu Dawkins (1982)]. The following list identifies the principal levels and entities of the genealogical hierarchy.

3.3.1 Molecular Level

This level of organization encompasses the genome—the entirety of genetic material, the nucleic acids. Information transfer is achieved by a template-based replication of entire chromosomes, so that individual genes replicate only in concert as a sequence of linked fragments. DNA has the peculiar property of being both a replicator and an informer: It serves as a template for its own synthesis and functions as a template for making proteins. As such, it serves as a source of information, as software for the assembly of interactive entities. DNA has an interactive role as well (such as associations with histones and transcription factors), but these interactions exclusively serve the purpose of storage and retrieval of genetic information.

3.3.2 Cellular Level

As genealogical entities, cells are reproductive individuals owing to different processes of cell division mechanisms. In eukaryotes, the division of the DNA-containing nucleus by mitosis or meiosis is distinct from cytokinesis, the division of the metabolic cytoplasm, and may be decoupled from it, such as in the production of multinucleate coenocytes or plasmodia in some protists and fungi. It is possible to distinguish the level of organelles, but only a few of them (mitochondria and chloroplasts) can replicate and they can do so only within a host cell. In prokaryotes, there is a single cell compartment and DNA replication is followed by binary fission. Cellular and organismal levels are conflated in unicellular organisms, so that cell division becomes the locus of reproduction at the cellular and organismal levels.

3.3.3 Organismal Level

From a genealogical perspective, a multicellular organism is an integrated whole consisting of multiple cell lineages that can survive only as parts of larger organisms (excepting the laboratory). Organismal replication includes sexual and asexual modes of reproduction. In sexually reproducing multicellular organisms, only a specialized part of the organism—the germ line and associated tissues and organs—is technically a part of the genealogical hierarchy, even though the genetic transmission is greatly affected by ontogeny (Buss 1983).

3.3.4 Population Level

At the population level, the information transfer is conducted by *demes*, or breeding clusters of organisms (Gilmour and Gregor 1939; Wright 1955). In some organisms, such as some eusocial insects, demes can be restricted to highly specialized reproductive castes, whereas other members of the population do not contribute to the reproductive process directly. Even though it is possible to distinguish a continuum of genealogical hierarchical levels for conspecific individuals with greater degree of sociality or complex spatial structure (e.g., families, clans, and metapopulations), the included individuals in all cases are the same kind of entities—individual organisms—that reside at the same hierarchical level.

Collectively, conspecific demes comprise a spatially distributed network of few highly productive demes (localized in "sources" habitats) and demes in which within-habitat reproduction is insufficient to balance mortality (distributed in "sink habitats") (Pulliam 1988). The persistence of demes depends on the dynamics of continuous local extinction and recolonization in the species-wide, sink-source network, making demes rather ephemeral entities that frequently separate and reunite over relatively brief time intervals (Futuyma 1989; Miller 2006).

3.3.5 Species Level

Species are segments of time-extended, population-level lineages demarcated by origin through lineage-splitting, or speciation events, and eventual demise through extinction. Typically, species are considered to be more inclusive population-level lineages, though the exact level of inclusiveness differs among authors. The temporal boundaries of species become less distinct at smaller timescales due to a gradual process of divergence that appears instantaneous at geological timescales. Species replication is the process of speciation, or cladogenesis, that ultimately results from the perturbation and partitioning of demic networks. In sexually reproducing organisms, speciation is typically followed by reproductive divergence that ensures that the new species remain discrete individuals by preventing hybridization (Dobzhansky 1937). Reproductive divergence can be due to continued selection for mate recognition characters within isolated demes (Paterson 1985), selection of characters genetically linked to reproductive characters, or by chance events. For a review of proposed species-level aggregate and emergent traits, see Jablonski (2008).

3.3.6 Monophyletic Taxon Level

A monophyletic taxon (clade) is a group of species that includes an ancestral species and all of its descendants. Clades arise and diversify as a result of continuous speciation events, thus representing a product of genealogical dynamics at the species level. The relative degree of inclusiveness of the resulting hierarchical structure is conventionally represented by ranks of the Linnaean classification. The ranks are merely convenient communication devices and lack biological significance (Ereshefsky 2001; Okasha 2011).

3.3.7 The Tree of Life Level

All life forms that evolved throughout the Phanerozoic aeon, all species—living and extinct—are presumed to descend from a single common ancestor. At this, highest, level of biological organization, the tree of life is a record of the history of the biosphere from the origin of life to the present.

3.4 The Economic Hierarchy

Biological individuals in the economic hierarchy are open systems that interact with their environment through energy and matter exchange. These interactions can be meaningfully represented and analyzed as complex networks using the theoretical apparatus of network theory. Because the environment can have biotic, abiotic, or both components, the literature on economic hierarchy invariably conflated biogeographical, geobiological, and biogeochemical aspects resulting in multiple irreconcilable ecological hierarchies (Miller 2001). A unique, internally consistent, and ontologically sound economic hierarchy can be constructed by restricting its members to solely biotic entities. From an epistemological standpoint, decoupling the biotic and abiotic ecological components is critical for elucidating the causal relationships between them. Boundaries in ecological hierarchies above individual levels are complex and difficult to identify precisely in practice.

3.4.1 Molecular Level

Proteins govern the economy of life at the molecular level. By their ability to catalyze reactions, and interact with other proteins and nucleic acids, they read and express genetic information and ultimately determine the phenotype at the cellular level. Interactions at the molecular level in biological systems do not occur outside cells but must have been instrumental in prebiotic chemistry (Michod 1983). Unquestionably, there are multiple levels in the functional hierarchy of control subsumed under the molecular level of biological organization. However, these elements reside at the same phenomenological level because their relationships are governed by the same principles of chemical interactions that occur at commensurable rates. At the molecular level, interactions are best described by a hierarchy of control, rather than that of nested composition, that can be represented as an integrated system of complex networks including transcriptional regulation, protein–protein interactions, cellular metabolism, and cell signaling (Albert 2005; Bray 2003; Zhu et al. 2007). Complex molecular interaction networks predominantly display scale-free or broad-tailed distribution and have been shown to be extremely stable in face of environmental and genetic perturbations under real and simulated conditions (Albert 2005; Almaas 2007; Barabási and Oltvai 2004; Giot et al. 2003; Jeong et al. 2000, 2001; Meiklejohn and Hartl 2002; Siegal et al. 2007; Stearns et al. 1995; Wagner and Fell 2001).

3.4.2 Cellular Level

Economic processes at the cellular level amount to maintenance of homeostasis of the cell through integrating metabolic processes, transport of substances across the plasma membrane, spatially structuring the internal environment using the cytoskeleton and membrane-bound compartments, and interactions with other cells.

3.4.3 Organismal Level

In an economic sense, individual organisms are highly integrated, cohesive wholes composed of interacting cells, structural units typically differentiated into functional modules (tissues and organs). The phenotype, a complex of emergent properties of the organism that includes morphology, physiology, behavior, and, in some instances, cognitive abilities, is derived epigenetically through development from the dynamics of gene regulatory networks (Villarreal et al. 2012). The enduring and robust organismal phenotypic stability ("canalization" of Waddington

1942; "epigenetic stability" of Wagner 1996) maintained over the life cycles of conspecific individuals is an emergent property of evolved, complex developmental genetic networks (Álvarez-Buylla et al. 2008; Bergman and Siegal 2003; Gibson and Wagner 2000; Scharloo 1991; Siegal and Bergman 2002).

3.4.4 Population Level

From the standpoint of economic interactions, populations are referred to as *avatars* (Damuth 1985). Avatars are more inclusive entities than demes: Whereas all members in a population interact with their environment, however differently, only a subset of individuals of reproductive age in a population comprises an interbreeding group. In eusocial insects, some castes are never reproductive, so their entire function is an economic one that indirectly benefits the reproductive outcome of a population as a whole. Differences in the economic processes within avatars may result in a population structure, where non-random associations among individuals may reflect age and sex structure, geographical distribution, pecking orders, territoriality, survivorship, and behavioral differences that can be expressed using the tools of network theory (Dyer and Nason 2004; Fewell 2003).

The most extensive population-level interactor corresponds to a metapopulation (Levins 1969), an instantaneous cross section of an entire species lineage, a geographic and ecological mosaic of contemporaneous avatars. The spatial distribution of avatars in a metapopulation is referred to as patchiness, a great variation in the continuum from near-contiguous ranges to highly isolated, disjunct patches. Moreover, the metapopulation structure may be hierarchical, where a large avatar consists of smaller, more strictly localized units, in which interactions among individuals are more uniformly distributed (Hanski and Gilpin 1991).

3.4.5 Biotic Assemblage Level

Biocenosis (Möbius 1877) is a highly integrated, discrete assemblage of nonconspecific avatars occupying a particular biotope (habitat). It roughly corresponds to the concept of *community* (Clements 1916) and represents a biotic component of an ecosystem, the network of interactions among organisms, and between organisms and their environment. Because the concept of the ecosystem includes both biotic and abiotic components, it cannot be consistently used in the context of a hierarchy composed of entirely biotic entities, if the logical structure of compositional nested organization is to be preserved: An association of avatars alone does not comprise an ecosystem. Moreover, it has been suggested that ecosystems are not temporally limited and, therefore, lack individuality (Maynard Smith and Szathmáry 1995; Szathmáry 1989). Decoupling the biotic and abiotic components of an ecosystem (i.e., biocenosis and an assemblage of ecological niches realized by avatars, respectively) allows for investigating their causal links.

Biocenoses are characterized by stable taxonomic composition and ecological interactions that fall into four fundamental types: trophic, topic, phoric, and fabricational (Beklemishev 1970). They differ widely in their complexity and typically exhibit a hierarchy of control of their functional subsystems that includes synusiae (Gams 1918), guilds (Root 1967; Simberloff and Dayan 1991), consortia (Masing 1981; Reinke 1872), and coenomes (Chernyshenko 2012). The complexity of the configuration of biocenosis boundaries precludes an easy demarcation of adjacent biocenoses.

The best characterized form of biotic interactions is trophic, or predator–prey relationships, typically referred to as food webs and represented by directed networks. Most examined real-world food webs share various topological properties (i.e., non-random degree distribution and short path length), but display a range in connectivity from power law to broad-scale degree distributions (Camacho et al. 2002; Dunne et al. 2002a; Montoya and Solé 2002; Solé and Montoya 2001; Williams et al. 2002). Despite differences, many food webs share common motifs, specific configurations of nodes not found in other kinds of networks (Milo et al. 2002). Other kinds of ecological interactions, such as plant–animal mutualistic networks, also show a variety of node connectivity distributions, including scale-free, truncated power law, or broad-scale regimes (Jordano et al. 2003).

From the explicitly hierarchical perspective, biocenoses can be conceptualized as hierarchical patch mosaics that can be perceived and studied as spatially nested patch hierarchies in the context of a landscape (Wu 1999; Wu and Loucks 1995). The nested structure of biotic assemblages most likely represents a continuum of levels. Recognizing discrete, nested hierarchical units—such as local, regional, and provincial assemblages—appears to be highly context-dependent, mirroring the difficulty of attributing significance to classificatory ranks of monophyletic taxa.

3.4.6 The Biosphere Level

The biosphere, or the global biocenosis—frequently referred to as Gaia (Lovelock 1972, 1979; Lovelock and Margulis 1974; Volk 1997)—has long been viewed as a self-regulating planetary system that affects Earth's geochemical, hydrological, and climatological dynamics [Vernadsky 1926 (1998)].

4 Evolution

4.1 The Domain of Evolution

In such a complex view of life that emphasizes interactions within and across levels as well as between the two hierarchies of life, everything becomes causally related to processes of biological evolution (directly or not, and to different extents). To understand the nature of the evolutionary process, we need to (1) provide a clear-cut definition of biological evolution and (2) identify entities of biological systems capable of evolving.

A widespread, commonplace notion equates evolution with any kind of historical development. By defining biological evolution as "descent with modification," Darwin made a sharp distinction between biological evolution and other processes of change (Darwin 1859). Biological evolution amounts to the fate of information that can be altered as it is transmitted from ancestors to descendants. Such definition does not limit evolution to processes that take place at the level at which the information ultimately resides (i.e., nucleic acids), but encompasses phenomena across levels of biological hierarchies that directly or indirectly channel, alter, or interrupt the flow of information through the entire timescale. Thus defined, evolution is restricted to living systems and different aspects of human cultural evolution (e.g., languages and material culture). Historical processes in non-living entities are deterministic, stereotypically repeated processes: Thus, related types of minerals or similar stars have analogous predecessors, not common ancestors.

4.2 Evolving Entities

What properties should entities have in order to evolve? Minimally, evolving entities must be able to (1) store information that can be potentially altered and (2) transmit the information through some sort of replication process by partitioning or copying of a preexisting entity. Thus, evolving entities must be individuals; that is, they have to persist for some duration of time in order to have any structure that can be transmitted and altered. These requirements also imply that replication is necessary but not sufficient for evolution: *Simple replicators* that produce descendants identical to ancestors cannot evolve (Fig. 6a), whereas *hereditary replicators* that generate progeny that might differ from its ancestral template can (Maynard Smith and Szathmáry 1995; Fig. 6b, c).

There are two modes of hereditary replication: proliferative (Fig. 6c) and non-proliferative (Fig. 6b). *Non-proliferative replication* results in an entity producing a single descendant, so that diversity remains the same over generations, even though the information content itself could have been altered, as in the case of a species giving rise to a descendant species. *Proliferative replication* generates variation in heritable attributes, offering the possibility of sorting of variants, so that only some of them persist and transmit information to the next generation. While both modes of replication occur in biological systems, proliferatively replicating biological individuals are the main focus of evolutionary biology, because only these processes ultimately result in biological diversity that characterizes life at all levels of biological organization. Non-proliferative replication is significant only as a means for linear modification of historical lineages established by proliferative replication. In the variational process of evolution, the properties of an ensemble change, not because individual elements change, but rather because of the action of some process sorting on preexisting variation within the ensemble.

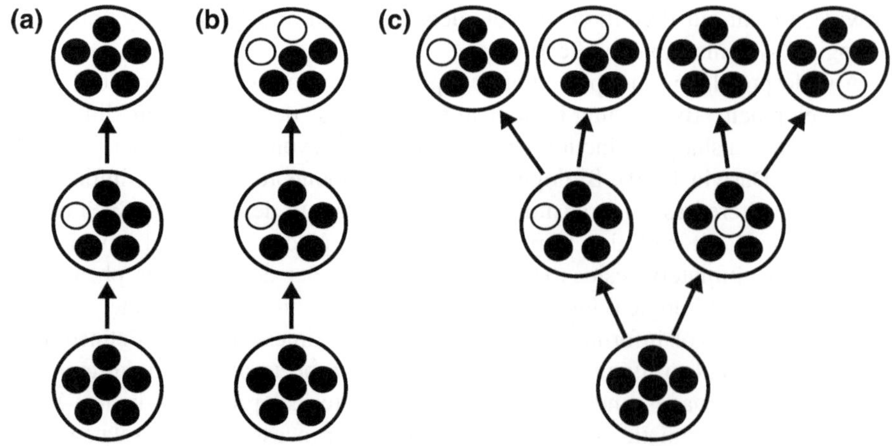

Fig. 6 Types of replication processes. **a** In simple replication, any variation that might occur is not transmitted to descendants and, consequently, simple replicator systems cannot evolve. **b**, **c** In hereditary replication, descendants that differ from their ancestors can pass on variation to their offspring, enabling such hereditary replicator systems to evolve. Non-proliferative hereditary replicators (**b**) produce a single descendant, resulting in a constant level of variation over generations, whereas proliferative hereditary replicators (**c**) continuously increase variation over generations. *Large circles* illustrate replicator entities; enclosed *filled circles* denote their component parts and *open circles* represent changes in a replicator's elements that produce variation among replicators

Paraphrasing the requirements for an evolving entity stated at the beginning of the section, the evolving individual must be (1) capable of generating variation in stored information amenable to sorting and, consequently, (2) must be relatively less spatiotemporally restricted than the lower-level constituents that display the variation.

Having established explicit criteria for evolution, we can identify where evolution can occur (at least in principle) in the dual hierarchical system and determine its upper and lower bounds. First, the levels containing evolving entities must contain individuals capable of proliferative replication. Second, there must be processes enabling sorting of variation at that level. As the information transfer is a defining feature of entities in the genealogical hierarchy, evolution is restricted to entities of the genealogical hierarchy, although not including all its levels. Even though the term evolution is used to describe a change in ecological entities (such as community or ecosystem evolution), ecological individuals do not reproduce. Undoubtedly, successional systems have histories, but they undergo changes as individuals, paralleling individual ontogeny (Clements 1916).

The minimal evolving unit is a nucleotide sequence. Changes at the level of individual nucleotides (i.e., substitutions, insertions, or deletions) have evolutionary significance only in the context of contiguous nucleotides that make the nucleic acid fragment a homology unit. The upper bound of evolution is at the level of the species. Even though supraspecific entities (monophyletic taxa) do show patterns of differential survival, they lack the capacity to replicate: Evolution above the species level is an epiphenomenon that results from processes that occur at lower levels.

4.3 Evolutionary Patterns

The history of life on Earth is characterized by a combination of unique and repeated events. An *evolutionary pattern* is an empirically discernible regularity in the history of a biological system. As such, the elements of a pattern recur in a predictable manner. A great many evolutionary patterns are recognized for classes of individuals at different levels of biological hierarchies, although some patterns appear to be isomorphic across levels (such as trends for increasing complexity; e.g., McShea and Brandon 2010).

There are two fundamental categories of historical patterns: diversity and disparity patterns. *Diversity patterns* describe the dynamics of change in the quantity of evolving individuals at a given level of a genealogical hierarchy. They are the product of the origin and death dynamics among individuals. Most studies have focused on three kinds of rates of evolution: rates of morphological evolution, rates of taxic evolution, and rates of genomic evolution (Raup 1987; Schoch 1986). Historical species-level diversity, or *taxic*, patterns (conceptualized as phylogenies) have long been considered a cornerstone of evolutionary theory (e.g., Alroy 2000; Gould 2002; McKinney 1990a; Fig. 7a). Phylogenetic patterns are the product of speciation and extinction dynamics and can be approached by causal analysis of rates of biological diversification. Since taxonomic ranks are arbitrary, the only meaningful approach to taxic diversity is through comparison of species-level diversity between sister groups (Cracraft 1984; Vrba 1980, 1984a). Taxic diversity patterns are typically studied by counting the distribution of taxa through time and typically represented by spindle diagrams (Gould et al. 1977, 1987; Raup et al. 1973; Stanley et al. 1981) or diversity curves (Newell 1952, 1967; Sepkoski 1978, 1993) and the shape of their profiles investigated for biological significance.

Disparity patterns describe the dynamics of change in the attributes (typically phenotypic) of evolving individuals at a given level (Fig. 7a). For example, it might be of interest to determine not only how many species existed at any time, but also how different they are phenotypically (what area of morphospace they occupied). Morphological disparity, or *transformational*, patterns are typically analyzed by quantitative approaches of morphometrics (Rohlf and Marcus 1993), theoretical morphology (McGhee 1998, 2007), and constructional morphology (Schmidt-Kittler and Vogel 1991; Seilacher 1970). Even though disparity and diversity can be investigated on their own right, empirical evidence suggests that patterns of diversity and disparity can be intimately linked.

Patterns of diversity and disparity vary in scale and mode. With regard to their scale, patterns can be *local* or *global* taxonomically and temporally. Taxonomically global patterns apply to many distantly related organisms up to and

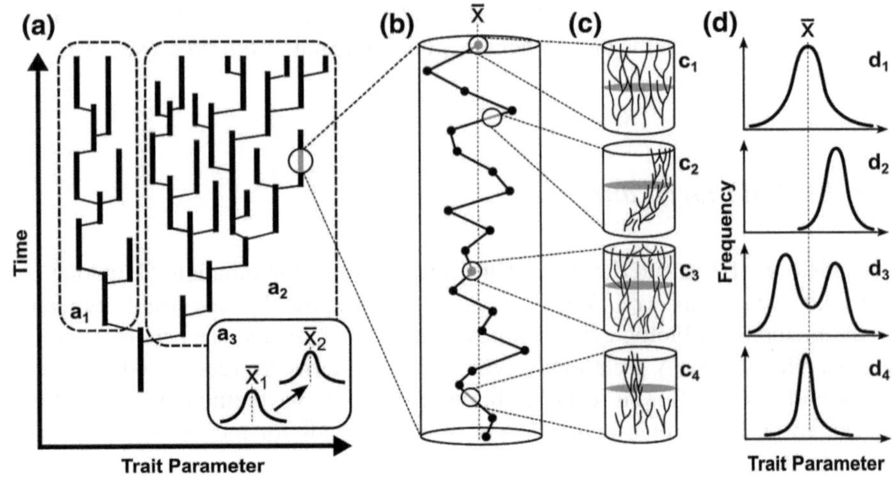

Fig. 7 Hypothetical examples of evolutionary patters at different hierarchical levels. At the level of monophyletic taxa (**a**), sister clades showing different diversity and disparity patterns: clade a_1 displays a low rate of taxic evolution and stasis in disparity; clade a_2 exhibits a high turnover rate accompanied by a passive directional trend in disparity (a_3). At the species level, species lineage remains in stasis, showing minimal oscillations around a stable mean (**b**), whereas at the population level, component demes (**c**) continually differentiate, merge, or become extinct, producing short-lived diversity and disparity patterns of stability (c_1, d_1), directional change (c_2, d_2), divergence (c_3, d_3), or decline (c_4, d_4). *Gray plane* in **c** indicates an instantaneous cross section of the species lineage described by a corresponding disparity profile in **d** (inspired by Eldredge and Gould 1972: Fig. 5–10 and Miller 2006: Fig. 3)

including all of life, whereas taxonomically local patterns are unique to a specific taxon. Temporally global patterns recur throughout the history of life, whereas temporally local patterns are unique to a specific period in the geological history.

The three principal modes of evolutionary patterns are stasis, trend, and random walk (Fig. 8). *Stasis* is a historical pattern characterized by little or no net change, typically showing minimal oscillations around a stable mean (Figs. 7b and 8a). An example of stasis is the taxonomically and temporally global disparity pattern of marked morphological stability, or equilibrium, displayed by many species in the fossil record (Eldredge et al. 2005).

A *trend* is sustained directional change in a statistic descriptor for some attribute observed over time (Figs. 7a and 8b). Trends display two kinds of dynamics: passive or active; the latter can be either strongly or weakly driven (McShea 1994). A classic example of a global trend is an increase in body size over time across evolutionary lineages ("Cope's Rule"; Kingsolver and Pfennig 2004; McKinney 1990b). The dynamics of driven trends occur primarily in one direction and apply to most component lineages, whereas passive trends are the net result of complex dynamics operating in different directions in different lineages or at different times. For many examples and a general discussion of evolutionary trends, see McNamara (1990, 2006).

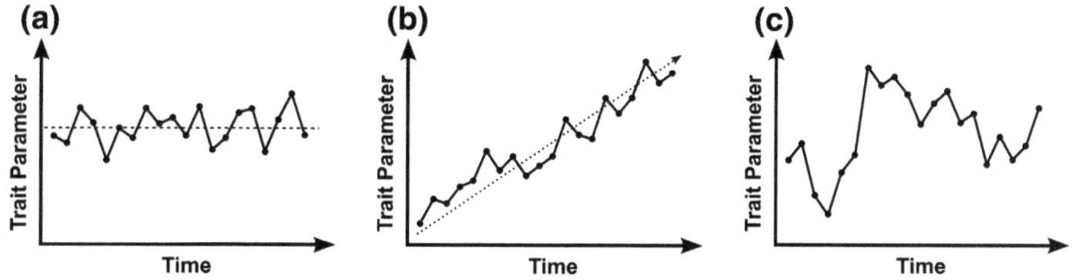

Fig. 8 Three principal modes of evolutionary patterns. **a** A stasis, showing minimal fluctuations around a stable mean (*dashed line*) that results in little or no net change; **b** a trend, illustrating sustained directional change (*dotted arrow*); and **c** a random walk, characterized by a series of independent steps that depart from the mean in either direction

A *random walk* is a pattern characterized by a succession of independent steps, where increases and decreases of a trait are equally probable (Fig. 8c). Both stasis and random walk patterns are not inherently directional but, unlike stasis, random walks depart from the mean and produce increasing evolutionary divergence over time. An example of a random walk is a global diversification (species origination and extinction) pattern in marine animals throughout the Phanerozoic aeon (Cornette and Lieberman 2004). Population-level dynamics may approach random walk in fluctuating environments and during biocenotic crises, but such microevolutionary patterns, as reflected by incomplete stratophenetic series in the fossil record, might rather reflect sampling bias and time averaging effects (e.g., Roopnarine 2003; Roopnarine et al. 1999).

5 Causes and Mechanisms of Biological Evolution

5.1 Hierarchical View of Evolutionary Processes

Evolution occurs at the nexus between the economic and genealogical hierarchies, where the dynamic interactions in the former are translated into a historical pattern of the latter. In other words, the interaction of individuals in the economic hierarchy with their environment is causally related to differential replication of genealogical individuals. A change in the information content at a given level has an effect on the diversity pattern of genealogical individuals: Variation can either increase through the origin of new variants, or decrease through sorting, where only a subset of variants persists. The origination of novelties at any given level does not necessarily entail sorting at the higher level; it does, however, provide raw material for sorting. These initiating conditions are necessary but insufficient for evolution. For instance, large changes in allele frequencies do not cause speciation and rampant species turnovers do not necessitate trends in monophyletic groups.

Sorting of variation is a consequence of differential birth and death rates among genealogical individuals differing in their properties. (The meaning of birth and death is different depending on the kinds of individuals: At the molecular and cellular level, replication processes result in the nominal death of a parent as it physically becomes part of its descendants, whereas at the organismal and species levels, parents typically remain alive, so that their durations, even if briefly, overlap with those of their descendants.) The dynamic shuffling of genealogical entities through sorting may result in the evolution of the more extended (more inclusive) genealogical individuals.

The origin and sorting of variation may be tightly coupled because variation at a focal level can emerge from sorting of lower-level individuals. Complex evolutionary patterns integrate variational dynamics of sorting processes that occur at different levels with their effects propagated indirectly to other levels within the genealogical hierarchy via downward and upward causation.

Hierarchy theory provides a theoretically and operationally unified framework for unraveling causal processes responsible for generating evolutionary patterns by identifying the involved individuals and their properties, hierarchical levels where these individuals reside, and their interactions within and across levels as well as between the two hierarchies. The outline of a formal approach to investigating causality is presented below:

- Identifying the focal level(s) in the genealogical hierarchy at which the pattern under study is manifest;
- Identifying entities and their specific attributes at that level pertinent to the pattern;
- Discerning whether these attributes are emergent or aggregate;
- Determining whether variation in these attributes is a product of sorting of lower-level attributes or by de novo introduction at the focal level;
- Identifying the level(s) and entities in the economic hierarchy that can potentially directly impact the fate of replicators by sorting the variants at the focal level in the genealogical hierarchy and/or by modifying their attributes (generating variation);
- Exploring cascading effects of upward and downward causation by more remote levels in the economic hierarchy that potentially indirectly affect variants at the focal level in the genealogical hierarchy.

5.2 Sorting of Variation

Sorting operates on individuals' traits, and there are two key requirements with regard to traits as subjects of sorting: (1) A trait exhibits little or no variation within an individual relative to the variation among individuals at the given focal level and (2) the differential origin and demise of individuals covary consistently across one or more higher-level individuals with that trait. Even though there are undoubtedly level-specific sorting processes, there are two general modes of sorting that are isomorphic across hierarchical levels: selection and drift.

Selection is a deterministic process that represents a special type of sorting resulting from non-random differences in fitness, or replication success. Gains in fitness produce *adaptations*, or attributes that ensure greater stability and persistence of an individual entity, typically envisioned as maximizations of the match between biological entities and their immediate environment. In Darwin's original formulation, selection ("natural selection") operates on phenotypic properties of individual organisms (Darwin 1859). The concept of selection has been subsequently extended to entities at other hierarchical levels (Dobzhansky 1937; Lewontin 1970; Wright 1931, 1953, 1980). Most generally, selection can be characterized as any process in which differential extinction and proliferation of interactors causes the differential perpetuation of the replicators that produced them (Hull 1988). For a recent, comprehensive review of selection, see Okasha (2006).

Drift is a non-deterministic sorting process owing to chance. Drift was originally proposed as a process of random fluctuations in allele frequencies due to sampling effects in finite populations ("genetic drift," Wright 1929, 1955). The neutral theory of molecular evolutions endowed the stochastic processes of genetic drift and mutation pressure with much greater significance than initially thought (Kimura 1968, 1983). Similar in principle, stochastic processes later were proposed to occur at other levels (e.g., Gould et al. 1977). In contrast to selection, drift operates on replicators directly, as the function of the latter is independent of ecological context (Hull 1988). The interplay of selection and drift can produce complex evolutionary patterns. For example, selection operating in different directions in multiple populations across species can result in stochastic dynamics for clades (Gould 2002; Raup 1981; Raup and Gould 1974).

5.2.1 Molecular Level

There is no evidence for specific, molecular-level processes of sorting of genetic variation that do not involve processes at higher levels (though such processes must have been instrumental in prebiotic evolution). The expression of molecular variation is downwardly constrained, or canalized, by the higher, phenotypic levels of the cell and organism.

5.2.2 Cellular Level

Sorting at the cellular level may arise by a number of different mechanisms. In unicellular organisms, where the cellular level of organization is coincident with that of the organism, natural selection and drift play major roles. In multicellular organisms, additional sorting processes involve a variety of cell–cell interactions during development in the soma (e.g., apoptosis, stem cell segregation). Sorting in the germ line includes a diversity of deterministic processes leading to intragenomic conflict (segregation distortion, meiotic drive, and maternal effect lethality) and stochastic events (e.g., random fertilization and transposition). Some of sorting processes at the cellular level leading to differential reproductive success of individual cells are analogous to natural selection: differential rates of cell division among different cell lineages or cell types, tissue repair, and neoplastic growth. If the latter becomes malignant and develops into cancer, the tumor cells are effectively outcompeting normal cells in the struggle for resources and space (Casás-Selves and DeGregori 2011). Many of these processes must occur simultaneously, but their relative frequencies are unknown.

5.2.3 Organismal Level

The individual has been historically regarded as the primary focus of sorting through natural and sexual selection, and genetic drift. Natural selection has long been recognized as the causal agent of organismal adaptation. In genetic terms, the theory of individual selection culminated in the Fundamental Theory of Natural Selection stating that the rate of change of population fitness is equal to the genetic variance in fitness (Fisher 1958). The theory was later modified for frequency-dependent and for multiple loci (reviewed by Lewontin 1970). The stochastic sorting at the level of the organism that cascades downward to the molecular level is interpreted as genetic drift, which is a lower-level description of population-level process expressed in terms of dynamics of allele frequencies.

5.2.4 Population Level

At the population level, variation among demes arises from upward causation by sorting of their component conspecific individual organisms and by density-dependent demographic processes at the focal level. *Kin selection* [generalized by Hamilton (1964a, b) and named by Maynard Smith (1964)], that enhances the reproductive success of an organism's relatives, even at a cost to the organism's own survival and reproduction, can bias genetic variation within a deme, but is unlikely to account for sorting among demes. Apart from stochastic processes that may be affecting small populations, there is no evidence for intrinsic population-level sorting processes that can result in the differential survival of demes. Thus, sorting at the population level is a consequence of downward causation deriving from the cascading effect of higher-level ecosystemic (biocenotic) dynamics in the economic hierarchy.

5.2.5 Species Level

Variation among species arises as a historical consequence of interdemic genealogical relationships described collectively by metapopulation dynamics extended through geological time. In Sewall Wright's formulation, species are constructed from a shifting balance of demes with different proliferative capacities

(Wright 1931, 1982). Species sorting is a consequence of differential speciation and extinction events within a monophyletic group. Differential births and deaths of species are epiphenomena resulting from interactions of a (meta) population with its environment. Species and, consequently, more inclusive monophyletic taxa, are exclusively genealogical units that do not have ecological roles and, consequently, cannot participate directly in sorting (Eldredge 1985a, 1989; Ghiselin 1981).

The punctuated equilibria model suggested a range of hypotheses explaining patterns of differential speciation and extinction rates in an implicit hierarchical framework (Eldredge and Gould 1972). One such possibility is *species selection*, or differential reproductive success among species within monophyletic groups favoring species that persist for long periods and speciate at higher rates (Gould and Eldredge 1977; Stanley 1975, 1979). An important distinction is made between selection operating on aggregate traits that ultimately reside at the organismal level (the Effect Hypothesis; Vrba 1980, 1987) and selection operating on emergent, heritable properties of species level (species selection sensu stricto) (Vrba 1984b, 1989; Vrba and Eldredge 1984; Vrba and Gould 1986; Grantham 1995, 2001). In strict-sense species selection, the focal level is the species, with downward causation influencing the frequencies of organismic traits among clades and upward causation shaping overall clade composition. The paucity of affirmative examples of true species selection operating at the species-lineage level suggests that patterns of species sorting are better accounted for by effect-macroevolution, the upward causation of sorting of lower-level entities (Eldredge 1995; Lieberman and Vrba 1995). *Species drift* (Gould 2002; or *phylogenetic drift*, Stanley 1979), a stochastic origination and persistence of species in the absence of environmental interaction, is another potential species sorting mechanism that might be most pronounced in clades with relatively small numbers of species.

5.2.6 Supraspecific Levels

Higher taxa do not replicate, because their origins reside at the species level and sorting of monophyletic groups is restricted to extinction dynamics. Mass extinctions are an example of sorting at higher level, where entire clades may become extinct as a result of environmental changes or contingent factors of large magnitude (e.g., volcanism or meteor impact). Such higher-level sorting events have a profound effect on the lower level by downwardly filtering variation retained at lower levels.

5.3 Process Integration Across Hierarchies

Stability emerges as an all-pervasive pattern in most biological systems: Individual cells persist as highly stable homeostatic systems, phenotypes of individual organisms are remarkably buffered against environmental and genetic variation, the variance within demes fluctuates little over time, and species remain in stasis for most of their duration. The overall stability of biological systems across levels of organization is not surprising given the general tendency of complex systems to acquire complex network architectures that ensure relative insensitivity to external perturbations (Albert and Barabási 2002; Dorogovtsev and Mendes 2002). The state of relative equilibrium is further reinforced by the nested hierarchy itself because stability in adjacent levels effectively limits the extent of the origin and sorting of variation at the focal level by reducing potentialities of initiating conditions (upward causation) and tightly canalizing the boundary conditions (downward causation), respectively. Such systems are expected to remain in equilibrium in the absence of external perturbations.

Evolution occurs as a response to perturbations that are sufficient to disrupt the steady state of the system, so that the buffering mechanisms preventing the cascading effects across hierarchies fail. Regaining the equilibrium in a new state constitutes the basis of evolutionary transitions resulting from the integration of processes affecting origin and sorting of variation across levels of the genealogical hierarchy brought about by extrinsic perturbations. Therefore, the key to the explanation of fundamental evolutionary patterns needs to be sought in factors that disrupt the stability of biological systems and the way the effects of such perturbations cascade up and down hierarchical levels. Genealogical individuals do not interact with their environment to the extent that economic entities do. It is the external factors influencing the economic dynamics of biological individuals at different levels that ultimately bring in motion sorting mechanisms that alter the landscape of biological diversity and disparity in subsequent generations.

5.4 Evolutionary Drivers

What are the external (extrinsic) factors strong enough to bring about evolutionary change? The ultimate evolutionary causes are abiotic environmental factors of cosmic and planetary scale that translate into climatic, lithological, and geochemical processes perturbing the stability of biological systems across levels. These factors either have a direct impact upon entities in the genealogical hierarchy or indirectly affect them via interacting with individuals in the economic hierarchy.

The nature of evolutionary response to an extrinsic environmental perturbation is determined by the scope, the scale, and the temporal mode of the impact. The *scope* refers to the range of levels in biological hierarchies affected directly by the perturbation. For example, gamma-ray bursts produce perturbations of wide scope because they deplete the ozone layer, causing a deleterious mutagenic effect at the molecular level due to a flux of ultraviolet radiation, while they can also produce acid rain and global cooling affecting global biocenotic processes (Melott et al. 2004). The emergence of the Central American Isthmus, completed 2.8 million years ago, is of comparatively narrower scope: It disrupted regional marine biocenoses, ultimately leading to divergence of geminate species (Lessios 2008).

The *scale* refers to the spatial extent of the impact that can range, for example, from factors influencing bacterial microhabitats of gut flora within individual organisms to factors affecting the global ecosystem, such as the Great Oxygenation Event that occurred 2.4 billion years ago.

External factors vary in the *temporal mode* of their action: They can be either *regular* (perpetually acting or periodic/recurring) or *contingent* (unique or aperiodic). Examples of the former include perpetual cosmic radiation, everlasting tectonic plate movement, and periodic Milankovitch cycles; examples of the latter—a mudslide destroying local microhabitats, volcanic eruptions causing collapse or major restructuring of biocenoses on an island, and the asteroid impact that precipitated the mass extinction at the Cretaceous–Paleogene boundary at around 66 million years ago.

The magnitude of evolutionary response is contingent upon the nature and spatiotemporal scale of environmental physical perturbations. Attempts were made to capture this relationship by a formal model. According to the *intermediate disturbance model of maximal speciation* (McKinney and Allmon 1995), at the intermediate levels of environmental disturbance, speciation rate is maximal due to a balance between population fragmentation and establishment of favorable conditions for isolate persistence. The *sloshing bucket model* (Eldredge 2003) is a more generalized framework rooted in the dual hierarchical model that postulates a direct correlation between the magnitude of disturbance and its evolutionary consequences that can be expressed as a series of levels progressing from virtually no disturbance, resulting in stasis, to massive perturbations affecting the global biota. According to the model, the higher the level of perturbation, the higher the level in the economic hierarchy at which its effects will be expressed and, consequently, the higher the level of the genealogical hierarchy at which the evolutionary pattern of change in diversity and disparity will be recorded.

5.5 Stability and Change in Biological Systems

While it might not be possible to construct a strict hierarchy of extrinsic drivers, their effect can be evaluated from a hierarchical perspective relative to their impact on evolutionary entities. To establish the significance of perturbation at multiple hierarchical levels, it is critical to consider the causes and effects of perturbations at each level along with their upwardly and downwardly cascading effects.

5.5.1 Molecular-Level Perturbations and Their Effects

Perturbations at the molecular level can affect either or both replicator and interactor entities at that level. If perturbations in the chemical balance are sufficient to disrupt metabolic and protein interaction networks, they might have a strong upward effect upon physiology at the cellular and organismal level to the extent of an organism's death. The limit of the upward effect would depend upon the magnitude of such perturbation and can have a cascading effect on world's biota. Examples of such occurrences include a tectonically driven nickel famine responsible for a global decline of anaerobic organisms at the end of the Archean aeon (Konhauser et al. 2009). Such an event, having directly impacted a particular lineage (or number of lineages) of the genealogical hierarchy (the methanogens), must have had ecological consequences for other organisms interacting with methanogens, thereby indirectly impacting their evolution at higher levels.

Direct perturbations of the genetic machinery can be environmentally induced via abiotic or biotic factors, or arise intrinsically due to random errors in DNA replication or repair. Although largely deleterious, mutations in DNA resulting from environmental radiation or chemicals are generally regarded as the ultimate source of novelty at the molecular level. Alteration of DNA can also result from sequence editing by viruses (Villarreal 2005) and concatenation via symbiosis (Maynard Smith and Szathmáry 1995). Duplication events followed by functional divergence of a duplicated gene are the principal source of new genes; small-scale mutations, such as point mutations and indels, are among the sources of new alleles and changes in non-coding regulatory sequences. Even though the mutation process is not entirely haphazard, as some parts of the genome are more likely to mutate or get repaired, mutations are random with respect to higher levels. In addition, heritable differences in gene expression pattern can be brought about by epigenetic processes (such as nucleotide methylation) that do not necessitate a physical change in the DNA sequence (Jablonka and Lamb 1995, 1998, 2005).

Because molecular genetic systems reside within living cells, the consequences of their perturbations must be viewed from the levels of the cell and the organism: Filtered to a higher level via differential gene expression, molecular genetic variation manifests at the cellular and organismal levels. In multicellular organisms, mutations in somatic cells are of little evolutionary consequence; however, tumors could have a role in the origin of new cell types, tissues, and organs (Kozlov 2014). It is mutations in the germ line of the parental generation that form the basis of phenotypic novelty at the level of individual organisms. In addition to its regulatory function as a genotype, the genome as a whole (the nucleotype) exerts an effect at the cellular level by virtue of its size and composition: for example, by influencing cell size (Gregory 2004; Gregory and Hebert 1999).

Translation of genetic changes into organismal phenotypes is channeled through a preexisting, inherited ontogenetic program that establishes organism-level boundary conditions limiting the range of potential phenotypes due to morphogenetic and phylogenetic constraints. This organism-level phenotypic stasis is an emergent property of molecular developmental networks that can be attained without direct selection for stability (Gibson and Wagner 2000; Scharloo 1991; Siegal and Bergman 2002). Models based on empirical data demonstrate that developmental networks are remarkably resilient to perturbations and show that the robustness resides in the topology of the network and the nature of the interactions within the system (Albert and Othmer 2003; Álvarez-Buylla et al. 2008; Siegal et al. 2007; von Dassow et al. 2000). Perturbations resulting in gene duplications and point mutations may lead to a preferential increase in the degree of highly connected proteins, further contributing to the overall stability of gene regulatory networks (Berg et al. 2004; Eisenberg and Levanon 2003; Wagner 2003). Furthermore, stochastic perturbations of gene regulatory networks contribute significantly to the emergence of stability in morphogenesis at the level of the organism (Álvarez-Buylla et al. 2008). In simulations, small populations that experience high mutation rates evolved to become particularly robust (Elena et al. 2007). As gene regulatory networks, metabolic networks are characterized by evolved redundancy and other topological features that make them robust to environmental fluctuations (Hintze and Adami 2008; Jeong et al. 2001; Ravasz et al. 2002).

Genetic variation is further buffered under normal conditions at the level of the deme and accumulates as hidden polymorphisms. Environmental perturbations that fall outside the normal range, such as temperature fluctuations, can functionally compromise the buffering capacity of genetic networks, resulting in the rapid increase of suppressed phenotypic variance (*evolutionary capacitance* sensu Bergman and Siegal 2003; Hermisson and Wagner 2004; Rutherford and Lindquist 1998). Such bursts of phenotypic variation provide raw material for sorting at the level of the deme.

Perturbation of developmental gene regulatory networks can result in a substantially reconfigured, albeit stable and robust, pattern of the network without non-functional intermediates and without creating new protein–protein interaction (Ingolia 2004). This emergent property of canalization in gene networks provides a mechanism for a rapid and stepwise phenotypic change (Cossins 1998), despite the fact that the accumulation of mutations contributing to hidden variation could have been a gradual and cumulative process. It is noteworthy that the stasis at the organismal level does not imply persistence of a particular morphology; instead, it suggests the fidelity of maintenance of the ontogenetic program as individuals go through life cycles frequently characterized by disparity of form ("semaphoronts" of Hennig 1950, 1966).

In addition to maintaining the fidelity of ontogenesis, topological features of genetic developmental networks exert an effect on the nature and direction of phenotypic change. The *cis*-regulatory elements, comprised of multiple binding sites for specific transcription factors, are the principal functional modules of gene network architectures that govern gene expression (Davidson 2001; Howard and Davidson 2004). The modules, composed of functionally quasi-autonomous gene regulatory subnetworks can be identified with specific biological functions (Hartwell et al. 1999; Wagner 2002). Modularity allows for certain parts of networks to be modified without affecting the integrity of ontogeny, so that different morphological features can evolve autonomously and over relatively brief periods of evolutionary time (Abouheif and Wray 2002; Alon 2003; von Dassow et al. 2000), with the magnitude of phenotypic effects depending on the specific components of the networks being altered (Davidson and Erwin 2006).

In summary, within a normal range of environmental conditions, molecular genetic networks buffer regular small-scale perturbations at the molecular levels, simultaneously increasing evolvability by accumulating hidden (unexpressed) variation at the molecular level and further enhancing robustness by evolving redundancy and interconnectedness. Such molecular-level dynamics is translated into relative phenotypic stability at the organismal and cellular levels. When the level of perturbations exceeds the threshold of the structural integrity and, consequently, the buffering capacity of genetic networks, the networks either collapse or become reconfigured, which is manifested at the higher levels as a rapid increase of phenotypic variation as a result of the release of previously accumulated hidden polymorphisms. Such a dynamical system, where the breakdown of the equilibrial state leads to a short-term episode of destabilization during which the system undergoes reorganization just to regain an equilibrium under altered circumstances, may possibly account for the empirical pattern of punctuated equilibria, characterized by a prolonged period of phenotypic stasis and a relatively brief burst of phenotypic change triggered by speciation as documented in the fossil record (Eldredge and Gould 1972; Hunt 2008; Mattila and Bokma 2008; Strotz and Allen 2013). Consistent with the results from studies on gene regulatory networks, population genetic models of punctuations indicate that the rapid transition between two states of genotype composition is accompanied by the increase in variance during punctuation while being preceded and followed by prolonged periods of low haplotypic variation, the dynamics independent of selection regime (Bergman and Feldman 2003).

5.5.2 Organism-Level Perturbations and Their Effects

The evolutionary fate of the organismal-level replicator, the germ line, depends upon economic success of the soma in the context of an individual organism's life. Variation in the phenotype (morphology, physiology, and behavior) of individual organisms is a synergetic effect of lower-level morphogenetic dynamics (upward causation), developmental and mechanical ("constructional" sensu Seilacher 1970) constraints acting at the focal level, and higher-level economic interactions with conspecific individuals and individuals of other species at higher levels (downward causation). The evolutionary effect of environmental perturbation at the level of individual organisms depends on the magnitude of perturbation: ranging from no evolutionary consequences (as in a trivial deleterious chance event of accidental death by a falling rock) to a substantial shift in the entire population in adaptive response to a climatic change mediated by natural selection. A classic example of environmentally driven evolutionary change resulting from selection operating on organismal traits is the evolution of jaws and beaks in Darwin's finches in response to the rainfall periodicity that results from cyclical waxing and waning of oceanic temperatures (Grant and Grant 2011). Such perturbations typically produce short-term, population-level anagenetic trends that over a greater period of time appear as fluctuations around a stable mean and contribute little to the evolution of higher genealogical entities.

5.5.3 Population-Level Perturbations and Their Effects

The introduction of phenotypic variation among organisms can result in the evolution of higher genealogical entities upon spreading and fixation of the mutation in a deme. As all the members of a deme are simultaneously members of avatars, demic sorting is a consequence of interactions of avatars with their environment. The avatar dynamic is driven by density-independent variation in resources and fluctuating physical factors, and density-dependent population reductions when resources are depleted. It is indirectly influenced by upward causation largely resulting from natural selection at the level of individual organisms and constrained by integration of biocenotic interactions among avatars within an ecosystem. Together, these processes can generate patterns of differential survival of demes but, as pointed out by Lewontin (1970), interdeme selection occurs under a restricted, although not exceptionally rare, set of circumstances and contributes little to evolution. Measured against the evolutionary timescale, in the context of equilibrium at the level of biocenosis, these localized, short-term processes are unlikely to yield significant evolutionary consequences in the context of the entire metapopulation, where net equilibrium is maintained by continuous local extinction, recolonization, and habitat tracking (Eldredge 1985b) of conspecific demes integrated by the network of genetic sources and sinks (Fig. 7b, c). With the exception of species consisting of spatially restricted, small populations, it is unlikely that drift can have a strong effect upon the fixation of mutations. These predictions are consistent with empirical evidence indicating that the rate of mutation fixation in natural populations is substantially lower than the potential rate (Maynard Smith 1978; Shaposhnikov 1965, 1978).

5.5.4 Biocenosis-Level Perturbations and Their Effects

Biocenoses are highly structured cybernetic systems of avatars interconnected by networks of energy and matter flow, exemplified by mutualistic relationships, trophic and host–parasite interactions, and competitive redistribution of resources. The resilience of biocenoses to perturbations emerges from the interplay of diverse ecological interavatar networks operating contemporaneously within ecosystems. Most examined real-world networks are characterized by non-random (predominantly broad-scale) degree distribution and short path length (Camacho et al. 2002; Dunne et al. 2002a; Jordano et al. 2003; Montoya and Solé 2002; Solé and Montoya 2001; Williams et al. 2002). These general structural features of ecological networks are implicated in their inherent stability. Food webs and mutualistic networks have been shown to preserve integrity even when subjected to strong perturbations (Dunne et al. 2002b; Montoya and Solé 2003; Olesen and Jordano 2002; Solé and Montoya 2001; Williams et al. 2002).

Sufficiently strong perturbations, capable of disrupting the buffering capacity of ecological networks, elicit differential responses related to avatar diversity and connectivity (Fig. 9a). Species-rich biocenoses with food webs characterized by skewed degree distributions are robust to random species removal, but become unstable, when removals target generalist or most-connected species. Species-poor biocenoses with food webs characterized by Poissonian degree distributions are highly susceptible to both random or targeted species removals (Montoya and

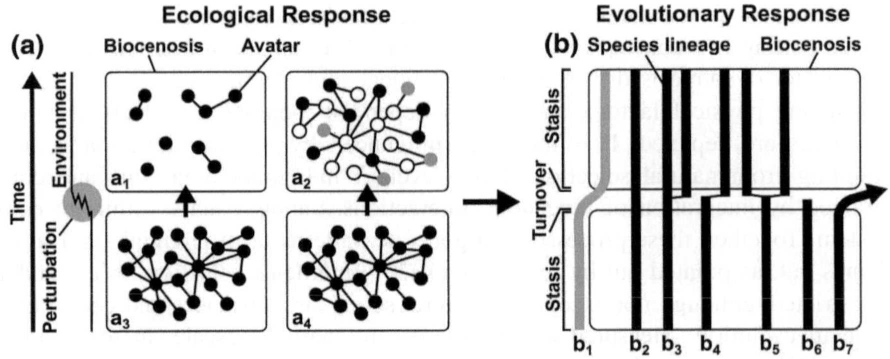

Fig. 9 Hypothetical models of ecological responses (**a**) to biocenosis-level perturbations and their evolutionary consequences (**b**). During prolonged periods of environmental stability, biocenoses behave as homeostatic avatar networks (a_{2-4}), resulting in a long-term taxonomic and morphological stability (coordinated stasis; **b**). Environmental perturbations capable of disrupting ecological networks can cause biocenotic collapse ($a_3 \rightarrow a_1$) or trigger a major reconfiguration of avatars, leading to a biocenotic replacement ($a_4 \rightarrow a_2$). Intensive ecological reconfiguration is accompanied by a variety of species-level responses, simultaneously affecting multiple lineages and concentrated in a relatively brief time interval (turnover pulse; **b**): invasion (b_1), persistence (b_2), speciation (b_{3-5}), extinction (b_6), and abandonment (b_7). The taxonomic composition of the emerging biocenosis (a_2) includes species that survives the crisis (*black circles*), invasions (*gray circles*), and newly evolved species (*white circles*)

Solé 2003). Highly connected avatars act as energy bottlenecks in food webs (Allesina and Bodini 2004) and, therefore, when selectively targeted, they cause large numbers of secondary extinctions (a number of coextinctions after species removal) resulting in food web fragmentation (Dunne et al. 2002b; Montoya and Solé 2003; Solé and Montoya 2001), ultimately producing a large detrimental effect on an overall community stability (Pimm 1980, 1991).

The analyses of food web dynamics support the hypothesis that biocenotic stability increases as the number of links in a food web grows due to an increasing number of paths through a species, dampening the impacts of its population fluctuations (MacArthur 1955). It must be emphasized that topological properties of trophic networks alone are not sufficient to capture the complexity of the dynamics and nonlinear biocenotic response to ecological perturbations involving avatar loss, calling for more realistic models that must take into account abundance, the nature and strength of interavatar interactions, and phylogenetic constraints (Cattin et al. 2004; Duffy 2002).

Highly non-equilibrial regimes of ecological remodeling are expected to endure for comparatively short periods of time until biocenoses reconstitute the balance and ecologically "lock" the locally optimal configuration of avatars in place. The reestablishment of network connectivity and increasing interdependence among avatars following a biocenotic crisis promotes species abundance within biocenoses (Anderson and Jensen 2005). The increased interconnectedness promotes further self-stabilization of the system. Thus, non-equilibrium and stochastic processes can underlie stability of ecological systems (e.g., Turner et al. 1993; Urban et al. 1987; Wu and Levin 1994).

Any response to a perturbation and transition to a new equilibrium requires that information about the perturbation to spread within the network. The mechanism of such rapid biocenotic response may be linked to the small-world property of ecological networks: Short path length in food webs suggests a potential for widespread and rapid dispersion of the effects of perturbation (Williams et al. 2002).

In the absence of disturbances in the physical environment, biocenoses behave as homeostatic systems, constraining interavatar (and, hence, demic) dynamics, which is reflected in persistent taxonomic stability and morphological stasis. This expectation is consistent with the predictions from ecological theoretical models that in a stable environment, the phenotypes of avatars within a biocenosis will remain in stasis for extensive duration (Stenseth and Maynard Smith 1984) and that highly integrated biocenoses effectively limit the establishment of durable populations of invasive species (Case 1990). The latter is substantiated by empirical data indicating that the collision of regional biotas does not typically result in formation of taxonomically mixed biocenoses (e.g., Zherikhin 1987).

Geographically co-localized avatars appear to frequently persist in stasis over extended intervals of geologic time ("coordinated stasis" of Brett and Baird 1995; Bonelli et al. 2006; Brett et al. 1996; Fig. 9b). Even though paleontological data do not allow for direct assessment of interavatar network dynamics, the observation that biocenoses frequently display analogous species composition when vastly separated geographically ["parallel communities" of Thorson (1957) or "homeomorphic paleocommunities" of Wallace (1978)] and across great expanses of time (Boucot 1975, 1978; Levinton and Bambach 1975; Walker and Laporte 1970) suggests that it is the ecological roles of the component taxa that are responsible for prolonged persistence of biocenoses.

Environmental perturbations capable of disrupting biocenotic networks (such as climate change or geological phenomena) are inextricably causally linked to evolutionary response at multiple levels and, most importantly, to speciation, due to a combination of two ensuing factors: They (1) trigger the expression of suppressed phenotypic organismal variance as a consequence of a compromised evolutionary capacitance at the molecular level and (2) provide a narrow window of opportunity for the spread and fixation of novel genotypes by releasing biocenotic control over demic dynamics during a brief interval of destabilization. High-level perturbation of entire metapopulation dynamics (conceptualized here as avatar networks) brought about by a biocenotic crisis has an upward effect of increasing speciation rates by simultaneously fragmenting metapopulations into isolated avatars (and, consequently, isolated demes) via patch extinction and by affecting their persistence and differentiation (Allmon 1992; McKinney and Allmon 1995).

Such high-scope biocenosis-level disturbances affect multiple lineages simultaneously, irrespective of their phylogenetic relationships (Fig. 9b). As documented by the fossil record, the empirical pattern of cross-lineage, synchronous, species-level transitions is referred to as a "turnover pulse" (Foote 2005; van Dam et al. 2006; Vrba 1985, 1993). A turnover pulse is a relatively rapid intensive period of taxonomic and ecological restructuring: A breakdown of ecological networks results in unchecked fluctuations in population demographics, largely driven by stochastic processes as the avatars, the former members of the collapsed biocenosis, opportunistically reconfigure their economic connections and spatial relationship. The factors affecting the taxonomic composition and the ecological configuration of the future biocenosis are influenced by a large number of factors, including phenotypic properties, behavior, stenotopic-eurytopic characteristics, dispersal ability, and population size. Controlled environmental perturbations in the laboratory have demonstrated a variety of context-dependent population responses that differentially affect population size and phenotypic variance, and might have nonlinear positive density-dependent effects in a variable environment (Benton et al. 2004).

Turnover pulses are the main engine that shapes the taxic patterns and trends of diversity and disparity. Even though it might be difficult or impossible to predict a specific species-level response due to stochasticity and nonlinear dynamics precipitated by perturbation, it generally falls into one of the following categories: extinction, speciation, or migration (habitat tracking) (Eldredge 1995; Miller 2002). In addition, substantial destabilization of biocenotic organization allows for successful invasion by non-native species that become important contributors to the emerging new biocenosis and also can induce a rapid morphological evolution in native species (Freeman and Byers 2006). The differential response to population-level perturbation agrees with the empirically derived correlation of taxic origination and extinction rates among clades in the fossil record (Stanley 1979, 1990): More geographically restricted, especially endemic, species are expected to speciate and become extinct at a higher rate than more geographically widespread species.

Extremely high levels of biocenotic perturbation may involve the entire global biota resulting in mass extinction events, where entire multiple lineages become extinct as a result of high-magnitude environmental factors (e.g., volcanism or meteor impact), such as the global Permian–Triassic (P–Tr) extinction, during which speciation was greatly depressed (Erwin 1993).

6 Epilogue

6.1 Summary

Incorporating insights from the hierarchy theory of evolution and from network theory provides a more complete theoretical framework for explaining complex patterns and processes of biological evolution. More specifically, network theory provides a description of interaction dynamics at all scales of biological organization and sheds light upon mechanisms of emergence in a hierarchical context. The proposed model of hierarchically nested networks of biological individuals offers promise for elucidating causal factors of metastability, displayed at all levels of biological hierarchies, and the factors responsible for its breakdown. The perpetual interplay of stasis and flux in network dynamics at all levels of life's hierarchies shapes its historical patterns and trends.

The prevalent pattern of stability in living systems across scale results from the structure of biological systems: Consisting of hierarchically nested, complex networks, they are extremely robust to extrinsic perturbations. The taxic patterns of diversity and disparity are the product of the synergetic outcome of complex network responses to partially temporally decoupled perturbations at different levels of the economic hierarchy and their cascading effects. Environmental perturbations contribute to both generating variation at the molecular level and facilitate the expression of hidden phenotypic variance by compromising evolutionary capacitance. The spread and fixation of novel genotypes, ultimately responsible for producing taxic evolutionary patterns, are enabled by temporary removal of control over population dynamics brought about by environmentally triggered disruption or destabilization of ecological networks at the level of biocenosis. Stochasticity and nonlinear dynamics characterizing the processes of the biocenosis in flux are among the principal factors behind the evolutionary contingency of life's history.

6.2 Future Perspectives

Defining a formal mathematical description of the theory developed here will facilitate quantitative empirical research and modeling of complex evolutionary phenomena in a hypothesis-testing framework. The mathematical approaches from the fields of graph theory, dynamical systems, fractal geometry, and chaos have been shown to be particularly effective in capturing important aspects of complex biological systems (e.g., Strogatz 2000). Facilitated by advances in computational technology, these methods have been fruitfully applied to studying different complex biological phenomena ranging from explaining allometry of plant vascular systems (West et al. 1999), causality in complex ecological dynamics (Sugihara et al. 2012), recovery following mass extinction (Solé et al. 2010), evolutionary responses of paleocommunities to ecological crisis (Roopnarine 2006, 2009), and the origins of life (Kauffman 1995). Mathematical description of hierarchical dynamics useful for modeling complex phenomena can be facilitated by object-oriented technology (Booch 2007). A few models have incorporated an explicitly hierarchical framework as an integrated modeling approach attempting to combine multiple-level-specific models into a global hierarchical model (Pavé 2006; Schmidt-Lainé and Pavé 2002). The application of these methods to evolutionary complexity is still in infancy and presents a wide field of opportunity for theoretical and methodological developments.

References

Abouheif E, Wray GA (2002) Evolution of the gene network underlying wing polyphenism in ants. Science 297:249–252

Albert R (2005) Scale-free networks in cell biology. J Cell Sci 118:4947–4957

Albert R, Barabási A-L (2002) Statistical mechanics of complex networks. Rev Mod Phys 74:47–97

Albert R, Jeong H, Barabási A-L (2000) Error and attack tolerance of complex networks. Nature 406:378–381
Albert R, Othmer HG (2003) The topology of the regulatory interactions predicts the expression pattern of the segment polarity genes in Drosophila melanogaster. J Theor Biol 223:1–18
Allen TFH (2008) Hierarchy theory in ecology. Encyclopedia of Ecology. Elsevier, Oxford, pp 1852–1857
Allen TFH, Starr TB (1982) Hierarchy: perspectives for ecological complexity. University of Chicago Press, Chicago
Allesina S, Bodini A (2004) Who dominates whom in the ecosystem? Energy flow bottlenecks and cascading extinctions. J Theor Biol 230:351–358
Allmon WD (1992) A causal analysis of stages in allopatric speciation. Oxf Surv Evol Biol 8:219–257
Almaas E (2007) Biological impacts and context of network theory. J Exp Biol 210:1548–1558
Alon U (2003) Biological networks: the tinkerer as an engineer. Science 301:1866–1867
Alroy J (2000) Understanding the dynamics of trends within evolving lineages. Paleobiology 26:319–329
Álvarez-Buylla ER, Chaos Á, Aldana M et al (2008) Floral morphogenesis: stochastic explorations of a gene network epigenetic landscape. PLoS ONE 3:e3626
Amaral LAN, Scala A, Barthélémy M, Stanley HE (2000) Classes of small-world networks. Proc Natl Acad Sci USA 97:11149–11152
Anderson PE, Jensen HJ (2005) Network properties, species abundance and evolution in a model of evolutionary ecology. J Theor Biol 232:551–558
Barabási A-L (2002) Linked: the new science of networks. Perseus, Cambridge, MA
Barabási A-L, Albert R (1999) Emergence of scaling in random networks. Science 286:509–512
Barabási A-L, Bonabeau E (2003) Scale-free networks. Sci Am 288:50–59
Barabasi A-L, Albert R, Jeong H (1999) Mean-field theory for scale-free random networks. Phys A 272:173–187
Barabási A-L, Oltvai ZN (2004) Network biology: understanding the cell's functional organization. Nat Rev Genet 5:101–114
Barthélémy M, Amaral LAN (1999) Small-world networks: evidence for a crossover picture. Phys Rev Lett 82:3180–3183
Beklemishev VN (1970) Biotsenologicheskie osnovy sravnitelnoj parazitologii. Nauka, Moskva
Benton TG, Cameron TC, Grant A (2004) Population responses to perturbations: predictions and responses from laboratory mite populations. J Anim Ecol 73:983–995
Berg J, Lässig M, Wagner A (2004) Structure and evolution of protein interaction networks: a statistical model for link dynamics and gene duplications. BMC Evol Biol 4:51
Bergman A, Feldman MW (2003) On the population genetics of punctuations. In: Crutchfield JP, Schuster P (eds) Evolutionary dynamics–exploring the interplay of selection, accident, neutrality, and function. Oxford University Press, Oxford, pp 81–100
Bergman A, Siegal ML (2003) Evolutionary capacitance as a general feature of complex gene networks. Nature 424:549–552
Bonelli JR, Brett CE, Miller AI, Bennington JB (2006) Testing for faunal stability across a regional biotic transition: quantifying stasis and variation among recurring coral-rich biofacies in the Middle Devonian Appalachian Basin. Paleobiology 32:20–37
Booch G (2007) Object-oriented analysis and design with applications. Addison-Wesley, Upper Saddle River, NJ
Boucot AJ (1975) Evolution and extinction rate controls. Elsevier, Amsterdam
Boucot AJ (1978) Community evolution and rates of cladogenesis. Evol Biol 11:545–655
Bray D (2003) Molecular networks: the top-down view. Science 301:1864–1865
Brett CE, Baird GC (1995) Coordinated stasis and evolutionary ecology of Silurian to Middle Devonian Faunas in the Appalachian Basin. In: Erwin DH, Anstey R (eds) New approaches to speciation in the fossil record. Columbia University Press, New York, pp 285–315
Brett CE, Ivany LC, Schopf KM (1996) Coordinated stasis: an overview. Palaeogeogr Palaeoclimatol Palaeoecol 127:1–20

Brogaard B (2004) Species as individuals. Biol Philos 19:223–242
Bunge M (1979) Treatise on basic philosophy. A world of systems, vol 4. D. Reidel, Dordrecht
Buss LW (1983) Evolution, development, and the units of selection. Proc Natl Acad Sci USA 80:1387–1391
Camacho J, Guimerà R, Amaral LAN (2002) Robust patterns in food web structure. Phys Rev Lett 88:228102
Casás-Selves M, DeGregori J (2011) How cancer shapes evolution and how evolution shapes cancer. Evo Edu Outreach 4:624–634
Case TJ (1990) Invasion resistance arises in strongly interacting species-rich model competition communities. Proc Natl Acad Sci USA 87:9610–9614
Cattin M-F, Bersier L-F, Banašek-Richter C et al (2004) Phylogenetic constraints and adaptation explain food-web structure. Nature 427:835–839
Chernyshenko SV (2012) Coenome model: elementary ecological cycle as a dynamical unit. In: Troitzsch KG, Möhring M, Lotzmann U (eds) 26th European conference on modelling and simulation, ECMS 2012, Koblenz, Germany, 29 May–1 June 2012. European Council for Modeling and Simulation, pp 143–149
Clements FE (1916) Plant succession: an analysis of the development of vegetation. Carnegie Institution of Washington, Washington
Cornette JL, Lieberman BS (2004) Random walks in the history of life. Proc Natl Acad Sci USA 101:187–191
Cossins A (1998) Cryptic clues revealed. Nature 396:309–310
Cracraft J (1984) Conceptual and methodological aspects of the study of evolutionary rates, with some comments on bradytely in birds. In: Eldredge N, Stanley SM (eds) Living fossils. Springer, New York, pp 95–104
Damuth J (1985) Selection among "species": a formulation in terms of natural functional units. Evolution 39:1132–1146
Darwin C (1859) On the origin of species by means of natural selection, or the preservation of favoured races in the struggle for life. J. Murray, London
Davidson EH (2001) Genomic regulatory systems: development and evolution. Academic Press, San Diego
Davidson EH, Erwin DH (2006) Gene regulatory networks and the evolution of animal body plans. Science 311:796–800
Dawkins R (1976) The selfish gene. Oxford University Press, New York
Dawkins R (1982) Replicators and vehicles. In: King's College Sociobiology Group eds. (ed) Current problems in sociobiology. Cambridge University Press, Cambridge, pp 45–64
Didion DM (2003) Relevant bounds on hierarchical levels in the description of mechanisms. Hist Philos Life Sci 25:5–25
Dobzhansky TG (1937) Genetics and the origin of species. Columbia University Press, New York
Dorogovtsev SN, Mendes JFF (2002) Evolution of networks. Adv Phys 51:1079–1187
Dorogovtsev SN, Mendes JFF (2003) Evolution of networks: from biological nets to the Internet and WWW. Oxford University Press Inc., New York
Duffy JE (2002) Biodiversity and ecosystem function: the consumer connection. Oikos 99:201–219
Dunne JA, Williams RJ, Martinez ND (2002a) Food-web structure and network theory: the role of connectance and size. Proc Natl Acad Sci USA 99:12917–12922
Dunne JA, Williams RJ, Martinez ND (2002b) Network structure and biodiversity loss in food webs: robustness increases with connectance. Ecol Lett 5:558–567
Dyer RJ, Nason JD (2004) Population graphs: the graph theoretic shape of genetic structure. Mol Ecol 13:1713–1727
Eisenberg E, Levanon EY (2003) Preferential attachment in the protein network evolution. Phys Rev Lett 91:138701
Eldredge N (1985a) Unfinished synthesis: biological hierarchies and modern evolutionary thought. Oxford University Press, New York

Eldredge N (1985b) Time frames: the rethinking of Darwinian evolution and the theory of punctuated equilibria. Simon and Schuster, New York
Eldredge N (1989) Macroevolutionary dynamics: species, niches, and adaptive peaks. McGraw-Hill, New York
Eldredge N (1995) Reinventing Darwin: the great debate at the high table of evolutionary theory. Wiley, New York
Eldredge N (2003) The sloshing bucket: how the physical realm controls evolution. In: Crutchfield JP, Schuster P (eds) Evolutionary dynamics–exploring the interplay of selection, accident, neutrality, and function. Oxford University Press, Oxford, pp 3–32
Eldredge N, Gould SJ (1972) Punctuated equilibria: an alternative to phyletic gradualism. In: Schopf TJM (ed) Models in paleobiology. Freeman Cooper, San Francisco, pp 82–115
Eldredge N, Salthe SN (1984) Hierarchy and evolution. Oxf Surv Evol Biol 1:184–208
Eldredge N, Thompson JD, Brakefield PM et al (2005) The dynamics of evolutionary stasis. Paleobiology 31:133–145
Elena SF, Wilke CO, Ofria C, Lenski RE (2007) Effects of population size and mutation rate on the evolution of mutational robustness. Evolution 61:666–674
Ereshefsky M (2001) The poverty of the Linnaean hierarchy: a philosophical study of biological taxonomy. Cambridge University Press, Cambridge
Erwin DH (1993) The great Paleozoic crisis: life and death in the Permian. Columbia University Press, New York
Fewell JH (2003) Social insect networks. Science 301:1867–1870
Fisher RA (1958) The genetical theory of natural selection, 2nd edn. Dover, New York
Foote M (2005) Pulsed origination and extinction in the marine realm. Paleobiology 31:6–20
Freeman AS, Byers JE (2006) Divergent induced responses to an invasive predator in marine mussel populations. Science 313:831–833
Futuyma DJ (1989) Speciational trends and the role of species in macroevolution. Am Nat 134:318–321
Gams H (1918) Prinzipienfragen der Vegetationsforschung: ein Beitrag zur Begriffsklärung und Methodik der Biocoenologie. Vierteljahresschr Naturf Gesellsch Zürich 63:293–493
Ghiselin MT (1974) A radical solution to the species problem. Syst Zool 23:536–544
Ghiselin MT (1981) Categories, life, and thinking. Behav Brain Sci 4:269–283
Gibson G, Wagner G (2000) Canalization in evolutionary genetics: a stabilizing theory? BioEssays 22:372–380
Gilbert W (1986) Origin of life: the RNA world. Nature 319:618
Gilmour JSL, Gregor JW (1939) Demes: a suggested new terminology. Nature 144:333–334
Giot L, Bader JS, Brouwer C et al (2003) A protein interaction map of *Drosophila melanogaster*. Science 302:1727–1736
Gould SJ (2002) The structure of evolutionary theory. The Belknap Press of Harvard University Press, Cambridge
Gould SJ, Eldredge N (1977) Punctuated equilibria: the tempo and mode of evolution reconsidered. Paleobiology 3:115–151
Gould SJ, Raup DM, Sepkoski JJ et al (1977) The shape of evolution; a comparison of real and random clades. Paleobiology 3:23–40
Gould SJ, Gilinsky NL, German RZ (1987) Asymmetry of lineages and the direction of evolutionary time. Science 236:1437–1441
Grant PR, Grant BR (2011) How and why species multiply: the radiation of Darwin's finches. Princeton University Press, Princeton
Grantham TA (1995) Hierarchical approaches to macroevolution: recent work on species selection and the "Effect hypothesis". Annu Rev Ecol Syst 26:301–321
Grantham TA (2001) Hierarchies in evolution. In: Briggs DEG, Crowther PR (eds) Palaeobiology II. Blackwell, Oxford, pp 188–192
Grantham T (2007) Is macroevolution more than successive rounds of microevolution? Palaeontology 50:75–85

Gregory TR (2004) Macroevolution, hierarchy theory, and the C-value enigma. Paleobiology 30:179–202
Gregory TR, Hebert PDN (1999) The modulation of DNA content: proximate causes and ultimate consequences. Genome Res 9:317–324
Grene M (1987) Hierarchies in biology. Am Sci 75:504–510
Hartwell LH, Hopfield JJ, Leibler S, Murray AW (1999) From molecular to modular cell biology. Nature 402:C47–C52
Hamilton WD (1964a) The genetical evolution of social behaviour. I. J Theor Biol 7:1–16
Hamilton WD (1964b) The genetical evolution of social behaviour. II. J Theor Biol 7:17–52
Hanski I, Gilpin M (1991) Metapopulation dynamics: brief history and conceptual domain. Biol J Linn Soc Lond 42:3–16
Hathaway RF (1969) Hierarchy and the definition of order in the letters of Pseudo-Dionysius: a study in the form and meaning of the Pseudo-Dionysian writings. Nijhoff, The Hague
Hennig W (1950) Grundzüge einer Theorie der phylogenetischen Systematik. Deutscher Zentralverlag, Berlin
Hennig W (1966) Phylogenetic systematics. University of Illinois Press, Urbana
Hermisson J, Wagner GP (2004) The population genetic theory of hidden variation and genetic robustness. Genetics 168:2271–2284
Hintze A, Adami C (2008) Evolution of complex modular biological networks. PLoS Comput Biol 4:e23
Howard ML, Davidson EH (2004) *cis*-Regulatory control circuits in development. Dev Biol 271:109–118
Hull D (1976) Are species really individuals? Syst Zool 25:174–191
Hull DL (1978) A matter of individuality. Philos Sci 45:335–360
Hull D (1980) Individuality and selection. Annu Rev Ecol Syst 11:311–322
Hull DL (1981) Units of evolution: a metaphysical essay. In: Jensen UJ, Harré R (eds) The philosophy of evolution. Harvester, London, pp 23–44
Hull DL (1988) Interactors versus vehicles. In: Plotkin HC (ed) The role of behavior in evolution. MIT Press, Cambridge, MA, pp 19–50
Hunt G (2008) Gradual or pulsed evolution: when should punctuational explanations be preferred? Paleobiology 34:360–377
Ingolia NT (2004) Topology and robustness in the *Drosophila* segment polarity network. PLoS Biol 2:e123
Jablonka E, Lamb MJ (1995) Epigenetic inheritance and evolution: the Lamarckian dimension. Oxford University Press, Oxford
Jablonka E, Lamb MJ (1998) Epigenetic inheritance in evolution. J Evol Biol 11:159–183
Jablonka E, Lamb MJ (2005) Evolution in four dimensions: genetic, epigenetic, behavioral, and symbolic variation in the history of life. MIT Press, Cambridge
Jablonski D (2008) Species selection: theory and data. Annu Rev Ecol Syst 39:501–524
Jacob F (1974) The logic of living systems: a history of heredity. Allen Lane, London
Jeong H, Tombor B, Albert R et al (2000) The large-scale organization of metabolic networks. Nature 407:651–654
Jeong H, Mason SP, Barabási A-L, Oltvai ZN (2001) Lethality and centrality in protein networks. Nature 411:41–42
Jordano P, Bascompte J, Olesen JM (2003) Invariant properties in coevolutionary networks of plant–animal interactions. Ecol Lett 6:69–81
Kauffman SA (1995) At home in the universe: the search for laws of self-organization and complexity. Oxford University Press, New York
Kimura M (1968) Evolutionary rate at the molecular level. Nature 217:624–626
Kimura M (1983) The neutral theory of molecular evolution. Cambridge University Press, Cambridge
Kingsolver JG, Pfennig DW (2004) Individual-level selection as a cause of Cope's rule of phyletic size increase. Evolution 58:1608–1612

Koestler A (1967) The ghost in the machine. Random House, New York
Koestler A (1978) Janus: a summing up. Random House, New York
Konhauser KO, Pecoits E, Lalonde SV et al (2009) Oceanic nickel depletion and a methanogen famine before the Great Oxidation Event. Nature 458:750–753
Kozlov AP (2014) Evolution by tumor neofunctionalization: the role of tumors in the origin of new cell types, tissues and organs. Academic Press, Oxford
Lane D (2006) Hierarchy, complexity, society. In: Pumain D (ed) Hierarchy in natural and social sciences. Springer, Dordrecht, pp 81–119
Lessios HA (2008) The Great American Schism: divergence of marine organisms after the rise of the Central American Isthmus. Annu Rev Ecol Evol Syst 39:63–91
Levins R (1969) Some demographic and genetic consequences of environmental heterogeneity for biological control. Bull Entomol Soc Am 15:237–240
Levins R (1970) Complex systems. In: Waddington CH (ed) Towards a theoretical biology. 3. Drafts. Aldine, Chicago, pp 73–88
Levinton JS, Bambach RK (1975) A comparative study of Silurian and recent deposit-feeding bivalve communities. Paleobiology 1:97–124
Lewontin RC (1970) The units of selection. Annu Rev Ecol Syst 1:1–18
Lieberman BS, Vrba ES (1995) Hierarchy theory, selection, and sorting. Bioscience 45:394–399
Lovelock JE (1972) Gaia as seen through the atmosphere. Atmos Environ 6:579–580
Lovelock JE (1979) Gaia: a new look at life on Earth. Oxford University Press, Oxford
Lovelock JE, Margulis L (1974) Atmospheric homeostasis by and for the biosphere: the Gaia hypothesis. Tellus 26:2–10
MacArthur R (1955) Fluctuations of animal populations and a measure of community stability. Ecology 36:533–536
Mandelbrot BB (1977) Fractals: form, chance, and dimension. W. H. Freeman, San Francisco
Mandelbrot BB (1982) The fractal geometry of nature. W. H. Freeman, New York
Masing V (1981) Consortia as elements of the functional structure of biocenoses. In: Laasimer L (ed) Anthropogenous changes in the plant cover of Estonia. Academy of Sciences of the Estonian SSR, Institute of Zoology and Botany, Tartu, pp 64–76
Mattila TM, Bokma F (2008) Extant mammal body masses suggest punctuated equilibrium. Proc R Soc Lond B Biol Sci 275:2195–2199
Maynard Smith J (1964) Group selection and kin selection. Nature 201:1145–1147
Maynard Smith J (1978) The evolution of sex. Cambridge University Press, Cambridge
Maynard Smith J, Szathmáry E (1995) The major transitions in evolution. W. H. Freeman, Oxford
McGhee GR (1998) Theoretical morphology: the concept and its applications. Columbia University Press, New York
McGhee GR (2007) The geometry of evolution: adaptive landscapes and theoretical morphospaces. Cambridge University Press, Cambridge
McKinney ML (1990a) Classifying and analyzing evolutionary trends. In: McNamara KJ (ed) evolutionary trends. Belhaven Press, London, pp 28–58
McKinney ML (1990b) Trends in body-size evolution. In: McNamara KJ (ed) Evolutionary trends. Belhaven Press, London, pp 75–118
McKinney ML, Allmon WD (1995) Metapopulations and disturbance: from patch dynamics to biodiversity dynamics. New approaches to speciation in the fossil record. Columbia University Press, New York, pp 123–183
McNamara KJ (1990) Evolutionary trends. Belhaven Press, London
McNamara KJ (2006) Evolutionary trends. In: eLS, Wiley. doi:10.1038/npg.els.0004136
McShea DW (1994) Mechanisms of large-scale evolutionary trends. Evolution 48:1747–1763
McShea DW, Brandon RN (2010) Biology's first law: the tendency for diversity and complexity to increase in evolutionary systems. University of Chicago Press, Chicago
Meiklejohn CD, Hartl DL (2002) A single mode of canalization. Trends Ecol Evol 17:468–473

Melott AL, Lieberman BS, Laird CM et al (2004) Did a gamma-ray burst initiate the late Ordovician mass extinction? Int J Astrobiol 3:55–61
Michod RE (1983) Population biology of the first replicators: on the origin of the genotype, phenotype and organism. Am Zool 23:5–14
Milgram S (1967) The small world problem. Psychol Today 1:61–67
Miller W III (2001) What's in a name? Ecologic entities and the marine paleoecologic record. In: Allmon WD, Bottjer DJ (eds) Evolutionary paleoecology: the ecological context of macroevolutionary change. Columbia University Press, New York, pp 15–33
Miller W III (2002) Regional ecosystems and the origin of species. Neues Jahrb Geol Palaontol Abh 225:137–156
Miller W III (2006) What every paleontologist should know about species: new concepts and questions. Neues Jahrb Geol Palaontol Abh 2006(9):557–576
Milo R, Shen-Orr S, Itzkovitz S et al (2002) Network motifs: simple building blocks of complex networks. Science 298:824–827
Möbius KA (1877) Die auster und die Austernwirtschaft. Verlag von Wiegandt: Hemple and Parey, Berlin
Montoya JM, Solé RV (2002) Small world patterns in food webs. J Theor Biol 214:405–412
Montoya JM, Solé RV (2003) Topological properties of food webs: from real data to community assembly models. Oikos 102:614–622
Newell ND (1952) Periodicity in invertebrate evolution. J Paleontol 26:371–385
Newell ND (1967) Revolutions in the history of life. In: Albritton CC Jr (ed) Uniformity and simplicity: a symposium on the principle of the uniformity of nature, Nov 1963. GSA Special Papers, No. 89. Geological Society of America, Boulder, CO, pp 63–89
Newman MEJ (2003) The structure and function of complex networks. SIAM Rev 45:167–256
Okasha S (2006) Evolution and the levels of selection. Clarendon Press, Oxford
Okasha S (2011) Biological ontology and hierarchical organization: a defense of rank freedom. In: Calcott B, Sterelny K (eds) The major transitions in evolution revisited. MIT Press, Cambridge, pp 53–64
Olesen JM, Jordano P (2002) Geographic patterns in plant-pollinator mutualistic networks. Ecology 83:2416–2424
Paterson HEH (1985) The recognition concept of species. Species and speciation. Transvaal Museum, Pretoria, pp 21–29
Pattee HH (1970) The problem of biological hierarchy. In: Waddington CH (ed) Towards a theoretical biology 3, drafts. Edinburgh University Press, Edinburgh, pp 117–136
Pattee HH (1973) Hierarchy theory: the challenge of complex systems. G. Braziller, New York
Pattee HH (1977) Dynamic and linguistic modes of complex systems. Int J Gen Syst 3:259–266
Pavé A (2006) Biological and ecological systems hierarchical organisation. In: Pumain D (ed) Hierarchy in natural and social sciences. Springer, Dordrecht, pp 39–70
Pimm SL (1980) Food web design and the effect of species deletion. Oikos 35:139
Pimm SL (1991) The balance of nature. University of Chicago Press, London
Polanyi M (1968) Life's irreducible structure: live mechanisms and information in DNA are boundary conditions with a sequence of boundaries above them. Science 160:1308–1312
Proulx SR, Promislow DEL, Phillips PC (2005) Network thinking in ecology and evolution. Trends Ecol Evol 20:345–353
Pulliam HR (1988) Sources, sinks, and population regulation. Am Nat 132:652–661
Raup DM (1981) Extinction: bad genes or bad luck? Acta Geol Hisp 16:25–33
Raup DM (1987) Major features of the fossil record and their implications for evolutionary rate studies. In: Campbell KSW, Day MF (eds) Rates of evolution. Allen and Unwin, London, pp 1–14
Raup DM, Gould SJ (1974) Stochastic simulation and evolution of morphology-towards a nomothetic paleontology. Syst Biol 23:305–322
Raup D, Gould S, Schopf T, Simberloff D (1973) Stochastic models of phylogeny and the evolution of diversity. J Geol 81:525–542

Ravasz E, Somera AL, Mongru DA et al (2002) Hierarchical organization of modularity in metabolic networks. Science 297:1551–1555
Reinke J (1872) Ueber die anatomischen Verhältnisse einiger Arten von Gunnera L. Nachrichten von der Königl Gesellschaft der Wissenschaften und der Georg- Augusts-Universität zu Göttingen 9:100–108
Rohlf FJ, Marcus LF (1993) A revolution in morphometrics. Trends Ecol Evol 8:129–132
Roopnarine PD (2003) Analysis of rates of morphologic evolution. Annu Rev Ecol Syst 34:605–632
Roopnarine PD (2006) Extinction cascades and catastrophe in ancient food webs. Paleobiology 32:1–19
Roopnarine PD (2009) Ecological modeling of paleocommunity food webs. In: Dietl GP, Flessa KW (eds) Conservation paleobiology: using the past to manage for the future. Paleontological Society, [S.l.], Boulder, CO, pp 195–220
Roopnarine PD, Byars G, Fitzgerald P (1999) Anagenetic evolution, stratophenetic patterns, and random walk models. Paleobiology 25:41–57
Root RB (1967) The niche exploitation pattern of the blue–gray gnatcatcher. Ecol Monogr 37:317–350
Rutherford SL, Lindquist S (1998) Hsp90 as a capacitor for morphological evolution. Nature 396:336–342
Salthe SN (1985) Evolving hierarchical systems: their structure and representation. Columbia University Press, New York
Scharloo W (1991) Canalization: genetic and developmental aspects. Annu Rev Ecol Syst 22:65–93
Schmidt-Lainé C, Pavé A (2002) Environnement: modélisation et modèles pour comprendre, agir et décider dans un contexte interdisciplinaire. Natures Sci Société 10:5–25
Schmidt-Kittler N, Vogel K (1991) Constructional morphology and evolution. Springer, Berlin
Schoch RM (1986) Phylogeny reconstruction in paleontology. Van Nostrand Reinhold, New York
Seilacher A (1970) Arbeitskonzept zur Konstruktions-Morphologie. Lethaia 3:393–396
Sepkoski JJ (1978) A kinetic model of Phanerozoic taxonomic diversity I. Analysis of marine orders. Paleobiology 4:223–251
Sepkoski JJ (1993) Ten years in the library; new data confirm paleontological patterns. Paleobiology 19:43–51
Shaposhnikov GK (1965) Morpfologicheskaya divergentsiya i konvergentsiya v eksperimente s tlyami (Homoptera, Aphidinea). Entomol Obozr 44:3–25
Shaposhnikov GK (1978) Dinamika klonov, populyatsij i vidov i evolyutsiya. Zh Obshch Biol 39:15–33
Siegal ML, Bergman A (2002) Waddington's canalization revisited: developmental stability and evolution. Proc Natl Acad Sci USA 99:10528–10532
Siegal ML, Promislow DEL, Bergman A (2007) Functional and evolutionary inference in gene networks: does topology matter? Genetica 129:83–103
Simberloff D, Dayan T (1991) The guild concept and the structure of ecological communities. Annu Rev Ecol Syst 22:115–143
Simon HA (1955) On a class of skew distribution functions. Biometrika 42:425–440
Simon HA (1962) The architecture of complexity. Proc Am Philos Soc 106:467–482
Simon HA (1973) The organization of complex systems. In: Pattee HH (ed) Hierarchy theory: the challenge of complex systems. G. Braziller, New York, pp 1–27
Solé RV, Montoya JM (2001) Complexity and fragility in ecological networks. Proc R Soc Lond B Biol Sci 268:2039–2045
Solé RV, Saldaña J, Montoya JM, Erwin DH (2010) Simple model of recovery dynamics after mass extinction. J Theor Biol 267:193–200
Stanley SM (1975) A theory of evolution above the species level. Proc Natl Acad Sci USA 72:646–650
Stanley SM (1979) Macroevolution: pattern and process. W. H. Freeman, San Francisco

Stanley SM (1990) Delayed recovery and the spacing of major extinctions. Paleobiology 16:401–414
Stanley SM, Signor PW, Lidgard S, Karr AF (1981) Natural clades differ from "random" clades: simulations and analyses. Paleobiology 7:115–127
Stearns SC, Kaiser M, Kawecki TJ (1995) The differential genetic and environmental canalization of fitness components in Drosophila melanogaster. J Evol Biol 8:539–557
Stenseth NC, Maynard Smith J (1984) Coevolution in ecosystems: Red Queen evolution or stasis? Evolution 38:870–880
Strogatz SH (2000) Nonlinear dynamics and chaos: with application to physics, biology, chemistry, and engineering. Westview press, Cambridge
Strogatz SH (2001) Exploring complex networks. Nature 410:268–276
Strotz LC, Allen AP (2013) Assessing the role of cladogenesis in macroevolution by integrating fossil and molecular evidence. Proc Natl Acad Sci USA 110:2904–2909
Sugihara G, May R, Ye H et al (2012) Detecting causality in complex ecosystems. Science 338:496–500
Szathmáry E (1989) The emergence, maintenance, and transitions of the earliest evolutionary units. Oxf Surv Evol Biol 6:169–205
Turner M, Romme W, Gardner R et al (1993) A revised concept of landscape equilibrium: disturbance and stability on scaled landscapes. Landsc Ecol 8:213–227
Thorson G (1957) Bottom communities (sublittoral or shallow shelf). In: Hedgpeth JW (ed) Treatise on marine ecology and paleoecology. 1. Ecology, vol 67. The Geological Society of America Memoirs, Boulder, CO, pp 461–534
Urban DL, O'Neill RV, Shugart HH (1987) Landscape ecology. Bioscience 37:119–127
Valentine JW, May CL (1996) Hierarchies in biology and paleontology. Paleobiology 22:23–33
van Dam JA, Abdul Aziz H, Álvarez Sierra MÁ et al (2006) Long-period astronomical forcing of mammal turnover. Nature 443:687–691
Vernadsky VI (1998) The biosphere. Copernicus, New York
Villarreal LP (2005) Viruses and the evolution of life. ASM Press, Washington, D.C.
Villarreal C, Padilla-Longoria P, Alvarez-Buylla ER (2012) General theory of genotype to phenotype mapping: derivation of epigenetic landscapes from N-node complex gene regulatory networks. Phys Rev Lett 109:118102
Volk T (1997) Gaia's body: toward a physiology of Earth. Springer, New York
von Dassow G, Meir E, Munro EM, Odell GM (2000) The segment polarity network is a robust developmental module. Nature 406:188–192
Vrba ES (1980) Evolution, species and fossils: how does life evolve? S Afr J Sci 76:61–84
Vrba ES (1984a) Evolutionary pattern and process in the sister-group Acelaphini-Aepycerotini (Mammalia: Bovidae). In: Eldredge N, Stanley S (eds) Living fossils. Springer, New York, pp 62–79
Vrba ES (1984b) What is species selection? Syst Zool 33:318–328
Vrba ES (1985) Environment and evolution: alternative causes of the temporal distribution of evolutionary events. S Afr J Sci 81:229–236
Vrba ES (1987) Ecology in relation to speciation rates: some case histories of Miocene-Recent mammal clades. Evol Ecol 1:283–300
Vrba ES (1989) Levels of selection and sorting with special reference to the species level. Oxf Surv Evol Biol 6:111–168
Vrba ES (1993) Turnover-pulses, the Red Queen, and related topics. Am J Sci 293(A):418–452
Vrba ES, Eldredge N (1984) Individuals, hierarchies and processes: toward a more complete evolutionary theory. Paleobiology 10:146–171
Vrba ES, Gould SJ (1986) The hierarchical expansion of sorting and selection: sorting and selection cannot be equated. Paleobiology 12:217–228
Waddington CH (1942) Canalization of development and the Inheritance of acquired characters. Nature 150:563–565
Wagner A (1996) Does evolutionary plasticity evolve? Evolution 50:1008–1023

Wagner A (2002) Estimating coarse gene network structure from large-scale gene perturbation data. Genome Res 12:309–315
Wagner A (2003) How the global structure of protein interaction networks evolves. Proc Biol Sci 270:457–466
Wagner A, Fell DA (2001) The small world inside large metabolic networks. Proc R Soc Lond B 268:1803–1810
Walker KR, Laporte LF (1970) Congruent fossil communities from Ordovician and Devonian carbonates of New York. J Paleontol 44:928–944
Wallace P (1978) Homeomorphy between Devonian brachiopod communities in France and Iowa. Lethaia 11:259–272
Watts DJ, Strogatz SH (1998) Collective dynamics of "small-world" networks. Nature 393:440–442
Weiss PA (1973) The science of life: the living system—a system for living. Futura, Mount Kisco
West GB, Brown JH, Enquist BJ (1999) A general model for the structure and allometry of plant vascular systems. Nature 400:664–667
Whyte LL (1969) Structural hierarchies: a challenging class of physical and biological problems. In: Whyte LL, Wilson AG, Wilson D (eds) Hierarchical structures. Elsevier, New York, pp 3–17
Williams RJ, Berlow EL, Dunne JA et al (2002) Two degrees of separation in complex food webs. Proc Natl Acad Sci USA 99:12913–12916
Wimsatt WC (1976) Reductionism, levels of organization, and the mind-body problem. In: Globus GG, Maxwell G, Savodnik I (eds) Consciousness and the brain: a scientific and philosophical inquiry. Plenum Press, New York, pp 205–267
Woodger JH (1929) Biological principles: a critical study. Kegan Paul, Trench, Trubner & Co., London
Wright S (1929) The evolution of dominance. Am Nat 63:556–561
Wright S (1931) Evolution in Mendelian populations. Genetics 16:97–159
Wright S (1953) Gene and organism. Am Nat 87:5–18
Wright S (1955) Classification of the factors of evolution. Cold Spring Harb Symp Quant Biol 20:16–24D
Wright S (1980) Genic and organismic selection. Evolution 34:825
Wright S (1982) The shifting balance theory and macroevolution. Annu Rev Genet 16:1–20
Wu J (1999) Hierarchy and scaling: extrapolating information along a scaling ladder. Can J Remote Sens 25:367–380
Wu J, Loucks OL (1995) From balance of nature to hierarchical patch dynamics: a paradigm shift in ecology. Q Rev Biol 70:439–466
Wu J, Levin SA (1994) A spatial patch dynamic modeling approach to pattern and process in an annual grassland. Ecol Monogr 64:447–464
Zherikhin VV (1987) Biotsenoticheskaya regulyatsiya evolyutsii. Paleontol J 1:3–12
Zhu X, Gerstein M, Snyder M (2007) Getting connected: analysis and principles of biological networks. Genes Dev 21:1010–1024

Uniting Micro- with Macroevolution into an Extended Synthesis: Reintegrating Life's Natural History into Evolution Studies

Nathalie Gontier

Abstract The Modern Synthesis explains the evolution of life at a mesolevel by identifying phenotype–environmental interactions as the locus of evolution and by identifying natural selection as the means by which evolution occurs. Both micro- and macroevolutionary schools of thought are post-synthetic attempts to evolutionize phenomena above and below organisms that have traditionally been conceived as non-living. Microevolutionary thought associates with the study of how genetic selection explains higher-order phenomena such as speciation and extinction, while macroevolutionary research fields understand species and higher taxa as biological individuals and they attribute evolutionary causation to biotic and abiotic factors that transcend genetic selection. The microreductionist and macroholistic research schools are characterized as two distinct epistemic cultures where the former favor mechanical explanations, while the latter favor historical explanations of the evolutionary process by identifying recurring patterns and trends in the evolution of life. I demonstrate that both cultures endorse radically different notions on time and explain how both perspectives can be unified by endorsing epistemic pluralism.

Keywords Microevolution · Macroevolution · Origin of life · Evolutionary biology · Sociocultural evolution · Natural history · Organicism · Biorealities · Units, levels and mechanisms of evolution · Major transitions · Hierarchy theory

> But how … shall we describe a process which nobody has seen performed, and of which no written history gives any account? This is only to be investigated, first, in examining the nature of those solid bodies, the history of which we want to know; and 2dly, in examining the natural operations of the globe, in order to see if there now actually exist such operations, as, from the nature of the solid bodies, appear to have been necessary to their formation. (Hutton, cited in Teggart 1916: 249)

N. Gontier (✉)
Applied Evolutionary Epistemology Lab, Centre for Philosophy of Science, Faculty of Science, University of Lisbon, Lisbon, Portugal
e-mail: nlgontier@fc.ul.pt

E. Serrelli and N. Gontier (eds.), *Macroevolution*,
Interdisciplinary Evolution Research 2, DOI 10.1007/978-3-319-15045-1_7

> ... We need a third gambit, one that admits history as a fair game to the scientific enterprise – but does so without exclusive attention to the individual facts of the matter. One that focusses on historical pattern. (Eldredge 1999: 10)

1 Introduction

When you look up micro or macro in a dictionary, the "micro" is defined as small or minute and "macro" as large or long. To exemplify the meaning of the words, many dictionaries will give you a lineup that goes from the micro to the meso to the macro. The terms stem from the ancient Greek mīkrós, mésos, and makrós and are loanwords borrowed from more ancient Indo-European (including Indo-Iranian) languages spoken by Vedic Indian and Mesopotamian peoples. Like their predecessors, Greek philosophers used the concepts as prefixes to form combinatorial words whereby the micro–meso–macro determined either the size or the temporal and chronological scale of the phenomenon denoted by the noun that followed in the combination.

Stated otherwise, the micro, meso, and macro refer to verbal, geometric, and arithmetic measurements of matter, space, and time. We still use the micro–meso–macro prefixes. Biochemists, for example, call DNA a macromolecule to indicate that it is a structure composed of smaller subunits; archaeologists divide the Paleolithic or Stone Age into a Lower Mesolithic and Upper Paleolithic period to give chronometric and spatial accounts of hominid tool technologies. Paleontologists use geological time to define visible eukaryotic multicellular life that evolved at the beginning of the Cambrian as transitioning from the Paleozoic or ancient life to the Mesozoic or middle life and the Cenozoic or more recent life.

The conceptualization of matter, space, and time brings us to the heart of cosmology, i.e., the study of the universe. Cosmologies are worldviews that function as epistemic systems that delineate the boundaries of the universe and everything in it. Over the ages, cosmologies have often been depicted in cosmographies such as ancient Greek Chains of Beings, Judeo-Christian Scala Naturae, or Far Eastern Wheels of Time that function as mappings of the world.

Aristotle, for example, distinguished between the realm of non-being (nothing), the physical realm of coming and becoming (the cosmos), and the metaphysical realm of being (roughly the aetiology or causes that underlie all order in the universe which he reified into an ultimate and final cause: the unmovable mover). The micro–meso–macro prefixes were exclusively used to measure the physical realm which these ancient peoples understood as a single cosmic whole that for the sake of analysis is dividable into embedded parts, i.e., the micro-, meso- and macrocosmos. Together, they form the great chain of being and becoming, and that chain of being and becoming, for Aristotle, follows metaphysical, causal laws that give teleological directionality to how change occurs.

Aristotle's chain of being as well as his more ultimate distinction into 3 realms was synthesized into Judeo-Christian cosmologies where the micro–meso–macro come in the form of a lineup of distinct and hierarchically structured layers of reality, respectively, made up of inanimate matter, animate (living) matter, and the celestial/divine. The latter ladder became reconceptualized in more scientific terms by Spencer (1862, 1876, 1915) who, following Hutton, distinguished a linear sequence that goes from the inorganic to the organic and superorganic. The micro or inorganic associates with chemistry and particle physics, the meso or organic with the life sciences which include the humanities as well as the earth sciences, and the macro or superorganic with astronomy and the planetary sciences. Early historical linguists, ethnologists, and sociologists adhered to a doctrine we now call organicism and understood languages, cultures, and societies as beings or living organisms, but Spencer reconceptualized these phenomena as extended organs of living beings that surpass the organic and thus also form part of the superorganic.

In all these hierarchical lineups, the living holds the middle between on the one hand the dead and on the other hand what transcends the living. In religious terms, that what transcends the living coincides with the supernatural and is therefore conceptualized as distinct in matter, space, and time. In scientific terms, what transcends the living can associate either with geology, physics, and astronomy, or, on a smaller scale, with sociocultural, linguistic, and political structures which equally surpass the individual or the collective. In the sciences, the organic has a more fluent connection with some aspects of the superorganic, because living beings can individually and collectively alter aspects of the superorganic, either ecologically speaking by altering the earth's surface and atmosphere, or in what regards the creation of new phenomena such as cultures and languages. However, the living cannot influence or alter star or planetary formations, or at least there is no evidence that proves we can, so eventually there is a rupture. In recent years, science has also made the divide between the inorganic and organic more fluent because we accept that living matter is made up of the same chemical compounds and physical particles. Life and inorganic matter both follow the same chemical and physical laws. Nonetheless, living matter evolves, and the evolution of life is considered to follow new rules which are formulated in the form of evolutionary mechanisms.

What is noteworthy is that none of these lineups converge with the actual history of the universe, which according to the current standard view goes from the Big Bang to star and planetary constellations, the evolution of life out of non-living matter on at least one of these planets, and the evolution of new phenomena out of life. It demonstrates that our ways to conceptualize the universe are outdated and biased toward more ancient cosmologies that no longer form part of our worldview. Huxley (1942, 1957), for example, when defining what he called "evolutionary humanism" tried to fix the problem and went from the cosmic to the biological to the psychosocial, but he was silenced by mainstream scientists, strangely so for being too fanatic.

Another observation drawn from investigating these scales is that life has always held the middle, but its exact position and beginning within a universal cosmic scale has been far from stable. Are populations, species, or higher taxa real? Are cultures, languages, and societal structures alive? These make up fundamental questions of Western science. Biology is traditionally defined as the science of life and evolution as the change that life undergoes through time. But definitions of life are diverse. Darwin (1859) understood biological organisms as individual living beings. He evolutionized the study of the living by measuring how well organisms cope in the struggle for existence and how able they are in reproducing fertile offspring. By emphasizing struggle for existence and reproductive success, organisms are subjected to a two-way selection process, one natural and one sexual. Success is measured by how well organismal variation helps in survival and attracting mates. Both are tested within the environment. Organismal variation can be reduced to hereditary variation, but it is the organism that has to struggle. And selection can be explained by environmental parameters, but it is the organism that is selected, either by other organisms or by the abiotic environment. So there is a duality that takes the issues to a micro- and macrolevel, but only the organism–environmental interface is considered real or relevant for study. Organisms collectively make up populations, but for Darwin, species are real in name only and in reality always "incipient" and prone to undergo further gradual change because of generational organismal–environmental interactions that eventually underlie descent with modification.

Evolutionary theory has therefore been mostly defined at a mesolevel. Following Darwin, the founders of the Modern Synthesis argued that the evolution of existing life occurs through the natural selection of adaptive organisms at the level of the environment as a consequence of the weeding out of maladaptive organisms that did not make it, either in the struggle for existence, or the battle between the sexes. Consequently, the unit of selection has traditionally been identified as the phenotype or living organism, and the level where selection occurs has been identified as the environment. In a very real sense, expanding evolutionary research toward a micro- and macrolevel implies an inclusion of the dead. It avers a recognition that the Modern Synthesis explains biogenesis but not abiogenesis and merely defines evolutionary research of existing life at a mesolevel, one that is inspired by older cosmologies.

Discussions on micro- and macroevolution reach their peak after the foundation of the Modern Synthesis and associate with advances made in fields such as on the one hand population and especially molecular genetics as well as organic biochemistry and, on the other, paleontology, ecology, systematics, and biophysics. These debates have been brought to us in the form of discussions on biological or genealogical and ecological hierarchies and what their epistemic entry points or levels of analysis are, major transitions, patterns, trends, evolutionary rates, and discussions on the mechanisms that cause variation, speciation, and (mass) extinction.

The debates invoke fundamental epistemological questions on spatiotemporal scales (material loci or units and levels as well as tempos of evolution), causation (mechanisms that underlie evolutionary change at these loci), and questions of reductionism and holism. Ultimately, they invoke an inquiry into how an evolutionary worldview necessitates a reconceptualization of matter, space, and time. Things can hardly become more fundamental, and herein lies the important contribution to science that both micro- and macro-oriented evolutionary scholars have brought forth.

We currently find ourselves in the midst of this fascinating cosmological revolution. Although evolutionary theory was first formulated over 150 years ago, we are still delineating the consequences of an evolutionary worldview and the impact it has on our epistemic frameworks that are themselves formulated by an evolving species.

My background lies in evolutionary epistemology as well as comparative anthropological science of intellectual history, what Nietzsche called the genealogy of thought. As such, my contribution here involves an epistemic and historical contextualization of the evolutionary cosmological issues that the micro- and macroevolutionary debates bring forth, as well as an anthropological delineation of two distinct scientific cultures that, although they share a common cultural ancestor, have evolved distinct scientific practices and intellectual ideas to study evolution. In so doing, I build on the following three books written by Niles Eldredge, *Time Frames* (1985), *Reinventing Darwin, The Great Debate at the High Table of Evolutionary Theory* (1995), and *The Pattern of Evolution* (1999). The present work reads both as a comment and as an elaboration of these ideas.

2 Microevolution

In 1946, over fifty evolutionary scholars from various biological fields united in the first meeting of the Society for the Study of Evolution (Eldredge 1999: 138–40; Smocovitis 1996). The Society was an outgrowth of a North American National Research Committee, founded a couple of years earlier with the goal to investigate shared evolutionary problems of genetics and paleontology. The societal members elected George Gaylord Simpson as their president, and Ernst Mayr functioned as secretary. During their meetings, the scholars established a common epistemic framework for eukaryotic phylogeny: Multicellular organisms undergo descent with modification because they evolve by means of natural selection. The established theoretical framework served as the foundation of the Modern Synthesis (Huxley 1942; Provine and Mayr 1980) that, according to Mayr, designates:

> … the general acceptance of two conclusions: gradual evolution can be explained in terms of small genetic changes ("mutations") and recombination, and the ordering of this genetic variation by natural selection; and the observed evolutionary phenomena, particularly macro-evolutionary processes and speciation, can be explained in a manner that is consistent with the known genetic mechanisms. (cited in Gould 1982: 382)

Contrary to their intellectual ancestors, these founders did not address origin of life questions, and they stayed clear from ontogenetic and sociocultural sciences. At the time, there were good reasons that justified these decisions. History was no longer considered a science (Popper 1957). Weismann's (1893) barrier caused for a rupture between ontogeny and phylogeny because scholars assumed that ontogenetically acquired characteristics do not feed back into the gene pool. Consequently, embryological and overall epigenetic research (Waddington 1942) was not considered a relevant subject area of evolutionary biology. The founders also refrained from sociocultural behavioral research because the subject was simply not considered to define their area of research. At the time, behaviorist schools considered behavior an outgrowth of individual and societal learning, and learned behavior in turn was an acquired behavior that did not feed back into the gene pool. In so far as learning is underlain by neurocognitive conditions, the subject belonged to the neurological and psychological sciences; in so far as learning requires teachers, it is a sociocultural group behavior that needs to be studied by anthropologists and sociologists. These fields developed altogether different epistemic frameworks and scientific practices to understand behavioral phenomena. The adoption of early social Darwinian theories had caused havoc in colonial and war times (Eliav-Feldon et al. 2009; Fanon and Philcox 2004), and many of the founders of the Modern Synthesis spent a lifetime eliminating vitalist and historicist (unilineal, orthogenetic, straight line) evolutionary ideas. Rosalind Franklin's work had led to the discovery of the double helix (Watson and Crick 1953), which later became understood as a "frozen accident" (Crick 1968), and such conceptualization temporarily eliminated discussion with cytologists and cell biologists who investigated heredity beyond the nucleated genome (Margulis 1998: 25). Bacteriology was altogether considered a medical science. Bacteria, viruses, and fungi were known to genetically infect multicellular life during ontogeny (Avery et al. 1944; Griffith 1928; Lederberg and Tatum 1946; Morse et al. 1956), and such infections can alter fitness (Haldane 1949; Lederberg 1999), but it was assumed that most infections did not enter the germ line and were thus negligible from an evolutionary point of view.

For an evolutionary scholar that wanted to work within the epistemic boundaries delineated by the Modern Synthesis, little else remained but the study of genes. The microevolutionary tradition developed by trying to incorporate lower- and higher-order phenomena on the scale, but at the same time reducing these phenomena to a genetic level of analysis. Adaptation became synonymous with fitness (how many times genes are passed on), and natural selection became understood as genetic selection (Williams 1966), because ultimately, these dead structures are the only "survivors" and "beneficiaries" (Lloyd 1988, 2000) of evolution. By understanding living organisms as temporary "vehicles" that ride "selfish genes" (Dawkins 1976), the micromolecular genetic level is considered more lasting in time and therefore more real, just because genes are "replicators" that have "fecundity, longevity, and copying fidelity" (Dawkins 1976).

Such an approach is interesting, because it demonstrates an epistemic struggle with evolutionary time and a subsequent search for generational continuity and stability which the founders as well as Darwin proclaimed a necessary feature in order for natural selection to work. By taking the differential distributions of genes over populations as point of departure, it implies an annihilation of the individual living organism which had always been understood as the unit of selection. It is here that we therefore find a first attempt to "evolutionize" the dead, i.e., to apply evolutionary theory to non-living phenomena.

I give two examples, research on kinship and research on sex. In these debates on the targets and beneficiaries of selection, attention was first drawn to those who from a genetic point of view do not benefit. Statistically speaking, kinship and altruism do not enhance an individual's fitness though other members of a population can benefit from such cooperative actions. When asking what such traits are good for, Wynne-Edwards (1962, 1986) raised the possibility that such traits benefit the group. But such higher-order selection was fiercely denied by Williams (1971) and Maynard Smith (1964), Maynard Smith and Price (1973). Instead, calculations began that enable such seemingly altruistic behavioral traits to become reduced to individual fitness via reciprocity or tit for tat (Axelrod 1981, Hamilton 1964). In times of competition, it pays to cooperate. Nonetheless, questions of benefit are a natural outgrowth of questions on adaptation. If selection favors the fit, then traits can only evolve when they are somehow adaptive or they would not be selected in the first place. So if the individual living organism does not benefit, it must be either something above or below the organism that benefits, the group or the genes themselves.

Another conundrum, passionately studied in flower power times, was the benefit of prolonged sexual immaturity, menopause and andropause, homosexuality, and meiotic sex which constrains "selfish genes." Meiotic sex (Edwards 1998; Fisher 1930; Hamilton 1967; Maynard Smith 1978; Becks and Alavi this volume) involves the acceptance of half of the nucleated genes from total strangers in neglect of the individual's own genes. Why would evolution have taken such a route when organisms reproduced by division wherein they passed on their complete genome? The most common answer given was that genetic recombination is beneficial because it increases variation and it enables genetic repair or the masking of maladaptive traits (which the fruit fly experimenters already showed to be mostly recessive) in *future* progeny. In particular, the first argument implies a higher-order population level. By examining how well genes fare in a population's gene pool, mate choice and sexual fitness become understood as a form of group selection (Van Valen 1975). Given that it takes two, meiotic sex annihilates the sexual individual as an independent organism in favor of the study of genetic (re-)combinations at a population level. What matters more is the differential distribution of genes over populations *through time*.

Neo-Darwinian theory was very successful in reducing many seemingly social behaviors to a genetic level, but an equally valid question, from an epistemological and historical point of view, is why, from all behaviors, it was particularly social and cooperative behavior that received so much attention. One reason is the following. Evolutionary theory itself evolved in the nineteenth century as an outgrowth of discussions on sociopolitical debates on the common goods of society and moral debates on human nature. Without divine law, scholars sought naturalistic, secular principles that enable individuals to bond into societal political structures. Hobbes, for example, saw humans and the whole of nature as fierce and violent. Rousseau endorsed that all are good-natured but corrupted by society. Pierre Joseph Preudhom developed his political ideas on mutualism which gave way to socialism. The ideas developed, with on the one side, the liberal camp that understood nature as "red in tooth and claw" and, on the other side, the socialists.

Darwin was part of the liberals and the struggle he assumed, for Spencer was one of the "fittest" in a world filled with "parasites." On the other side, you had the socialists and communitarians, one of them being Kropotkin who emphasized mutual aid (see Sapp 1994 for a discussion). These features were also found in nature, under the form of symbiosis, mutualism, and commensalism, but these ideas did not survive the synthesis, or better yet, they did, but in the form of parasitism and "master–slave" relationships that were studied as outcomes of behavioral features from within fields such as ecology. Because symbiotic behavior was understood as a type of socialism and studied from within ecology, the subject was not considered a hallmark of synthetic evolutionary studies. For Neo-Darwinians, who focused on the vertical transmission of nucleated genes, cooperation forms an anomaly. Another reason why especially sexual behavior received so much attention during the rise of microevolutionary studies is that eukaryotic sex forms a hallmark of natural and sexual selection theory (see Becks and Alavi this volume). If not for sex, traits cannot be passed on from one generation to the next. Finally, research on cooperation and social behavior in particular was also approved during the Darwin centennial, discussed in the next part.

2.1 Evolutionizing the Inorganic and Superorganic

Shortly after the Society for the Study of Evolution was founded back in 1946, the Darwin centennial was organized at the University of Chicago in 1959. Tax (1960) organized the event and invited the scholars whom Julian Huxley (1942) had called the founders of the Modern Synthesis (Theodosius Dobzhansky, E.B. Ford, Ernst Mayr, George G. Simpson, G. Ledyard Stebbins, Bernhard Rensch, and Sewall Wright), anthropologists, and behavioral scientists (including, among others, Daniel Axelrod, A. Irving Hallowell, Clyde Kluckhohn, Alfred Kroeber, Louis Leakey, Niko Tinbergen, Gordon Willey, and Leslie White), as well as religious leaders. The conference was organized around five panels: the origin of life, the evolution of life, man as an organism, the evolution of the mind, and social and cultural evolution, a lineup that Smocovitis (1999: 296–7) rightfully notes: "… follows both a logical sequence of the history of life on earth, and the logical ordering of knowledge: the physical sciences were followed by the biological sciences and the social sciences."

It is easy to deduce from the program that the purpose of the meeting was to investigate the impact an evolutionary worldview had on all the sciences, from the micro- to the macrolevel. Following Huxley (1942, 1957) who in turn was highly inspired by his grandfather Thomas Henry Huxley as well as Herbert Spencer, it was both a plea for "universal evolution" and an attempt to undo the "eclipse of Darwinism" that had originated from the 1880s onward due to advances in symbiogenesis, neo-Lamarckianism, epigenetics, and embryology. Waddington was the only disturbance in that regard. The population geneticists had mechanized natural selection by mathematizing selection as gene distribution studies, and the founders had put discontinuity and saltationist debates to rest, so it was now time to evolutionize the behavioral and sociocultural sciences.

In comparison, during the Chicago Darwin bicentennial that was chaired by the philosopher Robert Richards in 2009, the main themes were eco–evo–devo and History and Philosophy of Science, neither of which had been represented at the 1959 centennial. Back in 1959, the former schools were part of the eclipse that needed to be undone. The latter, philosophers of science, were altogether not engaged, with the exception of Nagel (1949, 1951) who, a couple of years earlier and from within logical positivism, had written on the use and abuse of teleology in biology, and Grace A. De Laguna, who wrote a positive review of the conference proceedings she characterized as having taken on a "fresh approach to the philosophical problem of ends in nature." In her abstract and by following Pittendrigh (1958), she summarizes the problem as follows:

> They accept 'teleonomy' only as descriptive, and neglect its significance for theory. The present thesis is that each of the three recognized phases of universal evolution: inorganic, organic, and post-organic initiated by the advent of man and his culture, is characterized and made possible by the emergence of a distinctive type of teleonomic organization. (De Laguna 1962: 117)

The centennial thus reinstated in a post-synthetic and post-war spirit the attempt to extend the principles of natural selection to phenomena studied within the behavioral, cognitive, and sociocultural sciences, in particular the "superorganic" goal-oriented behavior of the living (Axelrod 1981; Hamilton 1964; Tinbergen 1963; Trivers 1971; Williams 1966). Inspired by Pittendrigh and Williams' notions on teleonomy, sociobiologists (Wilson 1975) took things altogether to another level. Early sociobiologists assumed that behavioral and cognitive traits were ultimately underlain by genes, and their scientific work initiated an era where specific genes were postulated for language, cognition, and cultural traits. In short, such investigations reinstated evolutionary research on ontogeny from within a Neo-Darwinian framework (Campbell 1960, 1974; Hahlweg and Hooker 1989; Tinbergen 1963).

What about the inorganic? Darwin's *On the origin of species by means of natural selection* provides a theory of biogenesis, and it explains how existing life brings forth new life. He speculated that life evolved in a small little pond, a primordial soup, and that all living organisms are genealogically related through the blood line (Darwin 1859: 420–2) and thus brought forth from a single common ancestor (for a discussion, see Gontier 2011). But he did not go beyond this speculation to address the origin of life out of inorganic matter. This transition was mostly studied from within epigenetic circles (for a discussion, see Maienschein 2012), where scholars investigated abiogenesis by means of spontaneous generation instead of natural selection, to explain how the living evolved out of inanimate matter. The doctrine was called to halt by scholars such as Redi and Pasteur, who reintroduced more preformationist notions by defending that *Omne vivum* ex vivo, all life comes from preexisting life. This credo also formed the basis for cell theory, which takes the cell as the basic unit of life and argues that all life comes from cells (Mazzarello 1999). The view strengthened Darwin's evolutionary theory that all life shares a common origin wherefrom descent with modification follows (for a discussion, see Huxley 1870).

Thinking through the consequences of Darwinian evolution theory, and in the context of research on the origin of life, Haeckel (1866) and Huxley (1870), who coined the term abiogenesis and also denied it, were the first to universalize evolutionary theory toward the inorganic. Comparative research led Haeckel to theorize that "higher animals" had evolved from "protozoa" such as "monera" and subsequently diversified into kingdoms ("Protista," Animalia, and Plantae). He depicted his monophyletic reconstructions in the first evolutionary trees of life that in turn gave credibility to his recapitulation theory alternatively known as his biogenetic law (literally, the law of the genesis or historical origin of life). During the First World War, he published a work on crystals which he characterized as "anorganic life" (Haeckel 1917) wherein he came very close to the work so typical of biophysics.

Returning to the Chicago 1959 centennial, what is interesting, and very reflective of the time, is the differentiation the organizers made between origin and evolution of life problems which they examined in two distinct panels. Back then, origin questions were frowned upon as speculative and unscientific because the initial conditions can neither be repeated nor observed, though today progress is indeed being made in that regard. But such scientific demands equally go out to the future, which we can neither observe nor induce, yet prediction is valued as one of the highest epistemic goods of any scientific theory. Equally, "what is life" questions are also often dismissed as unscientific, while on the contrary, mechanical questions on how life evolved, and thus, identifying the mechanisms of evolution is considered science. The 1959 panel on the origin of life was composed of physicists, biochemists, geneticists, and ecologists who focused on the ecological conditions favorable to evolve eukaryotic life on earth, the biochemical aspects of the double helix, and the evolution of physiology, especially nerve wirings.

The structure of genetic material had been discovered 6 years earlier through X-ray crystallography, and topics included how genes mutate and how they underlie general animal physiology. Analyzing the panel, it altogether remains a strange combination of scholars and topics. Participants included the astrobiologist Harlow Shapley, a participant of the great debate on the size of the universe and its nebulae (galaxies), who also introduced the "liquid water belt theory" that defines habitable zones for life; Sir Charles Galton Darwin, descendent of Darwin and physicist involved in X-ray crystallography; Hermann Joseph Muller, a Nobel laureate known for his work on the impact radiation has on genetic mutations; the geneticist Theodosius Dobzhansky and biochemist Earl A. Evans; Ralph W. Gerard, a neurophysiologist and behavioral scientist working on the chemicoelectric wirings of the central nervous system; Clifford Ladd Prosser, a comparative physiologist; Georgii F. Gause, a Russian biologist known for his ecological competitive exclusion principle that states that only one of two similar species competing over the same niche will survive (the other one will go extinct); and Hans Gaffron, a biochemist working on photosynthesis in plants.

One can only interpret such a selection for a panel on the origin of life from within the bias there existed among the founders of the Modern Synthesis and their predecessors to understand the evolution of multicellular life and to understand that evolution exclusively by means of natural selection. Granted, great scientists such as Manfred Eigen (Eigen and Schuster 1977), Orgel (1973, 1992) and De Duve (1991) or the Schopf (1978) brothers and Fox and Dose (1972) were still coming of academic age (which on average takes a decade or two longer). But by then, Wallin (1927) had published his work on symbiogenesis, Pauling (1939) had already received the Nobel prize for his work on the nature of the chemical bond, Miller (1953) had performed his experiments on the primordial atmosphere of life that spontaneously generated amino acids, and Oparin's (1955) work had been translated and was well read within the Americas and Europe. Luria and Delbruck (1943), in discussions on Lamarckian versus Darwinian inheritance and the occurrence of mutations as random or directed toward environmental change, gave statistical evidence that bacteria mutate randomly and only subsequently become the target of selection. Avery et al. (1944), who pointed out that DNA and not proteins were the transforming factor during, what we now call, lateral gene transfer among strands of pneumococci were also not invited, and neither was Lederberg (1951, Zinder and Lederberg 1952), who discovered the mechanisms of transduction and bacterial conjugation and reintroduced the term abiogenesis before Carl Sagan did.

Perhaps funding ran out, but the composition of the 1959 Origin of Life panel does demonstrate a conscious selection of favoured topics as well as a lack of scholarly interaction between the founders of the Modern Synthesis and pioneers in areas such as abiogenesis, actual molecular genetics as opposed to theoretical population genetics, cytoplasmic heredity, and prokaryotic life. In their defense, it is true that such inclusion would not have involved a celebration of natural selection or Darwin per se. But in so far as they endorsed universal evolution and even accepted the possibility for selection to be directed as opposed to blind in what regards sociocultural evolution, much more so than later sociobiologists would, it remains strange why, besides selection, there appears to have been no room whatsoever for symbiogenesis or spontaneous generation (or self-organizing, autocatalytic complex adaptive systems as we call them today). Spontaneously generated, biochemical catalytic and autocatalytic systems show selectiveness, but much of this can be explained by electromagnetics, physics, and chemistry, rather than by struggle for existence and competition. This is one of the reasons why such views were not favoured at a time when biology was seeking its own field-specific and epistemic boundaries. Wright (1929, 1932) was there to defend drift, but he was part of the evolution panel, and drift, before Kimura (1968, 1983), was understood as a period of no selection rather than as a mechanism in and of itself.

Nonetheless, in relation to post-synthetic introductions of origin of life debates, and battles over which came first, genes (Eigen and Schuster 1977; Orgel 1973, 1992), proteins (Miller 1953; Miller and Urey 1959; Kauffman and Levin 1987; Kauffman 1995), or cells (De Duve 1991; Fox and Dose 1972; Oparin 1955), the inorganic eventually became evolutionized. Today, all known evolutionary theories (natural selection, symbiogenesis, spontaneous generation or self-organized complex adaptive systems theory, lateral gene transfer, drift, punctuated equilibria) are, to some extent, put to use to understand abiogenesis.

It was especially Eigen (1996: 547), who selectionized the inorganic or "prebiotic" by asking whether "'being alive' really [is] a necessary prerequisite of selective and evolutionary behavior." His theory on the autocatalytic hypercycle explains, from a selectionist point of view, the origin of autocatalytic RNA molecules and DNA compartmentalization into larger genomes. His background in physics and his adherence to Shannon's rising information theory even enabled him to turn natural selection into a physical law of functional optimization toward increasing complexity:

> That is how a physical interpretation of the Darwinian principle might sound. According to Darwin's principle, whatever is better adapted spreads out and displaces its less well adapted predecessor. Thus, complexity, built upon simplicity, has accumulated throughout biological evolution from the first single-celled organisms to human beings. Evolution as a whole is the steady generation of information - information that is written down in the genes of living organisms. ... Today we can apply our knowledge to molecular systems such as genes and the products of their translation. We can also investigate in a much more objective way the physical nature of the Darwinian principle: theoretically, by defining accurately the prerequisites and constraints, and experimentally, by exact control of experimental conditions. We find that the selection principle is neither a mystical axiom immanent in living matter nor a general tendency observable primarily in living processes. On the contrary, it is - like many of the known physical laws - a clear 'if-then' principle, that is, a principle according to which defined initial situations lead to deducible behavior patterns. It is thus analogous to the law of mass action, which regulates the attainment of the quantities of the components in a chemical equilibrium. (Eigen 1997: 17–8)

It is a complex theory, but the short version is that within genetic landscapes—to be read quite literally, in the context of (pre-)RNA worlds (Cech 1986; Gilbert 1986), not the wild type, but the most optimal variations of the wild type (the quasi-species) together *as a group* are the target of selection. In order for the quasi-species to overthrow the wild type (and thus to evolve by splitting), it needs to be more optimal. It is a bit like the "hill-climbing" metaphor Dawkins (1996) uses to describe what he calls "accumulative selection." Eigen (1996: 19):

> Selection contains an element of exact 'if-then' behavior. It has nothing to do with the tautological interpretation 'best adapted = selected'. 'Selection' could in principle just refer to any kind of preference. But here it means a particular kind of preference, which adheres unerringly to a single scale of values. Selection is based upon self-replication. It distinguishes sharply between competitors, it constructs a broad mutant spectrum on the basis of value, and in this way it organizes and steers the entire, complex system.

Selection becomes understood as replication of preferential information. Eigen and Dawkins' ideas find their roots in the works of Fisher (1930) and Wright (1929), with that very important difference that Eigen saw his ideas on natural selection as a physical law restricted to the early "life" of replicators (RNA and DNA molecules and viruses). Eigen therefore pointed out the limit of natural selection. As soon as there is a distinction between a genotype and a phenotype (and thus an established translation machinery), he argued that selection occurs at the level of the cell, which is different from how selection occurs at the level of replicators.

> The chief criticism of Darwin's idea was directed against its supposed claim to explain all of evolution. However, the development of life, from molecular systems to human beings, has passed through many stages of organisation, and, while some of these were Darwinian in nature, many were fundamentally different. Since the preservation of all living systems is based upon reproduction, selection plays a role at all levels. But selection is expressed in many different ways, sometimes as coexistence or even cooperation, and sometimes as competition and the often irreversible weeding-out of some forms of life. (Eigen 1996: 29–30)

Eigen's introduction of information theory and focus on increasing levels of complexity and optimization also inspired scholars such as Szathmáry (2002, 2006), and other researchers interested in the major transitions of life, as well as scholars that form part of the field of biosemiotics. For Eigen, for example, the origin of autocatalytic replicators, and their compartmentalization into a genome, was a transition that marked a "peculiar 'once-forever'—selection mechanism [...] a consequence of hypercyclic organization" (Eigen and Schuster 1977: 542). Oparin, for example, once asked why life does not originate de novo on a daily basis. The most popular answer in his time was that existing life takes up all resources, which is an ecological argument. The answer scholars like Eigen, Jacob (1977), or Monod (1971) gave is that it is so unlikely an event, that if it were to happen once, it will most definitely not happen twice due to the complexity involved. Life genealogically builds upon what exists, and what exists is the outcome of once-forever events, likely at the time, but often not repeatable thereafter. This tinkering argument is based upon an understanding of genes as carrying an increasingly complex informational code, and just because of its emphasis on information and increasing complexity, it is so popular with creationists (Hoyle 1981), whom assume that such tinkering requires an intelligent designer.

Nonetheless, if the origin of the genetic code was a "once-forever" event, another way of stating the latter is as follows: However the genome evolved, it is unique to that period. In so far as the situation of viruses resembles that realm, probably because they too are remnants of that particular phase (see, e.g., Villarreal 2004; Villarreal and Defillips 2000), they too can be studied as examples of hypercyclic organization. But beyond that, the law of selection stops and selection takes on different forms, sometimes as co-evolution, sometimes it is completely overthrown by other mechanisms and thus, itself, as a mechanism, not selected simply because the if-then conditions are not met. If correct, then such a view would imply a higher-order sorting or selection of evolutionary mechanisms. One of the challenges we are faced with is understanding how various and distinct evolutionary mechanisms work together to bring forth the phenomenon that evolution is.

Finally, ecology, for the founders as well as the organizers of the centennial, was by and large reduced to the study of genetic landscapes (Serrelli this volume), as well as the conditions for habitable life on earth. Research on both ontogeny and ecology (Odling-Smee 1988; Van Valen 1973, 1976), which are both areas of research that were first defined by Haeckel, became reintroduced in standard Neo-Darwinian theories from the 1960s and 1970s onward, partly because of advances in symbiosis and symbiogenesis (Margulis 1970, 1981) and partly due to advances in macroevolutionary fields, including biophysics. Eldredge (1985b, 1989, 1992), Gould (1977), Lewontin (1983, 2000), Gould and Lewontin (1979), and Stanley (1979; Eldredge and Stanley 1984), for example, were quite active in reinstating ecology, ontogeny, and biophysics as valid sub-branches of the Neo-Darwinian framework. I return to this later in the text.

Today, the movements evolved into, on the one hand, the new sociocultural evolutionary sciences, including evolutionary psychology, evolutionary linguistics, and evolutionary anthropology (for a review, see Gontier 2012a) and, on the other hand, the schools that associate with eco–evo–devo research. Eco–evo–devo today reinstates a mesolevel analysis between the biomolecular (micro) and ecological and systematic (macro) sciences. Within the sociocultural sciences, especially the work of Jablonka and Lamb (1989) and Oyama (1985) was foundational. Typical of the eco–evo–devo approach is that scholars try to define and map actual life history, by, e.g., correlating genes with organismal phenotypes and sociocultural, biotic, and abiotic environments. On the other hand, work on symbiosis and symbiogenesis and reticulate evolution in general remains poorly integrated in Neo-Darwinian theory.

2.2 *Universalizing Selection*

One of the reasons why the microevolutionary tradition has been so successful is that it has been able to "universalize" (Cziko 1995; Dawkins 1982a; Dennett 1995) natural selection theory into a common scientific language, not by searching for the law that natural selection can be (as Eigen had demonstrated), but the pattern that natural selection produces as well as the parameters that need to be present for natural selection to work. These parameters and patterns have been mostly formulated in the form of heuristics. Starting with the units and levels of selection debate (Brandon and Burian 1984), evolutionary epistemologists (Bradie 1986) and evolutionary biologists have identified "universal" units of selection such as the replicator (Dawkins 1982a), the interactor (Hull 1981), and the reproducer (Griesemer 2000), and they have been engaged in identifying the levels where selection occurs (Brandon 1982; Godfrey-Smith 2009; Okasha 2005). Such research relates to hierarchy theory and the quest for the major transitions, or patterns and trends of life that especially natural selection, understood as the selection of information, has brought forth in all domains of life (Maynard Smith and Szathmáry 1995; see Brandon and McShea 2010 for a critique).

Overall, the universalization of selection is characterized by abstracting a general template from natural selection that provides information on how the mechanism works (Gontier 2006a, b). This template can then be used as a heuristic to study the evolution of genes, organisms, culture, cognition, and language. In biology, Lewontin (1970: 1) has defined such a template or "logical skeleton" of natural selection as "phenotypic variation, differential fitness, and heritability of that fitness." And ten years earlier, Campbell (1960), the founder of evolutionary epistemology, argued that such a general formula of natural selection is "blind variation and selective retention." This means that as soon as one can identify something that varies blindly and is selectively retained, that something evolves by means of natural selection. In sum, the work by scholars engaged in universal selectionism has led to finding (1) more units and levels of selection and (2) the conditions under which natural selection operates, conditions they formulate in the form of a heuristic, logical skeleton, or formula.

Both approaches are today somewhat being synthesized into the new and rising field of Biosemiotics (Barbieri 2008a, b; Favereau 2005; Hoffmeyer 2008), where, as the name implies, the evolution of informational codes serves as a new means to delineate the major transitions of life from within a more historical perspective.

3 Macroevolution

In his *Structure of Scientific Revolutions*, Kuhn (1962) characterized scientific practice as a paradigm, defined in his preface as "scientific achievements that for a time provide model problems and solutions to a community of practitioners." Scientific development goes through distinct phases from preparadigm and paradigm formation phases, to phases of paradigm execution, and paradigm reformation or scientific revolutions. The latter result from how well scientific frameworks deal with anomalies that challenge the tenets of an accepted paradigm. Anomalies cause paradigm instability and scientific crisis. Either scientists are able to solve the anomalies from within the standard epistemic framework, or they develop another paradigm that either solves the issues or gives hope that it eventually will.

The historical developments of Darwinian and Neo-Darwinian evolutionary theories follow Kuhn's description of scientific revolutions quite accurately. Darwin (1859) did not find evidence for the fixity of species, which was a tenet in the standard paradigm of his time. Instead, his voyages had made him come to believe that species evolve. He knew that if true, his ideas would cause for a scientific revolution. Calling his *Origin of Species* "one long argument," Darwin (1859: 495) recognized that he had merely provided a hypothesis based upon generalizations of observable artificial selection in animal and plant breeding. For such a generalization to become a theory, rigorous scientific testing and proof was needed. In Chaps. 6 and 9, Darwin also pointed out anomalies or objections to his arguments. One of them was the lack of evidence for intermediate forms ("insensibly graded series," Eldredge 1995: 95) in the fossil record. Darwin countered the objection by arguing that not his hypothesis, but the fossil record was incomplete.

During the "eclipse of Darwinism," other anomalies were found (see Bowler 2003 for an overview). The rediscovery of Mendelian genetics and the introduction of mutation theory questioned Darwin's gemmules theory and his ideas on genetic blending. More discontinuous and saltational views emerged as a consequence (see Gould 1977, 2002 for an overview). These anomalies were solved from within the evolutionary paradigm by the theoretical population geneticists who synthesized natural selection theory with Mendelian heredity laws and who advanced a more gradual and continuous interpretation of genetic mutations (see Schwartz 1999 for an overview). The scholars provided mathematical and experimental proof for the theory of evolution and turned the hypothesis into a scientific theory, Neo-Darwinism, that formed the basis for paradigm selection by the founders of the Modern Synthesis.

Neo-Darwinist population geneticists are best regarded as executers of the program. They have been solving anomalies that phenomena such as altruistic and sociocultural behavior and eukaryotic sex pose to the Neo-Darwinian framework. By incorporating below and above organismal phenomena, they have had to make compromises to both Darwin's original theory of natural selection and the synthetic framework, but at the same time, they have been able to expand its scope and application range.

The macroevolutionary perspective as we know it today has been introduced from within the field of paleontology by Niles Eldredge and Stephen J. Gould in the early 1970s when the Modern Synthesis was well formed and on the expanding. Eldredge and Gould started out as executers of that particular program, but in so doing, they found anomalies that lead to crisis. Scholars in this tradition accept the lack of intermediates in the fossil record as a fact, they find evidence for periods of stasis, and they demonstrate that speciation and extinction are events that often occur rapidly. Scholars on both sides of the scale are currently investigating whether the Neo-Darwinian framework can overcome this crisis that research on stasis, speciation/extinction, and the lack of intermediates introduces, or whether the findings necessitate a scientific revolution.

When studying the evolution of trilobites, *Phacops rana* and *Phacops iowensis* which Eldredge (1971: 162) called "true biospecies," he did not find intermediate species. Instead, he encountered long periods of morphological stasis within the trilobites, and when variation in eye lenses did come along, older forms were rapidly replaced. Speciation was a rare event rather than a natural outcome of incessant and continuous anagenetic evolving which would imply intermediates (for a detailed description, see Eldredge 1985a). The speciation events that occurred appeared to have been driven by the spatiotemporal distributions of the species; geography, group size, and how ecological regions are populated by real individuals as opposed to how genes map onto genetic landscapes. New traits become fixed in marginal groups isolated from their parental species, and they split and rapidly replace the older population. In other words, Eldredge explained morphological change of trilobites by making use of Mayr's model of allopatric speciation by peripheral isolates and saw it as an anomaly to "phyletic gradualism" that endorsed speciation to occur through anagenesis where species "gradually and systematically evolve themselves out of existence" (Eldredge 1995: 69). For Eldredge, Mayr's model explained the gaps he found in successive strata. But he went further, by arguing that such a model annihilates the necessity to postulate intermediates, and furthermore encourages the idea that most species evolve through cladogenesis.

> The majority of species preserved in epeiric sediments show no change in species-specific characters throughout the interval of their stratigraphic occurrence, and the phyletic model is inapplicable to most of these elements of the fossil record. Instead, change in, or development of, species-specific characters are envisioned as occurring relatively rapidly in peripheral isolates. Morphological stability of epeiric species is attributed to stabilizing selection. (Eldredge 1971: 166)

At first, Eldredge explained morphological stasis as a form of stabilizing selection, and later and inspired by Vrba (1980, 1985a, b), Lieberman (2000; Lieberman et al. 1993, 1995, 2005, 2007), Lieberman and Dudgeon (1996), Lieberman et al. (1995, 2007), Eldredge et al. (2005) would understand it as an outcome of drift and habitat tracking. A parallel with Sewall Wright's shifting balance theory was first made by Thomas Schopf in the introduction to the first punctuated equilibria paper (Eldredge and Gould 1972). By denying that genetic selection plays sufficient role, drift and habitat tracking are recognized as independent mechanisms, not merely the result of a genetic level in evolution; rather, they add an ecological level, very much in the same way as Hubbel (2001) applies ecological drift to describe biodiversity and biogeography (Rosindell et al. 2011).

Returning to the 1970s, around the American Museum of Natural History and Columbia University, a group had been developing where Eldredge would become part of. Together with Stephen J. Gould, who found the same results in land snails, Eldredge published the consequences of their findings a year later (Eldredge and Gould 1972). Their conclusions can be summarized as follows:

1. Gaps in the fossil record are real, and when no intermediates are found, then there is no "insensibly graded fossil sequence" (Eldredge and Gould 1972: 87);
2. Inspired by Hennigian cladistics of genealogical or ancestral–descent relationships (Eldredge 1971: 157); in anticipation of Ghiselin's (1974) notion of species as biological individuals, they claimed that "biospecies" are real because they have stratigraphic and thus historical existence (a beginning, life span, and ending in time p. 92), and species demonstrate a type of stasis or "homeostasis" that surpasses "genetic revolution" (a term coined by Mayr, p. 114)—they never denied the occurrence of genetic mutations or the importance of genetic selection, what they claimed was that neither suffice to cause speciation;
3. Speciation is rare, but when it occurs, it occurs rapidly and species evolve by splitting because of allopatric speciation by peripheral isolates (p. 96);
4. The history of life is characterized by punctuated equilibria rather than by phyletic gradualism, and the former is a consequence of "the theory of allopatric speciation" and "implies that a lineage's history includes long periods of morphological stability, punctuated here and there by rapid events of speciation in isolated subpopulations" (pp. 108–10);
5. Trends are real (p. 110);
6. The Modern Synthesis is biased toward phyletic gradualism, but macroevolution (understood as the evolution of species or speciation) and also stasis of species cannot be fully explained by microevolutionary theory by which they meant genetic selection theory—the critique goes out to microevolutionary schools of thought much more than it goes out to Darwinism or general evolutionary theory as some creationists have falsely claimed.

In a Popperian sense, one anomaly rightfully disproves all, but it would only be the beginning of a long series of publications where scholars would confirm both their results and the pattern of stasis and rapid speciation in numerous lineages (reviewed in Gould and Eldredge 1977, 1993), including our own (Eldredge and Tattersall 1982; Gabrowski and Wood this volume). In a way, it was neither the trilobites or the snails, nor their theory of punctuated equilibria that made the difference. The major conclusion was that an exclusive focus on genetic distributions and population levels so typical of the microevolutionary perspective did not fully explain speciation and large-scale trends. Their work marked a beginning of a fresh breath of air for paleontologists who, ever since Darwin, were treated like the stepchildren of evolution. Eldredge and Gould's work justified other scholars to speak up against the claims of the Synthesis, in favor of what the data and the fossil record actually said. It marked a beginning of macroevolutionary thought as distinct from microevolution and a celebration of the work performed by paleontologists, systematists, and ecologists which had been eclipsed.

3.1 "Evolutionizing the Dead" by Means of Organicism

Like microevolutionary scholars, macroevolutionary scholars have also introduced an inclusion of "the dead" by emphasizing the evolutionary impact that phenomena traditionally not conceived as living have. Macroevolutionary scholars recognize species and higher taxa such as genera and phyla as historically real entities that occupy space and time. The species–genera distinction was introduced by Aristotle. In Judeo-Christian tradition, which was more influenced by Plato, it was believed that species were created by God according to fixed prototypes. Taxonomic classification, including Linnaean systematics, was nonetheless understood as a theoretical and logical discipline, where universals and particulars were distinguished based upon logical dichotomies. As theoretical frameworks, they showed the order of the universe, but like any universalia (theoretical concepts), their independent existence was questioned by nominalists and accepted by realists (for a discussion, see Gontier 2011). Darwin adhered to a nominalistic species concept, thereby claiming that species are not real, but Mayr (1942) introduced his biological species concept that defines species based upon sexual exclusion. He did so out of necessity because the gene population point of view annihilates the living sexual individual. And his theory of allopatric speciation due to peripheral isolates equally demarcated species in ecological terms. With Mayr, species become understood as bounded in space and time, due to sexual compatibility and geographical accessibility requirements. Eldredge and Gould took the reasoning further by "evolutionizing" and "geneologizing" not only species but also higher taxa. I give two examples, one from Eldredge's work and one from Gould's.

For Eldredge (1999: 22–3), species are "real entities, 'individuals' with births, histories, and deaths." They are real genealogical entities that occupy an equally real ecological space in the history of life. When Ghiselin (1974) introduced his ideas of species as biological individuals, an approach Eldredge adopted, Ghiselin quite literally, revitalized the organic notion of beings that was so typical of presynthetic evolutionary and historical thinkers of the nineteenth century. Back then, languages and cultures were considered real organic and organismic beings or natural kinds as philosophers call it. All have a birth, life history, and ending in time from which linguistic, cultural, and sociopolitical genealogies could be inferred. Because they were beings, they even reproduced, though not by replication, but by splitting (diffusion) or merging (cultural blending or acculturation, Gontier 2006c; Kressing et al. 2014). Inspired by linguistics and language genealogies, Darwin argued that all biological organisms are also genealogically related through what we call genetics and what he called the blood line (discussed in Gontier 2011). The discovery of genetic material and Weismann's barrier blocked such continuity, and anthropological fields would introduce notions of historical particularism instead, claiming that no genealogy of sociocultural knowledge can be drawn because every culture has its unique and particular history. The unique and particular history of biological species, in turn, was genealogized by Hennig who revitalized the idea of natural kinds, and this work was foundational for Ghiselin when he argued that species are biological individuals.

By building on both Hennig's cladistics that are a "truly 'scientific' way of deriving evolutionary histories and taxonomies" (Eldredge 1999: 8) and Ghiselin's work, Eldredge understands species as real historical entities, beings or individuals that occupy space and time. Species genealogies are more stable in time than the individual organisms that make them up. His research studies on stasis (Eldredge et al. 2005), living fossils (Eldredge and Stanley 1984), extinctions (Eldredge 1991a, b), and biodiversity (1992, 1997; Casetta and Marques da Silva this volume) are investigations into the geological life span, the actual natural history of species, both in what regards their origin, genealogy and ending, as well as in what regards their relation to the natural environment. As an ardent environmentalist, much of Eldredge's career has also focused on "biologizing" and "evolutionizing" the abiotic environment. The abiotic environment is relevant for understanding biological evolution, not as a level where genetic selection occurs, but as an evolutionary and causal factor that impacts the evolution of life.

> Echoing some of Darwin's strongest themes and denials, modern evolutionary biology emphasizes genes and biotic interactions at the expense of explicit links to the physical world. Ecology plays a muted role in evolutionary theory. And patterns in the history of life – patterns strongly linked with the physical history of the earth – still await full integration with evolutionary theory. (Eldredge 1999: 95)

That ecology is real becomes most clear in his distinction between an ecological/economic and genealogical hierarchy (a distinction he first made in 1985, but for an overview, see Eldredge 2008). That surely was not part of the Modern Synthesis, but it can be incorporated into a reworked version, in so far as Darwin himself attributed more causality to the environment than his predecessors did.

We also find this idea with Gould (1989: 228) who argued that "Evolutionary change (as opposed to mere variation) is produced by forces of natural selection arising from the external environment (both physical conditions and interactions with other organisms)." Nonetheless, Gould (1977) was more inspired by developmental and genetic studies. Contrary to Eldredge, who explained stasis as the outcome of stabilizing selection and later drift, Gould investigated the "homeostatic" life span of species as a possible consequence of ontogenetic constraints (Gould 1977) and genetic aging (Gould et al. 1987). By understanding eukaryotic genera and clades as real historical individuals, that have a beginning, life span, and ending in time, he studied that life span or genealogy by looking for trends in eukaryotic multicellular evolution that indicate a "new arrow of time to specify the direction of evolution" (Gould et al. 1987: 1437). In association with the MBL group (for overviews, see Sepkoski and Ruse 2009), Gould investigated clade typology and concluded that the evolution of marine invertebrates and terrestrial mammals, among others, portrays a maximal disparity (rather than adaptive radiation which he rejected as a notion, Gould et al. 1987: 1438) or "bottom-heaviness" at the beginning of the clade. This bottom-heaviness is followed by rapid decimation and subsequent species diversification within a few remaining lineages. Gould (1989: 304) interpreted this bottom-heaviness as:

> early experimentation and later standardization. Major lineages seem able to generate remarkable disparity of anatomical design at the outset of their history – early experimentation. Few of these designs survive an initial decimation, and later diversification occurs only within the restricted anatomical boundaries of these survivors – later standardization.

Dawkins had launched the notion that genes are the ultimate survivors of selection. Inspired by Lerner and Valentine, Gould asked whether genetic systems do age. By believing that "organisms as well as environments were different in Cambrian times, [...] the explosion and later quiescence owes as much to a change in organic potential as to an altered ecological status" (Gould 1989: 230). He assumed that early multicellular life, though morphologically different, had a "shared genetic heritage, now dissipated" (Ibid.: 231–2). Since the Middle Cambrian, a period in life that is well preserved in the Burgess Shale Formation currently in the Canadian Rockies of British Columbia, no new phyla or "body plans," as he called them, have evolved, for which he raised the possibility that the genetic systems had aged. Gould was a true visionary of the evo–devo schools that emerged with the discovery of the homeobox gene complex (Gehring 1992) and regulatory, homeotic genes in general (Davidson and Erwin 2009), when he wrote:

> I have no profound suggestions about the potential nature of this genetic 'aging', but simply ask that we consider such an alternative. Our exploding knowledge of development and the mechanisms of genetic action should provide, within a decade, the facts and ideas to flesh out this conception. [...] We cannot, for now, go beyond such crude and preliminary suggestions. (Gould 1989: 231)

Gould found the anomalies less commensurable with Neo-Darwinian theory, though he too emphasized repeatedly that he himself was a Darwinist, because Darwin himself was a pluralist. For Gould (1982: 383):

> The non-Darwinian implications of punctuated equilibrium lie in its suggestions for the explanation of evolutionary trends [...] not in the tempo of individual speciation events. Although punctuated equilibrium is a theory for a higher level of evolutionary change, and must therefore be agonistic with respect to the role of natural selection in speciation,

> the world that it proposes is quite different from that traditionally viewed by palaeontologists (and by Darwin himself) as the proper geological extension of Darwinism. The "gradualist-punctuationalist debate", the general label often applied to this disparate series of claims, may not be directed at the heart of natural selection, but it remains an important critique of the Darwinian tradition.

Because both Eldredge and Gould understand higher taxa as real, (mass) extinction and origination/speciation are equally real phenomena, and their history in time and space can become documented. Another logical outcome is research on both biological/genealogical and ecological hierarchies (Tëmkin and Eldredge this volume). In that regard, and as a part-time historian of science, Gould (1981) was more interested in debunking cultural biases of the scale, while Eldredge has been actively engaged in developing new hierarchical frameworks.

Finally, if species are real, it raises questions on species sorting. In fact, if all higher taxa, such as kingdoms or phyla, are real because they demarcate actual historical groups in space and time that can be delineated based upon ecological niches, geological strata, morphological layout, and mechanisms according to which they evolve, one could in principle ask whether there exists some type of sorting from Woese and Fox (1977) three domains downward. Eldredge and Gould kept it to questions of species sorting and developed the ideas further in collaboration with Vrba (1989), Vrba and Eldredge (1984), Vrba and Gould (1986). The rationale is that in so far as natural selection in combination with drift and abiotic factors causally influences speciation, it becomes natural to ask whether any or all somehow influence speciation to be differential and thus to show trends or arrows in time. Eldredge (1999: 23):

> If species are construed as real historical entities – with the aforementioned births, histories, and deaths – that in itself suggests the very real possibility that there are factors that bias those births and deaths. And that means that there very probably is a higher-level analogue to natural selection – species selection, or, as I have come to prefer, species sorting. The apparent directionality of trends, we suggested, might be the result of, say, differential species survival …

Species, as biological beings, reproduce, not by replication, but by splitting or cladogenesis, and as such also introduce a pattern of descent with modification that is in turn characterized by punctuated equilibria rather than phyletic gradualism. For Gould, the importance of punctuated equilibria had been that it lent insight into such trends:

> At the macroevolutionary level of trends, the theory of punctuated equilibrium proposes that established species generally do not change substantially in phenotype over a lifetime that may encompass many million years (stasis), and that most evolutionary change is concentrated in geologically instantaneous events of branching speciation. […] (Gould 1982: 383)

Returning to Kuhn, does macroevolutionary research call for a revolution? Gould was more revolutionary than Eldredge. For Eldredge (1985b), the Synthesis is merely "unfinished," and he takes: "… being called anti-Darwinian very personally. … I have always thought of myself as more or less a knee-jerk Neo-Darwinian" who has "felt that, with one or two major exceptions, my version of how the evolutionary process works lines up very well with Darwin's" (Eldredge 2006: 36).

He accepts, as Thomas Huxley did, that the theory of natural selection does not necessitate gradualness, and he understands his ideas as a geological extension of Darwinian theory, which is why he has tried to time speciation based upon geological strata. But the fact of the matter is that by understanding speciation as an outcome of ecological and geographical factors, much more so than as a consequence of genetic selection, he attributes causation to the abiotic environment rather than understanding it as the mere scene or level where genetic selection occurs.

Migration was one of the four elements of the Modern Synthesis, but if habitat tracking or niche construction have causal influence on the future course of evolution, then, as Lewontin (2000) pointed out, such selective behavior goes out from active organisms rather than selective environments that weed out passive individuals (also see Stigall this volume). The field of ecology was not part of the Modern Synthesis and became introduced post-synthetically, in both the micro- and macroevolutionary schools. Ecologically oriented scholars have done excellent work in finding causal factors that explain stasis, evolution, and extinction (Futuyma this volume; Bokma this volume) that go beyond genetic selection. These include drift, habitat-tracking, abiotic environmental changes, developmental constraints, and biophysical constraints as well as affordances. Such research has created a scientific necessity to integrate environmentalist and ecological approaches into the evolutionary discipline. Quite often, the environment provides causal non-genetic factors that influence how species originate, evolve, and die. Climate change, meteor impacts, soil, and atmosphere composition all can influence the evolutionary faith of species and higher taxa. Thermodynamic and other physical laws (Bohm 1980; Kauffman 1995; Prigogine 1980; Stanley 1979; Stanley this volume; Thompson 1917; von Bertalanffy 1928, 1940, 1950) also play crucial roles, as enablers as well as constrainers in both the formation of biochemical and anatomical form and its development and extinction. Ecology, systematics, and biophysics today are on the rise, and macroevolutionary scholars have most certainly contributed to their recognition by averring for a wider recognition of the importance of development, biophysics, and ecology in providing a fuller view on the evolution of life.

These ideas find their origin in the period designated as the eclipse of Darwinism and associate with presynthetic ideas, and these ideas also form part of the credo of scholars who call out for an extended evolutionary synthesis (Gontier 2010b; Pigliucci 2009; Pigliucci and Müller 2010). It is probably also the main reason why topics such as stasis and eco–evo–devo ranged so high at the 2009 Chicago bicentennial. As an audience participant, I found it nothing less than amazing how ideas associated with the "eclipse of Darwin," characterized as heretic by the founders of the Synthesis, and first reintroduced from within macroevolutionary streams of thought were presented as basic subject areas of Neo-Darwinism and approached from within universal Darwinian (Dawkins 1983; Dennett 1995) and universal selectionist (Czicko 1995) accounts.

3.2 Universalizing Punctuated Equilibria

The study of actual scientific practice teaches us that hypotheses and theories are tested for their use, not merely for their truth or corroboration to the truth, as Popper (1957) or logical empiricists assumed. Universal Darwinism and Universal Selection theory have proven to be useful to understand certain aspects of behavior. Punctuated equilibria theory has equally proven its use by designating a shared theoretical framework by which we can understand the history and evolution of natural kinds such as species and higher taxa. It is safe to say that many of its claims, including that species, extinctions, speciation, stasis, and gaps are real, are not only accepted as part of scientific language, but they are intensely studied as phenomena in need of explanation by scholars on both micro- and macrosides.

Punctuated equilibria theory and macroevolutionary research in general, and cladistic analysis in particular, have furthermore provided a universal methodological toolkit that is used to comprehend non-gradual vertical descent.

Just as the tenets of natural selection theory have been universalized across disciplinary boundaries and across the cosmic scale, so too the basic tenets of macroevolutionary outlooks in general and punctuated equilibria in particular have been universalized. Punctuated equilibria theory describes a pattern of evolution: Long periods of stasis are intermitted by short periods of rapid change (Eldredge and Gould 1972). Eldredge and Gould were among the first scholars within the field of paleontology to adopt cladistics which is an intrinsic genealogical method, and as such, they have been exemplar for macroevolutionary research of the living as well as the dead, both sub- and superorganically. Darwin argued that species share common genealogical descent, and as Eldredge rightfully notes, Hennig's cladistics method is the best scientific method available today to infer such genealogy or actual life history, because it is through such cladistic analysis that one can infer the patterns of evolution.

> Cladistics, in keeping with the original search for natural kinds, but imbued with an evolutionary perspective, boils down to a search for the actual distribution of the features, or 'characters', of organisms. It turns out that such characters are complexly, hierarchically nested just as are the taxa they define. (Eldredge 1999: 70)

The major breakthrough cladistics has brought forth is that it enables to infer genealogies of the actual life history of populations and species. But one can just as well use them to study the actual genealogy or "half-life" of chemical substances such as proteins or genes, as well as particular morphological and behavioral traits. These genealogies of natural kinds enable the building of phylogenetic taxonomies and hierarchies of the evolution of these specific phenomena. In short, traits can be morphological, but also genetic, behavioral, or cultural, and their origin, history, and ending can be mapped across time and space.

With the rise of whole-genome-sequencing techniques and our growing possibility to compute and study large data sets across time and space, fields such as meta- and comparative genomics are introducing macroevolutionary reconstructions of various aspects of life's history. The application of phylogenetic reconstruction methods that map the spread and distribution of genes has confirmed the pattern of punctuated equilibria in both extinct and extant taxa (Bokma 2002; Bokma and Monroe 2010; Pagel et al. 2006; Venditti and Pagel 2008, 2010).

Historical linguists have from the nineteenth century onward and in close parallel with evolutionary biologists attempted to reconstruct the natural genealogies of languages (Gontier 2011; Kressing et al. 2014). When Lorenz (1958) and Tinbergen (1963) first defined ethology, they wanted nothing less than to build a taxonomy of behavior and thus give the natural history or the genealogy of different behavioral traits by mapping when they first arose in time and how they evolved, spread, went extinct, or transitioned into something new. Equally, Nietzsche (1874) wanted nothing less than to launch the study of the genealogy of intellectual thought in our species, and when Campbell (1960, 1974) introduced the field of evolutionary epistemology, he wanted to map the evolution of knowledge from bacteria to humans. These ideas are now finally entering the sociocultural sciences. Today, anthropologists and archeologists are increasingly mapping the evolution of material culture by making use of phylogenetic and cladistics analyses. From within sub-branches of anthropology that are increasingly designated as macroarcheology and macroanthropology, scholars are reconstructing the beginning, evolution, and ending of "superorganic" traits across space and time.

Any and all such mappings by necessity take on a macroevolutionary approach, because we have to go beyond individual organisms, groups, and species, and focus on the genealogy or life history of the specific traits, i.e., genes, languages, or cultural artifacts themselves, beyond and across the beings that house them. And such research necessitates an investigation into their biogeography or how they populate the evolutionary landscape. Beyond gene pools, scholars are just beginning to map morphological, proteomic, viriomic, genetic, linguistic, and material culture pools. It are eventually such tree and network analyses that enable us to infer patterns of evolution, patterns that include stasis, branching, and merging. These patterns in turn give clues to the pace of evolution as well as the mechanisms that underlie the natural history or evolution of these phenomena.

Today, phylogenetic methodologies are applied to model macroevolutionary, ancestral–descent relationships of our kind (Eldredge and Tattersall 1982; Schwartz and Tattersall 2001; Wood this volume) as well as the cultural historical dispersal of material artifacts (Currie et al. 2010; Borgerhoff Mulder et al. 2006; Eldredge 2011; Eldredge and Tëmkin 2007; Lipo et al. 2006; O'Brien et al. 2002; Pagel and Meade 2006; Steele and Kandler 2010) and languages (Atkinson et al. 2008; Mace and Jordan 2011; Pagel et al. 2007). These phylogenetic models provide rigorous evidence for the pattern of punctuated equilibria in strands of linguistic and cultural evolution, as well as patterns of merging. Such research also necessitates new hierarchical thinking on cultural traits and behavioral phenomena, how they relate to their inventors, and how they transcend them. In sum:

> The macroevolutionary approach in archeology represents the most recent example in a long tradition of applying principles of biological evolution to the study of cultural change. Archeologists working within this paradigm see macroevolutionary theory as an effective

> response to the shortcomings of Neo-Darwinian biological evolution for studying cultural evolution. Rather than operating at the level of individual traits, macroevolutionary archeologists emphasize the role of hierarchical processes in cultural change. … Major cultural change, when it happens, is likely to be rapid, even revolutionary, with periods of rapid change separated by periods of relative stasis of actively maintained stability. (Zeder 2008: 1)

Concepts such as stasis, rapid turnovers, and constraints, or exaptations (Gould 1991; Gould and Vrba 1982) that form part of the technical language introduced by macroevolutionary scholars are also applied on a metaphorical level in the sociocultural sciences to describe certain aspects of linguistic and cultural evolution that cannot be accounted for by universal selectionism (d'Errico 2003; d'Errico and Stringer 2011; Tattersall 2014).

Universalizing punctuated equilibria, and thus developing a universal heuristic of the pattern, implies that we identify the mechanisms that underlie both the long periods of stasis and the short periods of rapid change. Several mechanisms can lead to stasis, including overall developmental, genetic, or epigenetic constraints (Carroll 2005; Davidson and Erwin 2006; Futuyma 2005, 2010 this volume; Gehring 1999; Gould 1977, 1989; Jablonka and Lamb 1995), stabilizing selection (Lieberman et al. 2005; Lieberman and Dudgeon 1996; Schmalhausen 1949) or drift (Bentley and O'Brien 2011; Bentley et al. 2004; Eldredge et al. 2005; Hubbel 2001; Kimura 1968, 1983; Masel et al. 2007). Numerous mechanisms can explain rapid change or speciation, such as symbiogenesis (Margulis 1970, 1999; Gontier 2007, 2012b), hybridization (Arnold 2006, 2008), lateral gene transfer (Gogarten et al. 2009), and virolution (Ryan 2006, 2009), the Baldwin effect (Baldwin 1896), or the Ratchet effect.

4 A Clash of Scientific Cultures

> Natural selection is not the beginning, as a physical force that stands above and beyond time, it is the fallout, the record of history. (Eldredge 1998: 96)

In the 1980s, Maynard Smith (1983, 1984) invited paleontologists to join the high table of evolutionary theory. The invitation was followed by polemic debates in various articles and books. One of those responses came from Niles Eldredge in his book *Reinventing Darwin: The Great Debate at the High Table of Evolutionary Theory*. In it, he characterized John Maynard Smith, Richard Dawkins, and George Williams as ultra-Darwinists sitting at one side of the table, and on the opposite side, he positioned Stephen Jay Gould, Elisabeth Vrba, Stanley Salthe, and himself, whom he characterized as naturalists.

Ultra-Darwinists were those scholars who adhered to a microevolutionary perspective by reducing the theory of evolution by means of natural selection to the *mechanical study* of how genes underlie evolutionary change at a population level. In other words, they had taken genetic change at a population level as normative and causal for evolutionary change at all ranks of life. Naturalists, on the other hand, studied evolutionary history from within a macroevolutionary perspective and endorsed more holistic epistemic stances that included paleontological, biogeographical, ecological, physical, and systematic areas of research. Their research had been descriptive and had focused on finding recurring patterns and trends in various ranks of life that surpass the population level.

The problem for the naturalists was that the *actual history* of how life changes does not converge with the normative framework of how life *should* change. At the heart of the debate lie radically opposing views on how evolutionary research has to be conducted, and whose epistemic stance, the microreductionist or macroholistic perspective was more scientific. While the ultra-Darwinists were concerned with detailing the underlying mechanism of evolution (which they associated exclusively with genetic selection), paleontologists were concerned with describing life's history. In other words, it was a clash of scientific cultures where different epistemic practices had evolved. Eldredge (1995: 2):

> What is really at stake is diametrically opposed suppositions of how evolutionary biology should be conducted. Maynard Smith put the geneticist's position succinctly: paleontology is about history. It is the paleontologist's job to elucidate what the evolutionary process has wrought over the last 3.5 billion years. But evolutionary theory is about how the evolutionary process works. And that, to a geneticist, means how genetic information, underlying the production of the physical appearance (and physiologies and behaviors) or organisms, comes to be modified over the course of time. What can a paleontologist, trapped under the dead hand of history, possibly have to say about the mechanics of genetic change? After all, science is supposed to concern itself with how things work.

Eldredge (1995: 5) blamed ultra-Darwinists a form of "physics envy," because "they seek to transform natural selection from a simple form of record-keeping, a filter that biases the distribution of genes between generations, to a more dynamic, active force that molds and shapes organic form as time goes by." By wanting to turn natural selection into a constant physical force, they endorsed a form of "extrapolationism" because they assumed that evolution by means of natural selection was a steady-state, lawlike phenomenon where gradual changes in gene frequencies suffice to explain evolution at all time and in all ranks of life. For the naturalists, the steady-state, lawlike theory of evolution by means of natural selection that had been developing did not converge with the actual history of life, and the geneticists' point of view could not simply be extrapolated to large-scale systems defined as populations, species, and ecosystems.

But on the other side of the table (Dawkins 1982b, 1983; Maynard Smith 1983), it was argued that natural history is merely descriptive and that an evolutionary theory should aspire to transcend narration of singular, non-repetitive, and unobservable events, by finding the laws or mechanisms, the underlying causes that bring forth evolution. For ultra-Darwinists, natural history research or the narration of actual historical origination and extinction events, in and of itself, cannot explain the causes and effects of evolution.

4.1 Delineating the Scientific

These debates are not unique to evolutionary biology. All sciences at one point or another have developed debates about whether their science should be primarily "descriptive" or "normative," for example, as philosophers of science, in turn inspired by moral philosophers, call it. Inspired by de Saussure, linguists, anthropologists, and sociologists, for example, know the debates as diachronic (historical and descriptive) over synchronic (structural and functional) research, and in physics, we find debates on Newtonian mechanics versus the new theoretical (bio) physics associated with relativity theory, quantum mechanics, chaos theory, and thermodynamics. The reason these debates are similar is they share common intellectual roots to which I turn in this section.

Should evolutionary science describe history or find its causes? And can science tackle origin questions (or extinction questions for that matter) in a scientific manner, or should science reduce itself to the study of repeatable and observable events? These questions associate with five basic and inter-related problems.

1. The problem of causality: Should science be historical and observational and perform a kind of record-keeping, or instead be aetiological, and focus on finding the underlying causes, the reasons why things are the way they are? The ultimate question here is: Can we go beyond record-keeping and find constancy in underlying mechanisms?
2. The problem of predictability: Should science explain what *is* (either mechanically or historically), or should it be able to explain what *ought* to be? In other words, can science make predictions on the future and inferences from the past?
3. The problem of uniformity or constancy over time: Does observation enable prediction and inferences on causality? Does what happens in the present immediately set the norm for the past as well as the future?
4. The problem of determinism or relativity: Is the universe determined by fixed laws, or is evolutionary change relative and dependable upon changing variables?
5. The problem of knowing: Can we, human observers and theoreticians, come to everlasting and objective truths, or merely develop temporary, local knowledge perspectives, from within a variety of epistemic stances?

The problems are indeed interrelated, because mechanisms can only be invoked when change, and that what causes it, is constant through time, or, in other words, determined by everlasting and unchanging laws. Predictability also depends upon the constancy of the mechanism over time. Favoring a normative view therefore implies a uniformitarian epistemic stance.

> So a connection has to be forged between uniformitarianism, gradualism, and reductionism: extrapolationism, the projection of commonly observed rates and processes as a prediction of what history *ought* to look like. (Eldredge 1999: 40)

In 1965, Gould wrote a paper titled *Is uniformitarian necessary?* wherein he distinguished between two types of uniformitarianism: substantive uniformitarianism that endorses "uniformity of rates of geological change" and methodological uniformitarianism that assumes "time and space invariance of natural laws." The pattern of punctuated equilibria proves that substantive uniformitarianism is false, and according to Gould (1965), methodological uniformitarianism, which assumes constancy of natural laws, is obsolete, most of all because laws are interspersed by other, often random events that function as disturbances to these laws. In other words, catastrophism, or the acceptance that abiotic random events such as meteor impacts can influence extinction and the future course of evolution of remaining life (e.g., Alverez et al. 1980), is valid.

Both Eldredge (1985b) and Gould (1987) spent a significant amount of their career fleshing out the origin of uniformitarian views and the notions on time that it brings forth. Uniformitarianism was introduced by Hutton, popularized by Lyell (1830, 1832, 1833), and endorsed by Whewell (1833). Uniformitarianists assume that adequate observational knowledge of the present should suffice to inductively infer the past and predict the future. But, these early scholars did so because they endorsed a lawlike physical view of nature in line with Newtonian mechanics that was inspired by Cartesian philosophy, where space, matter, and therefore also time are absolute and thus a priori determined, linear, constant, and irreversible. Such a stance enables uniformity and predictability toward the past as well as the future because laws are real, steady, and constant. These scholars started from the assumption that the universe, its laws, and our knowledge of them were the result of a benign God. As such, scientists became objective viewers that transcend nature and take on a God's-eye view (Pinxten 1997). This brings us right back to cosmology or better yet, the transition from studies on cosmo*gony* (the origin of the cosmos) to cosmo*logy* (the scientific study of its underlying causes, or aitia).

Newtonian physics is best known among philosophers and physicists for validating only Aristotle's *causa efficiens* which associates with the how question: how does matter move and how do phenomena function? It calls out for a mechanical approach, but it nonetheless assumes perfect, eternal, and harmonious functionality in the best of all possible worlds. In short, all these scholars, as firm believers of Judeo-Christian genesis so beautifully depicted in the *scala naturae* of their time, assumed that the cosmos followed perfect and constant, clocklike mechanical laws because they were, in deistic worldviews, the "hand of God" so to say. God did not have to intervene, like was the case in theistic worldviews, because nature follows predictable and teleological perfection.

Such views no longer form part of our worldview. As Eldredge put it:

> Most laws are descriptions of what happens given the existence of a set of conditions. Thus, while we might tend to think of gravity as some sort of process (gravity is one of the four basic 'forces' physicists investigate), Newton's and all subsequent 'laws' of gravity are simply generalizations about just what will happen given the existence of two hunks of matter of specified mass and distance apart. (Eldredge 1985b: 174)

With the rise of relativity theory, quantum mechanics, and chaos theory, we are necessitated to acknowledge that our cosmologies or epistemic frameworks are bounded by observational stances and probabilistic analyses, as well as colored by our cultural and theoretical frameworks. In our cosmologies, there is no God's-eye view, and there are no reifiable constant physical forces, or if there are, we do not know them. The best we can do is define the conditions of change of current and past events, generalize from there, and compare the generalizations, which brings forth a statistical analysis.

> Philosophers and scientists alike are fond of the idea that an evolutionary theory pertaining to any sort of system is unlikely to be established without a convincing mechanism, a well-articulated and corroborated process of change. … Darwin is still commonly supposed to have succeeded where all others before him had failed in establishing the credibility of evolution because he supplied a process that could plausibly account for the history of life: natural selection. … the contention involves whether pattern is preeminent over process in establishing credibility of an evolutionary theory.' (Eldredge 1998: 98)

The naturalists, mostly because of the importance they give to biophysics (Bohm 1980; Kauffman 1995; Prigogine 1980; Stanley this volume), have made the cosmological transition from a teleological to a relativist worldview and have allowed it to enter evolutionary science, while microevolutionary scholars endorse a more ambivalent view that tries to fit the old cosmologies into the new. I give two examples to explain this transition: the rise of natural history research and the clock metaphor.

Evolutionary science is an outgrowth of eighteenth- and nineteenth-century research on the natural history of the cosmos. Advances in physics had caused for a transition from a geocentric to a heliocentric worldview, which in turn had raised doubts about the divine order of the world which scholars had for centuries been depicting in scala naturae where the earth was traditionally placed in the center of the universe. From the Renaissance onward and culminating in the Enlightenment, scholars sought out the natural history of the universe (for an overview, see Duhem 1969, 1985).

Natural history research started in an age when travel stories of colonizers, gentleman scientists, and missionaries reported on foreign cultures, languages, political systems, and fauna and flora; when historical linguists were mapping the natural genealogies of the world's languages; when anthropologists *avant la lettre* were providing the first narrations of foreign cultures; and when moral philosophers such as Hobbes, Rousseau, Hume, Smith, and Mill debated the natural as opposed to divine condition of man as well as the status of "common goods," in "primordial" (which they associated with non-Western) as well as sociopolitical societies. In short, natural history research was characterized by encompassing attempts to narrate the natural history of the world and everything in it by referring to its natural "generation" or "development" in opposition to its divine origin.

The term evolution comes from the Latin noun ēvolūtiō (the act of reading, or unrolling a script, or narrating a history) and the verb ēvolvō (ex—out of and volvō—roll) and denotes development or unfolding. In this context, it was used in teleological, preformationist circles where especially medics and to some extent natural history students studied the growth of organismal form by trying to narrate its historical development through time. But it was equally used in this context, by cultural and linguistic scientists such as Auguste Schleicher or Max Müller; sociopolitical scientists such Auguste Comte and Herbert Spencer, Karl Marx, or Kropotkin; philosophers of history such as Hegel, Herder, and Nietzsche; or geologists such as Hutton, Lyell, or Bronn. In these circles, the study of the development of natural beings (languages, cultures, societies, political structures, or biological species) was therefore, more than anything, the study of natural, chronological sequences of change that beings go through, or stated differently, it is the study of historical change through time. From within the major languages used to develop the scientific worldview, i.e., ancient Greek, neo-Latin, English, German, and French, scholars first used terms such as "development," "entwicklung," and "dévelopmement," to refer to processes of historical change of sociopolitical, biological, or earthly form through time, and this development eventually became denoted as evolution.

These investigations into the natural history of the earth and its inhabitants raised doubts on the age of the globe, the fixity of natural kinds (languages, cultures, societies, and species), and whether or not we live in the best of all possible worlds. In sum, it questioned the traditional assumptions there existed on matter, space, and time. In particular, the latter gave way to two types of "conservation ethics": *musealisierung*, the nineteenth century is typified by the rise of the great natural history museums of France, London, Vienna, Brussels, and New York (see, e.g., Lübbe 1990); and patriotism and nationalism, the encounter of "the other" made especially European scholars emphasize and favor their own natural history. In particular in Prussia, works by Kant, Hegel, Fichte, Herder, and eventually Marx, Nietzsche, and Heidegger would take on the question of time in relation to history.

Before, the act of conducting historical research involved a narration of past events, and such narration was often biased toward Western history and cosmology. But the natural history students went further and wanted to find the patterns and deeper lying meaning of historical change as well as the ultimate nature of time. In short, they wanted to "scientify" history, which they did by postulating equally biased developmental and vitalist laws of generation and decay so typical of the nineteenth century. They understood time as directional. More specifically, time was directed toward a given end state, and development was understood as progression toward that end state. Such a view entails a continuation of teleological thinking.

Comte and Spencer, for example, assumed a steady progress from nomad to shepherd to industrialized and civilized man. Their analyses of time were inspired by religious and philosophical ideas on the world's beginning, duration, and eschatological ending. By adhering to a non-cyclical, linear, and teleological worldview, they assumed that form necessarily develops according to a given and fixed plan, if not of divine origin then through natural cause by means of natural (physical and developmental) laws. Social Darwinists eventually understood natural selection, interpreted as the survival of the fittest, as one such law. Natural history research induced a first attempt to find a deeper and higher pattern or order in history which brought forth the stadial, unilineal, orthogenic, and racial theories that eventually became used and abused to justify racism, imperialism, and hegemony, as well as the two world wars.

This type of historical research is today called historicism (Popper 1957), and after the havoc it had caused in the wars, and for good reason, such research became condemned. Instead of "morphologically" comparing cultures and placing them on a developmental scale, sociocultural and linguistic scientists averred for historical particularism and the incommensurability of cultures and languages, which brought forth synchronic instead of diachronic investigations. This means that comparative research became redefined, not as a historical discipline, but as a functionalist/structuralist discipline where cultural and linguistic "universals" are examined in their particular form in the variety of cultures, languages, and societal structures that exist. de Saussure (2006) in linguistics, Malinowski (1944) in anthropology, and Durkheim (1895) or Talcott Parsons (1975) in sociology averred for a synchronization of the sociocultural sciences where the organic and developing natural kinds of before became reinterpreted as closed systems underlain by functional operationalist, universal and determining dead structures that, in their operations, work similar to natural clocks and in accordance with steady laws. In philosophy, science was equally argued to progress and advance toward the truth, which is why they assumed that older theoretical frameworks could be reformulated in "more advanced," later-developed scientific frameworks through "bridge laws" that enable a harmonious synchronization of human thought toward knowledge. Such "synchronization" also entered the biological sciences, where mechanical, functionalist, and structuralist explanations of underlying causes became favored over historical narrations.

When the early natural history and evolutionary scholars were developing their ideas, they neither foresaw nor planned the havoc it would bring forth. It would be foolish to think that the terrors of colonialism and the world wars were brought forth by such research instead of economic, political, and religious reasons. Nonetheless, it would be equally foolish to deny that it contributed to the general *Zeitgeist* of the time and served as a justification for the crimes against humanity. We, who know the consequences of the early attempts to scientify the study of natural history, therefore have an obligation to investigate and try to understand what could have caused this early historical research to have gone so wrong. One of the reasons why historical research in the nineteenth century was typified by orthogenic views is that they lacked an evolutionary concept of time and instead continued to adhere to a teleological notion of time. Time is absolute, irreversible, lineal, progressive, and directed toward a final goal. And the reason matter attained its goal was because it was subjected to natural laws that gave directionality toward the end state. Eternal forces actualize inherent potential. That is why the missionaries, for example, did what they did. They assumed a favored position by claiming to have insight into the eschatological end state of the universe, and the "civilization" of "barbaric men," for them, was an act of charity wherein they helped actualize the potential of "underdeveloped" cultures.

To make this point more explicit, let us look into the second example, the metaphor of a clock. Hume (1739) really did hit the sore spot by treating matters of morality and matters of causality in the same way because both go from what *is* to what *ought* to be. We have seen the sun come up today, yesterday, and all the days of our lives as far as we can remember. Our ancestors who lived before us also saw the sun coming up on a daily basis, and we have scientific evidence that she did so from the moment earth was formed and started its orbit around it. We have good scientific theories today that explain why we perceive the sun as coming up and going down, and these theories predict that the orbits of the earth and sun will undergo no significant change for many ages to come. But in the end, the prediction that the sun will come up tomorrow is not 100 % guaranteed, and however probable the prediction is today, based upon the parameters available to us, it is *our expectation* that she will. And that expectation in the end is based upon uniformitarian, teleological, and clockwork notions of the universe. Hume on the other hand explained all predictions as well as moral prescriptions as human expectations and considered them the dominion of what we today call psychology.

Throughout the years, the problem has only intensified. The logical positivists held firm beliefs that our languages, be they expressed in logic, mathematics, or natural language, somehow relate to the world and that theories can be bridged because they are commensurable just because science progresses toward the truth. Wittgenstein provided proof based upon logic that we cannot demonstrate that language, mathematics included, refers to the world. Moreover, even if we did possess the perfect language to explain the state of affairs of the world, the mere fact of endorsing an evolutionary worldview necessitates that whatever we are able to predict today will only happen if the world does not change and if the parameters we define in our scientific language as valid and foundational remain so and thus do not alter at a different moment in time.

Western science had grown from assumptions that humans lived in a perfect world, that filled an absolute space, with an absolute and irreversible time, and humans were considered to have insight into the perfect mechanisms of the world, which they described in a perfect language that they had received from a perfect God. The age of the Renaissance and the age of Enlightenment were a celebration of this human capacity, a capacity that enabled to take on a "God's-eye view," and to become humanistic gods themselves, that know right from wrong, truth from falsehood, and future from the past.

Descartes, a firm believer, introduced mechanical thinking within the field of epistemology, and he was the first to introduce the idea that the natural world functions as a clock. He inspired Newton who inspired Paley, who inspired Dawkins, in thinking that we live in the best of all possible worlds because mechanical laws lead to an optimal and balanced cosmos. The clock metaphor is an argument for design that is based upon teleological arguments. In line with his mind–body distinction, Descartes fought animism by arguing that anatomically, organisms function perfectly according to physiological mechanisms that make their organs function as parts of a harmonious clock of which the sum of the parts defines the whole. The mechanical laws according to which they function so adequately for Descartes proved the existence of God. Paley, as an ardent adherent of intelligent design *avant la lettre*, would build upon these ideas, and this argument for optimalization is exactly what we find in Dawkins' (1995) work, who characterizes natural selection as a "blind watchmaker." Inspired by Fisher and the "hill-climbing metaphor" used to describe Sewall Wright's landscapes, and by introducing the idea of "accumulative selection," he has always had an ambiguity in what regards the predictive powers of selection:

> The theory of natural selection provides a mechanistic, causal account of how living things came to look as if they had been designed for a purpose. So overwhelming is the appearance of purposeful design that, [...] we still find it difficult indeed boringly pedantic, to refrain from teleological language when discussing adaptation. Bird's wings are obviously 'for' flying, [...]. (Dawkins 1982a: 161)

For Newton, time and space were absolute and irreversible, and truth was conceived to be singluar. William Whewell, another firm believer, attacked the standard historical thinking of his time that was characterized by narration and instead reinstated aetiology, the Greek study of causes. Whewell endorsed that inductive research enables generalizations on these causes. But he could only do so by accepting Hutton's and Lyell's principle of uniformity which is a direct consequence of living in a Newtonian steady-state universe where it finds its justification. In such a world, you have steady cause and steady effect, under all circumstances according to the same final or unchanging teleological forces. Empiricism might be observation based, it no less assumes, just like rationalist deductionism, that laws are constant. In fact, that was the main contribution scholars like Lyell and Whewell advanced: Insight into eternal laws comes from observation, not inborn categories of the mind. As Gould (2002: 102) put it:

> History presents two special problems: (1) frequent absence of evidence, given imperfections of preservation; and (2) uniqueness of sequences, unrepeatable in theory contingent complexity, and thereby distancing the data of history from such standard concepts as prediction, and experimentalization. We may epitomize the dilemma in the following way: many people define science as the study of causal processes. Past processes are, in principle, unobservable. We must therefore work by inference from results preserved in the historical record. We must study modern results produced by processes that can be directly observed and even manipulated by experiment – and we must then infer the causes of past results by their "sufficient similarity" ... which presents results. This procedure requires, as Mill ... and other philosophers recognized long ago, a methodological assumption of temporal invariance for laws of nature.

When Mayr (1961) was creating a discipline for evolutionary biology, he found himself troubled by cause and effect in biology. Ever since Newton, only the how question, the question about mechanism, was considered science. Aristotle's what and what for questions were abandoned because they introduce teleology, and that is exactly what the organizers of the Darwin centennial and microevolutionary biologists were reintroducing. Ernest Nagel, a logical positivist (one of those who believed we possess the perfect logical scientific language) and philosopher of physics, was a firm believer in reductionism and singular truth. Mayr was hesitant in adopting a genetic reductionist stance and argued instead for the unity of the genotype and phenotype. He distinguished between functional/operational (synchronic) and evolutionary (diachronic) biologists. The first study operational clocks (the physiology of organisms or "programs guaranteeing behavior that increases fitness"), and the latter, instead of asking "what for," ask "how come" which is a question directed toward the past that requires *historical analysis* (for a more elaborate discussion, see Gontier 2012a).

As such, Mayr revived the study of natural history and found a scientific way to get rid of the teleology. He validated the fact that biologists cannot and in fact need not make predictions about the future, in the same way physicists do. Because biology is a study of evolutionary history, it is directed toward the past, because it explains the present in a historical, sequential manner. Evolutionary biology is not a science that needs to predict the future, because the future of biological life is itself unpredictable. Mayr did excellent work in delineating biology as a scientific field in and of itself, a discipline that stands on its own two feet where physical laws cannot simply be bridged into evolutionary ones. Natural selection is a mechanism, not because it explains the future, but because it explains the past. Mayr never endorsed either teleology or teleonomy. Similar discussions were also held between biophysicists and classic physicists, in what regards the second law of thermodynamics, but these go beyond the scope of the present paper. For a good analysis, I refer the reader to the 1995 special issue in Physicalia Magazine, where Bricmont (1995) and Prigogine debated the problems of predictability in relation to matters of indeterminacy as brought forth by chaos theory (also in relation to the Sokal–Bricmont affair).

Returning to Eldredge' naturalists, these scholars conceive the clock to be more crooked. Lewontin (2000: 81), for example, in responding to Dawkins, said:

> In biology, this "what for" question is not the same as it is in the analysis of the parts of a motorcar or a clock. [...] In the case of the organism there are, of course, general functions such as motion, respiration and reproduction that are common, but there are many particular functions, peculiar to different life forms, that cannot be known in advance. In addition it is by no means true that every part serves a function. Many features of organisms are the epiphenomenal consequences of developmental changes or the functionless leftovers from remote ancestors.

We can easily see that it is a clash of epistemic cultures, one inspired by the old and one by the new. In particular, Lewontin's latter argument also makes implicit reference to Gould and Vrba's (1982) notion of exaptations, which is another means by which uniformitarianism can be fought. Naturalists take on a more relativistic view. In both science and philosophy, relativism carries bad connotations because "anything goes," but that need not be the case. What a relativist stance does most is mark a transition from epistemic monism toward epistemic pluralism. Stated differently, it rejects the notion of a single universe and a single truth, in favor of multiple explanations, and in the most fascinating scenario, it accepts ontological pluralism. What does that mean?

Fill a glass up to its exact middle. That glass is either half-full or half-empty. If you call it half-full, you might be an optimist, and if you call it half-empty, that might indicate you are a pessimist. Reality is that the glass is both half-full and half-empty, and whatever stance you prefer, it eventually lies in the eye of the beholder. Quantum mechanics has taught us the exact same thing in what regards waves and particles. And the micro–macroevolutionary debate has done the same for the study of the evolution of life; one can explain life by its genes, or by its natural history. It demonstrates first and foremost that a particular frame of reference underlies any and all scientific observation as well as theory formation, and that frame of reference is brought forth by either instruments or human observers and therefore also partially bounded by both. Science is human, and in reality, truth need not be singular for the glass is both half-full and half-empty.

We no longer live in a Newtonian world where space is absolute, time is irreversible, everything works in the form of a mechanical clock, and observers take on a God's-eye view to deduce the everlasting truths about the universe. All we have are our frames of reference, and these immediately call out for epistemic pluralism. We can scientifically justify a microstance as well as a macrostance, there is scientific proof for both, and we can compare both to one another and incorporate them into yet another framework. But we cannot transcend these frameworks. All we can do is develop as many frames of reference as possible. The latter entails a reconstruction of biorealities, work that was first taken on by the early natural history scholars and that today is well on its way to define actual scientific practices. I return to this latter point in the following section.

Simpson (1944), for example, when introducing the macroperspective in paleontology, was most of all thinking through the consequences that thermodynamics, relativity theory, and quantum physics have on our notions of evolution. Such branches of physics necessitate more pragmatic and operational, instrumental definitions of causality as well as science, which becomes understood as a practical and instrumental attempt to infer the conditions of change. Pragmatism and instrumentalism (other words used to denote relativism) were first introduced by scholars such as James (1907, 1909) and Duhem (1954). Teggart (1916, 1925), who inspired Eldredge in both his definition of punctuated equilibria and the epistemic stance it necessitates, was very much inspired by these scholars when he wrote that:

> 'As the sciences have developed,' William James says, 'the notion has gained ground that most, perhaps all, of our laws are only approximations.' 'Investigators have become accustomed to the notion that no theory is absolutely a transcript of reality, but that any of them may from some point of view be useful. Their great use is to summarize facts and to lead to new ones.' Obviously, then, it is not the function of science to gratify the desire of men to certainty. No scientific 'law' is to be regarded otherwise than as a 'working hypothesis' which has proved of value in organizing some phase of experience. 'The conception,' Bertrand Russell says, 'of the 'working hypothesis', provisional, approximate, and merely useful, has more and more pushed aside the comfortable eighteenth century conception of 'law of nature'. (Teggart 1925: 163)

Take something as simple as the syllogism, if A then B, so if you have A, then it is B, at all times and everywhere. It forms the core of cause and effect in Aristotle as well as Newton's world, but it is not part of ours. If the evolutionary sciences have taught us anything, it is that if A, that A sometimes evolves by means of natural selection, sometimes by drift, sometimes by symbiogenesis, sometimes by lateral gene transfer, sometimes due to external events in the abiotic world, sometimes just not at all, and most of the time through a combination of all the previous at different levels of an all but perfect agglomeration we humans call "organism." The point is that how evolution occurs is conditional upon the parameters, and these are not stable. Another way of saying this is that it is a logical necessity, within an evolutionary worldview, to assume that the means by which change occurs themselves are variable through time, and what caused evolution in the past might neither set the conditions for the present, nor for the future. The directionality we sometimes find in life is there because of historical reasons and might change in the future. It all depends upon the stability of the initial conditions, which themselves change in an evolving world.

Such a view does not at all call out for the end of science because "anything goes," and it merely marks the beginning of a new epistemic stance, one that is characterized by epistemic plurality and the reconstruction of biorealities:

epistemic referential frameworks of life that describes the genealogy of natural kinds, biological individuals. And these genealogies lend insight into the underlying mechanisms or heuristics, i.e., the specific conditions through which phenomena originate, change, and end.

Both the micro- and macroperspectives have, from within their own set of references, created new and inspiring referential frameworks, evolution according to genes, evolution according to environmental conditions. Others have developed evolutionary views according to lateral gene transfer, symbiogenesis, drift, physics, etc. The frameworks enable us to compare and as such approximate or corroborate not one but multiple biological realities and provide a fuller understanding of just how it is that evolution comes about. And such work, most of all, is brought about through reconstructions of natural history.

It is no coincidence that both the micro- and macroperspectives reintroduce a notion of organicism, by revitalizing the dead and investigating their natural history, according to genetic shifts in populations over time, or according to external factors that define the conditions of change. It is a return to the natural history scholars' epistemic framework, something which can only be celebrated.

5 Conclusion: Reintegrating History by Reconstructing Biorealities of the Dead

> Patterns in the history of life have been suggesting for at least a half century that there are regularly occurring sets of conditions that seem to control evolutionary activity – dampening it, often for longer periods, and triggering often rapid evolution at other times. Like ecological succession, evolution produces, not isolated events, but repeated pattern which hold clues to how the process works – specifically, how the physical world of matter-in-motion impacts the biological realm. (Eldredge 1998: 5)

What was the eclipse of Darwinism, really? Why it was one of the most illuminating periods in the reconstruction of the natural history of life. It marks a time that coincides with the introduction of the genies: morphogenesis, ontogenesis, phylogenesis, monogenesis, polygenesis, heterogenesis, embryogenesis, epigenesis, spermatogenesis, somitogenesis, symbiogenesis, psychogenesis, anthropogenesis, biogenesis, abiogenesis, and so on. Any and all terms defined new fields that set out to narrate the natural history of various aspects of life, by giving the natural genealogy of the parts that make up the wholes and by reconstructing their generation and decay through space and time. But, alas, without cladistics at their disposal, and still too much inspired by their teleological worldviews, most of the researchers, with a fair exception of the symbiologists, endorsed a form of orthogenesis, a straight-line evolution toward a predefined end state. There is no scientific proof for any of the orthogenic claims that were ever made, and with the foundation of the Modern Synthesis, the genies turned into logies where natural selection was the underlying mechanism that was assumed to explain all. Subsequently, synchronization followed.

Nonetheless, studies on the natural history of life remained an activity performed either in the margins or outside the Modern Synthesis. Scholars who studied symbiosis were one of the first to accept life's multilinearity and multidimensionality. The theory of symbiogenesis, a term first coined by Merzhkowsky (1905, 1910), from its onset, has explained how multicellular life historically evolved through symbiotic mergings and has rejected a single common origin from its very beginning. Such symbiotic mergings, in a very real sense, entail a form of time travel, where the old can merge into the new, but the new can equally merge with the old. Reticulate evolution brings forth multidirectionality and multidimensionality, concepts that also formed the hallmark of neo-Lamarckianism, and it defines studies on cytology, where scholars investigate the actual processes that occur in cellular life; ecology, which studies the natural and actual organism–organism and organism–environmental interactions; embryology, which studies the actual developments of organisms; and paleontology, which studies the actual natural history of species and higher taxa. All these fields are first and foremost historically and data driven rather than theory-driven sciences. Their scholars aim to explain real-life events and at present that implies a narration of past events much more so than it provides explanatory frameworks that point out their constant underlying conditions of change. Nonetheless, the patterns such researchers found can be applied and generalized outside their context of discovery. Symbiogenesis, for example, is not a unique event in the history of eukaryotic cells, it is a repeated mechanism whereby multiple life forms speciate (Brucker and Bordenstein 2013; Carrapiço and Rodrigues 2005; Carrapiço forthcoming; Gontier 2015). Scholars that associate with these fields are therefore increasingly calling out for a revision of the Modern Synthesis because their facts do not add up with the theorized claims of how life *should* evolve, while they can explain many real-life evolutionary events.

In contrast, the Modern Synthesis developed because of mathematical generalizations of theoretical rather than actual population genetics and how they behave in imaginary adaptive landscapes. Mathematics is also what enabled Neo-Darwinians to generalize and "universalize" natural selection theory to all domains of life, where prisoner dilemmas and other mathematical models enable theoretical scholars to model the evolution of life without so much as looking at its actual history. We must not forget that the Modern Synthesis itself was formulated before the genetic code was deciphered and before scholars could actually "see" or cladistically map the chemical structures and compounds that make up cells, organisms, and populations.

Either fields such as exobiology, cytology, epigenetics, paleontology, ecology, and embryology were excluded from the general framework of the Modern Synthesis, or their theoretical frameworks and empirical evidence only accumulated after the Modern Synthesis was formed. What unites all these diverse fields is that they have developed diverse techniques to enable better insight into the actual natural history that life has undergone and the various paths it has chosen. Large genome-sequencing techniques, proteomics, the viriome, the microbiome, metabolic pathways, behavior, and material culture, today, are studied phylogenetically, by mapping their actual historical and geographical dispersals across space and time. Equally, in fields such as biophysics, scholars are just beginning to map the natural history of the earth, by documenting the genealogy, the beginning, "life span," and ending of chemical cycles such as the nitrogen and water cycle (Volk 1995, 1998). Doolittle (1999, 2010) together with Dawkins one of the most fierce opponents of the Gaia hypothesis—that views our planet as a living organism with a beginning, life span, and ending in time—is now actively engaged in reconstructing actual cell histories by understanding them as biological individuals. The work indeed is noble, but it nonetheless makes use of the same kind of reasoning that scholars such as Margulis and Lovelock (1974) used when they understood earth as a natural being.

We currently find ourselves in the midst of a return to natural history, and we have finally found the instrumental tools that enable to reconstruct the evolutionary lineages of the dead. These reconstructions together shed new light on the evolution of the living. By decomposing the living into various sub- and superunits, and investigating the individual history of each and every one of them independently, we have reached an age of epistemic plurality (Gontier 2010a). We can describe organisms based upon the evolution of their genes, their microbiome, their viriome, their morpho- or embryogenesis, etc. And together, that brings forth a better understanding of various and distinct biorealities.

Scholars associated with the extended synthesis are today proving that besides natural selection, the conditions of change that define other evolutionary mechanisms can also be generalized toward other domains of life and applied beyond their original context of discovery. It is hard to think about evolutionary patterns without the mechanisms, the conditions of change, by which they are brought forth. Natural selection was mainly defined through the study of eukaryotic animal lineages and first focused on extant species in their natural habitat as well as extinct species as they present themselves in the fossil record. Today, when possible, comparative genetic research (as opposed to morphometric comparisons) lends insight into the exact ancestral–descendant relationships of animals in deep time, and as described above, natural selection has been "universalized" to nonbiological domains of life. Drift was first defined based upon research on the existence of neutral organismal traits (Darwin 1959). Later, it became defined in theoretical terms as a statistical property of theoretical population models, first at a molecular genetic level (Kimura 1968, 1976, 1983) and then at a species and above species as well as environmental level (Hubbel 2001). The mechanisms of hybridization (Arnold 2006) (divergence with gene flow or introgression) were mostly obtained from the study of plants and, albeit to a lesser extent, animals. Symbiogenesis (Wallin 1927; Kozo-Polyanski 1924; Margulis 1970, 1981) became defined based upon the study of organismal, symbiotic interactions as well as investigations into the evolutionary origins of the various organelles present in the eukaryotic cell. The mechanisms that underlie lateral gene transfer (transformation, transduction, conjugation) were first obtained from studies into how prokaryotes, viruses, protozoa, and fungi underlie infection and disease in organisms (Avery et al. 1944; Lederberg and Tatum 1946; Morse et al. 1956). Today, genealogical research enables generalizations of the underlying mechanisms to all domains of life, because whole-genome analyses enable scholars to calculate the exact amounts of species-specific and alien (acquired) genes (Doolittle 1999, 2010; Woese 1998). Both epigenetics and evo–devo investigate how evolution can be informed by specific post-genetic (e.g., post-translational changes in protein structures) and more general ontogenetically acquired characteristics (Jablonka and Lamb 1989).

The history of life is characterized by repeating patterns and trends. Evolutionary patterns are hard to describe without making reference to evolutionary theories that identify the pattern and seek out to explain the likelihoods or necessities by which it occurs. We are still very much in the dark about those conditions, but scholars have made good progress in what regards recurring patterns.

Evolution appears to move across a horizontal, a vertical, and a reticulate axis. Not all life forms remain extant, and whole types and all individuals go extinct. Not all life forms remain constant. That is as basic as it gets. To give more patterns, we need to refer to evolutionary theories such as natural selection, symbiogenesis, and lateral gene transfer.

Conditions of change are by and large defined in terms of internal and external, biotic and abiotic, and organismal and environmental parameters. The principles of conditional change are defined in terms of evolutionary mechanisms. Evolutionary mechanisms in turn are theoretical generalizations, working hypothesis on the natural history of various life strands through time. In other words, they are generalizations of how conditional change appears to have occurred in the past, and they can be tested for how well they are generalizable toward other phenomena beyond their context of discovery.

Evolution by drift, for example, gives a random or non-directional pattern. The outcome of natural selection is vertical descent with modification, and the emerging pattern can be either unilineal or bifurcating. Symbiogenesis or LGT gives reticulate, multidirectional patterns of horizontal mergings.

Based upon how widespread and persistent the evolutionary mechanism is across lineages, the emerging pattern can be argued to be stable if not permanent in time. Selection will always lead to diversification, drift will always be random, and symbiogenesis or reticulate evolution in general will always involve mergings. If either pattern seizes to emerge in time, the mechanism seizes to be active, or better yet, the conditions for that particular kind of change are not met. Patterns are thus causally explained by referring to postulated parameters that define conditional change. And that is eventually what an evolutionary mechanism is, it is not a constant force.

Moreover, these reconstructions again necessitate us to acknowledge that truth need not be singular. At one point in time, eukaryotes evolved according to symbiogenesis, at another point in time, they did so by means of natural selection, and at yet another period in time, they evolved by drift. Most of the time, they do not evolve at all, and when they do, they evolve by a combination of these means, simply because the organisms themselves are compositional structures. As such, we also need to investigate whether there is some higher-order sorting of mechanisms, and what would define the parameters of such sorting.

Evolutionary mechanisms are theoretical frameworks that define the parameters and conditions of change under well-defined circumstances. Genes evolve by means of natural selection if certain conditions are met, and symbiogenesis and drift equally occurs when certain well-defined conditions are met. When these conditions are not met, life does not evolve according to that pattern. What we need to do is find the conditions of change.

Evolution is spontaneous, random, and selective, and species come into being through splitting and merging, which can occur slow or fast. We can use a lot of fancy words to say things differently, but that is basically it. These are the heuristics or working hypotheses scholars have been working with, and the fancy words used to describe them imply a scientification of the problems. All have proven their use, in so far as all have been put into scientific theories and mathematical models that explain the data they aim to describe.

If time is relative because it is a measurement of movement of matter through space, and if how matter "moves" is variable, then it is more relative to repetition of initial conditions and thus more uncertain. Most of all, these theories are awaiting an epistemic framework that enables us to deal with the multiple dimensions of the evolutionary biorealities they have opened, not in the least by an evolutionary reconceptualization of our notion of time. Never has a generation of natural history scholars been studying actual historical material phenomena in time and space without a solid concept of either. Time is a hallmark of evolutionary science. Without time, there would be no origin or evolution of life, where through time living organisms change. Nonetheless, as natural and real as time appears to us, much of how we conceptualize time is the result of sociocultural constructs.

For the ancient Greeks and the Vedic Indians, time was cyclical and repetitive. Earth, fire, water, wind, and void/aether, the periodic table *avant la lettre*, made up the micro-, meso-, and macrocosmos, which were considered embedded structures. Much like Russian dolls, they made up inanimate, animate, and divine matter (a fire hearth, language, and the God Prometheus, for example). These beings underwent generation and decay, but eventually, time was cyclical, because without a second law of thermodynamics, the elements themselves did not undergo decay, but rather they kept their integrity which is why the world could cyclically start over and reincarnate. For the Vedic Indian as well as ancient Greek philosophers, time is not real. Aristotle, for example, defined time in terms of the motion of matter, and this concept was also foundational for Newton and the whole of Western science. Although time is absolute, because space and matter are, time itself is a derivative of matter in motion, which is why we use different "calendars" and time scales, such as geological time, sidereal or tropical zodiac time, or heat. Our switch from a geocentric to a heliocentric worldview did not affect our watches in any way. Nonetheless, if we or any other living species travel a couple of thousand miles, we most certainly experience a change in time, and biological individuals have evolved circadian rhythms that enable them to deal with them. Time, whatever it is, is most certainly real for the living.

Equally, an organismal body is made up of different moving elements, which each follows their own pace. The cells of our body form complex and differentiated structures such as bones, vascular systems, organs, neurons, and a brain.

These structures as well as the cells and the genes they are composed of differ in how they occupy space and time. Your skin cells rejuvenate at a monthly basis. The cells undergo mitosis, and the older ones die. Every 6 weeks, your lung cells do the same. Your organelles and the genetic material they contain replicate and regenerate at different time intervals than the cell's nucleus. All have a different life span, a different beginning, and ending in time. In so far as the structures that compose you have come into existence before, in previously existing generations, they are more constant in time than you. The compositional structure that is "you" only exists once in so far as your specific compositions never come into existence twice. You have a clear beginning, life span, and ending in time. But it takes numerous generations of living cells to repeat the structures that form your existence over a longer period in time. Equally, it takes a couple of days before all cells that make up your body die, and the microbial communities that inhabit you live on for quite some time.

Current research is evolutionizing all. We are evolving toward a recognition of the existence of multiple units, levels, mechanisms, and kinds of evolution, and it is the challenge of our and future generations to investigate how they together make up the multilinear and multidimensional biological realities that have evolved (Gontier 2010a).

Acknowledgments The work was written with the support of the John Templeton Foundation (grant ID 36288) and the Portuguese Fund for Scientific Research (grant ID SFRH/BPD/89195/2012). The author is grateful to Diederik Aerts, Luís Borda de Agua, Francis Heylighen, Octávio Mateus, and Emanuele Serrelli for useful comments to the text.

References

Alvarez LW, Alvarez W, Asaro F, Michel HV (1980) Extraterrestrial cause for the cretaceous-tertiary extinction. Science 208(4448):1095–1108. doi:10.1126/science.208.4448.1095
Arnold ML (2006) Evolution through genetic exchange. Oxford University Press, New York
Arnold ML (2008) Reticulate evolution and humans: origins and ecology. Oxford University Press, Oxford
Atkinson QD et al (2008) Languages evolve in punctuational bursts. Science 319(5863):588
Avery OT, Macleod CM, McCarty M (1944) Studies on the chemical nature of the substance inducing transformation of pneumonococcal types. Induction of transformation by a deoxyribonucleic acid fraction isolated from pneumococcus type III. J Exp Med 79:137–157
Axelrod R (1981) The evolution of cooperation. Science 211:1390–1396
Baldwin JM (1896) A new factor in evolution. Am Nat 30(354):441–451
Barbieri M (2008) Biosemiotics: a new understanding of life. Naturwissenschaften 95(7):577–599
Barbieri M (ed) (2008) The codes of life: the rules of macroevolution. Springer, Berlin
Bentley RA, O'Brien MJ (2011) The selectivity of social learning and the tempo of cultural evolution. J Evol Psychol 9:125–141
Bentley RA, Hahn MW, Shennan SJ (2004) Random drift and culture change. Philos Trans R Soc B 2711547:1443–1450
Bohm D (1980) Wholeness and the implicate order. Routledge, London
Bokma F (2002) Detection of punctuated equilibrium from molecular phylogenies. J Evol Biol 15:1048–1056

Bokma F, Monroe MJ (2010) Punctuated equilibrium in a neontological context. Theory Biosci 129(2–3):103–111
Borgerhoff Mulder M, Nunn CL, Towner M (2006) Macroevolutionary studies of cultural trait variation: the importance of transmission mode. Evol Anthropol 15:52–64
Bowler PJ (2003) Evolution: the history of an idea, 3rd revised and expanded edition. University of California Press, Berkeley
Bradie M (1986) Assessing evolutionary epistemology. Biol Philos 1(4):401–459
Brandon RN (1982) The levels of selection. In: Brandon RN, Burian RM (eds) Genes, organisms, populations: controversies over the units of selection 1984. MIT Press, Cambridge, pp 133–139
Brandon R, Burian Richard M (eds) (1984) Genes, organisms, population: controversies over the units of selection. MIT Press, Cambridge
Brandon RN, McShea D (2010) Biology's first law: the tendency for diversity and complexity to increase in evolutionary systems. University of Chicago Press, Chicago
Bricmont J (1995) Science of chaos or chaos in science? Physicalia Mag 17(3–4):159–208
Brucker RM, Bordenstein SR (2013) The capacious hologenome. Zoology 116(5):260–261
Campbell DT (1960) Blind variation and selective retention in creative thought as in other knowledge processes. Psychol Rev 67(6):380–400
Campbell DT (1974) Evolutionary epistemology. In: Schlipp PA (ed) The philosophy of Karl Popper, vol I. LaSalle, New York, pp 413–459
Carrapico F (forthcoming) In: Gontier N (ed) Reticulate evolution. Springer, Dordrecht
Carrapiço F, Rodrigues T (2005) Symbiogenesis and the early evolution of life. Proc SPIE 5906:59060R-1–59060R-4
Carroll SB (2005) Endless forms most beautiful: the new science of evo-devo. WW Norton, New York
Cech TR (1986) RNA as an enzyme. Sci Am 255(5):76–84
Crick FHC (1968) The origin of the genetic code. J Mol Biol 38:367–379
Currie TE, Greenhill SJ, Mace R (2010) Is horizontal transmission really a problem for phylogenetic comparative methods? A simulation study using continuous cultural traits. Philos Trans R Soc B 365(1559):3903–3912. doi:10.1098/rstb.2010.0014
Cziko G (1995) Without miracles: universal selection theory and the second Darwinian revolution. MIT Press, Cambridge
d'Errico F (2003) The invisible frontier: a multiple species model of the origin of behavioral modernity. Evol Anthropol 12:188–202
d'Errico F, Stringer CB (2011) Evolution, revolution or saltation scenario for the emergence of modern cultures? Phil Trans R Soc B 366:1060–1069
Darwin C (1859) On the origin of species by means of natural selection, or the preservation of favored races in the struggle for life. John Murray, London
Davidson EH, Erwin DH (2006) Gene regulatory networks and the evolution of animal body plans. Science 311:796–800
Davidson EH, Erwin DH (2009) Evolutionary innovation and stability in animal gene networks. Part B: molecular and developmental evolution. J Exp Zool 314B(3):182–186
Dawkins R (1976) The selfish gene. Oxford University Press, New York
Dawkins R (1982a) The extended phenotype. Oxford University Press, New York
Dawkins R (1982b) Replicators and vehicles. In: Brandon NR, Burian RM (eds) Genes, organisms, populations 1984. MIT Press, Cambridge, pp 161–179
Dawkins R (1983) Universal Darwinism. In: Hull DL, Ruse M (eds) The philosophy of biology. Oxford University Press, New York, pp 15–35
Dawkins R (1995) River out of Eden. Basic books
Dawkins R (1996) Climbing mount improbable. W. W. Norton and Company, New York
De Duve C (1991) Blueprint for a cell: the nature and origin of life. Neil Patterson Publishers, Burlington
De Laguna GA (1962) The role of teleonomy in evolution. Philos Sci 29(2):117–131
de Saussure F (2006) Writings in general linguistics. Oxford University Press, New York
Dennett D (1995) Darwin's dangerous idea. Simons and Schuster, New York
Doolittle FW (1999) Phylogenetic classification and the universal tree. Science 284:2124–2129. doi:10.1126/science.284.5423.2124

Doolittle FW (2010) The attempt on the tree of life: science, philosophy and politics. Biol Philos 25:455–473
Duhem P (1954) The aim and structure of physical theory, 2nd edn (1991). Princeton University Press, Princeton
Duhem P (1969) To save the phenomena, an essay on the idea of physical theory from Plato to Galileo. University of Chicago Press, Chicago
Duhem P (1985) Medieval cosmology: theories of infinity, place, time, void, and the plurality of worlds. University of Chicago Press, Chicago
Durkheim E (1895) The rules of sociological method, preface to the second edition (trans: Halls WD, 1982). The Free Press, New York
Edwards AWF (1998) Natural selection and the sex ratio: fisher's sources. Am Nat 151(6):564–569. doi:10.1086/286141
Eigen M (1996) Steps towards life: a perspective on evolution. Oxford University Press, New York
Eigen M, Schuster P (1977) The Hypercycle: a principle of natural self-organisation. Part A: emergence of the hypercycle. Naturwissenschaften 64:541–565
Eldredge N (1971) The allopatric model and phylogeny in Paleozoic invertebrates. Evolution 25(1):156–167
Eldredge N (1985a) Unfinished synthesis: biological hierarchies and modern evolutionary thought. Oxford University Press, New York
Eldredge N (1985b) Time frames. Simon and Schuster, New York
Eldredge N (1989) Macroevolutionary dynamics: species, niches and adaptive peaks. McGraw Hill, New York
Eldredge N (1991a) The miner's canary: extinctions past and present. Prentice Hall Books, New York
Eldredge N (1991b) Fossils: the evolution and extinction of species. Abrams, New York
Eldredge N (ed) (1992) Systematics, ecology and the biodiversity crisis. Columbia University Press, New York
Eldredge N (1995) Reinventing Darwin: the great debate at the high table of evolutionary theory. Wiley, New York
Eldredge N (1997) Dominion. University of California Press, Oakland
Eldredge N (1998) Life in the balance: humanity and the biodiversity crisis. Princeton University Press, Princeton
Eldredge N (1999) The pattern of evolution. W. H. Freeman and Co., New York
Eldredge N (2006) Confessions of a Darwinist. V Q Rev Spring 82(2):33–53
Eldredge N (2008) Hierarchies and the sloshing bucket: toward the unification of evolutionary biology. Evol Educ Outreach 11:10–15
Eldredge N (2011) Paleontology and cornets: thoughts on material culture. Evol Educ Outreach 4:264–373
Eldredge N, Gould SJ (1972) Punctuated equilibria: an alternative to phyletic gradualism. In: Schopf TJM (ed) Models in paleobiology, Freeman, Cooper and Co., New York, pp 82–115
Eldredge N, Salthe SN (1984) Hierarchy and evolution. Oxf Surv Evol Biol 1:184–208
Eldredge N, Stanley SM (eds) (1984) Living fossils. Springer, New York
Eldredge N, Tattersall I (1982) The myths of human evolution. Columbia University Press, New York
Eldredge N, Tëmkin I (2007) Phylogenetics and material cultural evolution. Curr Anthropol 48:146–153
Eldredge N, Thompson JN, Brakefield PM, Gavrilets S, Jablonski D, Jackson J, Lenski R, Lieberman B, McPeek M, Miller W (2005) The dynamics of evolutionary stasis. Paleobiology 312:133–145
Eliav-Feldon M, Isaac B, Ziegler J (2009) The origins of racism in the west. Cambridge University Press, Cambridge
Fanon F, Philcox R (2004) The wretched of the earth/Frantz Fanon; translated from the French by Richard Philcox; introductions by Jean-Paul Sartre and Homi K. Bhabha. Grove Press, New York

Favareau D (2005) Founding a world biosemiotics institution: the international society for biosemiotic studies. Sign Syst Stud 33(2):481–485
Fisher RA (1930) The genetical theory of natural selection. Clarendon Press, Oxford
Fox S, Dose K (1972) Molecular evolution and the origin of life. W.H. Freeman and Co, San Francisco
Futuyma D (2005) Evolution. Sinauer Associates, Sunderland
Futuyma D (2010) Evolutionary constraint and ecological consequences. Evolution 64(7):1865–1884. doi:10.1111/j.1558-5646.2010.00960.x
Gehring WJ (1992) The homeobox in perspective. Trends Biochem Sci 17(8):277–280. doi:10.1016/0968-00049290434-B
Gehring WJ (1999) Lifting the lid on the homeobox discovery. Nature 399:521–522
Ghiselin M (1974) A radical solution to the species problem. Syst Zool 23(4):536–544
Gilbert W (1986) Origin of life: the RNA world. Nature 319:618. doi:10.1038/319618a0
Godfrey-Smith P (2009) Darwinian populations and natural selection. Oxford University Press, New York
Gogarten M, Gogarten JP, Olendzenski L (eds) (2009) Horizontal gene transfer: genomes in flux. Humana Press, Springer
Gontier N (2006a) Evolutionary epistemology. In: Fieser J, Dowden B, Beebe J (eds) The internet encyclopedia of philosophy. http://www.iep.utm.edu/e/evo-epis.htm
Gontier N (2006b) Introduction to evolutionary epistemology, language and culture. In: Gontier N, Van Bendegem JP, Aerts D (eds) Evolutionary epistemology, language and culture: a non-adaptationist systems theoretical approach. Springer, Dordrecht, pp 1–26
Gontier N (2006c) Evolutionary epistemology and the origin and evolution of language: taking symbiogenesis seriously. In: Gontier N, Van Bendegem JP, Aerts D (eds) Evolutionary epistemology, language and culture: a non-adaptationist systems theoretical approach. Springer, Dordrecht, pp 195–226
Gontier N (2007) Universal symbiogenesis: a genuine alternative to universal selectionist accounts. Symbiosis 44:167–181
Gontier N (2010a) Evolutionary epistemology as a scientific method: a new look upon the units and levels of evolution debate. Theor Biosci 129(2–3):167–182
Gontier N (2010b) Darwin's legacy. Theor Biosci 129(2–3):77–87
Gontier N (2011) Depicting the tree of life: the philosophical and historical roots of evolutionary tree diagrams. Evol, Educ Outreach 4(3):515–538
Gontier N (2012a) Applied evolutionary epistemology: a new methodology to enhance interdisciplinary research between the human and natural sciences. Kairos 4:7–49
Gontier N (2012b) Introducing universal symbiogenesis. In: Pombo O et al. (eds) Special sciences and the unity of science. Springer, Dordrecht, pp 89–111 (Logic, Epistemology, and the Unity of Science 24). doi: 10.1007/978-94-007-2030-5
Gontier N (ed) (2015) Reticulate evolution. Springer, Dordrecht
Gould SJ (1965) Is uniformitarianism necessary? Am J Sci 263:223–228
Gould SJ (1977) Ontogeny and phylogeny. Belknap, Harvard University Press, New York
Gould SJ (1981) The mismeasure of man. W. W. Norton, New York
Gould SJ (1982) Punctuated equilibrium—a different way of seeing. New Sci 94(Apr. 15): 137–139
Gould SJ (1986/1989) Wonderful life. Penguin books, London
Gould SJ (1987) Time's arrow, time's cycle. Harvard University Press, Cambridge. ISBN 0-674-89198-8 online preview
Gould SJ (1991) Exaptation: a crucial tool for an evolutionary psychology. J Soc Issues 47(3):43
Gould SJ (2002) The structure of evolutionary theory. Harvard University Press, Cambridge
Gould SJ, Eldredge N (1977) Punctuated equilibria: the tempo and mode of evolution reconsidered. Paleobiology 3(2):115–151
Gould SJ, Eldredge N (1993) Punctuated equilibrium comes of age. Nature 366(6452):223–227. doi:10.1038/366223a0

Gould SJ, Lewontin RC (1979) The spandrels of san marco and the panglossian paradigm: a critique of the adaptationist programme. Proc Royal Soc Lond B 205:581–589
Gould SJ, Vrba ES (1982) Exaptation: a missing term in the science of form. Paleobiology 8:4–15
Gould SJ, Gilinsky NL, German RZ (1987) Asymmetry of lineages and the direction of evolutionary time. Science 236(4807):1437–1441
Griesemer J (2000) Development, culture and the units of inheritance. Philos Sci 67:S348–S368
Griffith F (1928) The significance of pneumococcal types. J Hyg 27(2):113–159. doi:10.1017/S0022172400031879
Haeckel E (1866) Generelle morphologie der organismen: allgemeine grundzüge der organischen formen-wissenschaft, mechanisch begründet durch die von c. Darwin reformirte decendenz-theorie. Georg Reimer, Berlin
Haeckel E (1917) Kristallseelen, studien über das Anorganische Leben. Alfred Kroner Verlag, Leipzig
Hahlweg K, Hooker CA (eds) (1989) Issues in evolutionary epistemology. State University of New York Press, Albany
Haldane JBS (1949) Disease and evolution. Ric Sci Suppl A 19:68–76
Hamilton WD (1964a) The evolution of social behavior. J Theor Biol 1:295–311
Hamilton WD (1964b) The genetic evolution of social behavior, I and II. J Theor Biol 7:1–52
Hamilton WD (1967) Extraordinary sex ratios. Science 156(3774):477–488. doi:10.1126/science.156.3774.477
Hoffmeyer J (2008) Biosemiotics: an examination into the signs of life and the life of signs. University of Scranton Press, Scranton
Hoyle F (1981) Hoyle on evolution. Nature 294(5837):105
Hubbell S (2001) The unified neutral theory of biodiversity. Princeton University Press, Princeton
Hull DL (1981) Units of evolution. In: Brandon NR, Burian RM (eds) Genes, organisms, populations, 1984. MIT Press, Cambridge, pp 142–159
Hume D (1739/2000) A treatise of human nature. Oxford University Press, Oxford
Huxley TH (1870) Biogenesis and abiogenesis. In: Huxley TH (2011) Collected essays, vol 8. Cambridge University Press, Cambridge, pp 229–271. doi: http://dx.doi.org/10.1017/CBO9781139149273.009
Huxley J (1942) Evolution: the modern synthesis. Allen and Unwin, London
Huxley J (1957) The three types of evolutionary process. Nature 180:454–455
Jablonka E, Lamb MJ (1989) The inheritance of acquired epigenetic variations. J Theor Biol 139(1):69–83
Jablonka E, Lamb MJ (1995) Epigenetic inheritance and evolution. Oxford University Press, Oxford
Jacob F (1977) Evolution and tinkering. Science 196:1161–1166
James W (1907) Pluralism, pragmatism, and instrumental truth. In: A pluralistic universe. Harvard University Press, New York, pp 321–324
James W (1909) Pragmatism: a new name for some old ways of thinking. Longmans Publishing Group, New York, pp 52–61
Kauffman SA (1995) At home in the universe. Penguin Books, London
Kauffman SA, Levin S (1987) Towards a general theory of adaptive walks on rugged landscapes. J Theor Biol 128:11–45
Kimura M (1968) Evolutionary rate at the molecular level. Nature 217(5129):624–626
Kimura M (1976) How genes evolve: a population geneticist's view. Ann Génét 19(3):153–168
Kimura M (1983) The neutral theory of molecular evolution. Cambridge University Press, Cambridge
Kozo-Polyansky B (1924) Symbiogenesis: a new principle of evolution, edited and translated by Fet V, Margulis L 2010. Introduction by Peter H Raven. Harvard University Press, Cambridge
Kressing F, Fangerau H, Krischel M (2014) The global phylogeny and its historical legacy—a critical review of a unified theory of human biological and linguistic co-evolution. Med Stud Int J Hist Philos Ethics Med Allied Sci 4:15–27. doi:10.1007/s12376_013_0081-8

Kuhn TS (1962) The structure of scientific revolutions. University of Chicago Press, Chicago
Lederberg J (1951) Papers in microbial genetics: bacteria and bacterial viruses. University of Wisconsin Press, Madison
Lederberg J (1999) Haldane JBS (1949) on infectious disease and evolution. Genetics 153:1–3
Lederberg J, Tatum EL (1946) Gene recombination in *E. coli*. Nature 158(4016):558
Lévi-Strauss C (1952) Race and history (UNESCO)
Lewontin R (1970) The units of selection. Annu Rev Ecol Syst 1:1–18
Lewontin R (1983) Gene, organism, and environment. In: Bendall DS (ed) Evolution from molecules to men. Cambridge University Press, Cambridge, pp 273–285
Lewontin R (2000) The triple helix: gene, organisms and environment. Harvard University Press, New York
Lieberman BS (2000) Paleobiogeography: using fossils to study global change, plate tectonics, and evolution. Kluwer Academic Publishers, New York
Lieberman BS, Dudgeon S (1996) An evaluation of stabilizing selection as a mechanism for stasis. Palaeogeogr Palaeoclimatol Palaeoecol 127:229–238
Lieberman BS, Vrba ES (2005) Stephen Jay Gould on species selection: 30 years of insight. Paleobiology 31:113–121
Lieberman BS, Allmon WD, Eldredge N (1993) Levels of selection and macroevolutionary patterns in the turritellid gastropods. Paleobiology 19:205–215
Lieberman BS, Brett CE, Eldredge N (1995) A study of stasis and change in two species lineages from the middle devonian of New York state. Paleobiology 21:15–27
Lieberman BS, Miller W III, Eldredge N (2007) Paleontological patterns, macroecological dynamics and the evolutionary process. Evol Biol 34:28–48
Lipo CP, O'Brien MJ, Collard M, Shennan SJ (eds) (2006) Mapping our ancestors: phylogenetic methods in anthropology and prehistory. Aldine Transaction, Hawthorne
Lloyd E (1988) The structure and confirmation of evolutionary theory. Greenwood Press, Westport
Lloyd E (2000) Units and levels of selection: an anatomy of the units of selection debates. In: Singh et al. (eds) Thinking about evolution: historical, philosophical, and political perspectives. Cambridge University Press, Cambridge, pp 267–291
Lorenz K (1958) The evolution of behavior. SCIAM 199(6):67–78
Luria SE, Delbrück M (1943) Mutations of bacteria from virus sensitivity to virus resistance. Genetics28(6):491–511
Lübbe H (1990) Zeit-Verhältnisse. Über die veränderte Gegenwart von Zukunft und Vergangenheit. In: Zeitphänomen Musealisierung, S 40–45
Lyell C (1830) Principles of geology, being an attempt to explain the former changes of the Earth's surface, by reference to causes now in operation, vol 1. John Murray, London
Lyell C (1832) Principles of geology, being an attempt to explain the former changes of the Earth's surface, by reference to causes now in operation, vol 2. John Murray, London
Lyell C (1833) Principles of geology, being an attempt to explain the former changes of the Earth's surface, byreference to causes now in operation. vol 3. John Murray, London
Mace R, Jordan F (2011) Macro-evolutionary studies of cultural diversity: a review of empirical studies of cultural transmission and cultural adaptation. Philos Trans R Soc Lond B Biol Sci 366:402–411
Maienschein J (2012) Epigenesis and preformationism. The stanford encyclopedia of philosophy. In: Zalta EN (ed). http://plato.stanford.edu/archives/spr2012/entries/epigenesis/. Retrieved 20 Feb 2013
Malinowski B (1944) A scientific theory of culture and others essays. The University of North Carolina Press, Chapel Hill
Margulis L (1970) Origin of eukaryotic cells. Yale University Press, New Haven
Margulis L (1981) Symbiosis in cell evolution. Freeman, New York
Margulis L (1998) The symbiotic planet: a new look at evolution. Orion Books, London, Phoenix
Margulis L, Lovelock JE (1974) Biological modulation of the Earth's atmosphere. Icarus 21:471–489

Masel J, King OD, Maughan H (2007) The loss of adaptive plasticity during long periods of environmental stasis. Am Nat 169(1):38–46. doi:10.1086/510212
Maynard Smith J (1964) Group selection and kin selection. Nature 201(4924):1145–1147
Maynard Smith J (1978) The evolution of sex. Cambridge University Press, Cambridge
Maynard Smith J (1983) The genetics of stasis and punctuations. Annu Rev Genet 17:12
Maynard Smith J (1984) Paleontology at the high table. Nature 309(5967):401–402. doi:10.1038/309401a0
Maynard Smith J, Price GR (1973) The logic of animal conflict. Nature 246(5427):15–18. doi:10.1038/246015a0
Maynard Smith J, Szathmáry E (1995) The major transitions in evolution. Oxford University Press, New York
Mayr E (1942) Systematics and the origin of species. Columbia University Press, New York
Mayr E (1961) Cause and effect in biology. Science 134(3489):1501–1506
Mazzarello P (1999) A unifying concept: the history of cell theory. Nat Cell Biol 1(1):E13–E15. doi:10.1038/8964
Merezhkowsky C (1905) Ueber natur und ursprung der chromatophoren im pflanzenreiche. Biologisches Zentralbl 25:595–596
Merezhkowsky C (1910) Theorie der zwei plasmaarten als grundlage der symbiogenesis, einer neuen lehre von der entstehung der organismen. Biol Zentralbl 30:278–288, 289–303, 321–347, 353–367
Miller SL (1953) A production of amino acids under possible primitive Earth conditions. Science 117:528–529
Miller S, Urey H (1959) Organic compound synthesis on the primitive Earth. Science 130(3370):245. doi:10.1126/science.130.3370.245
Monod J (1971) Chance and necessity. Alfred A. Knopf, New York
Morse ML, Lederberg EM, Lederberg J (1956) Transduction in Escherichia Coli K-12. Genetics 41(1):142–156
Nagel E (1949) The meaning of reduction in the natural sciences. In: Stauffer RC (ed) Science and civilization. University of Wisconsin Press, Madison, pp 99–135
Nagel E (1951) Mechanistic explanation and organismic biology. Philos Phenomenological Res 11:327–338
Nietzsche F (1874) Vom Nutzen und Nachteil der Historie für das Leben. Leipzig
O'Brien MJ et al (2002) Two issues in archaeological phylogenetics: taxon construction and outgroup selection. J Theor Biol 215(2):133–150
Odling-Smee FJ (1988) Niche constructing phenotypes. In: Plotkin HC (ed) The role of behavior in evolution. MIT Press, Cambridge, pp 73–132
Okasha S (2005) Multilevel selection and the major transitions in evolution. Phil Bio 72:1013–1025
Oparin A (1955) L'origine de la vie. Editions en langues étrangères, Moscow
Orgel LE (1973) The origins of life: molecules and natural selection. Wiley, New York
Orgel LE (1992) Molecular replication. Nature 358(6383):203–209
Oyama S (1985) The ontogeny of information: developmental systems and evolution. Cambridge University Press, Cambridge
Pagel M, Meade A (2006) Estimating rates of lexical replacement on phylogenetic trees of languages. In: Forster P, Renfrew C (eds) Phylogenetic methods and the prehistory of languages. McDonald institute Monographs, Cambridge, pp 173–182
Pagel M, Venditti C, Meade A (2006) Large punctuational contribution of speciation to evolutionary divergence at the molecular level. Science 314:119–121
Pagel M, Atkinson Q, Meade A (2007) Frequency of word-use predicts rates of lexical evolution throughout Indo-European history. Nature 449:717–720
Parsons T (1975) The present status of structural-functional theory in sociology. In: Parsons T (ed) Social systems and the evolution of action theory. The Free Press, New York

Pauling L (1939) The nature of the chemical bond and the structure of molecules and crystals. Cornell University Press, Ithaca
Pigliucci M (2009) An extended synthesis for evolutionary biology. the year in evolutionary biology. Ann NY Acad Sci 1168:218–228
Pigliucci M, Müller G (eds) (2010) Evolution: the extended synthesis. MIT Press, Cambridge
Pinxten R (1997) When the day breaks: essays in anthropology and philosophy. Lang Frankfurt am Main, Switzerland
Pittendrigh CS (1958) Behavior and evolution. In: Roe A, Simpson GG (eds) Behavior and evolution. Yale University Press, New Haven, pp 390–416
Popper K (1957) The poverty of historicism. Routledge, London
Prigogine I (1980) From being to becoming. Freeman, New York
Provine WB, Mayr E (eds) (1980) The Evolutionary synthesis: perspectives on the unification of biology. Harvard University Press, Cambridge
Rosindell J, Hubbell SP, Etienne RS (2011) The unified neutral theory of biodiversity and biogeography at age ten. Trends Ecol Evol 26(7):340–348
Ryan F (2006) Genomic creativity and natural selection. Biol J Linn Soc 88:655–672
Ryan F (2009) Virolution. Harper Collins, London
Sapp J (1994) Evolution by association: a history of symbiosis. Oxford University Press, New York
Schmalhausen II (1949) Factors of evolution: the theory of stabilizing selection. Blakiston, Oxford
Schopf W (1978) The evolution of the earliest cells. SCIAM 239:85–103
Schwartz J (1999) Sudden origins: fossils, genes, and the emergence of species. Wiley, New York
Schwartz JH, Tattersall I (2001) Extinct humans. Westview Press, USA
Sepkoski D, Ruse M (eds) (2009) The paleobiological revolution. University of Chicago Press, Chicago
Simpson GG (1944) Tempo and mode in evolution. Columbia University Press, New York
Smocovitis VB (1996) Unifying biology: the evolutionary synthesis and evolutionary biology. Princeton University Press, New York
Smocovitis VB (1999) The 1959 Darwin centennial celebration in America. In: Elliot C, Abir-Am P (eds) Commemorations of scientific grandeur, Osiris vol 14(1999), pp 274–323
Spencer H (1862/1958) First Principles. DeWitt Revolving Fund, New York
Spencer H (1876) The principles of sociology, vol 1 and 2. Appleton, New York
Spencer H (1915) Essays: scientific, political, and speculative. Appleton, New York
Stanley SM (1979) Macroevolution. W. H. Freeman, San Francisco
Steele J, Kandler A (2010) Language trees ≠ gene trees. Theory Biosci 129:223–233
Szathmáry E (2002) Units of evolution and units of life. In: Pályi G, Zucchi L, Caglioti L (eds) Fundamentals of life. Elsevier, Paris, pp 181–195
Szathmáry E (2006) The origin of replicators and reproducers. Phil Trans R Soc Lond B Biol Sci 361:1761–1776
Tattersall I (2014) Communication and human uniqueness. In: Gontier N, Pino M (eds) The evolution of socialcommunication in primates, Springer, Dordrecht, pp 219–229
Tax S (ed) (1960) Evolution after Darwin: the university of Chicago centennial, 3 volumes. University of Chicago Press, Chicago
Teggart FJ (1916) Prolegomena to history. Univ Calif Publ Hist 4(3):155–291
Teggart FJ (1925) Theory of history. Yale University Press, London
Thompson DW (1917) On growth and form. Cambridge University Press, Cambridge
Tinbergen N (1963) On aims and methods of ethology. Z Tierpsychologie 20:410–433
Trivers RL (1971) The evolution of reciprocal altruism. Q Rev Biol 46(1):35–57
Van Valen LA (1973) A new evolutionary law. Evol Theor 1:1–30
Van Valen LA (1975) Group selection, sex, and fossils. Evolution 29:87–94
Van Valen LA (1976) Ecological species, multispecies, and oaks. Taxon 25:233–239
Venditti C, Pagel M (2008) Speciation and bursts evolution. Evol Educ Outreach 13:274–280

Venditti C, Pagel M (2010) Speciation as an active force in evolution. Trends Ecol Evol 25:14–20
Villarreal LP (2004) Are viruses alive? Sciam 291(6):97–102
Villarreal LP, Defilipps V (2000) A hypothesis for DNA viruses as the origin of eukaryotic replication proteins. J Virol 74(15):7079–7084
Volk T (1995) Metapatterns across space, time, and mind. Columbia University Press, New York
Volk T (1998) Gaia's body: toward a physiology of the Earth. Copernicus Books. Springer, Berlin
Von Bertalanffy KW (1928) Kritische theorie der formbildung, borntraeger
Von Bertalanffy KW (1940) Vom molekül zur organismenwelt. Akademische Verlagsgesellschaft Athenaion, Potsdam
Von Bertalanffy L (1950) An outline of general systems theory. Br J Hist Sci 1(2):134–162
Vrba ES (1980) Evolution, species and fossils: how does life evolve? S Afr J Sci 76:61–84
Vrba ES (1985a) Ecological and adaptive changes associated with early hominid evolution. In: Delson E (ed) Ancestors: the hard evidence. A.R. Liss, New York, pp 63–71
Vrba ES (1985b) Environment and evolution: alternative causes of the temporal distribution of evolutionary events. S Afr J Sci 815:229–236
Vrba ES (1989) Levels of selection and sorting, with special reference to the species level. Oxf Surv Evol Biol 6:111–168
Vrba ES, Eldredge N (1984) Individuals, hierarchies and processes: towards a more complete evolutionary theory. Paleobiology 10:146–171
Vrba ES, Gould SJ (1986) The hierarchical expansion of sorting and selection: sorting and selection cannot be equated. Paleobiology 122:217–228
Waddington CH (1942) The epigenotype. Endeavour 1:18–20
Wallin IE (1927) Symbionticism and the origin of species. Williams and Wilkins Company, Baltimore
Watson JD, Crick FHC (1953) A structure for deoxyribose nucleic acid. Nature 171(4356):737–738
Weismann A (1893) The germ-plasm: a theory of heredity. Charles Scribner's Sons, New York
Whewell W (1833) Astronomy and general physics considered with reference to natural theology. W. Pickering, London
Williams GC (1966) Adaptation and natural selection. Princeton University Press, Princeton
Williams GC (ed) (1971) Group selection. Aldine Atherton, Chicago
Wilson EO (1975) Sociobiology: the new synthesis. Harvard University Press, Cambridge
Woese CR (1998) The universal ancestor. PNAS 95(12):6854–6859
Woese CR, Fox GE (1977) Phylogenetic structure of the prokaryotic domain: the primary kingdoms. PNAS 74(11):5088–5090
Wright S (1929) The evolution of dominance. Am Nat 63(689):556–561. doi:10.1086/280290
Wright S (1932) The roles of mutation, inbreeding, crossbreeding and selection in evolution. Proc 6th Int Conf Genet 1:356–366
Wynne-Edwards VC (1962) Animal dispersion in relation to social behavior. Oliver and Boyd, London
Wynne-Edwards VC (1986) Evolution through group selection. Blackwell Scientific, Oxford. ISBN 0-632-01541-1
Zeder MA (2008) The Neolithic macro-evolution: macroevolutionary theory and the study of cultural change. J Archeol Res 17:1–63. doi:10.1007/s10814-008-9025-3
Zinder ND, Lederberg J (1952) Genetic exchange in Salmonella. J Bacteriol 64(5):679–699

Part II
Evidencing Macroevolution with Case Studies

Using Microevolution to Explain the Macroevolutionary Observations for the Evolution of Sex

Lutz Becks and Yasaman Alavi

Abstract Macroevolution deals with large-scale and complex changes such as the rise of species, mass extinctions, and evolutionary trends. Microevolution describes evolutionary change within a species with changes in allele or genotype frequencies over short time periods. The separation of the two processes is somewhat arbitrary however, as macroevolution is the accumulation of a number of microevolutionary changes leading over time to large-scale changes. As a consequence, understanding macroevolution requires understanding microevolutionary processes. In this chapter we discuss this link between macro- and micro-evolution using the example of the evolution of sexual reproduction. Explaining the evolution of sexual reproduction is one of the most puzzling problems in evolutionary biology and despite ongoing research a general explanation for the evolution and maintenance of sex has not yet been widely confirmed. Sexual reproduction has been suggested to have evolved only once from asexual reproduction during the early stages of the eukaryote evolution. Today, we find that almost all animals and plants reproduce sexually rather than asexually, suggesting that sex is advantageous. There are, however, exemptions from the macroevolutionary observation of the ubiquity of sexual reproduction: asexual taxa are found across almost the whole phylogenetic tree in a twig-like pattern, and these asexual taxa are found in habitats where related sexual taxa are often absent. The evolution of sex, with all its superlatives such as 'the queen of evolutionary problems' and the 'evolutionary scandals' of the ancient asexuals, is probably one of the fields in evolutionary biology, where already early on macroevolutionary patterns were directly related to microevolutionary processes. Examples of the literature are reviewed here with an emphasis on the link between macro- and microevolution.

Keywords Evolution of sex • Geographic parthenogenesis • Phylogenetic distribution • Origin of sexual reproduction • Maintenance of sexual reproduction

L. Becks (✉)
Max Planck Institute for Evolutionary Biology, D-Plön, Germany
e-mail: lbecks@evolbio.mpg.de

Y. Alavi
Department of Zoology, The University of Melbourne, Melbourne, VIC, Australia

E. Serrelli and N. Gontier (eds.), *Macroevolution*,
Interdisciplinary Evolution Research 2, DOI 10.1007/978-3-319-15045-1_8

1 Sexual Versus Asexual Reproduction: The Big Mystery

Going back in evolutionary history, all taxa reproduced asexually. This probably changed around the time of the evolution of eukaryotes when sexual reproduction originated and became the dominant mode of reproduction (Cavalier-Smith 2002). Today, the vast majority of eukaryotes reproduce sexually, suggesting that sexual reproduction has a fitness advantage over asexual reproduction (Williams 1975; Smith 1978; Bell 1982). Asexuals among eukaryotes are few and tend to have a twig-like phylogenetic distribution: out of all ~42.000 vertebrates, there are only 74 asexual species described, and in some classes, as for example in birds and mammals, asexuals are absent. 20 % of fungi, less than 1 % of seed plants and ~10 % of ferns are asexual. Interestingly, these asexual taxa are often found in habitats where sexual-related taxa are not found: asexual reproduction is found in variable and unpredictable habitats, e.g., high altitudes and latitudes, xeric conditions and islands in contrast to sexual reproduction, which is more often associated with less variable habitats (Suomalainen 1950; Ghiselin 1974; Williams 1975; Glesener and Tilman 1978; Bell 1982). We discuss below mechanisms and processes, related to microevolution, that may create these patterns and how they might be used to explain the origin and maintenance of sex.

Before we start discussing the macroevolutionary observations and their connection to microevolutionary processes, we need to define what we mean by sexual and asexual reproduction as these terms are used differently throughout the literature and various intermediate modes exist combining aspects of asexual and sexual reproduction. Here, we use sexual reproduction (hereafter sex or sexuals) only when there is true sex including meiosis, syngamy (i.e., gamete fusion from different parents), followed by zygote formation and embryogenesis, resulting in a new generation containing a unique mix of parental genes (amphimixis and outcrossing). In our definition of sex, we therefore include two steps: segregation and/or recombination of alleles at each locus and the combination of genome information from two different parents. Automixis (or 'selfing') is a common mode of reproduction and represents a special case of sexual reproduction. Similar to amphimixis, meiosis and gamete fusion occurs, but both gametes come from the same individual (also referred to as uniparental sex) with differences in genetic and demographic consequences compared to amphimixis. In contrast, we refer to asexual reproduction (hereafter asex or asexuals) when meiotic divisions are absent or repressed (apomixis). While this distinction is probably true for many animals and plants as they are obligatory sexual (or asexual), there are other taxa that are able to switch between sexual and asexual reproduction (facultatively sexual). We exclude in our discussion parasexual mechanisms (or asymmetric sex) such as gene exchange in, e.g., bacteria through plasmid conjugation, transduction (exchange of genetic material mediated via phage or virus) and transformation (uptake of DNA from the environment), which usually do not result in reproduction. Our definition of sex and asex is thus focused on eukaryotes and reproduction. In the rest of this chapter, we refer to obligate sexuals (amphimixis) and obligate asexuals (apomixis) when not stated differently.

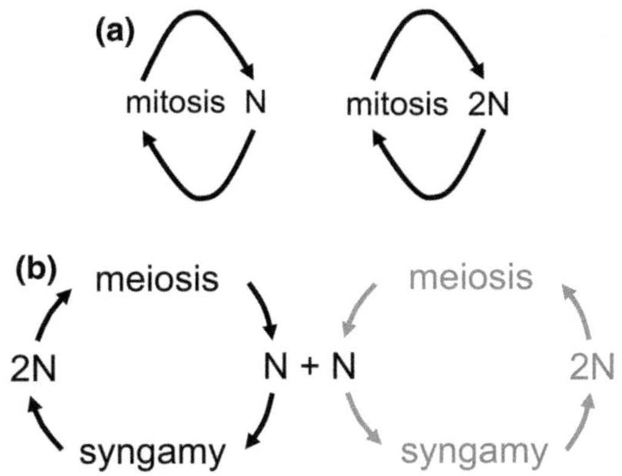

Fig. 1 Difference between asexual reproduction and sexual reproduction. Asexual reproduction (**a**) and sexual reproduction (**b**) differ in the cellular processes involved (mitosis, meiosis, syngamy) and whether the level of ploidy changes. Asexual reproduction: no change (stays haploid or diploid); sexual reproduction: alternation of diploid and haploid phase through meiosis and syngamy. Some sexual taxa stay in the haploid phase for most of their life cycle (e.g., the green alga *Chlamydomonas reinhardtii*), while others are most of their life at the diploid state (e.g., humans)

Sexual and asexual reproduction differ in respect of the cellular processes that are involved as well as their genetic consequences. Understanding these differences and the role they play in the shift from asexual to sexual reproduction in eukaryotes as well as the maintenance of sex (or less often asex) has proven to be difficult (Agrawal 2006; Otto 2009; Hartfield and Keightley 2012). This leaves us in the situation where we have some clear observations (e.g., that sex is dominant among eukaryotes), but no conclusive explanation general enough to explain these observations. Figure 1 shows the fundamental differences in terms of changes in ploidy levels, the cellular processes involved in sexual and asexual reproduction, and also indicates some further major distinctions between the two modes of reproduction. Asexual reproduction always produces clones through mitotic divisions, where the level of ploidy (number of chromosome sets, usually haploid or diploid, as well as polyploid) does not change between generations. In an asexually reproducing population, each individual produces offspring. Daughters are genetically identical to their mother, inheriting her unrecombined genome (with a low level of potential mutations) without changes of intra- and interlocus interactions (dominance and epistasis). This means any level of zygosity (which describes the genotype of a diploid organism at a single locus on the DNA; homozygosity and heterozygosity) is maintained in diploids and loci are in complete linkage (a measure for the association of alleles at two or more loci on a chromosome). Mutations are the only source of novelty with asexual reproduction. In contrast, sexual reproduction always involves meiosis, a reduction division causing a halved level of ploidy, which is restored by syngamy. In a sexually reproducing population, males and females are produced (or two different mating types) and sexually produced offspring are genetically different from their parents, getting half of their genome from one, the other half frowm the other parent (for automixis both gametes are produced by the same individual). With sexual reproduction novelty can be gained by creating new intra- and interloci associations through recombination and segregation (however, this is only true, when the mother's and father's genomes differ). A disadvantage is that with sexual reproduction, only half of the genome is passed on to the next generation (the so-called cost of meiosis (c.f., Lehtonen et al. 2012)).

If one difference between sex and asex is genetic mixing at the genetic level (through recombination and segregation) and the majority of eukaryotes reproduce sexually, can we than assume that sex is the dominant mode of reproduction because mixing is advantageous as it increases genetic variation? 'Depends,' as shuffling the genome can be detrimental or beneficial (Fig. 2). If for example previous rounds of selection created high-fitness genotypes, breaking up these advantageous gene associations through sex can result in low-fitness genotypes of the offspring (Lynch and Deng 1994; Pfrender and Lynch 2000) (Fig. 2b). This 'sex load' (or 'recombination load') has been demonstrated for example in facultative sexual rotifer populations (monogonont rotifers are cyclic parthenogens, switching between asexual and sexual reproduction) that were well adapted to two different conditions in the laboratory (Becks and Agrawal 2012). Sexually produced offspring in these populations had average fitness levels up to 3 times lower than that of asexually produced offspring. On the other hand, creating new genetic associations might be beneficial by increasing genetic variation, thus giving natural selection the possibility to be more effective (Fig. 2a). For example sex can combine beneficial alleles that arose independently in different genetic backgrounds through mutations and create new high-fitness genotypes (Fisher 1930 (reprinted by Dover, 1958); Muller 1932). These high-fitness genotypes will initially be rare, but increase disproportionally over time. In the same rotifer study mentioned above (Becks and Agrawal 2012), but in populations that adapted to novel environmental conditions, sexual reproduction did result in higher variance of fitness compared to asexual reproduction. Sex produced significantly more high-fitness genotypes whereas asexuals had a lower potential to adapt to environmental changes, resulting in a 1.5 times higher fitness of sexuals. This and other studies (Wolf et al. 1987; DaSilva and Bell 1996; Colegrave et al. 2002; Kaltz and Bell 2002) suggest that sex might be beneficial in changing environments where populations have to adapt to novel conditions. Note that our discussion here and in the rest of this chapter is on variation at the genetic level, and not the genomic level (e.g., variation in ploidy level). For the latter, Gorelick and Heng (2011) argue that sexual reproduction reduces genetic variation (always restores the level of ploidy).

Does this mean that asexual reproduction is better in constant environments? Probably not, as the absence of genetic mixing eventually results in the accumulation of deleterious mutations. Assuming a finite population and no back mutation, a fitness decline is irreversible without sex and the lack of purging deleterious mutations will eventually lead to extinction of the population through clonal decay (Muller 1964). Evidence for higher number of mutations accumulating in

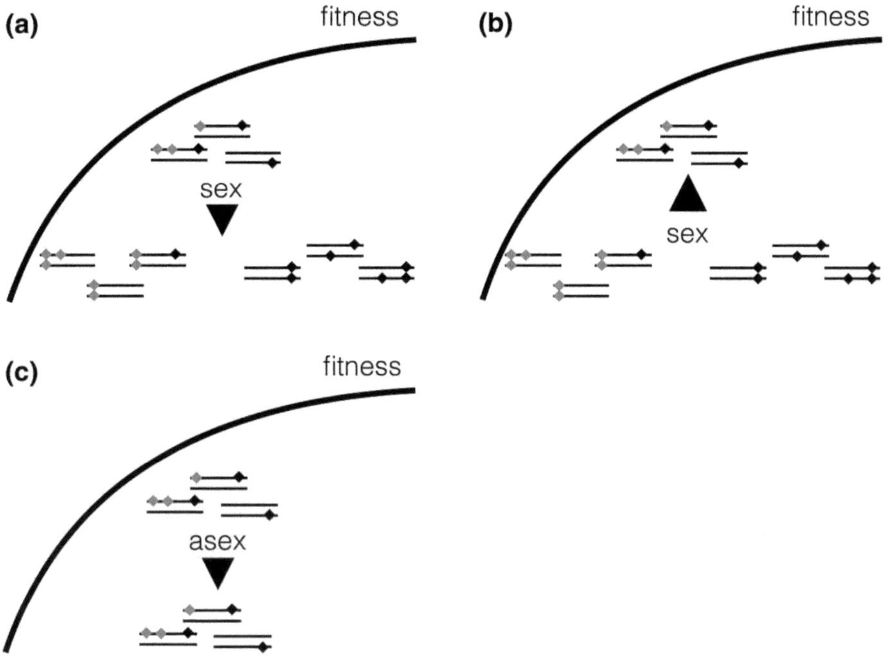

Fig. 2 Genetic effects of sexual (**a**, **b**) and asexual (**c**) reproduction shown for a population consisting of individuals (represented by two *horizontal lines*) carrying different combinations of beneficial (*black diamonds*) and deleterious mutations (*gray diamonds*). Distributions of individuals before and after reproduction are shown below a fitness *curve*: lower fitness at the *left end* and higher fitness at the *right* end of the *curve*. **a** The population reproduced sexually and the next generation (*below the arrow*) consist of individuals with new genetic associations due to recombination and segregation. Because the initial population has negative genetic associations (beneficial and deleterious mutations are more often found together than expected by chance), the new generation has a higher fitness variance than the parent generation (i.e., a greater distribution under fitness curve). In particular, sex results in high-fitness genotypes. In **b** the initial population has positive genetic associations (beneficial are more often found together with beneficial mutations and deleterious with deleterious mutations) and reproducing sexually will create an excess of the intermediate types, lowering the variance (the high- and low-fitness genotypes are lost). **c** No change in the genetic associations occurs with asexual reproduction. Figure modified from (Roze 2012)

asexuals (also referred to as Muller's Ratchet) is not clear; many studies demonstrated results confirming lower rates of mutation accumulation with genetic mixing (Normark and Moran 2000; Bruggeman et al. 2003; Paland and Lynch 2006). Other studies, however, failed to find such correlations between mutation accumulation and genetic mixing (e.g., Cutter and Payseur 2003). It appears that measuring and comparing the accumulation of deleterious mutations is difficult. One reason is probably that genotypes with deleterious mutations are too quickly removed from natural populations to be found in studies. Furthermore, even small amounts of sex (i.e., rare or unconventional meiosis) might be enough to eliminate high numbers of mutations (Charlesworth et al. 1993; Som and Reyer 2007). However, for many taxa that are currently considered to be asexual, the cellular mechanisms during reproduction and thus the genomic consequences are unknown. More studies aiming to discover the exact mechanisms will hopefully help to identify the extent and role of clonal decay in asexuals.

Is sex the better strategy in stable and varying environments because it prevents clonal decay? Most often not, as sex is costly in terms of sex load (see above) and other costs. Based on the notion that in asexually reproducing populations every individual (female) produces offspring and that in sexual population only half as many offspring per individual are produced (as males do not contribute to population growth), an asexually reproducing clone would outcompete and replace sexuals within a few generations (Smith 1978). Sexual reproduction pays this so-called twofold cost of males however only when we can make the assumption that everything else is equal besides the reproduction mode (i.e., the number of offspring produced per sexual and asexual female is equal and there is no other phenotypic difference between sexually and asexually produced offspring), males only contribute half of their genome and no other resources, and the population is outbred. The all-else-being-equal assumption for the twofold cost of sex makes it difficult to provide evidence for. Support comes from an experiment by Stelzer (2011) with obligate parthenogenetic and facultative parthenogenetic rotifers with the same genetic backgrounds (being 'equal'). In this experiment, the obligate asexuals outcompeted and successfully invaded a population of cyclic parthenogenes within a few generations (Stelzer 2011). While the genetic background of sexuals and asexuals was equal in these experiments, sexually and asexually produced offspring differed in at least one important way: sexual reproduction results in the formation of resting eggs from which new females hatch with some delay. Under the experimental conditions, sexually produced offspring did not hatch from these eggs and thus did not contribute to the population growth. Here, sexually reproducing individuals had to pay an even larger cost: producing males and resting eggs (note that the formation of resting eggs is essential for the survival of the species in nature as this is how they overwinter). Comparing population growth rates of obligate sexual and related obligate asexual snails (*Potamopyrgus antipodarum*) showed that the asexual snails indeed grew faster than their sexual counterparts (Jokela et al. 1997) but sexual and asexual snails differed in many other ways. Other studies failed to find differences in growth rates between sexual and asexuals (Kumpulainen et al. 2004). Besides these obvious demographic and genetic effects of sexual and asexual reproduction, other 'costs' of sex have been identified (Lehtonen et al. 2012). Sex is costly for example in terms of time needed to complete a sexual cycle compared to an asexual cycle. Sex can also be costly in terms of energy and risks (predation, transmission of diseases) and the highest cost obligate sexuals have to pay is when they do not find a mate. Although sexuals and asexuals probably always differ in some respect (besides the genetic mixing and the production of males), it is clear that the demographic differences and the other costs of sex might play a major role for the outcome of competition between sexuals and asexuals.

To summarize, sexual and asexual reproduction differ in many ways. Sex seems to be in most situations a bad choice, considering the immediate demographic and potential genetic costs. At the same time, sex might provide an opportunity to deal with long-term consequences such as eliminating deleterious mutations and a higher adaptability under changing conditions. The opposite seems to be the case when looking at asexual reproduction: asexuals have a higher extinction risk due to accumulations of deleterious mutations and clonal decay, as well as a lack of potential for adaptation to novel environmental conditions and speciation due to limited novelty. On shorter time scales, asexual reproduction appears to provide a clear demographic advantage. Most of the differences between sexual and asexual reproduction described above have effects on a microevolutionary scale (evolution within populations) and probably require population genetic explanations. How can we link these microevolutionary processes to observations on the macroevolutionary scale, i.e., between species and on a phylogenetic scale?

2 Phylogenetic Distribution

One of the macroevolutionary observations for the evolution of sex is its origin. Smith and Szathmary (1995) suggest in their book 'The major transitions in Evolution' several different stages for the evolution of eukaryotic sex: the evolution of a haploid–diploid cycle with endomitosis (mitosis taking place without dissolution of the nuclear membrane, and not followed by cytoplasmic division, resulting in doubling of the number of chromosomes within the nucleus), then a one-step meiosis followed by syngamy, and finally sex with syngamy and a two-step meiosis. One hypothesis for the origin of sex and this series of (micro-) evolutionary changes is based on the idea that sex increases fitness directly through DNA repair. Using an undamaged copy of the gene as a template, damaged double-stranded DNA can be repaired and the genome regenerated (Bernstein et al. 1985; Bernstein et al. 1987). Whilst appealing, this hypothesis does not necessarily apply to asexual diploids or polyploids. They already carry two or more copies of each gene (Otto and Lenormand 2002), and the evidence from experiments with prokaryotes provides only weak support for the DNA repair hypothesis (Redfield 1993; Engelmoer and Rozen 2011). However, conclusive evidence is still wanting for or against this hypothesis (see also the discussion in Smith and Szathmary 1995; Gorelick and Heng 2011).

Another direct hypothesis for sex is the transposable element mechanism, which suggest that sex facilitates transmission of selfish genes (Hickey 1982) that integrate randomly and often as several copies into the genome, potentially affecting the function of genes (e.g., Doolittle and Sapienza 1980; Finnegan 1989, 1992; McDonald 1993). Sex results in mixing of genes, thus allowing transposable elements to spread throughout the population while the 'selfish' transposable elements will be eliminated by selection in asexuals. At the same time, the host benefits from sexual reproduction by eliminating deleterious mutations/changes caused by the random integration of the elements. Hickey (1982) was the first to suggest that transposable elements drove the evolution of sex (origin) of their hosts. Evidence for this comes for example from screens for transposable elements across animal phyla (Arkhipova and Meselson 2000). While transposable elements were present in all tested sexual taxa, they could not be found in bdelloid rotifers, for which meiosis has not been described and are considered to be ancient asexuals (Hsu 1956; Welch and Meselson 2000). On the contrary, transposable elements have been found in higher numbers in mitotic parthenogenetic nematodes compared to their sexual relatives (Castagnone-Sereno and Danchin 2014). Here, the high number of transposable elements has been suggested to be beneficial as transposable elements allow overcoming the low adaptability of asexuals. However, it is not currently clear what role transposable elements have played in the origin of sexual reproduction.

The evolution back to asexual reproduction occurred several times independently (Fig. 3). There are several potential mechanisms for the origin of asexuality in eukaryotes. Asexual reproduction can result from disruption and/or rearrangement of genomes through hybridization (Vellend et al. 2007), deleterious mutations (Simon et al. 2003), or symbiotic microorganisms such as *Wolbachia* (Koivisto and Braig 2003). The shifts back to asexual reproduction must have happened recently, as the asexual taxa we find today usually have short lineages and are at the terminal branches in the tree of life, and they are also rare. The small number of asexual taxa appear to be young, not older than the Pleistocene (Williams 1975; Smith 1978; Bell 1982; Moritz and Heideman 1993; Howard and Lively 1998; Johnson and Bragg 1999; Pongratz et al. 2003), except for some 'ancient asexuals'. Ancient asexuals are defined to be without sex for more than 500,000 generations (Agren and Ericson 1996) and have been called an evolutionary scandal as they should not exist. Microevolutionary mechanisms (clonal decay and low adaptability of asexuals) can explain the young age of asexuals, pointing to the benefits of sexual reproduction. An important observation that supports the idea that sex prevents clonal decay and/or increases adaptability comes from recent studies on asexuals. Studying the genomics of ancient asexuals has revealed that they evolved different ways to avoid clonal decay, for example, through increased ploidy and higher number of gene copies (Welch and Meselson 2000) and horizontal gene transfer (Flot et al. 2013).

These patterns are intriguing and have often been discussed as a means of understanding the evolutionary mystery of the evolution of sex. Looking at evolutionary history and phylogenetic distribution, we see the large-scale pattern of sexual reproduction being almost ubiquitous among eukaryotes. Both processes, from asex to sex either due to transposable elements or the repair mechanism and back to asexual reproduction are evolutionary processes within populations and thus on microevolutionary scales. In recent years, alternative microevolutionary explanations have been postulated for the phylogenetic distribution of asexuals (Janko et al. 2008; Neiman et al. 2009; Schwander and Crespi 2009; Janko et al. 2011). These do not use higher extinction rates of asexuals due to mutational meltdown or low adaptability to explain the young age of asexuals but rather (i) low transition rates from sexuals to asexuals, (ii) asexual lineage turnover (i.e., parental

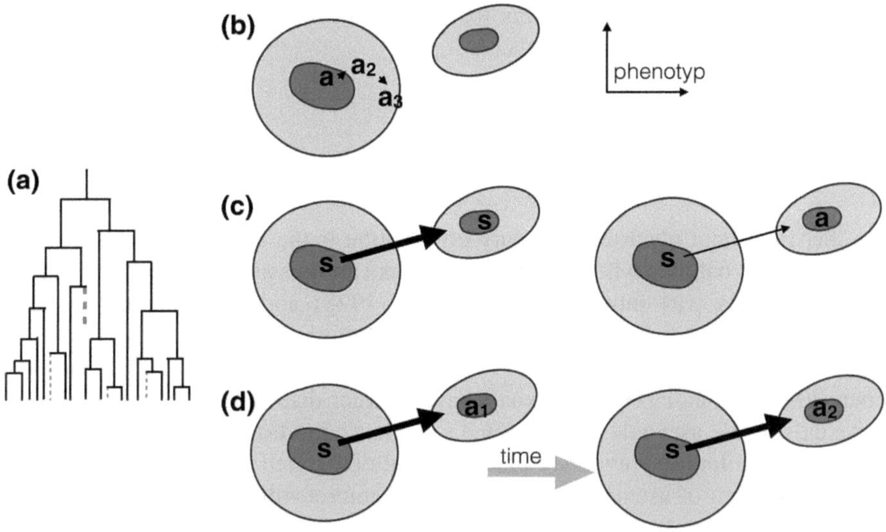

Fig. 3 Schematic of the twig-like distribution and young age of asexual lineages (**a**) and the suggested (micro)-evolutionary mechanism creating this distribution shown on adaptive landscapes (**b**–**c**). The potential underlying mechanisms are depicted on an adaptive landscape with fitness as a function of two different phenotypic traits (*right*) in **a**) and the differences in fitness indicated by different *shades* of gray representing higher fitness. Populations are shown as *s* (sexual) and *a* (asexual), and subscripts indicate changes over time. Initially, only one fitness peak is occupied. **a** Asexual lineages (*gray dashed lines*) evolved independently several times from sexual ancestors (*black lines*) and got eventually extinct (*wide gray dashed line*). **b** The traditional view is that the young age of asexuals results from clonal decay and thus high extinction rates. Over time, the accumulation of deleterious mutations moves the population from a fitness peak till the population goes extinct. Alternative hypothesis explain the young age and twig-like distribution of asexuals with **c** low transition rates from sexual to asexual reproduction: The open fitness peak is more likely to be occupied by a sexual species (*s*, *wide arrow*) than by an asexual species (*a*, *narrow arrow*). **d** A high lineage turnover of asexuals results in young ages of asexuals, because new asexual lineages (a_2) are recurrently created from related sexuals and the new lineages replace already existing asexual lineages (a_1) that got extinct by chance

sexual species continuously generate asexual lineages that replace older asexual lineages) and (iii) the fact that asexuals are often found in harsh environments and are more prone to become extinct in these environments (Fig. 3). Support for the traditional view (clonal decay, low adaptability) comes from studies that showed the accumulation of deleterious mutations with negative fitness effects in obligate asexual populations (Neiman et al. 2010; Henry et al. 2012). There is however also some evidence supporting alternative explanations. For example the asexual linage turnover hypothesis predicts that younger asexual lineages should be geographically closer to sexual lineages (Neiman et al. 2009; Schwander and Crespi 2009) and several studies support this prediction (Quattro et al. 1991; Law and Crespi 2002; Johnson 2006). As there are quantitative differences for the predictions made by the different models (clonal turnover and clonal decay), future work using phylogenetic and genomic data could help to identify the underlying reason and microevolutionary mechanisms for the macroevolutionary observation of the distribution of asexuals and sexuals.

3 Geographic Distribution

Another important macroevolutionary observation is the fact that asexual taxa are often found in habitats where related sexual taxa are missing. The term geographic parthenogenesis was introduced by Vandel (1928) and the general phylogeographic pattern has been described by several authors and for different taxonomic groups (Bell 1982; Bierzychudek 1985; Kearney 2003; van Dijk 2003; Haag and Ebert 2004; Hörandl 2009): Asexuals have a larger distributional range compared to sexuals, while asexuals tend to range to higher latitudes and altitudes, and asexuals tend to colonize previously glaciated and devastated areas. As with the phylogenetic pattern of asexuals, there are several microevolutionary explanations for these patterns (Fig. 4).

Geographic parthenogenesis seems to support the theories for the maintenance of sex that are based on species interactions and density dependency (see sect. 4 below). Species interactions are thought to be less and weaker in extreme environments such as higher altitudes and latitudes where sexuals are often missing. The absence of sexuals might result from the fact that costs of sex are not balanced when for example coevolving parasites are missing in the extreme habitats, suggesting that the benefits of sex derive from species interactions. Such mechanisms might be in action during interactions between antagonistic species such as host–parasite interactions (Hamilton 1980). The "Red Queen" hypothesis (Bell 1982), one of the most prominent models, suggests that species must constantly run away from coevolving parasites to keep their fitness constant. Coevolution of species imposes negative frequency-dependent selection, which can drive the evolution of sex (Jaenike 1978; Hamilton 1980). Hosts involved in coevolutionary interactions with parasites benefit from sexual reproduction by producing genetically heterogeneous offspring (e.g., Decaestecker et al. 2007; Morran et al. 2011). This is advantageous because rare genotypes might be resistant to the common parasite genotype. Consequently, the parasites are under selective pressure to overcome the common resistance strategies. Other explanations for the lack of sexuals in more extreme habitats are based on the differences in the costs of sex, which are considered to be greater in more variable habitats. The twofold cost of sex resulting from producing males may be reduced in constant environments such as low altitudes and latitudes. These stable environments might favor slower development, greater competitive ability, delayed reproduction, higher survival rates, and lower resource thresholds leading to efficiency. More variable environments such as high altitude and latitude habitats might favor rapid development, higher maximal growth rates, lower survival rates, and early reproduction leading to productivity. Because offspring quantity is relatively less important than offspring quality in stable environments than in more variable habitats, several authors have argued that the cost of

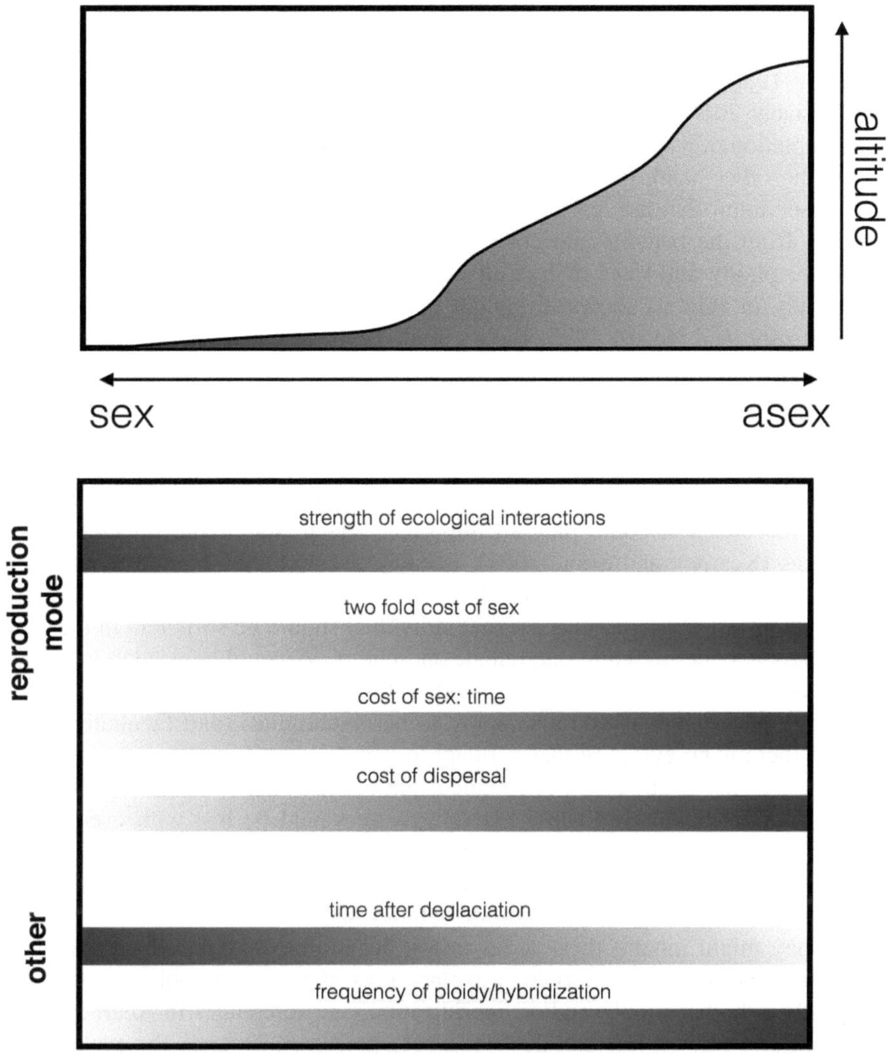

Fig. 4 Schematic of the geographic parthenogenesis (*top panel*) and (micro)-evolutionary processes (*bottom panel*) that have been suggested explaining the distribution of asexuals, which tend to inhabit extreme habitats, e.g., high altitudes and sexuals, which are often missing in the extreme habitats. Traditional explanations link the geographic distribution to benefits and costs of sex: sex can provide an advantage when coevolving with other species, e.g., parasites and because species interactions are less/weaker in extreme habitats, the costs of sex are not balanced under these conditions. Because the costs of sex are more easily balanced in stable habitats (here *low altitudes*), sexuals are less often found in the unstable high altitude habitats. Other explanations (reproduction mode independent) are based on higher levels of polyploidy and hybridization and the time after deglaciation (see text)

sex is reduced in stable environments or that the reproductive costs of sex can be more easily balanced by increases in quality, which have greater selective value (Abugov 1986; Lomnicki 2001; Doncaster et al. 2003; Olofsson and Lundberg 2007; Lively 2010). Similarly, other costs of sex, such as the longer time needed for completion of a sexual cycle, are most likely reduced in stable environments.

On the other hand, these explanations may be incorrect or even unnecessary. The observation that asexuals live in a boarder range of habitats might be independent from the benefits and costs of sexual reproduction but rather the result of higher ploidy and rates of hybridization found in these habitats. For example polyploidy in animals shows a similar pattern to the geographic parthenogenesis (Dufresne and Hebert 1995; Barata et al. 1996) and this factor may underlay the real reason for this pattern as higher ploidy often results in higher tolerances to environmental stressors compared to the diploid ancestors (e.g., Levin 1983). Kearny (2005) puts forward a similar argument. Hybrid advantages in new and open environments could be the primary reason for the geographic parthenogenesis (hybridization often involves a shift to asexual reproduction (Mallet 2007)) and asexual reproduction could play an important role in stabilizing selected hybrid genotypes (Kearney and Shine 2004). Because asexual reproduction is uniparental, it has also been suggested that the lower cost of dispersal of asexuals might explain their larger range . One asexual individual should be sufficient in theory to start a new population while one female and one male would be needed in sexuals (Baker and Stebbins 1965; van Dijk 2007; Hörandl 2008). The single individual advantage of asexuals would also apply to hermaphrodites (and facultative sexuals) but there is no evidence that hermaphrodites follow the pattern of geographic parthenogenesis. It has also been argued that asexuals can maintain heterozygosity of founder population while heterozygosity would be lost with every round of sexual reproduction (Beukeboom and Vrijenhoek 1998). Geographic parthenogenesis has also been linked to deglaciation. Because the ice retracted more recently from high altitude and latitude habitats and asexuals have lower dispersal costs, they might inhabit these areas faster, but might be replaced by sexuals in the future. Evidence for alternative explanations of the geographic parthenogenesis such as higher ploidy and/or hybridization (see references in Kearney 2003; Haag and Ebert 2004; Hörandl 2009) has been presented. Overall, the explanations dependent or independent of the consequences of sex consider microevolutionary processes and dynamics. From the different mechanisms discussed in this section, there is presently no clear proof, and future work testing the different mechanisms is needed.

4 The Maintenance of Sex

The maintenance of sex is another observation at the macroevolutionary level where in order to undrestand it, we need to look into microevolutionary mechanisms to understand this observation. The question, why sexual populations or taxa are not invaded and replaced by asexual taxa when sex comes at high costs is one of the central questions in the field on the evolution of sex. An enormous body of studies have presented various theoretical models and hypotheses. Most modern theories for the maintenance of sex are based on benefits of genetic mixing for the individual rather than for a group as some of the earlier explanations relied on. Intriguingly, all suggested theories depend on microevolutionary mechanisms, i.e., changes in allele or genotype frequencies within a population. Because sex affects genetic associations and it can be both beneficial and detrimental (see 1), it is necessary to find the conditions which constantly create genetic associations, and the conditions under which it is beneficial to break such associations apart. Most theories assume that the investment into sex is determined by a so-called modifier locus which determines the investment into sexual or asexual reproduction but has no direct effect on fitness (Nei 1967). Under this assumption, selection for (more) sex cannot work directly on the modifier and changes in the modifier frequencies are the result of hitchhiking with alleles at other loci that are positively selected; i.e., the fitness gain is greater than the costs of sex. Early theories suggested indirect advantages of sex because sex increases variability, which is required for selection (Weismann 1889, 1904), later confirmed by mathematical models (Felsenstein 1965, 1974; Smith 1988; Charlesworth 1993). Further progress revealed that genetic mixing does not always increase genetic variability, and when it does, it is not necessarily advantageous (Kondrashov 1993; Otto 2009). For sex to be beneficial in terms of increased genetic variation, the genetic associations need to be negative (negative linkage disequilibrium, LD) prior to sex such that beneficial and detrimental alleles are more often found together than expected by chance. In this case, breaking apart the associations will allow selection to act more efficiently in removing deleterious mutations (Mullers Ratchet) and/or creating beneficial gene combinations (Muller Fisher hypothesis; e.g., Fig. 2a). Two main forces, drift (stochastic changes in allele frequencies due to nonheritable variation in offspring number) and epistasis (intralocus interactions), can generate such associations (Felsenstein 1965, 1974; Kondrashov 1992). In presence of directional selection, drift can maintain negative LD in the population (stochastic effect). Negative epistasis also creates negative LD by creating a negative curvature in the fitness function (deterministic effects) (de Visser and Elena 2007; Otto 2009). The Fisher–Muller hypothesis, Muller's Ratchet (see 1, Fig. 3), and Hill–Robertson interference rely on drift coupled with directional selection to create negative associations and therefore benefits to sex by breaking up these associations. The Hill-Robertson model (Hill and Robertson 1966) shows that selection on one locus can interfere with selection on a second locus assuming linkage between the loci in a finite population.

Other hypotheses consider varying selection over time or space (Otto 2009). With constant change in the environment over time, previous genetic associations become unfavorable and breaking these associations down is beneficial (Salathe et al. 2009). However, for sex to be maintained under these conditions, fluctuations in epistasis should be high (Peters and Lively 1999; Gandon and Otto 2007; Peters and Lively 2007; Otto 2009). Such mechanisms might be in action during interactions between antagonistic species such as host–parasite interactions (Hamilton 1980; Bell 1982; Hamilton et al. 1990). Similarly, when selection varies over space through migration, the genetic associations may be unfavorable in the new environment and need to be broken down. In a spatially heterogeneous environment, sex will immediately increase the fitness of offspring (short-term benefit) and should be selected for (Pylkov et al. 1998; Agrawal 2009).

Previous empirical studies have, either directly or indirectly, tested the different conditions under which sex can be maintained. Most of these studies fall into one of the three general categories of correlative approaches, short- and long-term effects, and experimental evolution studies. All of these approaches have in common that they test microevolutionary mechanisms. The first group includes studies that correlate reproductive modes (sexual or asexual, selfing or outcrossing, recombinant or non-recombinant) with mutation accumulation (Normark and Moran 2000; Bruggeman et al. 2003; Cutter and Payseur 2003; Paland and Lynch 2006), or parasite/pathogen prevalence (Lively 1987; Jokela and Lively 1995; Killick et al. 2006; King et al. 2009; Elzinga et al. 2012; Verhoeven and Biere 2013). Empirical tests showed for example correlations between the prevalence of sex and the presence of parasites or pathogens, in line with the Red Queen hypothesis. One important study system examined in the recent years is the common freshwater snail *Potamopyrgus antipodarum* where phenotypically similar sexual and asexual forms coexist, allowing comparisons between sexuals and asexuals (Lively et al. 1998; Koskella and Lively 2009). In one study, King et al. (2009) cross-infected snails collected from shallow and deep water habitats with sympatric and allopatric parasites. They found evidence of local adaptation to parasites and showed that shallow waters are coevolutionary 'hot spots.' Most importantly, sexuals are more common in these 'hot spots' in comparison with the coevolutionary 'cold spots' where asexuals dominate. Again, other studies failed to show such correlations (Killick et al. 2006; Elzinga et al. 2012). In general, these kinds of studies are important because they inform us about some of the consequences of sexual and asexual reproduction even though they are not testing the microevolutionary mechanisms that maintain sex.

The second category includes studies that compare mean fitness and variance of fitness of sexually and asexually derived offspring to test for short- and long-term effects of sex (Barton 1995; Agrawal 2006). Since Nick Barton published his landmark paper on short- and long-term effects of sexual reproduction, it has become clear that the different theories for sex can be studied under this common population genetic framework. Both short- and long-term effects result from breaking up allele associations built by previous selection. In the presence of non-additive gene interactions (dominance or epistasis), sexual reproduction will have an immediate effect on the offspring (the mean fitness of the offspring). For example if selection favors heterozygotes, we will end up with an excess of heterozygotes. Sex will create less fit homozygotes and thus the average fitness of sexually produced offspring will be lower than that of the parents and asexually derived offspring. Whether sex has a long-term advantage, i.e., whether sex creates variation for selection to act on depends on the type of linkage disequilibrium. The balance between the sum of short- and long-term effects and the costs of sex determines whether sex is beneficial or not. Experimentally, we can measure short-term effects of sex by comparing the mean fitness of the sexual and asexually produced offspring when we can assume mating is random in the population. Comparing the variance in fitness between the two offspring types will give us the long-term effects of sex (Agrawal 2006). Several studies measured the effects of sex on adaptation over several generations (Zeyl and Bell 1997; Colegrave 2002; Kaltz and Bell 2002; Poon and Chao 2004; Cooper et al. 2005; Goddard et al. 2005; Becks and Agrawal 2012). The result of these studies suggest that sex could accelerate adaptation to new environments by increasing genetic variation and/or mean fitness of sexual offspring depending on population size (Colegrave 2002; Poon and Chao 2004) and the strength of selection (Kaltz and Bell 2002; Goddard et al. 2005; Becks and Agrawal 2012). One study using the green alga *Chlamydomonas reinhardtii* (Kaltz and Bell 2002) showed an increase in standardized variation in fitness after a few sexual episodes, supporting the Fisher-Muller hypothesis. Other studies measured mean and variance in fitness of sexuals and asexuals over one or a few generations using the general framework introduced by Barton (1995). Such studies have only been performed in a limited number of taxa, using *Drosophila* (Charlesworth and Charlesworth 1975), the monogonot rotifer *Brachoinus calyciforus* (Becks and Agrawal 2011), the planktonic crustacean *Daphnia pulicaria* (Lynch and Deng 1994; Pfrender and Lynch 2000; Allen and Lynch 2008), and the grass *Anthoxanthum odoratum* (Kelley et al. 1988). Both increases and decreases in mean fitness and/or fitness variance were reported in these studies. However, only the latter study was performed in the field, i.e., in the natural environment with all of the biotic and abiotic players. Thus, only the last offspring were produced in the same complex environment where genetic associations were built up before sex reshuffled them. This category of studies is particularly useful because it allows us to investigate the effect of different mechanisms involved in the maintenance of sex. It is, nevertheless, important to realize that the population genetic approach towards studying short- and long-term effects assumes that sexually and asexually produced offspring differ only in the reproductive mode, which might often not be valid (Meirmans et al. 2012).

The last category includes important studies aiming to test the mechanistic causes for the evolution of sex mainly through experimental evolution. Here, we consider those studies in which sexual and asexual individuals were in direct competition or studies that followed the evolution of the rates of sex (outcrossing or recombination) within populations. The role of sex in parasite resistance has been tested in *Tribolium castaneum* beetles (Kerstes et al. 2012), the nematode *Caenorhabditis elegans* (Morran et al. 2011), and a fresh water snail *Potamopyrgu antipodarum* (Koskella and Lively 2009). Morran et al. (2011) found that *C. elegans* populations benefited from outcrossing compared to selfing in the presence of coevolving pathogens, supporting the Red Queen hypothesis. One study tested and supported the hypothesis that higher rates of sex evolve in heterogeneous environments using *B. calyciflorus* (Becks and Agrawal 2010). Two experimental evolution studies on yeast found advantages to sex during adaptation to new environments by following the rate of recombination (Wolf et al. 1987) and competing sexual strain against asexual strains (Greig et al. 1998). However, the number of studies testing these mechanisms directly is limited and some of these studies use model systems where sexuals and asexuals do not coexist under natural conditions (e.g., Poon and Chao 2004).

In summary, all currently discussed mechanisms explaining the maintenance of sex are microevolutionary processes and rely on short- and long-term effects of sex that balance the costs of sex. The strength of the benefits and costs might differ depending on the environment and thus also explain the geographic pattern discussed above (see 3). Identifying the relevant mechanism or combination of mechanisms (West et al. 1999a, b) will need additional work.

5 Conclusion

Observations showing that sex is the dominant mode of reproduction but that there are phylogenetic and geographic exemptions have puzzled biologists since Charles Darwin, but we still struggle to explain them. It is intriguing that the origin and the maintenance of sexual (and asexual reproduction) in eukaryotes as well as the distributions of asexuals and sexuals are explained by microevolutionary mechanisms, while the observations are made on the macroevolutionary level. As a consequence, we can often derive predictions that are testable through theory and experiments. This will be important especially when using the macroevolutionary observations (such as the geographic parthenogenesis and the twig-like distribution of asexuals) to explain the general benefits of sex. In particular, one can use the microevolutionary predictions to test whether the general patterns observed in nature can be recreated under manipulated laboratory/field conditions. For example, in experiments with the green alga *Chlamydomonas reinhardtii*, obligate sexuality (and high genetic diversity) has been shown to be most effective when adapting to detrimental environments in comparison with facultative sexual and asexual populations (and low genetic diversity) (Lachapelle and Bell 2012). These results corroborate the lower adaptability of asexuals and their higher extinction risk in changing environments. Similarly, the general pattern of the geographic parthenogenesis could be recreated in rotifer populations; higher rates of sex evolved in populations in stable environments, while the rates of sex evolved to lower levels within populations from unstable environments (environments were manipulated in terms of population size; Becks and Agrawal 2013). However, further investigations are needed to (i) to test the different mechanisms explaining the macroevolutionary observations, (ii) to further develop clear predictions for models independent of the demographic and genetic consequences of sex and how they differ from the traditional models used to explain the geographic and phylogenetic distribution of asexuals and (iii) to understand the long-term effects of asexuality. This includes the consequences for the genome with rare or unconventional meiosis (Melters et al. 2012; Flot et al. 2013). With this information, we are starting to understand why sexual and asexual reproduction exist and why they follow the macroevolutionary patterns described above. Despite the challenges involved in studying evolutionary processes as complex as the evolution of sex, the field has and will continue linking microevolutionary processes to macroevolutionary observations.

Acknowledgments This work was supported and the German Research Foundation DFG (LB).

References

Abugov R (1986) Is there a cost for meiosis in life-history? J Theor Biol 116:613–623
Agrawal AF (2006) Evolution of sex: why do organisms shuffle their genotypes? Curr Biol 16:R696–R704
Agrawal AF (2009) Spatial heterogeneity and the evolution of sex in diploids. Am Nat 174:S54–S70
Agren J Ericson L (1996) Population structure and morph-specific fitness differences in tristylous Lythrum salicaria. Evolution 50(1):126–139
Allen DE, Lynch M (2008) Both costs and benefits of sex correlate with relative frequency of asexual reproduction in cyclically parthenogenic *Daphnia pulicaria* populations. Genetics 179:1497–1502
Arkhipova I, Meselson M (2000) Transposable elements in sexual and ancient asexual taxa. Proc Natl Acad Sci USA 97:14473–14477
Baker HG, Stebbins GL (1965) The genetics of colonizing species. Academic Press, New York
Barata C, Hontoria F et al (1996) Competition between sexual and parthenogenetic Artemia: temperature and strain effects. J Exp Mar Biol Ecol 196:313–328
Barton NH (1995a) A general-model for the evolution of recombination. Genet Res 65:123–144
Barton NH (1995b) Linkage and the limits to natural selection. Genetics 140:821–841
Becks L, Agrawal AF (2010) Higher rates of sex evolve in spatially heterogeneous environments. Nature 468:89–92
Becks L, Agrawal AF (2011) The effect of sex on the mean and variance of fitness in facultatively sexual rotifers. J Evol Biol 24:656–664
Becks L, Agrawal AF (2012) The evolution of sex is favoured during adaptation to new environments. PLoS Biol 10
Becks L, Agrawal AF (2013) Higher rates of sex evolve under K-selection. J Evol Biol 26:900–905
Bell G (1982) The masterpiece of nature: the evolution and genetics of sexuality. University of California Press, Berkeley
Bernstein H, Byerly HC et al (1985) Genetic damage, mutation, and the evolution of sex. Science 229:1277–1281
Bernstein H, Hopf FA et al (1987) The molecular basis of the evolution of sex. Adv Genet Mol Genet Med 24:323–370
Beukeboom LW, Vrijenhoek RC (1998) Evolutionary genetics and ecology of sperm-dependent parthenogenesis. J Evol Biol 11:755–782
Bierzychudek P (1985) Patterns in plant parthenogenesis. Experientia 41:1255–1264
Bruggeman J, Debets AJM et al (2003) Sex slows down the accumulation of deleterious mutations in the homothallic fungus *Aspergillus nidulans*. Genetics 164:479–485
Castagnone-Sereno P, Danchin EGJ (2014) Parasitic success without sex—the nematode experience. J Evol Biol 27:1323–1333
Cavalier-Smith T (2002) Origins of the machinery of recombination and sex. Heredity 88:125–141

Charlesworth B (1993) Directional selection and the evolution of sex and recombination. Genet Res 61:205–224
Charlesworth B, Charlesworth D (1975) Experiment on recombination load in *Drosophila melanogaster*. Genet Res 25:267–274
Charlesworth D, Morgan MT et al (1993) Mutation accumulation in finite outbreeding and inbreeding populations. Genet Res 61:39–56
Colegrave N (2002) Sex releases the speed limit on evolution. Nature 420:664–666
Colegrave N, Kaltz O et al (2002) The ecology and genetics of fitness in *Chlamydomonas*. VIII. The dynamics of adaptation to novel environments after a single episode of sex. Evolution 56:14–21
Cooper TF, Lenski RE et al (2005) Parasites and mutational load: an experimental test of a pluralistic theory for the evolution of sex. Proc R Soc B Biol Sci 272:311–317
Cutter AD, Payseur BA (2003) Rates of deleterious mutation and the evolution of sex in Caenorhabditis. J Evol Biol 16:812–822
DaSilva J, Bell G (1996) The ecology and genetics of fitness in *Chlamydomonas 7*. The effect of sex on the variance in fitness and mean fitness. Evolution 50:1705–1713
de Visser JAGM, Elena SF (2007) The evolution of sex: empirical insights into the roles of epistasis and drift. Nat Rev Genet 8:139–149
Decaestecker E, Gaba S et al. (2007). Host-parasite 'Red Queen' dynamics archived in pond sediment. Nature 450:870–U816
Doncaster CP, Pound GE et al (2003) Dynamics of regional coexistence for more or less equal competitors. J Anim Ecol 72:116–126
Doolittle WF, Sapienza C (1980) Selfish genes, the phenotype paradigm and genome evolution. Nature 284:601–603
Dufresne F, Hebert PDN (1995) Polyploidy and clonal diversity in an arctic cladoceran. Heredity 75:45–53
Elzinga JA, Chevasco V et al (2012) Low parasitism rates in parthenogenetic bagworm moths do not support the parasitoid hypothesis for sex. J Evol Biol 25:2547–2558
Engelmoer DJP, Rozen DE (2011) Competence increases survival during stress in Streptococcus pneumoniae. Evolution 65:3475–3485
Felsenstein J (1965) Effect of linkage on directional selection. Genetics 52:349–363
Felsenstein J (1974) Evolutionary advantage of recombination. Genetics 78:737–756
Finnegan DJ (1989) Eukaryotic transposable elements and genome evolution. Trends Genet 5:103–107
Finnegan DJ (1992) Transposable elements. Curr Opin Genet Dev 2:861–867
Fisher RA (1930) The genetical theory of natural selection. Clarendon Press, Oxford (reprinted by Dover 1958)
Flot J-F, Hespeels B et al (2013) Genomic evidence for ameiotic evolution in the bdelloid rotifer *Adineta vaga*. Nature 500:453–457
Gandon S, Otto SP (2007) The evolution of sex and recombination in response to abiotic or coevolutionary fluctuations in epistasis. Genetics 175:1835–1853
Ghiselin MT (1974) The economy of nature and the evolution of sex. University of California Press, Berkeley
Glesener RR, Tilman D (1978) Sexuality and components of environmental uncertainty: clues from geographic parthenogenesis in terrestrial animals. Am Nat 112:659–673
Goddard MR, Godfray HCJ et al (2005) Sex increases the efficacy of natural selection in experimental yeast populations. Nature 434:636–640
Gorelick R, Heng HHQ (2011) Sex reduces genetic variation: a multidisciplinary review. Evolution 65:1088–1098
Greig D, Borts RH et al (1998) The effect of sex on adaptation to high temperature in heterozygous and homozygous yeast. Proc R Soc B Biol Sci 265:1017–1023
Haag CR, Ebert D (2004) A new hypothesis to explain geographic parthenogenesis. Ann Zool Fenn 41:539–544

Hamilton WD (1980) Sex versus non-sex versus parasite. Oikos 35:282–290
Hamilton WD, Axelrod R et al (1990) Sexual reproduction as an adaptation to resist parasites (a review). Proc Natl Acad Sci USA 87:3566–3573
Hartfield M, Keightley PD (2012) Current hypotheses for the evolution of sex and recombination. Integr Zool 7:192–209
Henry L, Schwander T et al (2012) Deleterious mutation accumulation in asexual *Timema* stick insects. Mol Biol Evol 29:401–408
Hickey DA (1982) Selfish DNA—a sexually-transmitted nuclear parasite. Genetics 101:519–531
Hill WG, Robertson A (1966) Effects of linkage on limits to artificial selection. Genet Res 8:269–294
Howard RS, Lively CM (1998) The maintenance of sex by parasitism and mutation accumulationunder epistatic fitness functions. Evolution 52(2):604–610
Hörandl E (2008) Evolutionary implications of self-compatibility and reproductive fitness in the apomicitc *Ranunculus auricomus* polyplid complex (*Ranunculaceae*). Int J Plant Sci 169:1219–1228
Hörandl E (2009) Geographical parthenogenesis: opportunities for asexuality. In: Schon I, Martens K, VanDijk P (eds) Lost sex: the evolutionary biology of parthenogenesis. Springer, Dordrecht, pp 161–186
Hsu WS (1956) Oogenesis in the Bdelloidea rotifer *Philodina* roseola Ehrenberg. Cellule 59:281–296
Jaenike J (1978) An hypothesis to account for the maintenance of sex within populations. Evol Theory 3:191–194
Janko K, Drozd P et al (2011) Do clones degenerate over time? Explaining the genetic variability of asexuals through population genetic models. Biol Direct 6(1):17
Janko K, Drozd P et al (2008) Clonal turnover versus clonal decay: a null model for observed patterns of asexual longevity, diversity and distribution. Evolution 62:1264–1270
Johnson SG (2006) Geographic ranges, population structure, and ages of sexual and parthenogenetic snail lineages. Evolution 60:1417–1426
Johnson SG Bragg E (1999) Age and polyphyletic origins of hybrid and spontaneousparthenogenetic Campeloma (Gastropoda: Viviparidae) from the southeastern United States. Evolution53(6):1769–1781
Jokela J, Lively CM (1995) Spatial variation in infection by digenetic trematodes in a population of fresh-water snails (*Potamopyrgus antipodarum*). Oecologia 103:509–517
Jokela J, Lively CM et al (1997) Evidence for a cost of sex in the freshwater snail *Potamopyrgus antipodarum*. Ecology 78:452–460
Kaltz O, Bell G (2002) The ecology and genetics of fitness in *Chlamydomonas*. XII. Repeated sexual episodes increase rates of adaptation to novel environments. Evolution 56:1743–1753
Kearney M, Shine R (2004) Morphological and physiological correlates of hybrid parthenogenesis. Am Nat 164:803–813
Kearney MR (2003) Why is sex so unpopular in the Australian desert? Trends Ecol Evol 18(12):605–607
Kelley SE, Antonovics J et al (1988) A test of the short-term advantage of sexual reproduction. Nature 331:714–716
Kerstes NAG, Berenos C et al (2012) Antagonistic experimental coevolution with a parasite increases host recombination frequency. BMC Evol Biol 12
Killick SC, Carlsson AM et al (2006) Testing the pluralist approach to sex: the influence of environment on synergistic interactions between mutation load and parasitism in *Daphnia magna*. J Evol Biol 19:1603–1611
King KC, Delph LF et al (2009) The geographic mosaic of sex and the Red Queen. Curr Biol 19:1438–1441
Koivisto RKK, Braig HR (2003) Microorganisms and parthenogenesis. Biological Journal of the-Linnean Society 79:43–45
Kondrashov AS (1992) Classification of hypotheses on the advantage of amphimixis. In: Symposium on the evolution of sex, Blacksburg, Va

Kondrashov AS (1993) Classification of hypotheses on the advantage of amphimixis. J Hered 84:372–387
Koskella B, Lively CM (2009) Evidence for negative frequency-dependent selection during experimental coevolution of a freshwater snail and a sterilizing trematode. Evolution 63:2213–2221
Kumpulainen T, Grapputo A et al (2004) Parasites and sexual reproduction in psychid moths. Evolution 58:1511–1520
Lachapelle J, Bell G (2012) Evolutionary rescue of sexual and asexual populations in a deteriorating environment. Evolution 66:3508–3518
Law JH, Crespi BJ (2002) Recent and ancient asexuality in *Timema* walkingsticks. Evolution 56:1711–1717
Lehtonen J, Jennions MD et al (2012) The many costs of sex. Trends Ecol Evol 27:172–178
Levin DA (1983) Polyploidy and novelty in flowering plants. Am Nat 122:1–25
Lively CM (1987) Evidence from a New Zealand snail for the maintenance of sex by parasitism. Nature 328:519–521
Lively CM (2010) Parasite virulence, host life history, and the cost and benefits of sex. Ecology 91
Lively CM, Lyons EJ et al (1998) Environmental stress and the maintenance of sex in a freshwater snail. Evolution 52:1482–1486
Lomnicki A (2001) Carrying capacity, competition and maintenance of sexuality. Evol Ecol Res 3:603–610
Lynch M, Deng HW (1994) Genetic slippage in response to sex. Am Nat 144:242–261
Mallet J (2007) Hybrid speciation. Nature 446:279–283
Moritz C,m Heideman A (1993) The origin and evolution of parthenogenesis in Heteronotia binoei(Gekkonidae)—Reciprocal origins and diverse mitrochondrial DNA in western populations. Syst Biol 42(3):293–306
Pongratz N, Storhas M et al (2003) Phylogeography of competing sexual and parthenogenetic forms of afreshwater flatworm: patterns and explanations. BMC Evolutionary Biology 3
Welch DM, Meselson M (2000) Evidence for the evolution of bdelloid rotifers without sexual reproduction or genetic exchange. Science 288:1211–1215
Smith JM (1978). The evolution of sex. Cambridge University Press, Cambridge
Smith JM (1988) Selection for recombination in a polygenic model—the mechanism. Genet Res 51:59–63
Smith JM, Szathmary E (1995) The major transitions in evolution. Nature 374:227–232
McDonald JF (1993) Evolution and consequences of transposable elements. Curr Opin Genet Dev 3:855–864
Meirmans S, Meirmans PG et al (2012) The costs of sex: facing real-world complexities. Q Rev Biol 87:19–40
Melters DP, Paliulis LV et al (2012) Holocentric chromosomes: convergent evolution, meiotic adaptations, and genomic analysis. Chromosome Res 20:579–593
Morran LT, Schmidt OG et al (2011) Running with the red queen: host-parasite coevolution selects for biparental sex. Science 333:216–218
Muller HJ (1932) Some genetic aspects of sex. Am Nat 66:118–138
Muller HJ (1964) The relation of recombination to mutational advance. Mutat Res 1:2–9
Nei M (1967) Modification of linkage intensity by natural selection. Genetics 57:625
Neiman M, Hehman G et al (2010) Accelerated mutation accumulation in asexual lineages of a freshwater snail. Mol Biol Evol 27:954–963
Neiman M, Meirmans S et al (2009) What can asexual lineage age tell us about the maintenance of sex? In: Schlichting CD, Mousseau TA (eds) Year in evolutionary biology 2009, vol 1168. New York Academy of Sciences, New York, pp 185–200
Normark BB, Moran NA (2000) Testing for the accumulation of deleterious mutations in asexual eukaryote genomes using molecular sequences. J Nat Hist 34:1719–1729
Olofsson H, Lundberg P (2007) The twofold cost of sex unfolded. Evol Ecol Res 9:1119–1129
Otto SP (2009) The Evolutionary enigma of sex. Am Nat 174:S1–S14

Otto SP, Lenormand T (2002) Resolving the paradox of sex and recombination. Nat Rev Genet 3:252–261
Paland S, Lynch M (2006) Transitions to asexuality result in excess amino acid substitutions. Science 311:990–992
Peters AD, Lively CM (1999) The Red Queen and fluctuating epistasis: a population genetic analysis of antagonistic coevolution. Am Nat 154:393–405
Peters AD, Lively CM (2007) Short- and long-term benefits and detriments to recombination under antagonistic coevolution. J Evol Biol 20:1206–1217
Pfrender ME, Lynch M (2000) Quantitative genetic variation in *Daphnia* temporal changes in genetic architecture. Evolution 54:1502–1509
Poon A, Chao L (2004) Drift increases the advantage of sex in RNA bacteriophage Phi 6. Genetics 166:19–24
Pylkov KV, Zhivotovsky LA et al (1998) Migration versus mutation in the evolution of recombination under multilocus selection. Genet Res 71:247–256
Quattro JM, Avise JC et al (1991) Molecular evidence for multiple origins of hybridogenetic fish clones (*Peocilidae*, *Poeciliopsis*). Genetics 127:391–398
Redfield RJ (1993) Evolution of natural transformation—testing the DNA-repair hypothesis in *Bacillus subtilis* and *Haemophilus influenzae*. Genetics 133:755–761
Roze D (2012) Disentangling the benefits of sex. PLoS Biol 10
Salathe M, Kouyos RD et al (2009) On the causes of selection for recombination underlying the red queen hypothesis. Am Nat 174:S31–S42
Schwander T, Crespi BJ (2009) Twigs on the tree of life? Neutral and selective models for integrating macroevolutionary patterns with microevolutionary processes in the analysis of asexuality. Mol Ecol 18:28–42
Simon JC, Delmotte F et al. (2003) Phylogenetic relationships between parthenogens and theirsexual relatives: the possible routes to parthenogenesis in animals. Biol J Linn Soc 79:151–163
Som C, Reyer HU (2007) Hemiclonal reproduction slows down the speed of Muller's ratchet in the hybridogenetic frog rana esculenta. J Evol Biol 20:650–660
Stelzer C-P (2011) The cost of sex and competition between cyclical and obligate parthenogenetic rotifers. Am Nat 177:E43–E53
Suomalainen E (1950) Parthenogenesis in animals. Adv Genet 3:193–253
van Dijk PJ (2003) Ecological and evolutionary opportunities of apomixis: insights from *Taraxacum* and *Chondrilla*. Philos Trans R Soc London Ser B Biol Sci 358:1113–1120
van Dijk PJ (2007) Potential and realized costs of sex in dandelions, Taraxacum officinale s.l. In: Horandl E, Grossniklaus U, VanDijk PJ Sharbel TF (eds) Apomixis: evolution, mechanisms and perspectives. ARG Gantner Verlag KG, Lichtenstein, pp 215–233
Vandel A (1928) La parthénogénèse géographique contribution a l`étude biologique et cytologique de la parthénogénèse naturelle. Bull Biol France Belg 62:164–281
Vellend M, Harmon L et al. (2007) Effects of exotic species on evolutionary diversification. Trends in Ecology Evolution 22(9):481–488
Verhoeven KJF, Biere A (2013) Geographic parthenogenesis and plant-enemy interactions in the common dandelion. BMC Evol Biol 13
Weismann A (1889) The significance of sexual reproduction in the theory of natural selection. In: Poulton SSEB, Shipley AE (eds) Essays upon heredity and kindred biological problems. Clarendon Press, Oxford
Weismann A (1904) The evolution theory. Edward Arnold, London
West SA, Lively CM et al (1999a) A pluralist approach to sex and recombination. J Evol Biol 12:1003–1012
West SA, Lively CM et al (1999b) Sex may need more than one. J Evol Biol 12:1053–1055
Williams GC (1975) Sex and evolution. Princeton University Press, Princeton
Wolf HG, Wohrmann K et al (1987) Experimental-evidence for the adaptive value of sexual reproduction. Genetica 72:151–159
Zeyl C, Bell G (1997) The advantage of sex in evolving yeast populations. Nature 388:465–468

Speciation: Expanding the Role of Biogeography and Niche Breadth in Macroevolutionary Theory

Alycia L. Stigall

Abstract Understanding the processes that control speciation is critical to building a comprehensive macroevolutionary synthesis. A variety of theoretical constructs have been proposed to explain various differential speciation patterns observed in the fossil record, such as higher rates of speciation among specialist versus generalist taxa. Most of these explanations, however, rely on only one or two explanatory variables and may be overly simplistic. Developing a more complete understanding of speciation processes requires a broader synthesis of multiple explanatory factors including the role of external factors such as climatic and tectonics, impact of ecosystem-level processes, relative niche breadth, and relative stability of species' niches during environmental change (biotic and abiotic). This chapter explores the relationship between biogeography, ecological niches, and speciation in a series of case studies focused on Paleozoic (Late Ordovician and Late Devonian) shallow marine brachiopods and bivalves and Cenozoic (Neogene) horses of North America.

Keywords Ecological niche model • Phylogeny • Paleontology • Invasion • Ecology

1 Introduction

Macroevolution is typically defined as evolutionary change at or above the species level. Thus, speciation, the process by which new species evolve, could be considered a cornerstone of macroevolutionary theory. Developments in evolutionary theory over the past 60 years, including those detailed in this volume, indicate that an expansion of the classic Modern Synthesis is necessary to develop a more comprehensive and coherent macroevolutionary theory. In this contribution, I examine the

A.L. Stigall (✉)
Department of Geological Sciences and Ohio Center for Ecology and Evolutionary Studies, Ohio University, Athens, OH 45701, USA
e-mail: stigall@ohio.edu

E. Serrelli and N. Gontier (eds.), *Macroevolution*,
Interdisciplinary Evolution Research 2, DOI 10.1007/978-3-319-15045-1_9

impact of biogeographic processes and evolution of species niches on species formation using several examples of faunal turnover events in the fossil record. These examples highlight the importance of abiotic environmental factors, contingency, and the hierarchical structure of nature on speciation and emphasize need for these areas to be more explicitly included within a synthetic macroevolutionary theory.

1.1 Speciation in the Modern Synthesis

Architects of the Modern Synthesis, notably Ernst Mayr, considered the complexities of the speciation process, such as the importance of geographic barriers in promoting genetic isolation (e.g., Mayr 1942). However, the stated equivalence of macroevolution with microevolutionary processes in the original formulation of the Modern Synthesis (e.g., Dobzhansky 1937) has resulted in an overly simplistic conception of speciation; wherein, speciation is typically viewed as an outcome of population genetics processes with direct adaptive benefits that is divorced from external drivers (Coyne and Orr 2004; Allmon 2013). A consequence is that the importance of speciation has been diminished in modern evolutionary theory.

Speciation lies at the intersection of microevolution and macroevolution, because incipient species are reduced to single populations at the time of speciation. In fact, it is this reduction to a single population that makes speciation such a critical process to understand from a macroevolutionary perspective because incipient species interact simultaneously within both the genealogical and economic hierarchies. Indeed, the initial papers describing punctuated equilibria focused on this point explicitly. Eldredge and Gould (1972) applied Mayr's (1942) conceptual framework for geographic isolation to identify a process, distinct from the microevolutionary explanations inherent in the Modern Synthesis, from which macroevolutionary patterns could derive. Furthermore, macroevolution provides the unique opportunity to study speciation patterns and mechanisms across multiple clades in deep time, which studying snapshots of the speciation in modern organisms cannot accomplish.

Within this context, species are defined as groups of organisms that maintain genetic continuity by interbreeding among members of the group but that are distinct from other reproductive groups. The genetic integrity of a species persists through geologic time, and thus, species have temporal as well as spatial attributes (cf. Evolutionary Species Concept of Wiley 1978). Speciation is the separation of a set of organisms into a newly isolated reproductive unit that is discrete from the ancestral species (Coyne and Orr 2004). Speciation, therefore, relates to a unique event in geologic time that transpired at a specific location within a specific lineage of organisms. Speciation is a process, as discussed in Chaps. 1 and 4 (*Emanuele: Update as appropriate based on final chapter structure*), which typically occurs over some interval of time, usually less than 10,000 years (Eldredge and Gould 1972; Coyne and Orr 2004; Eldredge et al. 2005). During this interval, the ancestral and incipient species accumulate increasing amounts of genetic, morphological, behavioral, etc., differences. Eventually, the two entities may be considered distinct species under a variety of species definitions (cf. De Queiroz 2007). In the fossil record, species are recognized on the basis of morphological differences, but these morphological characters are epistemological tools to recognize the genetically independent entities inherent in ontological species definitions (Allmon 2013). Indeed, it is the process and causes of reproductive isolation that are important to examine when attempting to elucidate the process of speciation itself (Stigall 2013).

1.2 Speciation in Time and Space

The reproductive isolation that initiates speciation can occur when populations have allopatric (separated in geographic space), sympatric (have overlapping geographic ranges), or parapatric (overlap only along a border) distributions. Among modern taxa, allopatric speciation is considered to be much more common than the other styles (Brooks and McLennan 2002). Fortunately, because allopatric speciation results from populations becoming differently adapted to separated environments, it has high potential to result in structural morphological changes that could be fossilized (Eldredge and Gould 1972; Gould and Eldredge 1977; Coyne and Orr 2004). In contrast, sympatric and parapatric speciation typically result from behavioral or rapid genetic changes (e.g., host race switching, chromosomal duplication) (Coyne and Orr 2004) which are less likely to generate morphological changes perceivable in the fossil record. Certainly, notably instances of sympatric speciation involving fossilizable structural changes exist, such as cichlid fishes of African rift lakes, but the prevalence of such examples in modern taxa is controversial (Gavrilets and Losos 2009). Consequently, the discussion of speciation in this chapter addresses only allopatric speciation.

Allopatric speciation occurs via two primary mechanisms: (1) *vicariance*, in which the ancestral population is passively divided by the formation of a barrier into two or more large subpopulations, all of which may diverge to form new species, and (2) *dispersal*, in which a subpopulation actively moves away from the ancestral population and establishes a geographically isolated population that subsequently diverges from the ancestral population (Fig. 1) (Stigall 2010a; also referred to as Allopatry Model I and Model II by Wiley and Mayden 1985). Because vicariance and dispersal are characterized by discrete biogeographic patterns relative to the geographic ranges of daughter and ancestral species, it is possible to identify vicariance and dispersal events in fossil taxa where evolutionary relationships are known (Fig. 1) (Lieberman 2000).

Like every species, every individual speciation event is unique. Each speciation event occurred at a particular geographic location during a discrete interval of time within a specific evolutionary lineage. Therefore, by analyzing the pattern of speciation within and across taxa, it is possible to constrain the role of biogeography, ecology, and phylogenetic history on the development of macroevolutionary patterns (Stigall 2008). The three case studies presented in this chapter provide a

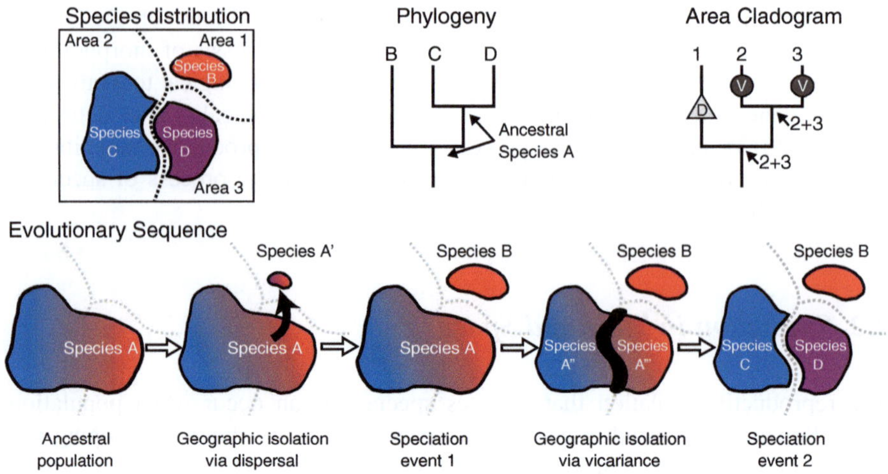

Fig. 1 Schematic representation of the relationship between geographic distribution, phylogenetic relationships, and speciation processes. Species distribution panel illustrates the observed distribution of three species (*Species B*, *C*, and *D*) among three areas (*Areas 1*, *2*, and *3*) which are separated by barriers indicated by *dotted lines*. Phylogeny panel indicates the evolutionary relationships among these three species and the hypothetical ancestor, *Species A*. In the area cladogram panel, the names of the terminal species are replaced with their areas of occurrence. Ancestral distributions are determined using the Fitch optimization outlined in Lieberman (2000). Speciation mode is identified as dispersal (*D*) when the daughter species occupies an area different or additional to the ancestral distribution, whereas vicariance speciation (*V*) is identified when the daughter species occupies a subset of the ancestral range. Evolutionary sequence panel illustrates the geographic steps through which evolution proceeded in this clade

framework to examine the controls on speciation dynamics during three very different intervals of geologic time: the shallow marine seas of the Ordovician and Devonian periods and the grassy plains of the Neogene.

2 Case Studies: Speciation in Action

2.1 Late Devonian Biodiversity Crisis

The first case study examines speciation dynamics of shallow marine organisms, mainly brachiopods and bivalves that inhabited the epicontinental seas of Laurentia (modern North America) between 390 and 360 million years ago. This interval included the Late Devonian biodiversity crisis (~375 Ma), which has been ranked as the third largest interval of ecosystem collapse in Earth history (McGhee et al. 2004, 2013). During the crisis interval, the stromatoporoid sponge/tabulate coral reef ecosystem was obliterated and many previously dominant clades, including atrypid brachiopods, lost their prominence in the marine

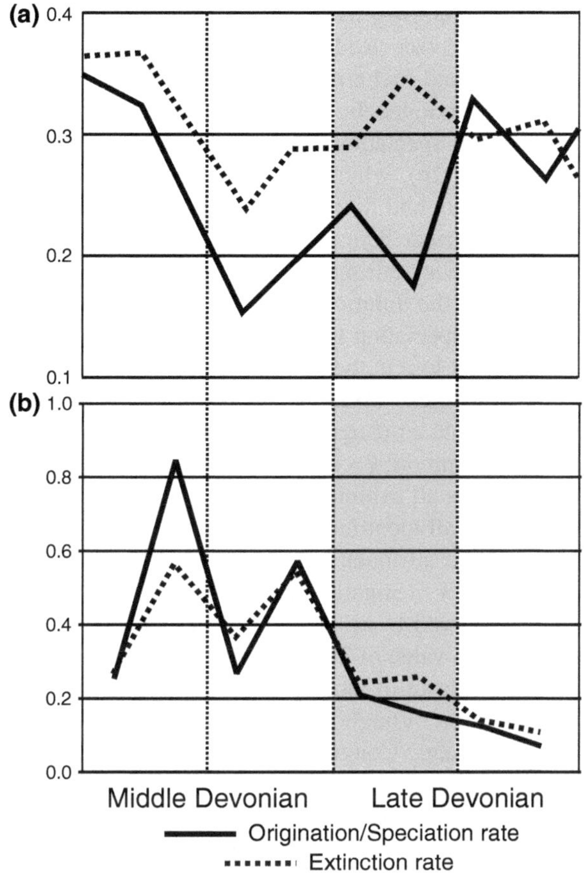

Fig. 2 Comparison of extinction versus origination/speciation during the Middle to Late Devonian interval. Late Devonian biodiversity crisis interval is indicated in gray. **a** Proportion of global generic extinction or origination per interval. *Modified* from Bambach et al. (2004). **b** Average instantaneous rates of species extinction and speciation for the brachiopod genera *Schizophoria* and *Floweria* and the bivalve *Leptodesma* (*Leiopteria*). *Modified* from Stigall (2010a). Similar patterns occur at both taxonomic levels: during the crisis interval, extinction rates are lower than occurs during the Middle Devonian background interval, and origination/speciation rates are substantially reduced

environment (McGhee 2013). Although extinction rate was somewhat elevated during the crisis interval, extinction rate did not statistically rise to the level of "mass extinction" (Bambach et al. 2004). Instead, the biodiversity crisis was caused by a dramatic reduction in speciation rates which, when combined with moderately elevated extinction levels, caused fundamental restructuring of the ecosystem (Fig. 2a) (McGhee et al. 2013; McGhee 1996; Stigall 2012b). Examining the reasons for speciation decline during the Late Devonian, therefore, can provide insight into macroevolutionary controls on speciation.

2.1.1 Speciation Mode and Rate

Late Devonian speciation dynamics have been assessed using a series of phylogenetically informed analyses to examine rate and mode of speciation across a range of Devonian marine taxa. Four genera from three phyla were subjected to species-level phylogenetic analysis: two rhynchonelliform brachiopods, *Floweria* and *Schizophoria* (*Schizophoria*) (see Stigall Rode 2005); a bivalve, *Leptodesma* (*Leiopteria*) (see Rode 2004); and a predatory crustacean order, Archaeostraca (see Rode and Lieberman 2002). This represents a combined dataset of 89 species and 148 cladogenetic events available for speciation analysis. Phylogenetically constrained speciation rates were calculated for the brachiopod and bivalve lineages (Fig. 2b), which confirms the dramatic decline in speciation rate in these clades before and during the biodiversity crisis interval (Stigall 2010a). Statistical analyses further demonstrated that both speciation and extinction rates observed in these clades during the biodiversity crisis interval are lower than the average rate during the duration of each clade (Stigall 2010a), which further indicates that depressed speciation rates, not elevated extinction rates, were the primary case of biodiversity loss in these lineages.

Among these taxa, the prevalence of speciation by vicariance and dispersal varied (Table 1) (Stigall 2010a). Within *Floweria*, dispersal and vicariance were equally common; each comprised 50 % of speciation events. However, the three other clades all exhibited higher levels of speciation by dispersal, ranging from 68 to 89 % of all identifiable speciation events. Combined only 28 % of all speciation events were attributable to vicariance versus 72 % to dispersal (standard deviation = 16 %) (Stigall 2012b). These values are almost diametrically opposed to those observed from analyses of the modern fauna; Brooks and McLellan (2002) recovered a value of 74 % (sd = 35 %) speciation by vicariance in a review of speciation analyses of modern taxa. Vicariance has also been the more common mode of speciation throughout geologic time based on comparative studies of Paleozoic trilobites (e.g., Congreve and Lieberman 2008, 2010; Lieberman 1997, 2003; Lieberman and Eldredge 1996), which exhibit an average of 54 % (sd = 16 %) speciation by vicariance (Stigall 2010a).

Therefore, the rates of vicariant speciation in the Late Devonian clades are anomalously low compared to other intervals in Earth history. Because speciation by vicariance typically occurs with a higher frequency than dispersal, the near elimination of this mode of speciation has the potential to reduce the total speciation rate by 50 % or more. The substantial reduction in frequency of vicariant speciation during this interval is certainly a causal factor in the biodiversity crisis (Stigall 2012b). Reduced vicariance explains the precipitous decline in total speciation rate that characterized the Frasnian Stage—both in these taxa and the fauna as a whole (Fig. 2). Consequently, speciation depression in the Late Devonian can only be understood by considering which factors would preferentially suppress speciation by *vicariance*, or fragmenting of ancestral populations to form new species.

2.1.2 Cause of Speciation Depression

Because the factors that prevented vicariance did so across the entire fauna, internal microevolutionary factors alone cannot explain this pattern; however, a series of external factors, most notably sea level changes and the global spread of invasive species, provide mechanisms for vicariance reduction (Stigall 2012b). During

Table 1 Comparison of primary controls on speciation in each of the three case studies

	Speciation patterns	Study group		Primary controls: environmental factors		Secondary controls: inherent factors	
		Taxon	Lifestyle	Abiotic	Biotic	Niche breadth	Niche evolution
Late Devonian biodiversity crisis	Depressed during invasion interval, suppression of vicariant speciation	Brachiopods and bivalves	Marine, sessile suspension feeders	Orogenic pulses and sea level changes, particularly transgressions	Global species invasions (competition)	Preferential survival of generalist versus specialist species throughout crisis	Not analyzed
Late Ordovician Richmondian invasion	Depressed during invasion, resumes once community stabilizes	Brachiopods, mollusks, arthropods, crinoids, corals	Marine, sessile suspension feeders, predators, and detritovores	Sea level change; rapid transgression versus gradual regressions	Basin-scale species invasions (competition)	Preferential survival of generalist versus specialist during invasion; specialists return after community stabilizes	Common only during rapid environmental changes; more pronounced in specialist than generalist species
Miocene equinid radiation	High during initial environmental change; lower during subsequent gradual change	Horses	Terrestrial, vagile herbivores	Tetonic uplift and global climatic cooling	Expansion of grasslands, shift from patchy to continuous habitats	N/A; all equinids considered to be specialist taxa	Only one species tested, shows niche evolution (~50 %) between Middle and Late Miocene

Modified from Stigall (2013)

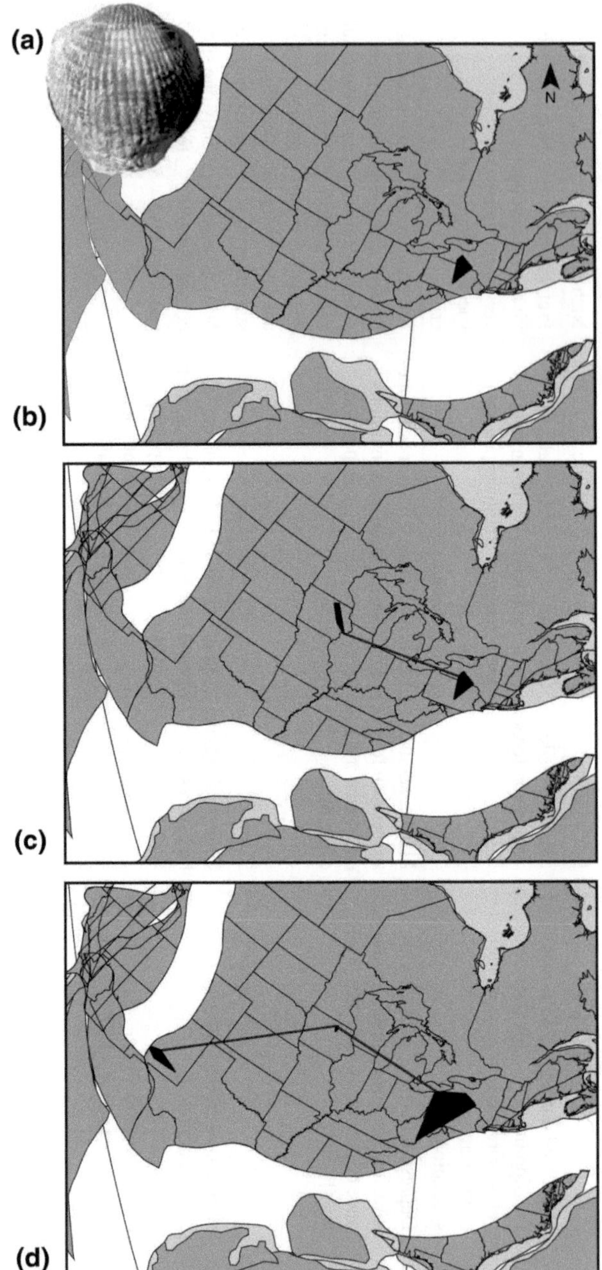
(a)
N
(b)
(c)
(d)

◀ **Fig. 3** Example of interbasinal species invasion events in the Late Devonian brachiopod species, *Pseudatrypa devoniana* (**a**), 0.75×. (**b**) During the earliest Late Devonian, *P. devoniana* occupied its ancestral basin in the Appalachian region; (**c**) this species invaded the Iowa basin during the early Late Devonian; (**d**) and subsequently invaded the New Mexico basin during the middle Late Devonian. The invasions depicted in (*b*) and (*c*) correspond to the sea level rises indicated by the *second* and *third* arrows in Fig. 4. *Modified* from Rode and Lieberman (2004)

the Frasnian Stage of the Late Devonian, a series of sea level rises facilitated the colonization of previously isolated tectonic basins by a new set of species, in a manner analogous to modern invasive species (Rode and Lieberman 2004). Using geographic information systems (GIS) to generate temporally restricted geographic range maps, Rode and Lieberman (2004), were able to characterize the geographic expansion and contraction of 341 species of the most common brachiopod and bivalve genera. In this study, they were able to identify discrete events of interbasinal species invasion (Fig. 3), characterize the distribution of invasion events within the fauna through time (Fig. 4), and calculate geographic range size for species through time (Fig. 4). Peaks in invasion intensity corresponded to episodes of sea level rise and were moderated by regional tectonic processes (Rode and Lieberman 2004). Mean geographic range size increased, and aggregate number of species invasion events increased during the study interval, in clear contrast with the monotonic decline in speciation rates (compare Figs. 2 and 4).

The observed spread of Devonian invasive species facilitated a shift in ecosystem dynamics and depressed speciation. The combination of overall range expansion and frequent invasive events prohibited sustained geographic isolation, thereby impeding the primary requirement for vicariant speciation. In addition, there was preferential extinction of narrowly distributed species and preferential survival of broadly ranging or invasive species across the biodiversity crisis interval (Rode and Lieberman 2004). Furthermore, species that had participated in interbasinal invasions preferentially survived the crisis interval. Analysis of paleoenvironmental distribution of these species indicates that the narrowly ranging species were typically ecological specialists, whereas the surviving broadly ranging and invasive species were ecological generalists (Rode and Lieberman 2004). This pattern of larger geographic ranges or invasive success of generalist taxa is well-documented among modern organisms (e.g., Fernández and Vrba 2005; Botts et al. 2013; Thuiller et al. 2005; Brown et al. 1996; Lockwood et al. 2009).

The preferential survival of generalist species is an important observation from a macroevolutionary perspective. Ecological niche breadth has been an important character on which species sorting may occur (Vrba 1984; Jablonski 2008). Specifically, lineages of ecological specialists tend to have higher diversification rates than generalist clades (Stanley 1990; Jablonski 2008; Vrba 1987; Jackson 1974; Kammer et al. 1997); therefore, their preferential loss from the Devonian ecosystem would have further exacerbated speciation depression. Ultimately, the series of species invasions and the subsequent differential survivorship patterns produced a fauna that was dominated by broadly ranging ecological generalist taxa, producing the notably cosmopolitan fauna (cf. McGhee 1996) of the latest Devonian.

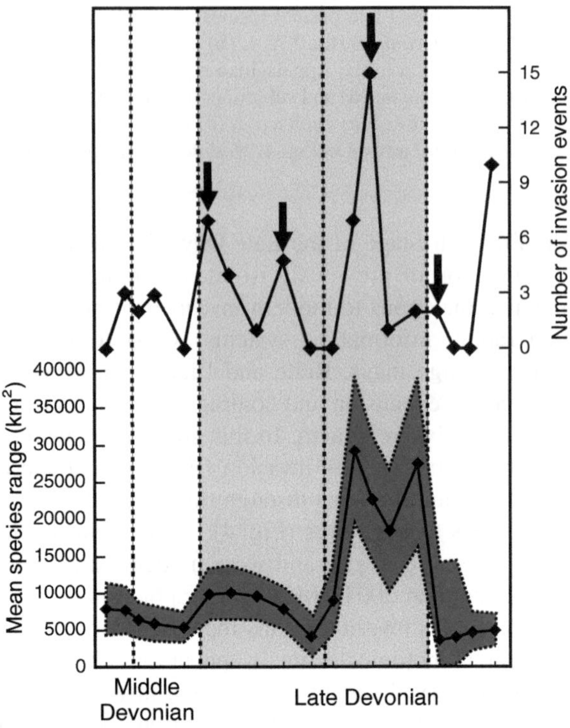

Fig. 4 Middle through Late Devonian species geographic range sizes and invasion intensity. *Arrows* indicate episodes of sea level rise. Both the mean geographic range size for brachiopod and bivalve species and interbasinal invasion intensity increase during Late Devonian crisis interval (*shaded gray*) compared with Middle Devonian background levels. *Modified* from Rode and Lieberman (2004)

2.1.3 Late Devonian Synthesis

The depression in speciation that defined the Late Devonian biodiversity crisis resulted from a combination of biogeographic, environmental, and ecological factors (Table 2). The primary elements of geographic range expansion and widespread interbasinal species invasions, and the dominance of generalist over specialist species produced an environment that essentially prohibited successful formation of new species via vicariant speciation. A combination of overall range expansion and frequent invasive events would have prohibited sustained geographic isolation, thereby impeding the primary requirement for vicariant speciation. The initial phase

Table 2 Comparison of speciation mode in marine invertebrates through the Phanerozoic

Clade	Percent speciation by vicariance (%)	Percent speciation by dispersal (%)
Late Devonian crustaceans	32	68
Late Devonian brachiopods and bivalves	27	73
Cambrian and Ordovician trilobites	54	46
Early to Middle Devonian trilobites	54	46
Modern fauna	74	26

Modified from Stigall (2010a)

of vicariant speciation, in which the ancestral species is fragmented into incipient species with smaller populations, was unsustainable during the invasive regime and incipient species became extinct rather than developing into viable new species.

The suppression of Late Devonian speciation is not attributable to microevolutionary processes alone. Competition or other biotic interactions may have played a role in promoting extinction of incipient species. However, the overarching controls on the aggregate pattern of speciation reduction occurred at higher hierarchical scales. Fluctuations in sea level that facilitated species invasions occurred on timescales exceeding the life span of individual organisms. The frequency of species invasions and the greater relative abundance of generalist species in the environment relate to ecosystem and community structure, both entities in economic hierarchies (cf. O'Neill et al. 1986; Eldredge 1996) that are outside the scope typically considered in the standard Modern Synthesis. It is only by integrating the disparate areas of biogeography, ecology, and environmental change that a synthetic understanding of Late Devonian speciation dynamics can be achieved.

2.2 Late Ordovician Richmondian Invasion

The second case study also focuses on macroevolutionary patterns of marine benthos (mainly brachiopods) in Paleozoic seas during an interval characterized by species invasions and fluctuating sea levels. However, this example focuses on species-level responses to a regional invasion event, the Richmondian Invasion, during an interval of community-level ecological overturn in the Late Ordovician strata (Cincinnatian Series; ~450 mya) deposited in a shallow marine basin centered on what is today Cincinnati, Ohio, USA. Strata of the Cincinnatian Series have been divided into six depositional sequences (C1 to C6), each of which represents a cycle of sea level rise and fall (Holland and Patzkowsky 1996) (Fig. 5). The preserved record of these sequences includes a shallowing upward succession that indicates relative infilling of the basin through time. These cyclical sea level changes are overprinted by the onset of the Richmondian Invasion during the C4 sequence. This regional invasion event introduced more than 50 genera into the basin, which included members of all major marine clades and trophic groups (Holland 1997). Notably, some invaders belonged to major clades of animals that had been absent from the region for up to 5 million years prior (e.g., rugose corals, rhynchonellid brachiopods) (Stigall 2010b).

The Richmondian introductions substantially impacted the established community structure, species survivorship, and speciation patterns (Meyer and Davis 2009; Stigall 2010b). Detailed gradient ecology and biofacies analyses by Holland and Patzkowsky (2007) indicated that the pre-invasion communities (C1 to C3 sequences) exhibited highly similar community structure, but the onshore–offshore gradient was reduced and faunal differentiation became diminished during the invasive regime (C4 sequence). Furthermore, well-defined biofacies

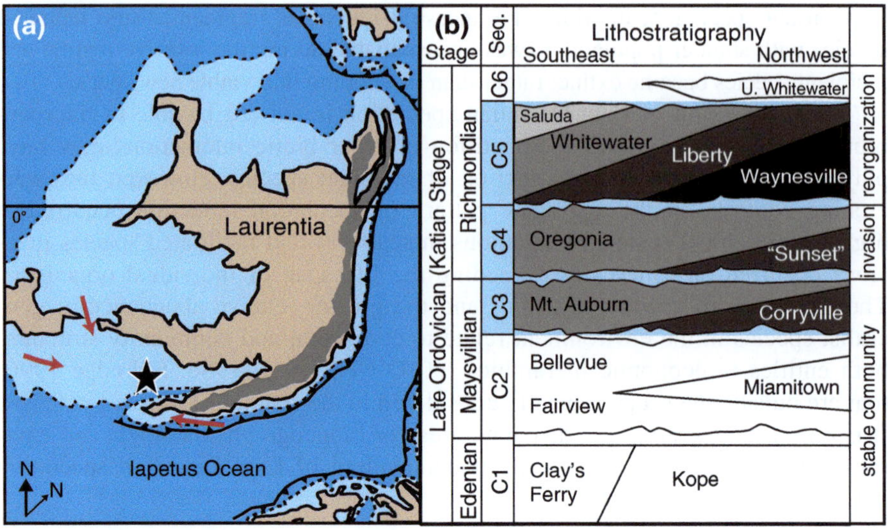

Fig. 5 Geologic framework of the Late Ordovician case study. **a** Location of the Cincinnati region (*star*) within Laurentia. *Arrows* indicate established pathways of migration during the Richmondian Invasion following Wright and Stigall (2013a); additional migration routes may have operated. **b** Ecological changes within a chrono-, litho- and sequence-stratigraphic framework of the Cincinnatian series, *modified* from Malizia and Stigall (2011). The Richmondian invasion occurred primarily during the *C4* sequence. The case study examines the *C3*, *C4*, and *C5* sequences. *Modified* from Stigall and Brame (2014)

were re-established during the C5 sequence, but these biofacies differed significantly from the pre-invasion structure due to increased habitat partitioning and the ecological dominance of many invader taxa (Holland and Patzkowsky 2007; Patzkowsky and Holland 2007). These community analyses focused on the genus level and form the framework for the detailed investigation of species-level responses discussed below.

2.2.1 Range Size, Survivorship, and Speciation Mode

The ecological changes that occurred across the Richmondian Invasion have clear biogeographic and macroevolutionary correlates. Stigall (2010b) examined biogeographic and survivorship patterns in Cincinnatian brachiopod species using GIS methods. In this study, species were divided into four groups: species that were native to the region and became extinct before or during the invasion, native species that persisted through the invasion interval, species that evolved from native ancestors during or after the invasion interval, and extrabasinal invaders. The emergent patterns resemble those discussed in the Devonian case study as: (1) broadly ranging species preferentially survived the invasive interval, whereas narrowly ranging species were more likely to become extinct; (2) both native survivors and extrabasinal

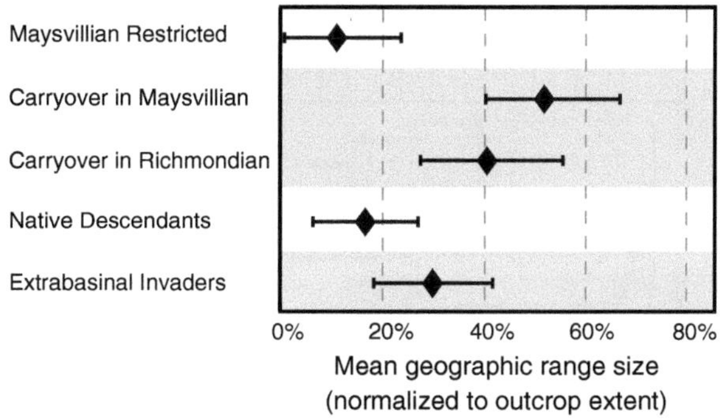

Fig. 6 Comparison of geographic range sizes Cincinnatian brachiopods by species group. Extrabasinal invaders and native species that carryover between the pre-invasion to post-invasion intervals have statistically larger ranges than either the native species that do not survive through the invasion (Maysvillian restricted) or those new species that form post-invasion from native ancestors (Native descendants). Individual 95 % confidence intervals for mean geographic range are based on pooled standard deviation. *Modified* from Stigall (2010b)

invaders occupied large geographic areas; (3) formation of new species was severely depressed during the invasion interval (Fig. 6) (Stigall 2010a).

Phylogenetic biogeographic analysis of speciation mode in three genera of Late Ordovician rhynchonelliform brachiopods, *Glyptorthis, Hebertella,* and *Plaesiomys,* provides additional insight into speciation dynamics during the invasion interval (Fig. 7). Speciation by vicariance was rare in *Plaesiomys* and *Hebertella* (0 and 29 %, respectively) but was more common in *Glyptorthis* (74 %) (Wright and Stigall 2013a). When placed in a stratigraphic context, it is apparent that the primary interval of vicariance predates the invasive regime; only two of the twenty-two vicariance events occurred after the C4 sequence (Fig. 7) (Wright and Stigall 2013a). Notably, the extremely low level of vicariance during the Richmondian Invasion parallels that of the Late Devonian (Stigall 2012a).

The similarity of speciation mode and biogeographic patterns suggests that the underlying mechanism—differential survival of specialist and generalist taxa across the invasion—driving diversity dynamics is similar during the Devonian and Ordovician case studies. Specifically, species sorting based on niche breadth exerted a primary control on speciation rates and outcomes. In the case of the Richmondian Invasion, however, the ecological dominance of generalist taxa persisted for only about 1 million years (throughout the C5 sequence) (Stigall 2010b). Even during the interval of adjustment, community diversity remained high due to niche partitioning (Patzkowsky and Holland 2007; Tyler and Leighton 2011; Brame and Stigall 2014). Once speciation resumed in the post-invasion C6 sequence, new specialist taxa were formed (Fig. 6) (Stigall 2010b), which promoted a biodiversity recovery rather than the decline of the Late Devonian. The different outcomes of the Late Devonian and Late Ordovician case studies likely derive from the global versus regional scope of the invasion impacts.

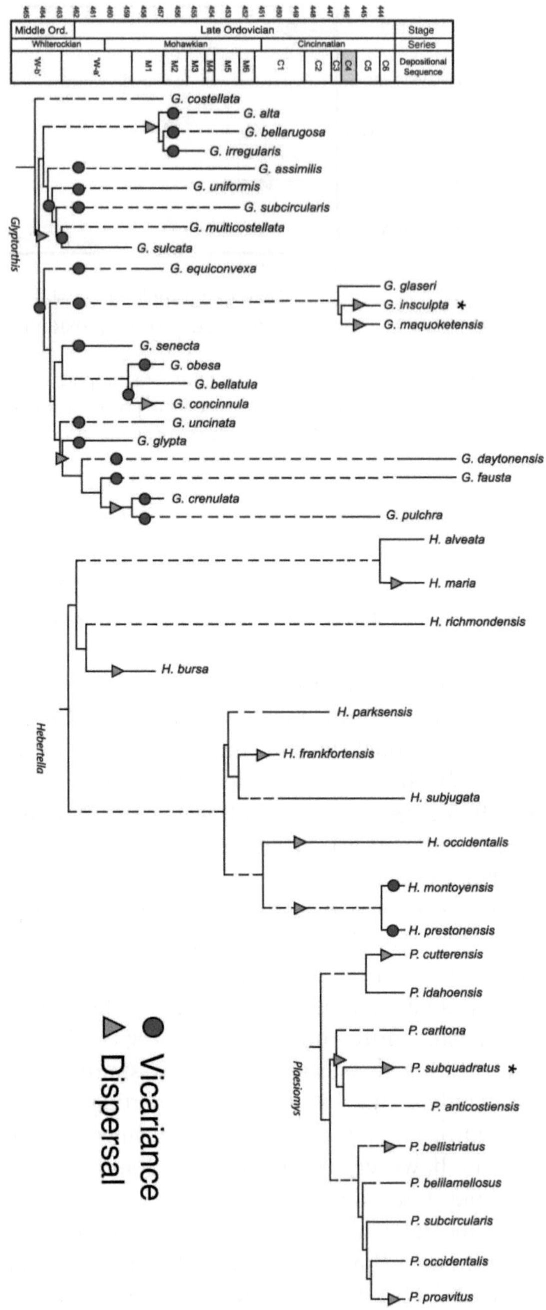

Fig. 7 Distribution of speciation events by mode within of the brachiopod genera *Glyptorthis, Hebertella, and Plaesiomys* mapped onto species-level phylograms from Wright and Stigall (2013b, 2014) at the level of depositional sequence. The Richmondian invasion occurred during the C4 sequence (*shaded gray*). *Modified* from Wright and Stigall (2013a)

2.2.2 Niche Stability Analyses

To fully characterize the impact of invasive species on the speciation process, it is important to more fully constrain aspects of population-level change that occur within populations and how the changes impact the propensity of populations to persist, go extinct, or subdivide to produce incipient species. Accordingly, ecological niche models were developed to assess whether native species responded to the influx of invaders by maintaining the parameters of their ecological niches and tracking their preferred habitats laterally (= niche conservatism or habitat tracking) or by undergoing shifts in dimensions their ecological niches (= niche evolution) (overview in Stigall 2012c). In this technique, species' ecological niches are estimated from a suite of environmental parameters associated with a set of known species occurrence points (Peterson et al. 2011). From these data, an equation is generated to describe the species' niche. This equation is then projected back into geographic space to generate potential geographic distribution map for the species (Peterson et al. 2011). To assess Late Ordovician niche stability in geographic space, the niche model created for a taxon in one time slice was projected onto the environmental parameters of the following time slice to generate a model of the geographic distribution that would be occupied by that taxon if it had conserved its niche completely between time slices (Fig. 8) (Stigall 2012c). This forward projection model was overlain on the geographic distribution model generated from the data collected for the second time slice, and the relative degree of spatial similarity was calculated (Brame and Stigall 2014; Malizia and Stigall 2011) (Fig. 8). Niche stability was also assessed by calculating the similarity of the percent contributions of each environmental variable to niche model equations for a taxon in consecutive time slices (Brame and Stigall 2014; Stigall 2011, 2012c).

A series of niche stability studies of this type have been undertaken to examine niche stability within brachiopod species (i.e., Stigall 2011, 2012c; Malizia and Stigall 2011; Walls and Stigall 2011, 2012; Dudei and Stigall 2010) as well as mollusk, trilobite, coral, and crinoid genera (Brame and Stigall 2014). The primary results are statistically similar for all taxa, regardless of autecology, clade membership, or taxonomic rank (Stigall and Brame 2014). Niche stability in geographic space is highest in the pre-invasion interval and becomes statistically reduced following the onset of the Richmondian Invasion (Fig. 9) (Stigall 2012c, 2014), which indicates that although species are able to track their preferred habitat laterally during the gradual sea level fluctuations of the pre-invasion interval, this ability is diminished in the presence of the extrabasinal invaders. Intriguingly, there is a concomitant statistical increase in stability of the contribution of the environmental parameters of species' niches through time (Brame and Stigall 2014; Stigall and Brame 2014; Stigall 2011). The increase in parameter similarity contrasts with the contemporaneous decrease in overall geographic niche stability (Fig. 9).

Shifting niche expression in geographic space concurrent with increasing parameter similarity requires a reduction of the initial ecological niche that resulted in reduced variability of niche parameters (Fig. 10) (Brame and Stigall 2014). Reduction in environmental niche parameters could occur through a variety of

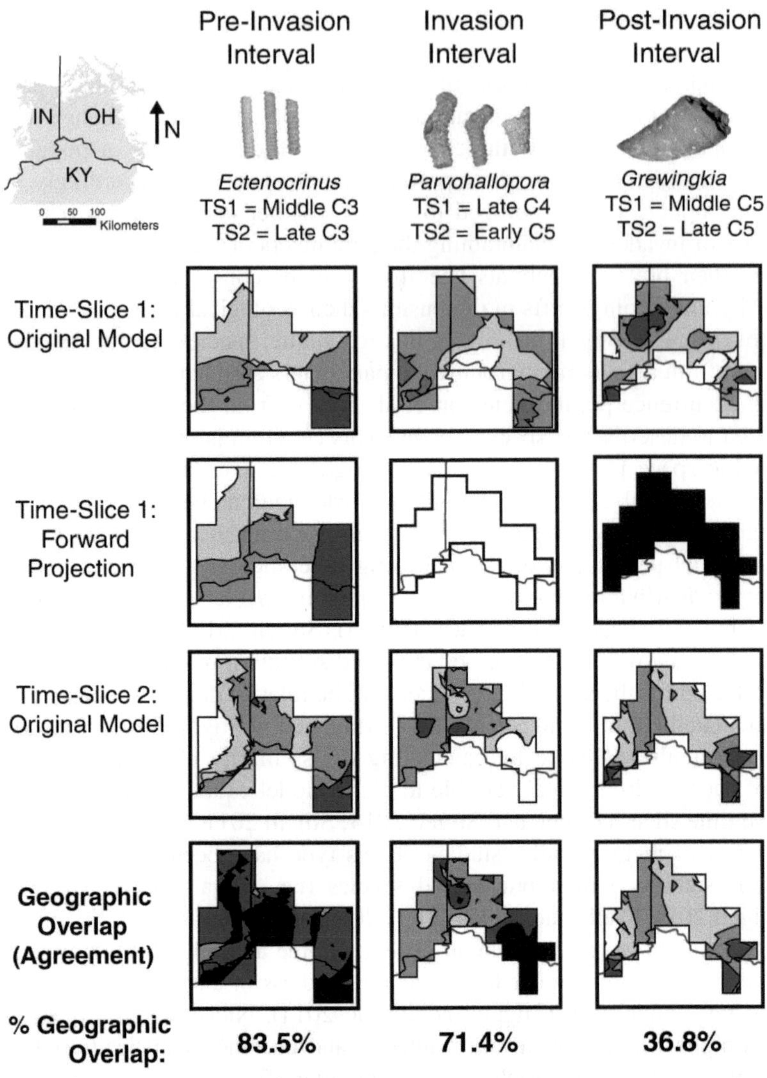

Fig. 8 Geographic-distribution models representative of typical patterns during the pre-invasion, invasion, and post-invasion intervals. On distribution maps, *dark gray* indicates high probability/agreement, and *light gray* indicates low probability/agreement. From the pre-invasion interval to the post-invasion interval, weighted percent geographic overlap decreases and model parameter similarity increases, a pattern also present within the entire fauna. *Modified* from Brame and Stigall (2014)

processes: direct product of natural selection (= adaptation), byproduct of selection (= exaptation) (Gould 2002), spatial sorting (Shine et al. 2011), or plastic ecophenotypic expression (cf. Levin 2003). Because the focal taxa exhibited niche stability across a depositional sequence that included changes in sea level and geographic expression of habitat during the pre-invasion interval, the post-invasion

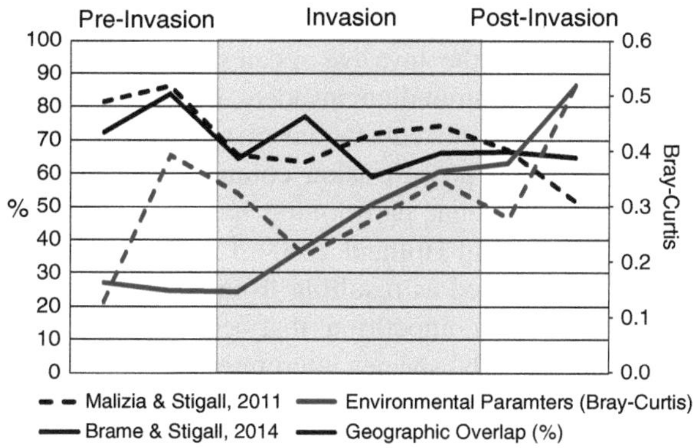

Fig. 9 Summary of the weighted percent geographic overlap and environmental parameter similarity data through time. Patterns are congruent between the species-level brachiopod data of Malizia and Stigall (2011) and the cross-faunal genus-level data of Brame and Stigall (2014). The geographic overlap percentage declines significantly between the pre-invasion and invasion intervals, whereas environmental parameter similarity statistically increases between the pre-invasion and invasion intervals. *Modified* from Stigall (2014)

Fig. 10 Relationship between niche evolution in environmental parameter and geographic space. The *left* column displays the initial environmental tolerance of a species along environmental gradient (*dark* to *light gray*) and the resulting geographic distribution. The column on the *right* indicates how contraction of the initial niche can result in high niche similarity but low geographic similarity. *Modified* from Brame and Stigall (2014)

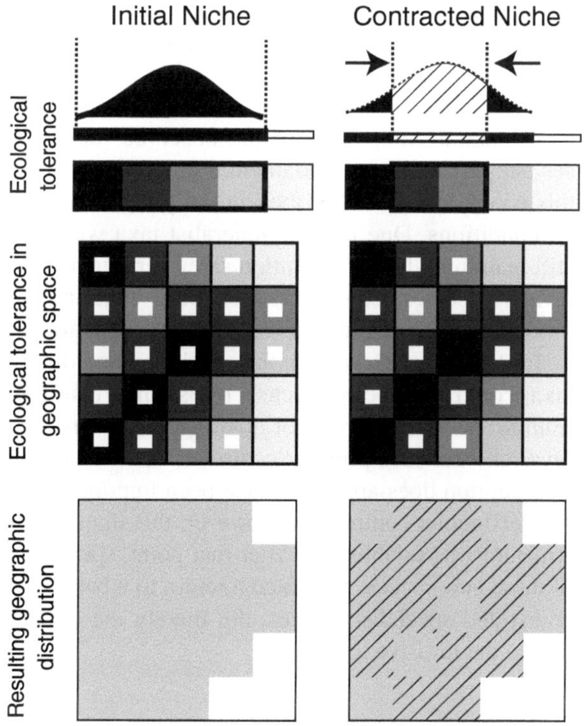

niche evolution can be attributed to changes produced within the environment related to the introduction of the invasive species rather than other environmental factors. Although the Richmondian invaders included a set of species from all trophic levels, there were no substantial changes in predation or mutualistic structures in the pre- and post-invasion communities; however, evidence for increased competition and habitat partitioning has been documented (Tyler and Leighton 2011; Patzkowsky and Holland 2007). Consequently, the primary driver for niche evolution is interpreted as resulting from adaptive response to the biotic changes, principally increased competition, that occurred during and following the Richmondian Invasions. Notably, species interpreted as ecological generalists (e.g., *Rafinesquina alternata*) were those most able to successfully habitat track initially and maintain their niche parameters through time (Malizia and Stigall 2011; Brame and Stigall 2014). The enhanced stability of generalist taxa within the invasion and post-invasion intervals supports the conclusions from the speciation analyses that eurytopy confers an adaptive advantage during invasive regimes.

2.2.3 Late Ordovician Synthesis

External factors (relative sea level and extrabasinal species invasions) exerted fundamental control over the macroevolutionary patterns that developed among the Late Ordovician fauna of the Cincinnati basin (Table 2). Primary aspects of community restructuring included differential loss of specialist taxa, 1 million year speciation gap, reduced vicariance, and shift to ecosystems dominated by extrabasinal taxa. This suite of characteristics indicates that species selection, notably preferential survival of generalist species and loss of ecological specialists, was a substantial component of the observed faunal change. Niche modeling analyses suggest that one fundamental key to the persistence of the generalist taxa in this ecosystem was that they initially could utilize a broad suite of environmental conditions. Due to this, generalist taxa were able to constrict their ranges and still maintain viable population sizes participate while partitioning habitat with the invasive species. Ecological specialists with narrowly defined niches could not do this effectively, resulting in very small population sizes, and ultimately extinction.

Like the Late Devonian, the speciation and diversity patterns of the Cincinnatian taxa could not be predicted by simple extrapolation within a neo-darwinian framework. The impact of contingent factors that facilitated the Richmondian Invasion exerted a primary control on Cincinnatian niche evolution and speciation. Competition does appear to have been important in determining the structure of the post-extinction community; however, this impact lessened once the new community structure was established. After that point, speciation resumed and the development of these new species promoted a return to a balanced ecosystem with a blend of generalist and specialist species, not merely the ecological generalists that dominated during the invasive regime.

2.3 Neogene Radiation of the Equinae

The third case study contrasts with the first two by examining speciation patterns and processes during a radiation, rather than a crisis interval, and by focusing on a clade of terrestrial mammals instead of marine invertebrates. Specifically, this case study examines the Miocene to Early Pliocene radiation of horse species (subfamily Equinae) in North America. The diversification of this clade has classically been referred to as an adaptive radiation under the interpretation that the rapid speciation occurred via adaptive response to the contemporaneous climatic and vegetation changes (Simpson 1944; MacFadden 1984; Hulbert 1993). Historically, the major Miocene radiation of the Equinae has been attributed to morphological changes (e.g., increased hypsodonty, reduction of digits) considered to be adaptations to the coincident emergence of grasslands (Simpson 1951; Stebbins 1981). However, other studies have demonstrated that the dentition of a horse species is not always conclusive evidence for the type of vegetation in their diet (Eronen et al. 2010; Fortelius 1985), and Strömberg (2006) advised against using tooth morphology alone to reconstruct habitat change due to inconclusive evidence regarding whether hypsodonty was an adaptive characteristic. Consequently, the extent that speciation in this clade represents an adaptive rather than exaptive radiation has been questioned (Lieberman 2012). Because detailed phylogenetic hypotheses and a dense fossil record exist for Miocene horses of North America, this radiation can be analyzed in detail to identify whether internal (adaptive) or external (e.g., climatic, tectonic) factors were more influential in promoting diversification.

2.3.1 Speciation Mode

The evolutionary relationships between equinid species have been examined repeatedly (e.g., Hulbert 1993; MacFadden 1984; Prado and Alberdi 1996), and well-supported phylogenetic hypotheses exist for members of this clade. Analysis of speciation mode in this clade by Maguire and Stigall (2008) recovered 81 % of inter-regional speciation events were attributable to dispersal. Contrary to the previous case studies, frequent dispersal in horses coincided with high diversification rates (e.g., Hulbert 1993). The link between dispersal and diversification stems from the biology and behavioral ecology of horses; these animals were vagile, migrational herbivores. Consequently, dispersal events were frequent and could have generated isolated populations and subsequent speciation events. Notably, ecological niche modeling analyses of this clade demonstrated a correlation between increased geographic discontinuity within species ranges and higher speciation rates (Maguire and Stigall 2009). This pattern suggests that vicariance was operating at the basinal, rather than regional, level within this clade. Because the analysis of Maguire and Stigall (2008) analyzed large geographic regions, basin-scale vicariant speciation would appear as no change in geographic region between ancestor and descendant nodes, a condition that characterizes most of the internal nodes on the optimized area cladogram (Fig. 11).

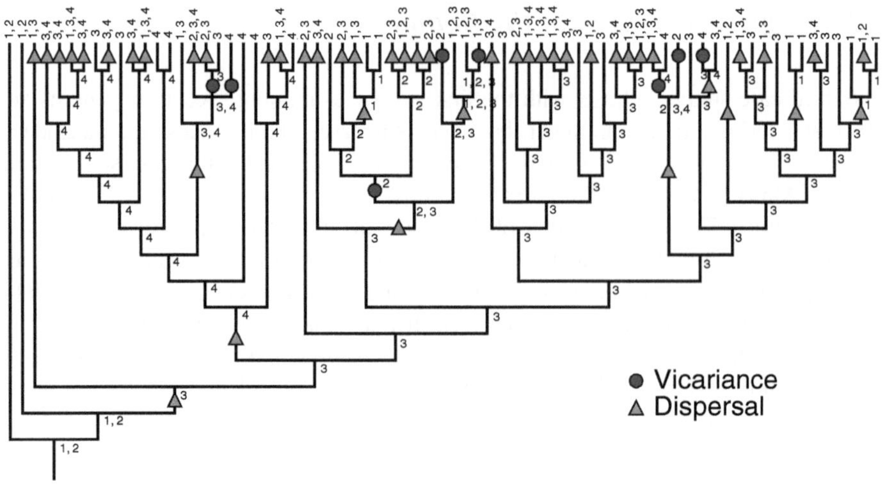

Fig. 11 Distribution of speciation events by mode within the subfamily Equinae mapped onto species-level area cladogram. *Modified* from Maguire and Stigall (2008)

Although organismal biology may have predisposed species to dispersal-based speciation, external processes, notably climatic and tectonic changes—both external, continent factors—have been shown to have substantial impact on the overall radiation of this clade (Maguire and Stigall 2008). A Lieberman-modified Brooks Parsimony Analysis of equinid species generated general area cladograms topologies consistent with geological events, such as uplift of the Rocky Mountains, and climatic conditions, such as the change from a warm and moist to cool and arid climate during the Miocene (Maguire and Stigall 2008). Congruence between the geodispersal and vicariance area cladograms indicates that the cyclical processes, notably climatic changes during the Miocene, were important in structuring diversification patterns and area relationships during the equinid radiation (Maguire and Stigall 2008).

2.3.2 Habitat Patchiness and Speciation

Reconstruction of species geographic ranges using ecological niche modeling indicated that the relative patchiness of habitats within regions provided an additional control on speciation (Maguire and Stigall 2009). Habitat patchiness developed as the previously widespread forests became fragmented due to the spread of grasslands during the Middle Miocene, but declined once grasslands were widespread in the Late Miocene and Early Pliocene (Fig. 12). Speciation was highest during intervals of environmental change. In particular, early to Middle Miocene climatic and tectonic events promoted the development of a landscape characterized by patchy environmental distribution. During this interval, horse species comprised several isolated populations and exhibited frequent speciation events, related to vicariance within and dispersal between basins. Once climate stabilized

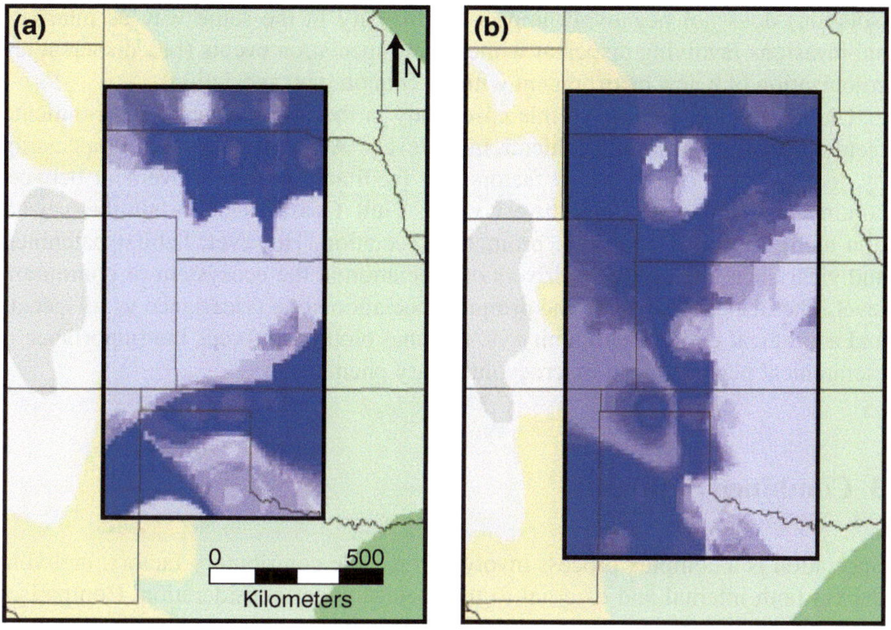

Fig. 12 Example of fragmented and continuous geographic ranges derived from ecological niche models of Neogene horses. Fragmented predicted range for *Pseudhipparion gratum* in the early to Middle Miocene (**a**) and continuous predicted range for *Nannippus lenticularis* in the late Miocene (**b**). *Light gray* indicates low probability of occurrence, and *dark gray* indicates high probability of occurrence. *Modified* from Maguire and Stigall (2009)

and habitats became more homogenous, speciation rate in horses declined (Maguire and Stigall 2009). Consequently, speciation was promoted by environmental change and habitat patchiness but retarded by environmental uniformity and climatic stability.

2.3.3 Neogene Synthesis

The Neogene radiation of the horses provides insight into some factors that promote speciation (Table 2). Speciation was highest during early to Middle Miocene climatic and tectonic events that promoted the development of a landscape characterized by patchy environmental distribution. Once climate stabilized and habitats became more homogenous, speciation rate in horses declined. Consequently, speciation was promoted by environmental change and habitat patchiness but retarded by environmental uniformity and climatic stability. Finally, although dispersal was common in this clade, it did not result in the speciation depression observed during the Late Ordovician or Late Devonian. This may suggest that speciation via dispersal (i.e., speciation within a new environment after active geographic isolation) does not negatively impact biodiversity in the same way as interbasinal invasions involving dispersal without new speciation events (i.e., dispersal and colonization of a new environment without concomitant speciation).

One of the key aspects of this case study is the importance of environmental factors operating at multiple hierarchical levels. At the regional level, speciation by dispersal was common and factors that facilitated species movement between continental regions (eg., Southwest versus Gulf Coast), such as similar vegetation along dispersal corridors, promoted speciation. However, habitat patchiness and vicariance were primary drivers of speciation at the ecosystem or community level. The disparity between the primary speciation style (vicariance vs. dispersal) and ecological entity (community vs. regional biota) reinforces the importance of hierarchical perspective to macroevolutionary phenomena.

3 Combined Synthesis

Speciation is a complex process involving multiple contributory factors, including drivers both internal and external to the species under consideration. Comparison of speciation dynamics among the three case studies indicates that critical determinants of speciation processes can include biotic and abiotic influences (Table 2). Although the details of each case study are unique, these examples share several broad features. Significantly, the primary controls on speciation in all three case studies were external, environmental factors (Table 3). Abiotic factors (tectonics, sea level, climate change) were the dominant controls on speciation, but external biotic factors (species invasions, ecosystem turnover, habitat patchiness vs. continuity) were also prominent. Notably, these external biotic factors relate to entities in the economic hierarchy that occur at higher levels than the organismal/population level where microevolutionary influences occur. Due to the discrete nature of information flow in the genealogical and economic hierarchies, the impacts of ecosystem or community-level change must be filtered to the genealogical hierarchy via organisms or populations (where deme = avatar) before resulting in the gene flow breakdown required for speciation. This process termed the "sloshing

Table 3 Summary of the impacts of internal and external parameters on speciation

	Promote	Inhibit
Internal	Populations spatially isolated	Populations spatially continuous
	Specialist ecology	Generalist ecology
	Vagile or high organismal mobility (dispersal potential)	Sessile or limited movement within organism life span (low dispersal potential)
External	Rapid or abrupt environmental changes	Slow and gradual environmental change
	Tectonic factors or sea level changes that cyclically join and separate of basins	Widespread species invasions of ecological generalists

Modified from Stigall (2013)

bucket" by Eldredge (2003, 2007) is an important theoretical construct in macroevolutionary theory (Pievani and Serrelli 2013), but has been less widely explored within case studies, such as those herein. The consistent role of external environmental parameters evident form these examples, however, suggests that the sloshing bucket may represent a fundamental control on biodiversity dynamics.

The secondary controls on speciation across the three case studies also have commonalities. The relative breadth of a species' niche impacts species survival and adaptability during intervals of environmental change. Species with broad ecological niches display reduced level of niche evolution and higher levels of persistence across disturbance intervals. This indicates that species-level emergent characters may play a key role in speciation dynamics and argues for the importance of species selection in determining diversity patterns through time.

The Modern Synthesis primarily emphasized evolutionary patterns and processes affected by gene flow and its interruption over short timelines. Consequently, the Modern Synthesis cannot incorporate the fundamental ecological and spatial attributes of species and their environments completely. The three case studies presented herein emphasize the centrality of large-scale environmental changes acting over long time scales in driving speciation and thus major pattern in biodiversity and thereby relegating microevolutionary processes to a more limited role in speciation initiation. The explicit incorporation of additional hierarchical levels and the interplay between the economic and genealogical hierarchies provides a dynamic perspective to analyze speciation. In particular, incorporation of a broader evolutionary theory with explicit consideration of abiotic environmental changes, species selection, details of species geographic ranges and niche breadth, and migration of species will provide a robust framework to assess cross-faunal speciation dynamics.

Acknowledgements Thanks to Emanuele Serrelli for constructive comments on an earlier version of this paper and to the editors for the invitation to submit to this volume. Many students, particularly Hannah Brame, Nicole Dudei, Kaitlin Clare Maguire, Richard Malizia, Robert Swisher, Bradley Walls, and Davey Wright contributed to the development of the datasets described in the case studies. This research was supported by NSF EAR-0922067 and is a contribution to International Geoscience Programme (IGCP) 591.

Appendix

Biofacies: A body of sediment characterized by a particular suite of fossils
Chronostratigraphy: Correlation of rock units based on intervals of equivalent time
Depositional or stratigraphic sequence: A cohesive cycle of sedimentary deposition; including a cycle from low sea level (lowstand systems tract), to rising sea level (transgressive systems tract), to high sea level (highstand systems tract), to falling sea level (falling stage systems tract). Depositional sequences are separated by unconformities (often erosional surfaces) or correlative conformities.

Ecological niche modeling (ENM): A computational method for estimating the boundaries of a species' ecological niche based on the correspondence of a set of known species occurrence points with a series of environmental variables.
Gradient ecology: An empirical analytical method used to relate abundances of species in a community to environmental gradients by ordination techniques
Hyspodonty: Development of high-crowned teeth with enamel that extends beyond the gum line. This type of dentition provides extra tooth material for extensive wear of siliceous grasses.
Lithostratigraphy: Correlation of rock units based on intervals of similar rock properties
Sequence stratigraphy: Correlation of rock units based on equivalent cycles of sedimentary deposition, i.e., depositional sequences.

References

Allmon WD (2013) Species, speciation and palaeontology up to the modern synthesis: persistent themes and unanswered questions. Palaeontology 56(6):1199–1223. doi:10.1111/pala.12054

Bambach RK, Knoll AH, Wang SC (2004) Origination, extinction, and mass depletions of marine diversity. Paleobiology 30(4):522–542

Botts EA, Erasmus BFN, Alexander GJ, Lawlor J (2013) Small range size and narrow niche breadth predict range contractions in South African frogs. Glob Ecol Biogeogr 22(5):567–576. doi:10.1111/geb.12027

Brame H-MR, Stigall AL (2014) Controls on niche stability in geologic time: congruent responses to biotic and abiotic environmental changes among Cincinnatian (Late Ordovician) marine invertebrates. Paleobiology 40(1):70–90 doi:10.1666/13035

Brooks DR, McLennan DA (2002) The nature of diversity: an evolutionary voyage of discovery. University of Chicago Press, Chicago

Brown JH, Stevens GC, Kaufman DM (1996) The geographic range: size, shape, boundaries, and internal structure. Annu Rev Ecol Syst 27:597–623

Congreve CR, Lieberman BS (2008) Phylogenetic and biogeographic analysis of Ordovician homalonotid trilobites. Open Paleontol J 1:24–32

Congreve CR, Lieberman BS (2010) Phylogenetic and biogeographic analysis of deiphonine trilobites. J Paleontol 84(1):128–136

Coyne JA, Orr HA (2004) Speciation. Sinauer Associates, Sunderland

De Queiroz K (2007) Species concepts and species delimitation. Syst Biol 56(6):879–886. doi:10.1080/10635150701701083

Dobzhansky T (1937) Genetics and the origin of species. Columbia University Press, New York

Dudei NL, Stigall AL (2010) Using ecological niche modeling to assess biogeographic and niche response of brachiopod species to the Richmondian invasion (Late Ordovician) in the Cincinnati Arch. Palaeogeogr Palaeoclimatol Palaeoecol 296(1–2):28–43. doi:10.1016/j.palaeo.2010.06.012

Eldredge N (1996) Hierarchies in macroevolution. In: Jablonski D, Erwin DH, Lipps JH (eds) Evolutionary paleobiology. Chicago University Press, Chicago, pp 42–61

Eldredge N (2003) The sloshing bucket: how the physical realm controls evolution. In: Crutchfield J, Schuster P (eds) Evolutionary dynamics: exploring the interplay of selection, accident, neutrality, and function: exploring the interplay of selection, accident, neutrality, and function. Oxford University Press, New York, pp 3–32

Eldredge N (2007) Hierarchies and the sloshing bucket: toward the unification of evolutionary biology. Evol Educ Outreach 1(1):10–15. doi:10.1007/s12052-007-0007-6

Eldredge N, Gould SJ (1972) Punctuated equilibria: an alternative to phyletic gradualism. In: Schopf TJM (ed) Models in paleobiology. Freeman, Cooper, and Co., San Fransisco, pp 82–115
Eldredge N, Thompson JN, Brakefield PM, Gavrilets S, Jablonski D, Jackson JBC, Lenski RE, Lieberman BS, McPeek MA, Miller WI (2005) The dynamics of evolutionary stasis. Paleobiology 31(2):133–145
Eronen JT, Evans AR, Fortelius M, Jernvall J (2010) The impact of regional climate on the evolution of mammals: a case study using fossil horses. Evolution 64(2):398–408
Fernández MH, Vrba ES (2005) Body size, biomic specialization and range size of African large mammals. J Biogeogr 32(7):1243–1256
Fortelius M (1985) Ungulate cheek teeth: developmental, functional, and evolutionary interrelations. Acta Zool Fennica 180:1–76
Gavrilets S, Losos JB (2009) Adaptive radiation: contrasting theory with data. Science 323:732–737
Gould SJ (2002) The structure of evolutionary theory. Harvard University Press, Cambridge
Gould SJ, Eldredge N (1977) Punctuated equilibria: the tempo and mode of evolution reconsidered. Paleobiology 3(2):115–151
Holland SM (1997) Using time/environment analysis to recognize faunal events in the Upper Ordovician of the Cincinnati Arch. In: Brett CE, Baird GC (eds) Paleontological event horizons: ecological and evolutionary implications. Columbia University Press, New York, pp 309–334
Holland SM, Patzkowsky ME (1996) Sequence stratigraphy and long-term paleoceanographic change in the Middle and Upper Ordovician of the Eastern United States. Geol Soc Am Spec Pap 306:117–129
Holland SM, Patzkowsky ME (2007) Gradient ecology of a biotic invasion: biofacies of the type Cincinnatian series (Upper Ordovician), Cincinnati, Ohio Region. U.S. Palaios 22(4):392–407. doi:10.2110/palo.2006.p06-066r
Hulbert RC Jr (1993) Taxonomic evolution in North American neogene horses (subfamily Equinae): the rise and fall of and adaptive radiation. Paleobiology 19(2):216–234
Jablonski D (2008) Species selection: theory and data. Annu Rev Ecol Evol Syst 39:501–524
Jackson JBC (1974) Biogeographic consequences of eurytopy and stenotopy among marine bivalves and their evolutionary significance. Am Nat 108:541–560
Kammer TW, Baumiller TK, Ausich WI (1997) Species longevity as a function of niche breadth: evidence from fossil crinoids. Geology 25(3):219–222
Levin LA (2003) Oxygen minimum zone bethos: adaptation and community response to hypoxia. Oceanogr Mar Biol Annu Rev 41:1–45
Lieberman BS (1997) Early Cambrian paleogeography and tectonic history: a biogeographic approach. Geology 25(11):1039–1042
Lieberman BS (2000) Paleobiogeography: using fossils to study global change, plate tectonics, and evolution. Top Geobiol 1:1–208
Lieberman BS (2003) Biogeography of the Trilobita during the Cambrian radiation; deducing geological processes from Trilobite evolution. In: Lane PD, Siveter DJ, Fortey RA (eds) Trilobites and their relatives, vol 70. Special Papers in Palaeontology, pp 59–72
Lieberman BS (2012) Adaptive radiations in the context of macroevolutionary theory: a paleontological perspective. Evol Biol 39(2):181–191
Lieberman BS, Eldredge N (1996) Trilobite biogeography in the Middle Devonian: geological processes and analytical methods. Paleobiology, pp 66–79
Lockwood J, Hoopes M, Marchetti M (2009) Invasion ecology. Wiley, New York
MacFadden BJ (1984) Systematics and phylogeny of Hipparion, Neohipparion, Nannippus, and Cormohipparion (Mammalia, Equidae) from the Miocene and Pliocene of the new world. Bull Am Museum Nat Hist 179:1–195
Maguire KC, Stigall AL (2008) Paleobiogeography of Miocene equinae of North America: a phylogenetic biogeographic analysis of the relative roles of climate, vicariance, and dispersal. Palaeogeogr Palaeoclimatol Palaeoecol 267(3–4):175–184. doi:10.1016/j.palaeo.2008.06.014

Maguire KC, Stigall AL (2009) Using ecological niche modeling for quantitative biogeographic analysis: a case study of Miocene and Pliocene equinae in the great plains. Paleobiology 35(4):587–611doi:10.1666/0094-8373-35.4.587

Malizia RW, Stigall AL (2011) Niche stability in Late Ordovician articulated brachiopod species before, during, and after the Richmondian invasion. Palaeogeogr Palaeoclimatol Palaeoecol 311(3–4):154–170. doi:10.1016/j.palaeo.2011.08.017

Mayr E (1942) Systematics and the origin of species from the viewpoint of a zoologist. Columbia University Press, New York

McGhee GR (1996) The Late Devonian mass extinctions: the Frasnian/Famennian crisis. Columbia University Press, New York

McGhee GR (2013) The legacy of the Devonian extinctions: when the invasion of land failed. Columbia University Press, New York

McGhee GR, Clapham ME, Sheehan PM, Bottjer DJ, Droser ML (2013) A new ecological-severity ranking of major Phanerozoic biodiversity crises. Palaeogeogr Palaeoclimatol Palaeoecol 370:260–270. doi:10.1016/j.palaeo.2012.12.019

McGhee GR Jr, Sheehan PM, Bottjer DJ, Droser ML (2004) Ecological ranking of Phanerozoic biodiversity crises: ecological and taxonomic severities are decoupled. Palaeogeogr Palaeoclimatol Palaeoecol 211(3):289–297

Meyer DL, Davis RA (2009) A sea without fish: life in the Ordovician Sea of the Cincinnati Region. Indiana University Press, Bloomington

O'Neill RV, DeAngelis DL, Waide JB, Allen TFH (1986) A hierarchical concept of ecosystems. Princeton University Press, Princeton

Patzkowsky ME, Holland SM (2007) Diversity partitioning of a Late Ordovician marine biotic invasion: controls on diversity in regional ecosystems. Paleobiology 33(2):295–309

Peterson AT, Soberón J, Pearson RG, Anderson RP, Martínez-Meyer E, Nakamura M, Araújo MB (2011) Ecological niches and geographic distributions. Princeton University Press, Princeton

Pievani T, Serrelli E (2013) Bucket thinking: the future framework for evolutionary explanation. Contrastes Revista internacional de filosofia - Suplementos 18:389–405

Prado JL, Alberdi MT (1996) A cladistic analysis of the horses of the tribe Equini. Palaeontology 39(3):663–680

Rode AL (2004) Phylogenetic revision of *Leptodesma* (*Leiopteria*) (Devonian: Bivalvia). Postilla 229:1–28

Rode AL, Lieberman BS (2002) Phylogenetic and biogeographic analysis of Devonian phyllocarid crustaceans. J Paleontol 76(2):271–286

Rode AL, Lieberman BS (2004) Using GIS to unlock the interactions between biogeography, environment, and evolution in Middle and Late Devonian brachiopods and bivalves. Palaeogeogr Palaeoclimatol Palaeoecol 211(3–4):345–359. doi:10.1016/j.palaeo.2004.05.013

Shine R, Brown GP, Phillips BL (2011) An evolutionary process that assembles phenotypes through space rather than through time. Proc Natl Acad Sci US A 108(14):5708–5711

Simpson GG (1944) Tempo and mode in evolution, vol 15. Columbia University Press, New York

Simpson GG (1951) Horses: The story of the horse family in the modern world and through sixty million years of history. Oxford University Press, New York

Stanley SM (1990) The general correlation between rate of specation and rate of extinction: fortuitous causal linkages. In: Ross RM, Allmon WD (eds) Causes of evolution: a paleontological perspective. University of Chicago Press, Chicago, pp 103–127

Stebbins GL (1981) Coevolution of grasses and herbivores. Ann Mo Bot Gard 68:75–86

Stigall AL (2008) Tracking species in space and time: assessing the relationships between paleobiogeography, paleoecology, and macroevolution. Paleontol Soc Pap 14:139–154 doi:10.1111/pala.12003

Stigall AL (2010a) Invasive species and biodiversity crises: testing the link in the Late Devonian. PLoS One 5(12):e15584. doi:10.1371/journal.pone.0015584.g001

Stigall AL (2010b) Using GIS to assess the biogeographic impact of species invasions on native brachiopods during the Richmondian invasion in the type-Cincinnatian (Late Ordovician, Cincinnati region). Palaeontol Electronica 13(1):5A–19

Stigall AL (2011) Application of niche modelling to analyse biogeographic patterns in Palaeozoic brachiopods: evaluating niche stability in deep time. Memoirs Assoc Australas Palaeontologists 41:229–255

Stigall AL (2012a) Invasive species and evolution. Evol Educ Outreach 5(4):526–533. doi:10.1007/s12052-012-0410-5

Stigall AL (2012b) Speciation collapse and invasive species dynamics during the Late Devonian "Mass Extinction". GSA Today 22(1):4–9. doi:10.1130/g128a.1

Stigall AL (2012c) Using ecological niche modelling to evaluate niche stability in deep time. J Biogeogr 39(4):772–781. doi:10.1111/j.1365-2699.2011.02651.x

Stigall AL (2013) Analysing links between biogeography, niche stability and speciation: the impact of complex feedbacks on macroevolutionary patterns. Palaeontology 56(6):1225–1238. doi:10.1111/pala.12003

Stigall AL (2014) When and how do species achieve niche stability over long time scales? Ecography 37:1123–1132. doi:10.1111/ecog.00719

Stigall AL, Brame H-MR (2014) Relating environmental change and species stability in Late Ordovician seas. GFF 136(1):249–253. doi:10.1080/11035897.2013.852619

Stigall Rode AL (2005) Systematic revision of the Middle and Late Devonian brachiopods *Schizophoria* (*Schizophoria*) and '*Schuchertella*' from North America. J Syst Paleontol 3(2):133–167. doi:10.1017/s1477201905001537

Strömberg CA (2006) Evolution of hypsodonty in equids: testing a hypothesis of adaptation. Paleobiology 32(2):236–258

Thuiller W, Lavorel S, Araujo MB (2005) Niche properties and geographical extent as predictors of species sensitivity to climate change. Glob Ecol Biogeogr 14(4):347–357. doi:10.1111/j.1466-822X.2005.00162.x

Tyler CL, Leighton LR (2011) Detecting competition in the fossil record: support for character displacement among Ordovician brachiopods. Palaeogeogr Palaeoclimatol Palaeoecol 307(1–4):205–217. doi:10.1016/j.palaeo.2011.05.020

Vrba ES (1984) What is species selection? Syst Zool 33(3):318–328

Vrba ES (1987) Ecology in relation to speciation rates: some case histories of miocene-recent mammal clades. Evol Ecol 1(4):283–300

Walls BJ, Stigall AL (2011) Analyzing niche stability and biogeography of Late Ordovician brachiopod species using ecological niche modeling. Palaeogeogr Palaeoclimatol Palaeoecol 299(1–2):15–29. doi:10.1016/j.palaeo.2010.10.024

Walls BJ, Stigall AL (2012) A field-based analysis of the accuracy of niche models applied to the fossil record. Paleontol Contrib 6:1–12

Wiley EO (1978) The evolutionary species concept reconsidered. Syst Zool 27:19–26

Wiley EO, Mayden RL (1985) Species and speciation in phylogenetic systematics, with examples from the North American fish fauna. Annals Mo Bot Gardens 72:596–635

Wright DF, Stigall AL (2013a) Geologic drivers of Late Ordovician faunal change in Laurentia: investigating links between tectonics, speciation, and biotic invasions. PLoS One 8(7):e68353. doi:10.1371/journal.pone.0068353

Wright DF, Stigall AL (2013b) Phylogenetic revision of the Late Ordovician orthid brachiopod genera *Plaesiomys* and *Hebertella* from North America. J Paleontol 87(6):1107–1128

Wright DF, Stigall AL (2014) Species-level phylogenetic revision of the Ordovician orthide brachiopod *Glyptorthis* from North America. J Syst Palaeontol 12:893–906. doi:10.1080/14772019.2013.839584

Morphological Misfits and the Architecture of Development

Alessandro Minelli

Abstract Morphological misfits can be defined as a miscellaneous class of numerically marginal, often lately discovered taxa, deviating in one or more dramatic aspects from the structural organization of their closest relatives. Morphological misfits are a widely diverse set of taxa which according to their anatomical, ontogenetic and phylogenetic nature may offer the opportunity for a multiplicity of case studies in evolutionary developmental biology. Anatomically, there are modular misfits such as the paussinae beetles, with their extravagant antennae borne on a quite usual beetle body, and systemic misfits such as *Wolffia* among the flowering plants, reduced to a minute blob of green matter, and the also bloblike parasitic crustacean *Sacculina* among the animals. The former, but not the latter, are suggestive of developmental modularity. Ontogenetically, there are one-phase only misfits, such as blepharicerid midges (aberrant larvae but conventional midge-shaped adults), and whole-life-cycle misfits such as cycliophorans. The former, but not the latter, would suggest an evolutionary independence of developmental stages. Some misfits have diverged recently from their morphologically conservative closest relatives (e.g. the duckweeds (Lemnoideae) from conventional Araceae), while others, such as chaetognaths and *Welwitschia*, are the sole members of lineages which diverged very deeply in time from their closest known or putative relatives. The former, but not the latter, can provide obvious opportunities for investigating character evolvability.

Keywords Developmental modularity · Developmental repatterning · Evolutionary developmental biology · Evolvability · Heterochrony · Heterometry · Morphological misfit · Heterotopy · Non-systemic metamorphosis · Synorganization

A. Minelli (✉)
Department of Biology, University of Padua, Padua, Italy
e-mail: alessandro.minelli@unipd.it

E. Serrelli and N. Gontier (eds.), *Macroevolution*,
Interdisciplinary Evolution Research 2, DOI 10.1007/978-3-319-15045-1_10

The history of exploration of organism's diversity on Earth is punctuated by an unceasing series of discoveries of 'odd' creatures that have repeatedly defeated our efforts to establish a satisfactory and stable classification of the living world. Exemplary, in this respect, was the discovery of the platypus (Eco 1999), whose duck bill appeared as an unnatural oddity in a fur-covered quadruped, not to mention the later ascertained oviparity of an animal feeding the newborn with milk. Two hundred years later, zoologists have been confronted with a diversity of newly dredged carnivorous sponges, including the extraordinary *Chondrocladia lyra* (Lee et al. 2012) whose body—vaguely similar to a stemless crinoid—comprises five threadlike horizontal arms radiating from a clump of rhizoids, each arm bearing a series of long vertical branches, each of which bears, in turn, a blob of sperm on the tip, apparently used as a lure for tiny prey. Eventually, in the last two decades of the XX century, three new 'phyla' (Loricifera: Kristensen 1983; Cycliophora: Funch and Kristensen 1995; Micrognathozoa: Kristensen and Funch 2000) were erected to accommodate newly discovered animals that could not fit into any of the major animal groups known at the date.

Bell (1991, 2008) introduced the term *morphological misfit* to denote those plant forms "that cannot as yet sensibly be accommodated in traditional descriptions … that is, misfits, for the moment, to botanical discipline, not misfits for a successful existence" (Bell 2008, p. 247). This colloquial term can be used also in zoology, as many examples in this chapter will show.

A major reason for our embarrassment in the face of these unusual organisms is their numerical marginality, as in the case of the few carnivorous sponges summing up to little more than one thousandth of the total diversity of Porifera or the duckweeds (Lemnaceae), common water plants well known to layperson in several parts of the world, but just 34 species, i.e. 0.01 % of the flowering plants, and the monotremes, i.e. the platypus and the echidnas, just four species, less than 0.1 % of living mammal species.

Morphological uniqueness is frequently a symptom of peculiar adaptations, which may in turn suggest adaptive dead ends, such as the mandible of the extinct cartilagineous fish *Helicoprion*, prolonged into a spirally coiled saw with impressively large teeth (Lebedev 2009), or the only two leaves of the gymnosperm *Welwitschia mirabilis*, which continue growing at the base while the distal end progressively withers, all along the individual's life, which may last a century or more.

Less obvious is the adaptive significance of the unique (although inconspicuous) flowers of *Lacandonia* (Márquez-Guzmán et al. 1989) and *Trithuria*, which apparently bear female parts external to the male ones, opposite to all other flowers with distinct floral whorls, unless they should be interpreted as compact, miniaturized inflorescences, with unisexual male flowers surrounded by unisexual female ones (Rudall et al. 2009).

The discovery (or rediscovery and reinterpretation) of plant and animal misfits makes often breaking news and even finds a space in popular collections of illustrations, e.g. in stamps, as in the case of *Helicoprion* in an improbable stamp of Equatorial Guinea or the tiny 'worm' *Limnognathia maerskii*, the only known species of the Micrognathozoa, in a stamp of its homeland Greenland. As biodiversity is currently a fashionable issue in the media, many spectacular discoveries are presently flagged in popular websites, e.g. those of the Natural History Museum, London, although only a fraction of the species appearing there deserves to be called a misfit.

The characterization of a living species as a morphological misfit translates generally into difficulties with its placement in the classification, at least in so far as the latter is based on morphological characters. Morphological oddity means that we acknowledge problems in establishing homologies between the newly discovered misfit and the 'normal' animals or plants for which we have already provided accommodation in the system of living beings. To some extent, these circumstances re-open an old case. During the first season of comparative anatomy, in the early decades of the XIX century, Cuvier (1816) divided the animal kingdom into four major groups (*embranchments*)—the vertebrates, the molluscs, the articulates and the radiates. Cuvier placed in the same embranchment those animals among which he was able to trace correspondences (we would say, homologies) between anatomical parts, whereas he placed in separate embranchments those forms between which he failed to recognize obviously corresponding parts. Although very seldom explicitly acknowledged (but see Minelli 1993; Conway Morris 1995), the phyla of modern classifications (for instance, the 30^+ phyla currently recognized in most classifications of the Metazoa) are similarly based on the difficulty in tracing homologies between the representatives of different phyla, compared to the much higher number of homologies confidently recognized among the members of the same phylum. No wonder, therefore, that the first description of *Nanaloricus mysticus*, *Symbion pandora* and *Limnognathia maerskii* corresponded to the proposal of three new phylum-level taxa—Loricifera, Cycliophora and Micrognathozoa, respectively. In other cases, lesser degrees of morphological oddity did not suggest establishing new phyla; nevertheless, a number of newly discovered misfits have been placed, initially at least, in new classes or orders specifically created for them. For example, a new echinoderm class, Concentricycloidea, was created for *Xyloplax medusiformis*, an unusual disc-shaped animal discovered in the 80s in the seas of New Zealand (Baker et al. 1986; its real affinities, placing it within the sea stars, were ascertained later, cf. Janies and Mooi 1998; Janies et al. 2011), and several new orders or classes of crustaceans (Remipedia: Yaeger 1981; Tantulocarida: Boxshall and Lincoln 1983; Mictacea: Bowman et al. 1985) were established in the last half of the XX century for newly discovered species.

Eventually, however, most of these misfits turn out not to deserve such a special taxonomic treatment. In other words, their morphological oddity is not necessarily a proof of a long separation of their lineage from the closest extant relatives. The advent of molecular systematics has dramatically improved our chances of tracing these organisms' affinities. Most interesting, the consequences of this reallocation are not limited to phylogeny and classification, but create a comparative context within which interesting questions concerning the evolution of morphology, and the underlying developmental processes, can be addressed. This is the main intended message of this chapter. However, before coming to that, I must fix two points.

The first is that some little groups whose unusual morphology has characteristically caused big problems to taxonomists relying on morphological evidence are still largely groups *incertae sedis*, despite the application of the modern tools of molecular phylogenetics. Within the Metazoa, the most critical example is provided by the mainly planktonic arrow worms (Chaetognatha), about which we can only say that they likely belong to the protostomians, but no further precision can be safely added as to their affinities (e.g. Marlétaz et al. 2006; Matus et al. 2006; Harzsch and Müller 2007; Helmkampf et al. 2008; Edgecombe et al. 2011).

The second point is that the oddities of morphological misfits, although extremely diverse and in principle unpredictable in terms of both taxonomic distribution and anatomical extent, allow nevertheless a classification into three main classes.

A first class is represented by *misfits by reduction*. These are organisms whose bodily organization is very simple, lacking many of the specialized parts or organs otherwise found in the 'normal' members of the group to which they belong. In many cases, their reduction appears to be the effect of progenesis, that is, of reproductive maturity reached at a stage morphologically similar to an embryonic or larval stage of their relatives (e.g. Gould 1977; Westheide 1987). Reduction, however, is often accompanied by the expression of novel traits, for which there is no equivalent among the close relatives of the misfit. These are the cases where the contribution of molecular systematics is most important in establishing the correct phylogenetic placement of the morphological misfits. Three extraordinary examples of strong morphological simplification accompanied by the development of unusual specializations have been revealed among the Cnidaria. All members of the phylum, as traditionally circumscribed, are either medusae or polyps (solitary such as the hydra and the sea anemones or colonial such as the corals), the two forms often alternating within a species' life cycle. Recently, three different kinds of misfits have been recognized as strongly modified cnidarians. One is the wormlike *Buddenbrockia*, a parasite of freshwater bryozoans (see Jiménez-Guri et al. 2007), another is *Polypodium*, a parasite of sturgeon's eggs, which is reduced to an irregular blob of jelly with a number of threadlike projections (see Siddall et al. 1995; Zrzavý et al. 1998; Siddall and Whiting 1999), and the third is the Myxozoa, a whole group (more than one thousand species) of organisms classified in the past within the protozoans, with a complex life cycle including two forms that are so different as to be classified in the past in two different protozoan classes (see Wolf and Markiw 1984; Siddall et al. 1995).

A second class of misfits is characterized by peculiarities in the *building blocks* by which they are formed. This is the case of the Loricifera. These little marine animals have unusually small size (adults are in the order of a quarter of millimetre); nevertheless, their anatomical complexity is very high. Indeed, they are formed by a high number of cells, but these are extremely small, more in the range of bacterial cell sizes than in the range of animal cell sizes. The cells of loriciferans can be regarded as truly miniaturized (Kristensen 1991), to the extent that they have lost the organelles of which they could dispose: loriciferans are the only metazoans whose cells lack mitochondria (Danovaro et al. 2010).

A third class includes misfits by *synorganization*. Here, body parts that are well recognizable in 'normal' animals or plants appear instead indistinctly separated, be this because of primary lack of separation, or because of secondary fusion. An example is provided by the extraordinary antennae of many paussine beetles, where the majority of articles (completely separated and reciprocally articulated in all other beetles, including the less specialized among the paussines) are fused to form a leaf-like blade. No wonder that these beetles have been long separated into a distinct family Paussidae, before being recognized as a specialized subclade of the ground beetles (Carabidae); their precise relationships within this larger clade, however, are still unclear (Di Giulio et al. 2003; Beutel et al. 2007).

This tentative classification of misfits is far from being a futile exercise. Classifications are often a necessary step in organizing knowledge, at a time when a theory-based approach is not yet at hand (Wilkins and Ebach 2013). A corresponding exercise proved useful in animal and human teratology, especially under the consequently systematic approach of Geoffroy Saint-Hilaire (1832–1837). In his case, the very possibility to offer a classification of monsters was a proof of their lawfulness. Eventually, the criteria used in diagnosing monster 'taxa' would turn into as many windows into developmental processes, each kind of anomaly pointing to a specific developmental step that, in the monster, has been somehow perturbed or bypassed. Eventually, this proved to be a 'teratological dissection' of development. This has not changed with the advent of developmental genetics, except for the level of analysis and the corresponding experimental tools that can be employed nowadays.

In the case of morphological misfits, a classification such as the one suggested above indicates some pathways (or, at least, some scenarios) along which developmental schedules may have evolved, eventually resulting in the unusual phenotypes I am discussing here.

I will briefly present in the following some evo-devo perspectives on morphological misfits, in terms of some of the main concepts of this discipline, namely modularity (structural as well as ontogenetic), heterochrony, heterotopy, heterometry and evolvability.

1 Morphological Misfits and Structural Modularity

Departure of morphological misfits from the conventional structural plan of a major clade of which they represent an offshoot is sometimes systemic, sometimes limited to one or a few body parts. These two types of misfits open very different views on developmental events.

Among the flowering plants, classic examples of systemic misfits are the riverweeds and the duckweeds. The riverweeds (family Podostemaceae) are aquatic plants of tropical regions, whose flowers clearly suggest affinities to, e.g., the St. John's wort family, Hypericaceae, whereas the vegetative structures deviate—conspicuously to exceptionally—not only in relation to their closest relatives, but also in comparison with all other flowering plants. In the most derived forms, none of the conventional vegetative parts of angiosperms is recognizable and the whole takes a form more reminiscent of an alga than of a plant articulated into roots, stem, branches and leaves. Gene expression patterns studied by Katayama et al. (2010, 2013) in two Podostemoideae reveal an indeterminate stem capped by something like a determinate "leaf", with new leaves/branches developing endogenously from the base of the "leaf".

In the case of duckweeds, which belong to a completely different plant group, currently regarded as a basally splitting lineage within the arum family, Araceae (Cusimano et al. 2011; Henriquez et al. 2014), the whole photosynthetic structure is reduced to an irregular lens- or disc-shaped body, or a small cluster of similar, floating bodies, out of which one or more roots sprout out inferiorly, while the flowers, absolutely inconspicuous and only rarely occurring, are reduced to the essential reproductive parts, a tiny ovary in the female flowers and a rudimentary stamen in the male one. The final stage of this trend is *Wolffia arrhiza*, a subspherical grain of green matter without any root or other projection, perhaps 1 mm in diameter.

Systemic misfits defy our efforts to dissect the complexity of an organism's structure into morphological, developmental or evolutionary modules, at least to the extent that we are guided by the conventional vocabulary we have produced to describe features and behaviour of 'normal' organisms'. In the duckweeds and in the most derived riverweeds, we would search in vain for parts strictly homologous to branches or leaves. These morphological misfits are indeed textbook examples of continuum morphology, a view of plant morphology championed by Sattler and Rutishauser (see Rutishauser and Sattler 1985, 1987, 1989, 1997; Sattler 1988, 1992, 1996; Sattler and Jeune 1992; Rutishauser 1995; Sattler and Rutishauser 1997; Rutishauser et al. 2008).

In other misfits, however, most of the body is structurally comparable to the organization in typical representatives of higher clades to which the misfit belongs, the strongly divergent features being limited to one or a few body parts. This circumstance reveals an evolutionary modularity which may rest on a similarly articulated developmental modularity. This hypothetical co-extension of evolutionary and developmental modules should encourage detailed study from either point of view, because it represents a far from common occurrence. Usually, in fact, the mapping of evolutionary modules onto developmental ones is far from straightforward. Consider, for example, the giraffe. In this lineage, we can identify several body units that have been quite obviously targets of directional selection: the neck with its unusually elongated cervical vertebrae and the associated musculature, the correspondingly long blood vessels serving the brain, the elongated trachea and oesophagus, etc.; the circulatory system, with a powerful heart and specialized arterial valves; and the legs, among which the anterior pair is now longer than the posterior pair, contrary to a nearly universal trend in the whole of mammals. The adaptive significance of these peculiar features of the giraffe neck, circulatory system and limbs marks these body parts as fairly well-circumscribed evolutionary modules. None of them, however, can be considered as a largely independent developmental module.

Things are possibly different in the case of structurally modular misfits, for example the paussine beetles mentioned above, whose peculiar antennae, borne on an otherwise 'normal' body, are suggestive of a developmental evolution driven by changes in the genetic control of the developmental processes patterning the antenna. Insect antennae, indeed, are prone to extraordinary diversification that remains strictly circumscribed to this modular unit. This should perhaps not be that surprising: each pair of appendages of the arthropod head, that is, the antennae as well as the mandibles, maxillae, chelicerae, pedipalps, etc., has its precise identity, consequent to the evolution of a differential genetic control (partly due to the combinatorial Hox code; cf., e.g. Carroll 1995); thus, it should not be a surprise that any of these pairs of appendages evolves quite independently from the others, provided that its function and the function of other appendages are not impaired. In particular, we should expect—as in fact we find—more 'evolutionary freedom' in the antenna than in the mouthparts, as the antenna is not controlled by the Hox genes, as all the following appendages are instead. Right for the same reason, we should not expect that just one pair of appendages within a series of usually identical limbs will start evolving in a strongly diverging way, without any visible correlative change in all other limb pairs, those in front of it and those behind it. However, this is exactly what we observe in a peculiar class of morphological misfits, the males of the vast majority of millipede species, those classified as the Helminthomorpha.

The number of walking legs in helminthomorph millipedes varies from 32 to as many as 375 pairs (Fusco 2005). Basically, all leg pairs are morphologically identical, except for the smaller size of the first pair or the first few pairs. This is also true of male juveniles, the segment number of which increases with each moult, together with the number of completely articulated leg pairs. At later stages, however, one or two pairs of legs will undergo a unique metamorphosis. This deep structural modification is very strictly and precisely localized and does not affect the other leg pairs, both those in front and those posterior to the affected zone, this feature being captured by the term *non-systemic metamorphosis* (Drago et al. 2008, 2011). In these arthropods, one or two pairs of legs are replaced in the adult by specialized sexual appendages, the gonopods, used by males as claspers or to transfer sperm.

In two major subgroups of millipedes, the Polydesmida and the Callipodida, only the leg pair 8 is replaced by gonopods, as in most Spirostreptida, but in the latter, the leg pair 9 is completely atrophied, or nearly so (Demange 1967). In the remaining clades, two pairs of legs (8 and 9) are eventually transformed into gonopods; in many Chordeumatida, less dramatic modifications may additionally affect legs 7 and 10 (Blower 1985). Only in Colobognatha the leg pairs modified into gonopods are pairs 9 and 10 (Hoffman 1982). With few exceptions, gonopods differ dramatically from walking legs, being articulated only at the joint with the corresponding body ring and presenting a fantastic diversity of shapes, very often branched, coiled, spiny and mostly too complex to be adequately described in words.

Transformation of walking legs into gonopod is often abrupt, although accomplished in steps across several moults. As an example, I summarize here the events as recently described by Drago et al. (2011) in *Nopoiulus kochii*. During stadia III and IV, leg pairs 8 and 9 are identical to ordinary walking legs, but at stadium V, these legs are usually replaced by two pairs of inconspicuous gonopod primordia, to be subsequently changed into two pairs of complex and bulky gonopods following another moult. This non-systemic metamorphosis is not strictly fixed in time with respect to the progression of post-embryonic stadia: in some males, changes are delayed by one stadium in relation to the just described schedule (Kheirallah et al. 2000; Drago et al. 2011). The dramatic changes seen externally in the shape of these appendages are accompanied by not less dramatic changes in the internal structure of the ring to which these appendages belong. The gonopod apodemes, that is the skeletal elements to which the gonopods articulate, are much bulkier than those of the walking legs, and the corresponding muscles are also much more conspicuous. As a consequence, much less space is locally available to accommodate the other organs. This affects the ventral nerve cord and the digestive tract especially: at the level of the gonopods, these organs are displaced dorsally, with a consequent dramatic reduction in the lumen of the gut.

From the point of view of morphological misfits, male millipedes, with their fantastic diversity of gonopods, are non-systemic misfits. The non-systemic metamorphosis they undergo reveals the existence of developmental modules, only one of which is the theatre of this dramatic change. Due to the segmental architecture of arthropods and, specifically, to the articulation of the millipede trunk in a series of morphological units (diplosegments), each of which bears two pairs of legs (except for the legless collum and the following 3, rarely 4 rings, with one leg pair each), one might expect the gonopodal module to correspond to one ring, but this is a wrong expectation, for two distinct reasons. First, as said before, in Polydesmida and Callipodida, only one leg pair is eventually transformed into gonopods. As a consequence, in these millipedes, only the anterior half of trunk ring VII corresponds to the gonopodal module, whereas the posterior half belongs to the post-gonopodal module with appendages all in the form of walking legs. Second, and more compelling, in the Colobognatha, the two pairs of gonopods belong to two distinct rings, i.e. to the posterior half of ring VII and the anterior half of ring VIII, respectively. Thus, the 'misfit module' in the millipede trunk does not correspond to a segmental (or diplosegmental) module, but to a short bit of trunk length marked by other than segment-specifying genes. It has been suggested indeed that this marker could be one of the Hox genes, *Abdominal-B* (Drago et al. 2008). What matters in the present context, anyway, is that the position at which legs will be transformed into gonopods, often only after several months of post-embryonic development, must have been specified during embryonic development and, most likely, before full segmentation of the relevant trunk region has been fully specified. This is suggested indeed by the segmental positioning of extra pairs of gonopods in recently described teratological specimens (Akkari et al. 2014).

One more unusual phenomenon stressing the modularity of the millipede trunk, in respect to the differentiation of these unique appendages, is *periodomorphosis*, a developmental schedule during which gonopod differentiation is temporarily reversed. This phenomenon has been recorded in a few species where a mature male, following a reproductive season, moults to an 'intercalary' stadium with dedifferentiated gonopods, only eventually to moult again, to give rise to a second reproductive stadium with newly differentiated gonopods (Verhoeff 1923; Sahli 1990).

2 Ontogenetic (Stage) Modularity

In the case of some morphological misfits, the whole life cycle is represented by oddly shaped stages, and the sequence of stages can also strongly depart from the norm. This is the case of the Cycliophora. In reconstructing the still imperfectly known, but obviously very complex life cycle of these tiny animals that live epizoically on the appendages of the Norwegian lobster, zoologists were forced to introduce new terms such as the Pandora larva and the Prometheus larva, because the terminology available for the other animals seemed not to offer adequate labels for their unique stages (Obst and Funch 2003).

In other cases, however, strong morphological singularities are limited to a phase, mainly one stage only, within a life cycle. For example, the adults of the blepharocerids do not deviate significantly from the midgelike organization of most 'primitive' dipterans despite the fact that this family split quite long ago (Early Jurassic, ca. 200 million years BP) from their closest extant relatives (Tanyderidae + Psychodidae) (Bertone et al. 2008). In contrast to the broadly conservative aspect of the adults, however, the larvae of blepharicerids are among the oddest among the dipterans, and insects generally. Their body is divided by well-marked constrictions into six 'macrosegments'—each of which bears a large sucker ventrally—plus a less well-marked terminal division devoid of sucker, representing the abdominal segments 7–10, virtually fused together. Macrosegments II–VI correspond to conventional abdominal segments 2–6, respectively, whereas macrosegment I results from the fusion of all head and thorax segments plus abdominal segment 1 (Hogue 1981).

Opposite is the case of cirripeds such as true barnacles (*Balanus*) and goose barnacles (*Lepas*). Here, the larval stages—the nauplius, in particular—have the general morphology of conventional crustacean larvae, whereas the adult deviates so conspicuously from the crustacean body plan, that the real affinities of these animals were discovered only when their metamorphosis was observed and described (Thompson 1830, 1835). Similar is the history of *Sacculina carcini*. The early suggestion by Cavolini (1787), based on observing typical crustacean larvae being apparently generated by the non-descript, saclike parasitic adult, only gained substance one full century later, when Delage (1884) reconstructed this animal's life cycle in satisfactory detail and eventually recognized *Sacculina* as a crustacean, although deserving placement in a new order of its own (Kentrogonida) due to its being an extremely specialized morphological misfit.

From an evo-devo perspective, stage misfits such as blepharocerids and cirripeds offer food for thought in addressing the question whether individual life stages should be regarded as homologues, i.e. as temporal (developmental) modules subjected to autonomous evolution. This thesis has been defended, in recent years, by Scholtz (2005, 2008). Prima facie, the circumstantial evidence provided by the stage misfits discussed here seems to support this view, but a word of caution is needed. Ontogenetic stages, and body parts alike, are likely to be, in the vast majority of cases, the compound outcome of only partially overlapping developmental modules and functional/evolutionary modules. We cannot determine a priori how far a developmental unit (a stage) or an anatomical unit (an organ, or a body part such as a body segment) is to be interpreted in terms of modularity of developmental processes, or in terms of modularity of the phenotype as target of natural selection. The combined effects of ontogeny and evolution are clearly shown by all those patterns of change that Arthur (2011) has chosen to present collectively under the suggestive term of *developmental repatterning*. These patterns are the subject of the next section.

3 Heterochrony, Heterotopy and Heterometry

Heterochrony, heterotopy and heterometry are three of the four kinds of developmental repatterning (=developmental reprogramming sensu Arthur 2000, 2002) denoting evolutionary changes that have occurred through modifications in the temporal, positional and quantitative (metric or meristic) aspects, respectively, of the ontogenetic production of individual body parts. A fourth kind, heterotypy, would encompass those changes where a body part is eventually replaced by a radically different structure. Of course, such a change can produce obvious morphological misfits, as in the case of millipede walking legs replaced by gonopods, but I prefer to keep such radical changes—of which we have already seen examples—under a separate heading. On a first sight at least, the other three kinds of change seem to result from a minor, purely quantitative tuning of developmental process; nevertheless, their outcome is often so conspicuous that the resulting phenotypes may easily fall into the broad class of the morphological misfits. This is most obvious when the 'reprogramming' is pushed to such an extreme that a structural or functional limit is reached, beyond which a major reorganization is eventually required.

An extreme example of *heterometry* is offered by the female of the nematode *Sphaerularia bombi*, a parasite of bumblebees. In this tiny worm, up to 3 mm long, the mature reproductive system is monstrously hypertrophic, eventually becoming much bigger than the worm itself. The latter's epidermis, and especially the tough cuticle, cannot be extended to adapt to the rapidly increasing volume of the reproductive organs; therefore, a prolapse of the huge uterus occurs, and the uterus continues growing outside the worm's body, until the latter ends up as a tiny external appendage of its own reproductive organs (Poinar and van der Laan 1972).

Heterochrony as a source of morphological novelty has been studied extensively. Gould's (1977) book on *Ontogeny and Phylogeny* provided both a detailed overview of previous work on this subject and a stimulus to further research, as witnessed by major works such as McNamara (1986, 1995), McKinney (1988), McKinney and McNamara (1991), followed by a more diversified approach, so diversified indeed that Gould (2000) felt eventually obliged to take distance from what he regarded as an unjustified application of the term to phenomena, or patterns, other than those which he exclusively wanted to call heterochrony. Gould's criticisms notwithstanding, the study of heterochrony has gained new momentum since then, especially through the sequence heterochrony approach of Smith (2001, 2002, 2003), Jeffery et al. (2002a, b, 2005) and others. These modern approaches have not simply demonstrated the developmental modularity of animals with high structural complexity such as the vertebrates, but have additionally stimulated the development of quantitative tools for analysing it.

Heterotopy, that is, a change in the spatial localization of a body part in relation to the remaining of the body, is best illustrated by an example, the marine worm *Sipunculus*. This is a free-living creature with robust, nearly cylindrical body. Prior to closer inspection, we would expect it to have mouth and anus, respectively, at either end of the body; however, while the mouth is actually apical, the anus opens laterally, in the proximity of the mouth. This external oddity is the visible aspect of an internal feature, the unusual spatial arrangement of the digestive tube. The gut of *Sipunculus* is much longer than the body; thus, following a first segment running straight from the mouth towards the opposite body end, it coils onto itself in the anterior direction and then moves again in the posterior direction, to finally move back again, with further coils, until it eventually opens to the exterior through the anteriorly localized anus. This arrangement shows the existence of two independently patterned modules: the external, strongly muscular body sack, with its obvious anteroposterior polarity corresponding to its cylindrical shape, and the internal gut, whose spatial arrangement is simply constrained by the space provided by the body cavity, within which it follows a spatial pattern independent from the geometry of the body sack (Minelli 2003).

4 Concluding Remarks: the Evolvability of Misfits

A correct phylogenetic positioning of morphological misfits is the first obligate step towards a tentative interpretation of their evolution. An important, further step is estimating how old is their divergence from their ‘normal’ relatives. The isolated phylogenetic position of chaetognaths and, to a lesser degree, of a plant-like *Welwitschia* does not seem to offer interesting insights about the evolvability of their unique structures. Many other misfits, however, are certainly much more recent. Detailed studies are mostly lacking, but a guess is often possible about the age of their origin, considering the phylogenetic position of these taxa. Recently evolved misfits deserve to be the target of detailed studies, aimed at determining how major, but eventually successful, departures from ‘normal’ phenotypes can be accomplished in relatively few steps. There are perhaps so many ways to become a viable misfit as there are misfits around us, but this expectation, perhaps, will be demolished after a careful study of several cases. This study will be anyway a rewarding contribution to our understanding of the intricacies of the genotype $\rightarrow$ phenotype map.

Acknowledgments I am very grateful to Emanuele Serrelli for kindly inviting me to contribute to this volume and to Wallace Arthur for his sensible suggestions on my MS.

References

Akkari N, Enghoff E, Minelli A (2014) Segmentation of millipede trunk as suggested by a homeotic mutant with six extra pairs of gonopods. Front Zool 11:6

Arthur W (2000) The concept of developmental reprogramming and the quest for an inclusive theory of evolutionary mechanisms. Evol Dev 2:49–57

Arthur W (2002) The emerging conceptual framework of evolutionary developmental biology. Nature 415:757–764

Arthur W (2011) Evolution: a developmental approach. Wiley, Chichester

Baker AN, Rowe FWE, Clark HES (1986) A new class of Echinodermata from New Zealand. Nature 321:862–864

Bell A (1991) Plant form: an illustrated guide to flowering plant morphology. Oxford University Press, Oxford

Bell A (2008) Plant form: an illustrated guide to flowering plant morphology. New edition. Timber Press, Portland

Bertone MA, Courtney GW, Wiegmann BM (2008) Phylogenetics and temporal diversification of the earliest true flies (Insecta: Diptera) based on multiple nuclear genes. Syst Entomol 33:668–687

Beutel RG, Ribera I, Bininda-Emonds ORP (2007) A genus-level supertree of Adephaga (Coleoptera). Org Divers Evol 7:255–269

Blower JG (1985) Millipedes. Synopses of the British fauna (NS) No. 35. EJ Brill/Dr W Backhuys, London

Bowman TE, Garner SP, Hessler RR, Iliffe TM, Sanders HL (1985) Mictacea, a new order of Crustacea Peracarida. J Crustac Biol 5:74–78

Boxshall GA, Lincoln RJ (1983) Tantulocarida, a new class of Crustacea ectoparasitic on other crustaceans. J Crustac Biol 3:1–16

Carroll SB (1995) Homeotic genes and the evolution of arthropods and chordates. Nature 376:479–485

Cavolini F (1787) Sulla generazione dei pesci e dei granchi. Napoli

Conway Morris S (1995) A new phylum from the lobster's lips. Nature 378:661–662

Cusimano N, Bogner J, Mayo SJ, Boyce PC, Wong SY, Hesse M, Hetterscheid WLA, Keating RC, French JC (2011) Relationships within the Araceae: comparison of morphological patterns with molecular phylogenies. Am J Bot 98:654–668

Cuvier G (1816) Le Règne Animal distribué d'après son organisation pour servir de base à l'histoire naturelle des animaux et d'introduction à l'anatomie comparée. Deterville, Paris

Danovaro R, Dell'Anno A, Pusceddu A, Gambi C, Heiner I, Kristensen RM (2010) The first metazoa living in permanently anoxic conditions. BMC Biol 8:30

Delage Y (1884) Evolution de la sacculine (*Sacculina carcini* Thomps.), crustacé endoparasite de l'ordre nouveau des Kentrogonides. Archives de zoologie expérimentale et générale (2)2:417–736

Demange JM (1967) Recherches sur la segmentation du tronc des Chilopodes et des Diplopodes Chilognathes (Myriapodes). Mémoires du Muséum national d'Histoire naturelle, Paris (NS) A44:1–188

Di Giulio A, Fattorini S, Kaupp A, Vigna Taglianti A, Nagel P (2003) Review of competing hypotheses of phylogenetic relationships of Paussinae (Coleoptera: Carabidae) based on larval characters. Syst Entomol 28:509–537

Drago L, Fusco G, Minelli A (2008) Non-systemic metamorphosis in male millipede appendages: long delayed, reversible effect of an early localized positional marker? Front Zool 5:5

Drago L, Fusco G, Garollo E, Minelli A (2011) Structural aspects of leg-to-gonopod metamorphosis in male helminthomorph millipedes (Diplopoda). Front Zool 8:19

Eco U (1999) Kant and the platypus—essays on language and cognition. Secker & Warburg, London

Edgecombe GD, Giribet G, Dunn CW, Hejnol A, Kristensen RM, Neves RC, Rouse GW, Worsaae K, Sørensen MV (2011) Higher-level metazoan relationships: recent progress and remaining questions. Org Divers Evol 11:151–172

Funch P, Kristensen RM (1995) Cycliophora is a new phylum with affinities to Entoprocta and Ectoprocta. Nature 378:711–714
Fusco G (2005) Trunk segment numbers and sequential segmentation in myriapods. Evol Dev 7:608–617
Geoffroy Saint-Hilaire I (1832–1837) Histoire générale et particulière des anomalies de l'organisation chez l'homme et les animaux. Ballière, Paris (1, 1832; 2, 1836; 3, 1836; 4, 1837)
Gould SJ (1977) Ontogeny and phylogeny. Belknap Press, Cambridge
Gould SJ (2000) Of coiled oysters and big brains: how to rescue the terminology of heterochrony, now gone astray. Evol Dev 2:241–248
Harzsch S, Müller CHG (2007) A new look at the ventral nerve centre of *Sagitta*: implications for the phylogenetic position of Chaetognatha (arrow worms) and the evolution of the bilaterian nervous system. Front Zool 4:14
Helmkampf M, Bruchhaus I, Hausdorf B (2008) Multigene analysis of lophophorate and chaetognath phylogenetic relationships. Mol Phylogenet Evol 46:206–214
Henriquez CL, Arias T, Pires JC, Croat TB, Schaal BA (2014) Phylogenomics of the plant family Araceae. Mol Phylogenet Evol 75:91–102
Hoffman RL (1982) Diplopoda. In: Parker SP (ed) Synopsis and classification of living organisms, vol 2. McGraw-Hill, New York, pp 689–724
Hogue CL (1981) Blephariceridae. In: McAlpine JF, Peterson BV, Shewell GE, Teskey HJ, Vockeroth JR, Wood DM (eds) Manual of nearctic Diptera, vol 1. Research Branch Agriculture Canada, Ottawa, pp 191–197
Janies D, Mooi R (1998) *Xyloplax* is an asteroid. In: Candia Carnevali MD, Bonasoro F (eds) Echinoderm research 1998. Balkema, Rotterdam, pp 311–316
Janies D, Voight JR, Daly M (2011) Echinoderm phylogeny including *Xyloplax*, a progenetic asteroid. Syst Biol 60:420–438
Jeffery JE, Bininda-Emonds ORP, Coates MI, Richardson MK (2002a) Analyzing evolutionary patterns in amniote embryonic development. Evol Dev 4:292–302
Jeffery JE, Richardson MK, Coates MI, Bininda-Emonds ORP (2002b) Analyzing developmental sequences within a phylogenetic framework. Syst Biol 51:478–491
Jeffery JE, Bininda-Emonds ORP, Coates MI, Richardson MK (2005) A new technique for identifying sequence heterochrony. Syst Biol 54:230–240
Jiménez-Guri E, Okamura B, Holland PWH (2007) Origin and evolution of a myxozoan worm. Integr Comp Biol 47:752–758
Katayama N, Koi S, Kato M (2010) Expression of *SHOOT MERISTEMLESS*, *WUSCHEL*, and *ASYMMETRIC LEAVES1* homologs in the shoots of Podostemaceae: implications for the evolution of novel shoot organogenesis. Plant Cell 22:2131–2140
Katayama N, Kato M, Yamada T (2013) Origin and development of the cryptic shoot meristem in *Zeylanidium lichenoides* (Podostemaceae). Am J Bot 100:635–646
Kheirallah A-M, Aly A-NH, Abdel-Wahed NY (2000) Anamorphosis and life-history of the millipede *Nopoiulus kochii* (Gervais, 1847), new for Egypt. Zool Middle East 21:159–168
Kristensen RM (1983) Loricifera, a new phylum with Aschelminthes characters from the meiobenthos. Zeitschrift für Zoologische Systematik und Evolutionsforschung 21:163–180
Kristensen RM (1991) Loricifera. In: Harrison FW, Ruppert EE (eds) Microscopic anatomy of invertebrates, vol 4. Aschelminthes. Wiley, New York, pp 351–375
Kristensen RM, Funch P (2000) Micrognathozoa: a new class with complicated jaws like those of Rotifera and Gnathostomulida. J Morphol 246:1–49
Lebedev OA (2009) A new specimen of *Helicoprion* Karpinsky, 1899 from Kazakhstanian Cisurals and a new reconstruction of its tooth whorl position and function. Acta Zoolog 90(Supp 1):171–182
Lee WL, Reiswig HM, Austin WC, Lundsten L (2012) An extraordinary new carnivorous sponge, *Chondrocladia lyra*, in the new subgenus *Symmetrocladia* (Demospongiae, Cladorhizidae), from off of northern California, USA. Invertebr Biol 131:259–284
Marlétaz F, Martin E, Perez Y, Papillon D, Caubit X, Fasano L, Dossat C, Wincker P, Weissenbach J, Le Parco Y (2006) Chaetognath phylogenomics: a protostome with deuterostomes-like development. Curr Biol 16:R577–R578

Márquez-Guzmán J, Engleman M, Martínez-Mena A, Martínez E, Ramos CH (1989) Anatomía reproductiva de *Lacandonia schismatica* (Lacandoniaceae). Ann Mo Bot Gard 76:124–127
Matus DQ, Copley RR, Dunn CW, Hejnol A, Eccleston H, Halanych KM, Martindale MQ, Telford MJ (2006) Broad taxon and gene sampling indicate that chaetognaths are protostomes. Curr Biol 16:R575–R576
McKinney ML, McNamara KJ (1991) Heterochrony. The evolution of ontogeny. Plenum Press, New York
McKinney ML (ed) (1988) Heterochrony in evolution: a multidisciplinary approach. Plenum Press, New York
McNamara KJ (1986) A guide to the nomenclature of heterochrony. J Paleontol 60:4–13
McNamara KJ (ed) (1995) Evolutionary change and heterochrony. Wiley, Chichester
Minelli A (1993) Biological systematics. The state of the art. Chapman & Hall, London
Minelli A (2003) The development of animal form. Cambridge University Press, Cambridge
Obst M, Funch P (2003) Dwarf male of *Symbion pandora* (Cycliophora). J Morphol 255:261–278
Poinar GO, van der Laan PA (1972) Morphology and life history of *Sphaerularia bombi*. Nematologica 18:239–252
Rudall PJ, Remizowa MV, Prenner G, Prychid CJ, Tuckett RE, Sokoloff DD (2009) Non-flowers near the base of extant angiosperms? Spatiotemporal arrangement of organs in reproductive units of Hydatellaceae, and its bearing on the origin of the flower. Am J Bot 96:67–82
Rutishauser R (1995) Developmental patterns of leaves in Podostemaceae as compared to more typical flowering plants: saltational evolution and fuzzy morphology. Can J Bot 73:1305–1317
Rutishauser R, Sattler R (1985) Complementarity and heuristic value of contrasting models in structural botany. I. General considerations. Botanische Jahrbücher für Systematik, Pflanzengeschichte und Pflanzengeographie 107:415–455
Rutishauser R, Sattler R (1987) Complementarity and heuristic value of contrasting models in structural botany. II. Case study on leafwhorls: *Equisetum* and *Ceratophyllum*. Botanische Jahrbücher für Systematik, Pflanzengeschichte und Pflanzengeographie 109:227–255
Rutishauser R, Sattler R (1989) Complementarity and heuristic value of contrasting models in structural botany. III. Case study on shoot-like "leaves" and leaf-like "shoots" in *Utricularia macrorhiza* and *U. purpurea* (Lentibulariaceae). Botanische Jahrbücher für Systematik, Pflanzengeschichte und Pflanzengeographie 111:121–137
Rutishauser R, Sattler R (1997) Expression of shoot processes in leaf development of *Polemonium caeruleum* as compared to other dicotyledons. Botanische Jahrbücher für Systematik, Pflanzengeschichte und Pflanzengeographie 119:563–582
Rutishauser R, Grob V, Pfeifer E (2008) Plants are used to having identity crises. In: Minelli A, Fusco G (eds) Evolving pathways. Key themes in evolutionary developmental biology. Cambridge University Press, Cambridge, pp 190–210
Sahli F (1990) On post-adult moults in Julida (Myriapoda, Diplopoda). Why periodomorphosis and intercalaries occur in males? In: Minelli A (ed) Proceedings of the 7th international congress of myriapodology. Brill, Leiden, pp 135–156
Sattler R (1992) Process morphology: structural dynamics in development and evolution. Can J Bot 70:708–714
Sattler R, Jeune B (1992) Multivariate analysis confirms the continuum view of plant form. Ann Bot 69(249):262
Sattler R, Rutishauser R (1997) The fundamental relevance of morphology and morphogenesis to plant research. Ann Bot 80:571–582
Sattler R (1996) Classical morphology and continuum morphology: opposition and continuum. Ann Bot 78:577–581
Sattler R (1988) Homeosis in plants. Am J Bot 75:1606–1617
Scholtz G (2005) Homology and ontogeny: pattern and process in comparative developmental biology. Theory Biosci 124:121–143
Scholtz G (2008) On comparisons and causes in evolutionary developmental biology. In: Minelli A, Fusco G (eds) Evolving pathways. Cambridge University Press, Cambridge, pp 144–159

Siddall ME, Whiting MF (1999) Long-branch abstractions. Cladistics 15:9–24
Siddall ME, Martin DS, Bridge D, Desser SS, Cone DK (1995) The demise of a phylum of protists: phylogeny of Myxozoa and other parasitic Cnidaria. J Parasitol 81:961–967
Smith KK (2001) Heterochrony revisited: the evolution of developmental sequences. Biol J Linn Soc 73:169–186
Smith KK (2002) Sequence heterochrony and the evolution of development. J Morphol 252:82–97
Smith KK (2003) Time's arrow: heterochrony and the evolution of development. Int J Dev Biol 47:613–621
Thompson JV (1830) On the Cirripedes or Barnacles; demonstrating their deceptive character; the extraordinary metamorphosis they undergo, and the class of animals to which they indisputably belong. In: Thompson JV (ed) Zoological Researches, and illustrations; or, natural history of nondescript or imperfectly known animals, vol 1(1). King and Ridings, Cork, pp 69–82
Thompson JV (1835) Discovery of the metamorphosis in the second type of the cirripedes, viz. the lepades, completing the natural history of these singular animals, and confirming their affinity with the Crustacea. Philos Trans R Soc 126:355–358
Verhoeff KW (1923) Periodomorphose. Zoologischer Anzeiger 56(233–238):241–254
Westheide W (1987) Progenesis as a principle in meiofauna evolution. J Nat Hist 21:843–854
Wilkins JS, Ebach MC (2013) The nature of classification: relationships and kinds in the natural sciences. Palgrave Macmillan, Basingstoke
Wolf K, Markiw ME (1984) Biology contravenes taxonomy in the Myxozoa: new discoveries show alternation of invertebrate and vertebrate hosts. Science 225:1449–1452
Yaeger J (1981) Remipedia, a new class of Crustacea from a marine cave in the Bahamas. J Crustac Biol 1:328–333
Zrzavý J, Mihulka S, Kepka P, Bezdek A, Tietz D (1998) Phylogeny of the Metazoa based on morphological and 18S ribosomal DNA evidence. Cladistics 14:249–285

Macroevolution in and Around the Hominin Clade

Bernard Wood and Mark Grabowski

Abstract In this review, we discuss the criteria for recognizing species and genera within the fossil record in general, and within the hominin clade in particular. We review the grade concept, suggest how taxa within the hominin clade can be divided into grades, and define the grade categories. We discuss the difficulties with studying macroevolution in the hominin clade but suggest that at least one trait, brain size, may provide insight into the tempo and mode of evolution. We also review evidence suggesting that stasis is the dominant signal in two early hominin taxa that have substantial and well-dated fossil records. We discuss the role of evolutionary forces in forming macroevolutionary patterns and find that while natural selection appears to be the dominant force, some well-known interspecific and intraspecific differences in hominins may have been the result of random genetic drift. Lastly, we suggest that homoplasy makes generating reliable hypotheses about relationships among early hominins more difficult than most researchers are willing to admit.

Keywords Clade · Grade · Hominin · Macroevolution · Mode · Selection · Tempo

1 Preamble

In his 1944 book *Tempo and mode in evolution,* Simpson uses the criterion of population continuity to distinguish micro- and macroevolution (Simpson 1944). He suggests that microevolution refers to "changes within potentially continuous populations" whose details can be revealed by "genetic experimentation" (ibid, p. 97).

B. Wood (✉) · M. Grabowski
Center for Advanced Study of Hominid Paleobiology,
George Washington University, Washington D.C., USA
e-mail: bernardawood@gmail.com

E. Serrelli and N. Gontier (eds.), *Macroevolution*,
Interdisciplinary Evolution Research 2, DOI 10.1007/978-3-319-15045-1_11

In contrast, he suggests that macroevolution involves the "rise and divergence of taxonomic groups that are at or near the minimum level of genetic discontinuity (species and genera)" (ibid, p. 98). Simpson (1953) later simplifies this by suggesting that macroevolution involves historical changes "from species upwards," whereas microevolution refers to historical change "within species" (ibid, p. 338).

Simpson credits Goldschmidt with introducing the term macroevolution (Simpson 1944, p. 97), but this is not correct. While it is true that Goldschmidt (1940) uses microevolution and macroevolution as the two major subheadings for his book *The material basis of evolution*, it seems that Dobzhansky (1937) introduced macroevolution into the English language 3 years before Goldschmidt's book was published. Dobzhansky, however, did not coin the term macroevolution. That distinction apparently rests with his teacher, the Russian geneticist, Filipčenko (aka Filipchenko or Philiptchenko) who used the Russian equivalent of macroevolution in 1934 in a text entitled *Genetics of soft wheats*.

Whereas Simpson (1944) had focused on genetic continuity as a criterion to distinguish micro- and macroevolution, Dobzhansky (1937) stressed the importance of temporal distinctions. Specifically, he referred to the differences between longer-term "macroevolutionary changes that require time on a geological scale" and shorter-term "microevolutionary processes" that are observable "within the span of a human lifetime" (ibid, p. 12). Levinton (2001) suggests that macroevolution is the sum of the processes that generate the "character-state transitions that diagnose evolutionary differences of major taxonomic rank" (ibid, p. 2), but Hallam's (1989) "evolution at and above the species level" (ibid, p. 59) is a more typical contemporary definition of macroevolution.

If the species category is used as the definition of what is, or is not, macroevolution, it raises problems for anyone reviewing that topic in the context of human evolution. This is because taxonomic hypotheses about the hominin clade run the gamut from those that recognize relatively few species (e.g., Wolpoff 1994) to those that are much more speciose (e.g., Wood 2010). Irrespective of their strengths and weaknesses, if the species is the rubicon that divides macroevolution from microevolution, then the type of taxonomic hypothesis that is adopted will have profound implications for what is included in a review of macroevolution in the hominin clade. This is because more inclusive interpretations of hominin species (i.e., "lumping" hypotheses) will result in substantial amounts of phenotypic evolution (e.g., an increase in brain size from *c*.600 cm^3 to *c*.>1,300 cm^3) being regarded as intraspecific, and if the definition of macroevolution is "evolution at and above the species level," then these changes would be regarded as microevolutionary and would be *outside* the purview of a review of macroevolution. In contrast, more exclusive interpretations of hominin species (i.e., "splitting" hypotheses) suggest that most phenotypic evolution within the hominin clade took place at the time of speciation, and thus, its discussion would be *within* the bailiwick of a review that focuses on macroevolution.

Thus, instead of using "evolution at and above the species level" as the definition of macroevolution, we follow Dobzhansky (1937), and especially Eldredge (1989), who suggested that macroevolution always connotes "large-scale phenotypic evolutionary change" (ibid, p. vii). We assume the following "broad-brush" distinction that macroevolution is what you can learn about evolution from the fossil record. So with respect to macroevolution in the hominin clade, we interpreted our remit as "what can be learned about human evolution from the hominin fossil record."

In this review, we focus on the hominin clade and consider the following questions. What is its comparative context? What are the criteria for recognizing species and genera within the hominin clade? Can the taxa within the hominin clade be usefully divided into grades, and if so how should they be defined? What evidence is there about the tempo and mode of evolution within the hominin clade? Are morphological trends in the hominin clade the result of selection, or can they be explained by random drift? Lastly, we consider what is known about the relationships among its constituent taxa. Inevitably, there is overlap between these questions, but they provide a structure for the task allotted to us of reviewing macroevolution in the hominin clade. Some of the topics included in the questions set out above have been addressed in our publications, so where appropriate, we point the reader to those publications rather than simply repeating arguments made elsewhere. We do not cite references listed in those publications.

2 Context

Recent attempts to use gross morphological evidence to generate hypotheses about higher primate relationships (e.g., Gibbs et al. 2002; Diogo and Wood 2011) have confirmed the close relationship between modern humans and the African apes suggested just over 150 years ago by Huxley (1863). During the first half of the twentieth century, the focus of the search for evidence about higher primate relationships shifted from evidence about gross morphology to evidence about the morphology of molecules (e.g., Grünbaum 1902; Nuttall et al. 1904). In the 1960s, two molecules, hemoglobin (Zuckerkandl et al. 1960) and albumin (Goodman 1963), were used to investigate the relationships among higher primates, and these studies concluded that chimpanzees were more closely related to modern humans than to gorillas. Sarich and Wilson (1967) came to a similar conclusion, and later, King and Wilson (1975) suggested that 99 % of the amino acid sequences of chimpanzee and modern human proteins were identical.

Initial attempts to compare the DNA of higher primates were crude (e.g., Caccone and Powell 1989); however, sequencing methods rapidly replaced hybridization as the preferred method for generating hypotheses about the relationships among extant hominoid taxa, and the number of sequence-based studies increases year by year (see Bradley 2008; Arnold et al. 2010; Perelman et al. 2011 and Prado-Martinez et al. 2014 for reviews). When DNA differences were calibrated using what was then the best paleontological evidence for the split between apes and Old World monkeys, it was predicted that the hypothetical ancestor of modern humans and chimpanzees/bonobos lived between *c*.8 and *c*.5 million years ago (Ma) (Bradley 2008). However, these predictions are likely to yield different results in light of the recent discovery of the Oligocene catarrhine *Rukwapithecus fleaglei* that may be a basal hominoid (Stevens et al. 2014). Langergraber et al. (2012) used comparative data about generation times and estimates of mutation rates and concluded that the date of the *Pan–Homo* split is probably closer to 8 than to 5 Ma, but the results of a recent analysis of a larger data set (Prado-Martinez et al. 2014) that used different assumptions suggest that it is closer to 5 Ma.

Whole genomes can now be sequenced with acceptable levels of coverage, and in the last few years, researchers have published good-quality draft sequences of the genomes of the chimpanzee (TCSAC 2005), orangutan (Locke et al. 2011), gorilla (Scally et al. 2012), and bonobo. Scally et al. (2012) sampled two western lowland and one eastern lowland gorillas and showed that when considering the entire genome, the greatest number of similarities is between modern humans and chimpanzees, but in 30 % of the genome, gorillas are closer to modern humans and chimpanzees than they are to each other. This phenomenon is known as incomplete lineage sorting (ILS). The Prüfer et al. (2012) study showed that bonobos and common chimpanzees are 99.7 % alike, yet 98.7 % of the bonobo genome resembles that of modern humans. Prüfer et al. (2012) also found evidence of ILS in their study to the extent that *c*.3 % of the modern human genome is more closely related to bonobos or to common chimpanzees than bonobos and common chimpanzees are to each other, and they suggest that 25 % of all genes contain evidence of ILS. That said, a recent comparative study of 79 great ape genomes representing all six species emphasized that the presence of genetically distinct populations within each great ape species (Prado-Martinez et al. 2014) confirms that despite the effects of ILS, chimpanzees and bonobos are more closely related to modern humans than they are to gorillas. Thus, the comparative context of the hominin clade is the one set out in Fig. 1.

3 Criteria for Including Taxa Within the Hominin Clade

The reasons for including the *c*.7 Ma remains assigned to *Sahelanthropus tchadensis* (Brunet et al. 2002; Guy et al. 2005), the *c*.6 Ma remains assigned to *Orrorin tugenensis* (Senut et al. 2001), the *c*.5.8–5.2 Ma remains assigned to *Ardipithecus kaddaba* (Haile-Selassie 2001, 2004), and the *c*.4.5–4.4 Ma remains assigned to *Ardipithecus ramidus* (White et al. 1994, 2009; White 2010) in the hominin clade, differ according to what anatomical regions are represented. However, three common lines of evidence run through the claims for the hominin status of these taxa. The first involves a reduction in size and a change in morphology of the canines, which is linked with the partial or complete loss of upper canine/P_3 honing and reduced canine sexual dimorphism. The second involves the location and orientation of the foramen magnum and inferences about posture and gait. The third involves features of the pelvis and other preserved postcranial elements that imply a dependence on bipedalism. In each case, the assumption is that these character complexes and their inferred behaviors are *only* seen in the hominin clade.

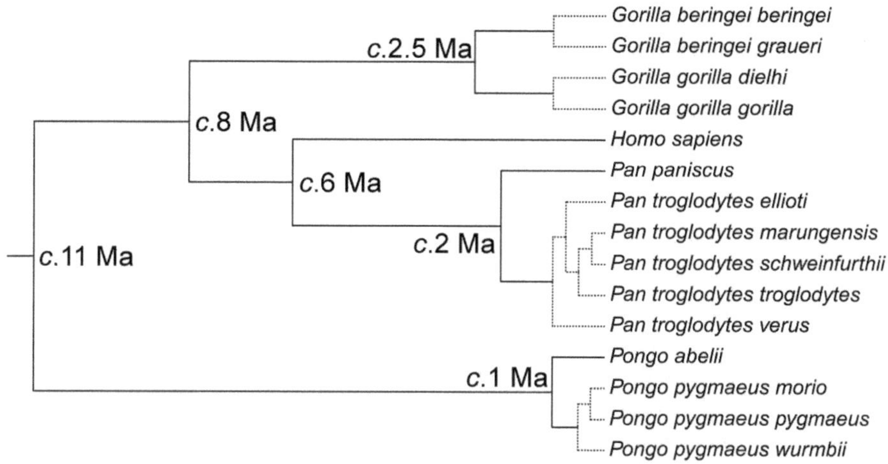

Fig. 1 Current consensus of the phylogenetic relationships and splitting times within the great ape clade. The only Asian great ape, the orangutan (*Pongo*), which is likely to have split off from the African great apes *c*.11 million years, diverged into the Bornean (*Pongo pygmaeus*) and Sumatran (*Pongo abelii*) orangs *c*.1 million years ago. There have been two major and two minor splits in the African ape clade. The first major splitting event, the one leading to gorillas, occurred *c*.8 million years ago. The second, leading to modern humans, occurred *c*.6 million years ago. The split within gorillas, into mountain (*Gorilla beringei*) and lowland (*Gorilla gorilla*), occurred *c*.2.5 million years ago. The split within chimpanzees occurred *c*.2 million years ago when the Congo River divided the ancestral chimpanzee population into bonobos (*Pan paniscus*) to the south and common chimpanzees (*Pan troglodytes*) to the north. The details of the subspecies, along with the timing of any splits, are more conjectural. Figure courtesy of Adam Gordon. Evidence for the phylogenetic relationships within the extant great ape genera is drawn from a variety of sources (*Pan*: Groves 2005; Gonder et al. 2011; *Gorilla*: Groves 2001; Scally et al. 2012; *Pongo*: Brandon-Jones et al. 2004; Singleton et al. 2004; Locke et al. 2011; Prado-Martinez et al. 2014)

The canine morphology that *Ar. ramidus* and *S. tchadensis* share with later hominins is the most convincing evidence to support their hominin status. But it is important to recognize that during the Late Miocene, a number of Eurasian hominids (e.g., *Oreopithecus*, *Ouranopithecus*, and *Gigantopithecus*) also developed smaller canines and a reduction in canine–premolar honing. Presumably, these were parallel responses linked to analogous shifts in dietary behavior and there is no a priori reason to exclude the possibility that a similar behavioral and phenotypic response could have occurred in at least one extinct African hominid clade.

The anteriorly positioned and more horizontal foramen magnum seen in modern humans and later hominins compared to the extant great apes has been assumed to relate to the upright posture and bipedal locomotion of the former. However, comparisons with other primates suggest that these features may also be linked with differences in head carriage and relative brain size rather than uniquely with bipedalism (Strait 2001) and the differences in the position and orientation of the foramen magnum seen in bonobos and chimpanzees, and the overlap between the morphology of bonobos and that of *Sahelanthropus* and *Ardipithecus* suggests that we should exercise caution before assuming that a relatively anteriorly positioned and more horizontal foramen magnum is linked exclusively with the adoption of habitual bipedalism.

The postcranial evidence for bipedalism in *Ardipithecus kadabba* mainly involves the morphology of a proximal pedal phalanx (presumed to belong to *Ar. kadabba,* but from an older geological horizon and with no associated craniodental remains), whereas in *O. tugenensis*, the evidence mainly involves the morphology of the proximal femur. The case for the femur being that of a committed biped is much stronger than the case for the pedal phalanx. The claim that *Ar. ramidus* was a biped is mainly based on highly speculative inferences about the presence of lumbar lordosis and on a few features of the pelvis and foot, but the claims are either based on questionable reconstructions, or they involve characters whose link to habitual bipedalism has yet to be convincingly demonstrated.

Researchers that support hominin status for *S. tchadensis*, *O. tugenensis*, *Ar. kaddaba*, and *Ar. ramidus* do so on the assumption that within the great apes, canine honing and bipedalism are *confined* to the hominin clade. We believe that their assumption is a logical fallacy. For even if all hominins are bipedal and lack canine honing, the converse proposition–that among the great apes bipedalism and the loss of canine honing are confined to the hominin clade–is not a logical corollary.

How strong are the cases for each of the four taxa being hominins? The argument for including *Ar. kaddaba* in the hominin clade at the present time is a particularly weak one. Its teeth are apelike, and because of the sparse fossil record, there is not enough evidence to be sure it is a committed biped. As for *O. tugenensis*, although the external morphology of the proximal femur is consistent with it being bipedal, the evidence from the internal morphology of the femoral neck is equivocal. The morphological evidence that *S. tchadensis* and *Ar. ramidus* should be included in the hominin clade is stronger, but is not compelling for either taxon. In addition, their age is against them being hominins. In the case of *S. tchadensis,* if the more recent splitting *c*.5 Ma times are correct, then if it is *c*.7 Ma it is too early for it to be the stem hominin. In the case of *Ar. ramidus*, if both it and the *c*.4.2 Ma *Australopithecus anamensis* are lineal ancestors of later hominins, as its discoverers claim, then there is simply too little time for the cranial and postcranial morphology of the former to evolve into the latter. Also, if the 3.4 Ma foot with an abducted hallux from the Burtele locality at Woranso-Mille belongs to *Ar. ramidus*, then the "ancestral" scenario is even less likely. Thus, for these reasons, one of us has referred to *S. tchadensis*, *O. tugenensis*, *Ar. kaddaba*, and *Ar. ramidus* as "possible hominins" (e.g., Wood 2010) and this is how we refer to them in this review.

4 Hominin Alpha Taxonomy

The definition of taxonomic categories is a vexed issue. With respect to the species category, Smith (2009) usefully divides contemporary species concepts into *process related* and *pattern related*, with the former emphasizing the processes involved in the generation and maintenance of species, while the latter emphasizes the methods used for recognizing species in the fossil record. The three main concepts in the process category are the biological species concept (BSC), the evolutionary species concept (ESC), and the recognition species concept (RSC). The ESC was an attempt by Simpson (1961) to add a temporal dimension to the BSC; thus, he suggested that under the ESC, a species is "an ancestral-descendant sequence of populations evolving separately from others and with its own evolutionary role and tendencies." Some use the term chronospecies to refer to a segment of the type of evolving lineage implied in the ESC definition of a species. Such segments are considered separate species because the fossil sample across time is deemed to exceed the degree or the pattern of variation that it would be reasonable to find within closely related, living species. The third concept in the process-related category, the recognition species concept, instead of emphasizing reproductive isolation, emphasizes the process that promotes interbreeding. Paterson (1985) refers to this as the "specific mate recognition system" (or SMRS), and as long as a species' SMRS signal fossilizes, the RSC can potentially be applied to the fossil record.

Given the twin impediments of having no direct evidence about interbreeding, and with only fragments of the hard tissue skeleton and the dentition as evidence, how are species recognized in the hominin fossil record? There are two main pattern-based species concepts, the phenetic species concept (PeSC) and the phylogenetic species concept (PySC). The PeSC gives equal weight to all aspects of the phenotype by assembling a matrix of characters and then uses multivariate analysis to detect clusters of individual specimens that share similar phenotypes. The PySC differs from the PeSC by emphasizing only the diagnostic aspects of the phenotype. According to Nixon and Wheeler (1990), a species defined under the PySC is "the smallest aggregation of populations diagnosable by a unique combination of character states."

In practice, most human evolution researchers use a version of the PySC in the sense that they search for the smallest cluster of individual organisms that is "diagnosable" on the basis of the preserved morphology. Because the hominin fossil record consists primarily of craniodental remains, most diagnoses of early hominin taxa inevitably emphasize craniodental morphology. Thus, using this evidence, paleoanthropologists must decide whether a collection of hominin fossils spanning several hundred thousand years consists of several samples of the same taxon, or samples of different taxa. When making these judgments, researchers should strive to neither grossly underestimate, nor extravagantly overestimate, the actual number of species represented in the hominin fossil record.

One of the many factors that paleoanthropologists must take into account in addition to the time represented in their sample is that the fossil record is predominantly confined to remains of hard tissues (i.e., bones and teeth). We know from living animals that many uncontested species are difficult to distinguish using bones and teeth (e.g., *Cercopithecus* species—see Manaster 1979); thus, there are sound, logical reasons to suspect that a hard tissue-bound fossil record is always likely to underestimate the number of species. Furthermore, if a punctuated equilibrium model of evolution is adopted along with a branching or cladogenetic interpretation of the fossil record (see below), then researchers will tend to divide the hominin fossil record into more rather than fewer species. Conversely, researchers who favor a phyletic gradualism model, which implies an anagenetic interpretation of evolution and emphasizes morphological continuity, will tend to resolve the hominin fossil record into fewer, more inclusive, longer-lived species that are more likely to show substantial changes in morphology through time.

Eldredge (1993) made a proposal about how to view the species category that is both intuitive and appealing. He suggested that species, like individuals, have a history. The history of any species begins at the point of speciation when it and its sister taxon (or taxa) arise from a common ancestor and ends when it becomes extinct or becomes the common ancestor of daughter taxa. Eldredge also acknowledges the reality that the morphological characteristics of a living species or of an evolutionary lineage are never uniformly distributed across its range, and like Sewall Wright, Eldredge is prepared to recognize the existence of distinctive local populations or demes. Related demes would share the same SMRS, but Eldredge suggests that their morphological distinctiveness could in some cases justify them being regarded as separate species. He also acknowledges that the same logic could be applied to lineage chronospecies on the basis that the number of cladogenetic events during evolutionary history is more likely to be underestimated than overestimated. Thus, within the fossil record, it may be possible to identify several paleospecies (sensu Cain, 1954) within the equivalent of a neontological BSC/RSC-type species. For many reasons, some of which are set out above, in this review, we use a relatively speciose taxonomic hypothesis (Table 1).

The genus is even more an elusive taxonomic category than the species, but for various reasons, we accept the proposal that a genus should be both a clade and a grade. To qualify as a clade, the prospective genus must consist of all the members of a monophyletic group, no more and no less. But not all of the species in the same grade have to be in the same genus, for a grade may contain species belonging to more than one monophyletic group. We have divided the species we recognize in the hominin fossil record into genera, but because we are generally skeptical about our ability to recognize subclades within the hominin clade, our genus-level distinctions are based more on evidence about grade distinctions (see below) than on hypotheses about relationships.

5 Differences Between Modern Humans and Chimpanzees/Bonobos

The features that set modern humans apart from chimpanzees and bonobos, and which can be tracked using a hard tissue-bound fossil record, are to do with cranio-dental morphology, axial and postcranial morphology, and life history.

With respect to dental morphology, chimpanzees and bonobos have larger canine and incisor teeth than modern humans, but if the size of the premolar and molar teeth is related to body mass, then the chewing teeth of chimpanzees/bonobos and modern humans are similar in relative size. However, the jaws of a modern human skull are generally, but not in all cases, smaller and lighter than those of chimpanzees and bonobos.

Table 1 The "old" taxonomy below reflects the pre-molecular consensus that chimpanzees and bonobos were more closely related to gorillas than to modern humans

Old
Superfamily Hominoidea (hominoids)
Family Hylobatidae (hylobatids)
Genus *Hylobates*
Family Pongidae (pongids)
Genus *Pongo*
Genus *Gorilla*
Genus *Pan*
Family Hominidae (hominids)
Subfamily Australopithecinae (possible and archaic hominins)
Genus *Ardipithecus*
Genus *Australopithecus*
Genus *Kenyanthropus*
Genus *Orrorin*
Genus *Paranthropus*
Genus *Sahelanthropus*
Subfamily Homininae (hominines)
Genus *Homo*
New
Superfamily Hominoidea (hominoids)
Family Hylobatidae (hylobatids)
Genus *Hylobates*
Family Hominidae (hominids)
Subfamily Ponginae
Genus *Pongo* (pongines)
Subfamily Gorillinae
Genus *Gorilla* (gorillines)
Subfamily Homininae (hominines)
Tribe Panini
Genus *Pan* (panins)
Tribe Hominini (hominins)
Subtribe Australopithecina (possible and archaic hominins)
Genus *Ardipithecus*
***Ardipithecus ramidus* (White et al., 1994) White et al., 1995**
***Ardipithecus kaddaba* HaileSelassie, 2001**
Genus *Australopithecus*
***Australopithecus africanus* Dart, 1925**
***Australopithecus afarensis* Johanson, 1978**
***Australopithecus anamensis* Leakey et al., 1995**
***Australopithecus bahrelghazali* Brunet et al., 1996**
***Australopithecus garhi* Asfaw et al., 1999**
***Australopithecus sediba* Berger et al., 2010**
Genus *Kenyanthropus*

(continued)

Table 1 (continued)

***Kenyanthropus platyops* Leakey et al., 2001**
Genus *Orrorin*
***Orrorin tugenensis* Senut et al., 2001**
Genus *Paranthropus*
***Paranthropus robustus* Broom, 1938**
***Paranthropus boisei* Leakey, 1959; Robinson, 1960**
***Paranthropus aethiopicus* Arambourg, 1968**
Genus *Sahelanthropus*
***Sahelanthropus tchadensis* Brunet et al., 2002**
Subtribe Hominina (hominans)
Genus *Homo*
Homo sapiens Linnaeus, 1745
***Homo neanderthalensis* King, 1864**
***Homo erectus* Dubois, 1893; Weidenreich, 1940**
***Homo heidelbergensis* Schoetensack, 1908**
***Homo habilis* Leakey, Tobias and Napier, 1964**
***Homo rudolfensis* (Alexeev, 1986) sensu Wood, 1992**
***Homo antecessor* Bermúdez de Castro et al., 1997**
***Homo floresiensis* Brown et al., 2004**

In this taxonomy, modern humans, and all of the taxa thought to be more closely related to modern humans than to any other living taxon, are distinguished at the level of the family as the Hominidae. The "new" taxonomic hypothesis set out above is one of several ways that researchers reflect the overwhelming molecular and morphological evidence that modern humans and chimpanzees and bonobos are more closely related to each other than chimpanzees and bonobos are to gorillas. In this taxonomy, modern humans, and all of the taxa thought to be more closely related to modern humans than to any other living taxon, are distinguished at the level of the tribe as the Hominini. Some researchers consider even this level of distinction too much and they prefer to reduce hominins to a subtribe as the Hominina. In this second, "new" taxonomy, we list fossil hominin species under each genus in the order they were established

With respect to the cranium, modern human brains are not just absolutely larger than those of chimpanzees/bonobos, but they are also larger relative to body mass. The modern human cranium has a relatively smaller face, and the cranium is more evenly balanced on the vertebral column. The foramen magnum is close to the middle of the cranial base in modern humans, whereas in common chimpanzees, it is situated more posteriorly, although bonobos have a more anterior foramen magnum than do common chimpanzees.

With regard to the axial skeleton, the chest is differently shaped in modern humans and in chimpanzees/bonobos. The thorax of chimpanzees/bonobos widens toward the base to accommodate their relatively large gut. The thorax of modern humans is uniform in width from top to bottom, and flatter from front to back, with the shoulder blades rotated around to the back so that they lie closer to the vertebral column. With respect to the vertebral column, there is a difference between modern humans and chimpanzees/bonobos in how thoracic and lumbar vertebrae contribute to trunk length. In modern humans, the dominant modal pattern for thoracic and lumbar vertebrae is 12:5, whereas in *P. troglodytes* and

P. paniscus, it is 13:4, with *P. troglodytes* averaging 13.1 thoracic and 3.6 lumbar vertebrae and *P. paniscus* averaging 13.4 thoracic and 3.8 lumbar vertebrae.

Postcranially, the longer and more mobile modern human thumb enables it to meet the tips of the fingers to make a precise "pinch" grip. In addition to the evident differences in the structure of the hand, it is likely that a neurological control component (e.g., motor unit size) accounts for the differences in dexterity between modern humans and chimpanzees/bonobos. Modern human adult locomotion is almost exclusively bipedal and thus contrasts with the predominantly quadrupedal locomotion of chimpanzees and bonobos. These differences are reflected in the morphology of the pelvic girdle and lower back, knee, ankle, and foot and in the disposition of the muscles connecting the lower limb to the pelvis and trunk. The modern human pelvis is arranged so that the body can be held upright with the body mass being supported on the hind limbs alone. The upper limbs of modern humans are relatively shorter than those of chimpanzees and bonobos, whereas the legs of modern humans are relatively longer than those of chimpanzees/bonobos. There are also differences in the foot, with the modern human foot creating a more stable platform than it does in chimpanzees and bonobos.

In addition, there are differences in the rate that the body grows and in the order in which structures appear. Modern humans reach maturity more slowly than do chimpanzees and bonobos, they erupt their teeth in a different order, and in modern humans, the milk, or deciduous, molars wear out before the adult molars have erupted.

6 Reconstructing Hypothetical Common Ancestors

The task of paleoanthropologists is to use the fossil record to try and trace the evolutionary history of the differences reviewed above back into the tree of life. This task is made more difficult because we can be sure that the differences between the earliest hominins and the Late Miocene ancestors of chimpanzees and bonobos are likely to have been more subtle and difficult to discern than the differences between modern humans and chimpanzees/bonobos. Some of the distinctive features of modern humans, such as those linked with obligate bipedalism, can be traced back a long way. Others, such as the relatively diminutive jaws and chewing teeth of modern humans, were acquired more recently and thus cannot be used to tell the difference between early hominins and potential ape ancestors. There is also reasonably sound evidence that at least two early hominin genera, *Australopithecus* and *Paranthropus*, had absolutely and relatively larger chewing teeth than did pre-modern *Homo*. Thus, even though absolutely and relatively larger chewing teeth (known as postcanine megadontia) may have been an important derived feature of early hominins, this trait has been reversed in the later stages of human evolution. Presently, we do not have sufficient information about the earliest stages of hominin evolution, or about fossil apes, to tell whether postcanine megadontia is confined to hominins.

So, given all of these caveats, how do we go about telling a *c*.5-6 million years old hominin from an early panin, or from a taxon that belongs to a closely related clade that has no living representative? The presumption based on parsimony is that taxa at the base of the panin clade (stem panins) would show little change from the presumed morphology of the hypothetical common ancestor of chimpanzees/bonobos and modern humans. Thus, they would have had projecting faces, jaws with relatively small chewing teeth, large, sexually dimorphic, honed canine teeth, and a locomotor system adapted for arboreal quadrupedalism. In contrast, taxa at the base of the hominin clade (stem hominins) would have been distinguished by cranial and other skeletal adaptations to a predominantly upright posture, and skeletal and other adaptations for a locomotor strategy that includes substantial bouts of bipedalism. These features would be combined with a masticatory apparatus that combines relatively large chewing teeth and modest-sized canines.

7 The Case for Grades Within the Hominin Clade

The reconstructions of the hypothetical common ancestors set out above are all working hypotheses that need to be reviewed and tested as appropriate evidence is uncovered and new methods for reconstructing ancestral states are developed. In the meantime, can we detect any trends in cranial, dental, and postcranial morphology that allow the extinct hominin taxa to be sorted into informal groupings that reflect their adaptation? In other words, can we sort them into grades?

Taxa, including extinct hominins, are put in the same grade if they share morphology that suggests they eat the same sorts of foods and share the same posture and mode of locomotion; no store is set by how they came by those behaviors. The judgment about how different two diets or two locomotor strategies have to be before the taxa being scrutinized are considered to belong to different grades is a subjective one, but until we can be sure we have access to ways of generating reliable hypotheses about the relationships among hominin taxa (i.e., about the nature of the subclades within the hominin clade), the grade concept helps sort taxa into broad functional categories. We use five grades in this review, "anatomically modern *Homo*" (the grade that includes modern humans), "pre-modern *Homo*," "transitional hominins," "archaic hominins," and "megadont and hyper-megadont archaic hominins" (Fig. 2). In the following sections, we describe each grade in terms of its characteristic regional morphology (e.g., brain volume, tooth morphology, limb proportions, and postcranial morphology). Within each grade, we describe the species in the historical order the taxa were recognized, not according to their estimated first appearance datum.

7.1 Anatomically Modern Homo

This grade includes hominin fossil evidence that is not significantly different from the morphology found in at least one regional population of modern humans. Presently, the earliest evidence of anatomically modern human

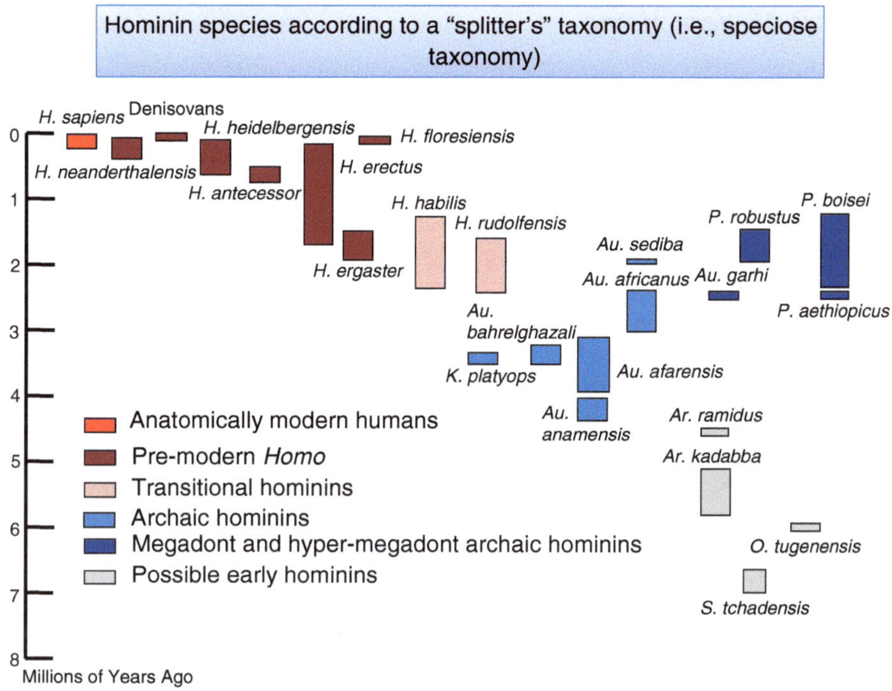

Fig. 2 Taxa recognized in a speciose hominin taxonomy sorted into six grades. In the case of taxa where there are well-dated horizons at several sites, the height of the column reflects current evidence about the earliest and the most recent fossil evidence of that taxon. For taxa known from a well-dated horizon at a single site, the height of the column reflects the age of that horizon. For some taxa, the height of the column is a reflection of uncertainty about the age of that taxon. The height of the column for the Denisovans reflects the age of the only known fossil evidence

morphology in the fossil record comes from Omo Kibish in Ethiopia. Elsewhere in Africa, there is evidence of crania (e.g., Jebel Irhoud from North Africa, Laetoli 18 from East Africa, and Florisbad and the Cave of Hearths from southern Africa) that are generally more robust and archaic looking than those of anatomically modern humans, yet they are not archaic or derived enough to justify being allocated to *Homo heidelbergensis* or to *Homo neanderthalensis*. The gradual and incremental nature of the morphological change between *Homo heidelbergensis* and anatomically modern humans makes setting the boundary between these two taxa challenging, but variation in the later *Homo* fossil record is too great to be accommodated in a single taxon (Mounier et al. 2009). Researchers who make a distinction between subrecent and living modern humans and fossils such as Florisbad and Laetoli 18 do so by formally by referring the latter specimens to a separate species, *Homo helmei* Dreyer, 1935, or informally by referring to them as "archaic *Homo sapiens.*"

7.2 Pre-Modern Homo

This very broad grade grouping includes Pleistocene *Homo* taxa that lack the distinctive size and shape of the modern human cranium and the gracility of the modern human postcranial skeleton, but all of the included taxa have postcranial morphology that is consistent with obligate bipedalism. The teeth are generally larger and the jaws more robust than those of anatomically modern *Homo*. What makes this a particularly broad grouping is the wide range of absolute brain size (*c*.600–*c*.>1,300 cm^3).

The first fossil taxon to be recognized in the pre-modern *Homo* grade is *Homo neanderthalensis* King 1864, whose temporal range is *c*.200–28 ka (but if the Sima de los Huesos material is included, then it is *c*.>400–28 ka). The type specimen, the Neanderthal 1 skeleton, was found in 1856 at the Kleine Feldhofer Grotte in Elberfeld, Germany, and fossil evidence for *H. neanderthalensis* has since been found in Europe as well as in the Near East, the Levant and Western and Central Asia. The distinctive features of the cranium of *H. neanderthalensis* include thick, double-arched brow ridges, a face that projects anteriorly in the midline, a large nose, laterally projecting and rounded parietal bones, and a rounded, posteriorly projecting occipital bone. Mandibular and dental features include a retromolar space, distinctively high incidences of some non-metrical mandibular and dental traits, and thinner tooth enamel than in modern humans. The average endocranial volume of *H. neanderthalensis* is larger than that of living modern humans. Postcranially, *H. neanderthalensis* individuals were stout with a broad rib cage, a long clavicle, a wide pelvis, and limb bones that are generally robust, with large joint surface areas. The distal extremities tend to be short compared to most modern *Homo sapiens*, and the generally well-marked muscle attachments and robust long bone shafts point to a strenuous lifestyle. Some researchers restrict the *H. neanderthalensis* hypodigm to fossils from Europe and the Near East that used to be referred to as "Classic" Neanderthals, but others interpret the taxon more inclusively and include fossil evidence that is generally older and less distinctive (e.g., Steinheim, Swanscombe and from the Sima de los Huesos). The first DNA recovered from a fossil hominin was from the type specimen of *H. neanderthalensis*.

The next fossil hominin taxon in the pre-modern *Homo* grade to be discovered was *Homo erectus* (Dubois 1893) Weidenreich 1940. Its temporal range is *c*.1.8 Ma–*c*.30 ka. The initial discovery at Kedung Brubus was made in 1890, but the type specimen was recovered in 1891 from Trinil. *Homo erectus* is known from sites in Indonesia (e.g., Trinil, Sangiran, Sambungmachan), China (e.g., Zhoukoudian, Lantian), Africa (e.g., Olduvai Gorge, Melka-Kunturé), and possibly the Caucasus (Dmanisi). The fossil record of *H. erectus* is dominated by cranial remains, and while there is some postcranial evidence (mainly femora), there are very few hand and foot fossils. *Homo erectus* crania have a low vault, a continuous supraorbital torus, a sharply angulated occipital region, and relatively thick inner and outer tables of the cranial vault. The body of the mandible is more robust than that of *H. sapiens*, it lacks a chin, and the mandibular tooth crowns are generally larger and the roots of the premolars more complex than those of modern humans. The limb proportions of *H. erectus* are similar to those of modern humans, but the shafts of the long bones of the lower limb are flatter (the femur from front to back and the tibia from side to side) relative to those of modern humans. Overall, the cortical bone of *H. erectus* is thicker than that in modern humans. All of the dental and cranial evidence points to a modern humanlike diet for *H. erectus*, and the postcranial elements are consistent with an upright posture and obligate bipedalism. Those who support *Homo ergaster* Groves and Mazák 1975 as a separate species point to features that are more primitive (e.g., mandibular premolar root and crown morphology and, vault and cranial base morphology) than *H. erectus*. However, most researchers are not convinced that there are sufficient consistent differences between the hypodigms of *H. ergaster* and *H. erectus* to justify the former being a separate species.

After *H. erectus*, the next taxon recognized within the pre-modern *Homo* grade was *Homo heidelbergensis* Schoetensack 1908. Although the type specimen, Mauer 1, was an adult mandible found in 1907 in a sand quarry near Heidelberg, Germany, *c*.600–100 ka fossils from sites in Europe (e.g., Petralona), the Near East (e.g., Zuttiyeh), Africa (e.g., Kabwe, Bodo), China (e.g., Dali, Jinniushan, Xujiayao, Yunxian), and possibly India (Hathnora) have been included in *H. heidelbergensis*. What sets this material apart from *H. sapiens* and *H. neanderthalensis* is its cranial morphology and robusticity of the postcranial skeleton. Some *H. heidelbergensis* specimens have endocranial volumes as large as those of modern humans, but they are always more robustly built with a thickened occipital region and a projecting face and with large separate ridges above the orbits. Researchers who see the African part of this hypodigm as distinctive refer it to a separate species, *Homo rhodesiensis*. Researchers who interpret the European component of the *H. heidelbergensis* hypodigm (e.g., Sima de los Huesos) as already showing signs of *H. neanderthalensis* autapomorphies would sink it into the latter taxon.

The taxon *Homo antecessor* Bermúdez de Castro et al. 1997 was introduced for hominins recovered from the Gran Dolina site at Atapuerca, Spain. The researchers who found the remains claim that the combination of a modern humanlike facial morphology with large and relatively primitive tooth crowns and roots is not seen in *H. heidelbergensis*, and they see *H. antecessor* and not *H. heidelbergensis* as the likely recent common ancestor of *H. neanderthalensis* and *H. sapiens*.

The most recent taxon to be added to pre-modern *Homo* is *Homo floresiensis* Brown et al. 2004. It is currently only known from Liang Bua, a cave in Flores. Its published temporal range is *c*.74–17 ka, but it may be closer to 100 ka. The initial discovery and type specimen is LB1, an associated partial adult skeleton, but a second associated skeleton, and close to a hundred separate fossils representing up to 10 individuals have subsequently been recovered. This hominin displays a unique combination of early *Homo*-like cranial and dental morphology, a hitherto unknown suite of pelvic and femoral features, a small brain (*c*.417 cm^3), a small body mass (25–30 kg), and small stature (1 m). When it was first described, researchers interpreted it as *Homo erectus*, or a *Homo erectus*-like taxon that had undergone endemic dwarfing; however, more recently, researchers have suggested that it could be a dwarfed *Homo habilis*-like transitional hominin.

7.3 Transitional Hominins

For the purposes of this review, *H. habilis* and *H. rudolfensis* are retained within *Homo*, but they are treated separately from the pre-modern *Homo* grade. This is because the fossils assigned to these taxa show a mix of morphology, some of which is seen in pre-modern *Homo* and some in archaic hominins.

The taxon *Homo habilis* Leakey, Tobias and Napier 1964 was introduced for fossils recovered from Olduvai Gorge, Tanzania. The rest of the *H. habilis* hypodigm consists of other fossils found at Olduvai Gorge and of fossils from Ethiopia (Omo Shungura and Hadar) and Kenya (Koobi Fora and perhaps Chemeron), and researchers have claimed that there is also evidence of *H. habilis* in southern Africa at Sterkfontein, Swartkrans, and Drimolen. The *H. habilis* hypodigm consists of mostly cranial and dental evidence. The endocranial volume of *H. habilis* ranges from *c*.500 cm^3 to *c*.700 cm^3—but a reassessment of the endocranial volume of OH 7 suggests that it may be closer to 800 cm^3. All *H. habilis* crania are wider at the base than across the vault, but the face is broadest in its upper part. The only postcranial fossils that can be assigned to *H. habilis* with confidence are the postcranial bones associated with the type specimen, OH 7, and the associated skeleton, OH 62: isolated postcranial bones from Olduvai Gorge assigned to *H. habilis* (e.g., OH 10) could also belong to *P. boisei.* If OH 62 is representative of *H. habilis*, the skeletal evidence suggests that its limb proportions and locomotion and carpal bones were archaic hominin-like, and the curvature and well-developed muscle markings on the phalanges of OH 7 indicate that *H. habilis* was capable of powerful grasping. The size of the mandible and postcanine teeth suggests that the diet of *H. habilis* was as mechanically demanding as that of archaic hominins. The inference that *H. habilis* used spoken language is based on links between endocranial morphology and language comprehension and production that are no longer supported by comparative evidence.The temporal range of *H. habilis* would be c.2.4–1.6 Ma.

Some researchers suggest that the transitional hominin grade contains a second taxon, *Homo rudolfensis* (Alexeev, 1986) *sensu* Wood 1992, but not all researchers are convinced that the scale and nature of the variation within early *Homo* justifies the recognition of two taxa. The temporal range of *H. rudolfensis* would be *c*.2.0-1.8 Ma., and members of the proposed hypodigm include the lectotype, the KNM-ER 1470 cranium from Koobi Fora, and other fossils recovered from Koobi Fora (e.g., KNM-ER 1482, 1801, 1590, 3732, 60000, 62000, 62003). Compared to *H. habilis*, the absolute size of the brain case in *H. rudolfensis* is a little greater, and its face is widest in its mid-part, whereas the face of *H. habilis* is widest superiorly, and the dental arcades are differently shaped. Despite the mean absolute size of the *H. rudolfensis* brain (*c*.725 cm^3), when it is related to estimates of body mass based on orbit size, the brain is not substantially larger than that of the archaic hominins. At present, no postcranial remains can be reliably linked with *H. rudolfensis*. As with *H. habilis*, the size of the mandible and postcanine teeth suggests that its diet made similar mechanical demands as that of the archaic hominins.

7.4 Archaic Hominins

This grade includes all the unambiguously hominin taxa not included in *Homo* and *Paranthropus*. All archaic hominins, no matter what their absolute size is, have relatively larger chewing teeth and a more primitive postcranial skeleton than pre-modern *Homo*. They were all likely to be predominantly bipedal, but unlike pre-modern *Homo*, the anatomy of their upper limb suggests that they were still effective and regular climbers. What is known of the life history of archaic hominins suggests that it is more like that of the extant apes than modern humans.

The first taxon to be recognized in this grade was *Australopithecus africanus* Dart 1925. The type specimen, Taung 1, a juvenile skull with a partial natural endocast, was recovered in 1924 from the limeworks at Taung (formerly Taungs), now in South Africa. Most of the other fossil evidence for *Au. africanus* comes from two caves, Sterkfontein and Makapansgat, with other evidence coming from Gladysvale cave. Its temporal range is *c*.3–2.4 Ma. The cranium, mandible, and the dentition are well sampled, but the postcranial skeleton, and particularly the axial skeleton, is less well represented in the fossil record, and many of the fossils that do exist have been crushed and deformed by rocks falling on the bones before they were fully fossilized. The picture that has emerged from morphological and functional analyses suggests that although *Au. africanus* was capable of walking bipedally, it was probably more arboreally adapted (i.e., it was a facultative and not an obligate biped) than other archaic hominin taxa such as *Australopithecus afarensis*. It had relatively large chewing teeth, and apart from the reduced canines, the skull is relatively apelike. Its mean endocranial volume is *c*.460 cm^3. The Sterkfontein evidence suggests that males and females of *Au. africanus* differed substantially in body size, but probably not to the degree they did in *Au. afarensis*.

The taxon *Australopithecus afarensis* Johanson et al. 1978 is only known from East Africa, unless *Australopithecus bahrelghazali* from Chad proves to be a conspecific. The type specimen is an adult mandible, LH 4, recovered in 1974 from Laetoli, Tanzania, but the largest contribution to the *Au. afarensis* hypodigm comes from Hadar in Ethiopia and from other Ethiopian (Belohdelie, Brown Sands, Dikika, Fejej, Maka, White Sands, and Woranso-Mille) and Kenyan (Allia Bay, Koobi Fora, Tabarin and West Turkana) sites. The temporal range of *Au. afarensis* is *c*.3.8–3 Ma (*c*.4 -3 Ma if the presence of *Au. afarensis* is confirmed at Belohdelie and Fejej). The *Au. afarensis* hypodigm includes a well-preserved skull, other skulls, partial and fragmented crania, many lower jaws, sufficient limb bones to be able to estimate stature and body mass, and a specimen, A.L.-288, that preserves just less than half of the skeleton of a small adult female. Most body mass estimates range from *c*.30 to 45 kg, and the endocranial volume of *Au. afarensis* is *c*.400–550 cm^3. It has smaller incisors than those of extant chimps/bonobos, but its premolars and molars are relatively larger. Comparative evidence suggests that the forelimbs of A.L.-288 are substantially longer than those of a modern human of similar stature. The discovery at Laetoli of several trails of fossil footprints provided very graphic direct evidence that at least one contemporary hominin, presumably *Au. afarensis*, but possibly *Kenyanthropus platyops*, was capable of bipedal locomotion, but the Laetoli prints are less modern humanlike than *c.*1.5 Ma footprints from Koobi Fora presumed to be made by a pre-modern *Homo*. The upper limb of *Au. afarensis*, especially the hand and the shoulder girdle, retains morphology that most likely reflects a significant element of climbing. Although a recent study argues that sexual dimorphism in this taxon is relatively poorly developed, most researchers interpret it as showing substantial sexual dimorphism.

The taxon *Au. anamensis* (Leakey et al. 1995) is presently restricted to East Africa. The type specimen, KNM-KP 29281, was recovered in 1994 from Kanapoi, Kenya. Other sites contributing to the hypodigm are Allia Bay, also in Kenya, and the Middle Awash study area, Ethiopia. The temporal range of *Au. anamensis* is *c.*4.2–3.9 Ma. The fossil evidence consists of jaws, teeth, and postcranial elements from the upper and lower limbs. Most of the differences between *Au. anamensis* and *Au. afarensis* relate to details of the dentition. In some respects, the teeth of *Au. anamensis* are more primitive than those of *Au. afarensis* (e.g., the asymmetry of the premolar crowns and the relatively simple crowns of the deciduous first mandibular molars), but in others (e.g., the low cross-sectional profiles and bulging sides of the molar crowns), they show some similarities to *Paranthropus*. The upper limb remains are similar to those of *Au. afarensis*, and a tibia attributed to *Au. anamensis* has features associated with bipedality.

The taxon *Australopithecus bahrelghazali* Brunet et al. 1996 is most likely a regional variant of *Au. afarensis*, but the Chad discovery is significant because it substantially extends the geographical range of early hominins and reminds us that important events in human evolution (e.g., speciation, extinction) may have been taking place well away from the very small (relative to the size of the African continent) percentage of the land surface of Africa that is sampled by the existing early hominin sites.

The penultimate archaic hominin taxon to be recognized is *Kenyanthropus platyops* Leakey et al. 2001. The type specimen, KNM-WT 40000, a *c.*3.5–3.3 Ma relatively complete but distorted cranium, was found in 1999 at Lomekwi, West Turkana, Kenya. The main reasons Meave Leakey and her colleagues did not assign this material to *Au. afarensis* are its reduced subnasal prognathism, anteriorly situated zygomatic root, flat and vertically orientated malar region, relatively small but thick-enameled molars, and the unusually small M^1 compared to the size of the P^4 and M^3. Despite this unique combination of facial and dental morphology, some suggest that the new taxon is not justified because they claim that KNM-WT 40000 is a distorted *Au. afarensis* cranium, but this explanation is not consistent with the shape of the face and the small size of the postcanine teeth.

The most recent archaic hominin taxon to be recognized is *Australopithecus sediba* Berger et al. 2010 which was recovered from Malapa cave in the Blaauwbank Valley in southern Africa. The initial discoveries consisted of two associated skeletons: MH1, a juvenile, was made the holotype and MH2, an adult, the paratype. Although the lower limb of *Au. sediba* is like that of other archaic hominins, Berger et al. (2010) claim that aspects of its cranial (e.g., more globular neurocranium, gracile face), mandibular (e.g., more vertical symphyseal profile, a weak *mentum osseum*), dental (e.g., simple canine crown, small anterior and postcanine tooth crowns), and pelvic morphology (e.g., acetabulocristal buttress, expanded ilium and short ischium) are only shared with early and later *Homo* taxa. But the immaturity of one of the skeletons (MH1) plus many overall similarities to *Au. africanus* suggests that the Malapa hominins may sample *Au. africanus* at a later stage of its evolution than the existing samples from Sterkfontein and Makapansgat. The demonstration that the Malapa sample differs from the Sterkfontein and Makapansgat samples does not exclude the possibility that the three samples were drawn from the same fossil taxon, but the finding that its stable carbon isotope and phytolith signatures are predominantly C_3 does suggest that the diet of the Malapa hominins was not like that of *Au. africanus* (Henry et al. 2012).

7.5 Megadont and Hyper-Megadont Archaic Hominins

This grade includes hominin taxa conventionally included in the genus *Paranthropus*, plus *Australopithecus garhi*. As the term megadont suggests, the criterion for inclusion in this grade is large tooth size, specifically the size of the postcanine dentition. This increase is both in absolute and in relative (e.g., in relation to the anterior dentition and to estimates of body mass) terms. The genus *Paranthropus*, into which *Zinjanthropus* and *Paraustralopithecus* are subsumed, was reintroduced when cladistic analyses suggested that the first three species discussed in this section most likely formed a clade. The postcanine teeth of *Paranthropus robustus* are not much larger than those of *Au. africanus*, but those of the East African taxa in this grade are substantially larger; hence, they are referred to as hyper-megadont. The enamel of all of the taxa in this grade is thick; the enamel of the two *Paranthropus* taxa from East Africa is exceptionally thick.

The taxon *Paranthropus robustus* Broom 1938 was established to accommodate an associated skeleton, TM 1517, recovered in 1938 from the southern African site of Kromdraai B. The other sites that contribute to the *P. robustus* hypodigm, Swartkrans, Gondolin, Drimolen, and Cooper's caves, are all situated in the Blaauwbank Valley near Johannesburg, South Africa. The dentition is well represented in the hypodigm of *P. robustus*, but many of the cranial remains are crushed or distorted and the postcranial skeleton is not well represented. Research at Drimolen was only initiated in 1992, yet already more than 80 hominin specimens (many of them otherwise rare juvenile specimens) have been recovered and it promises to be a rich source of evidence about *P. robustus*. The temporal range of the taxon is *c*.2.0–1.5 Ma. The brain, face, and chewing teeth of *P. robustus* are on average larger than those of *Au. africanus*, yet the incisor teeth are smaller. The morphology of the pelvis and the hip joint is much like that of *Au. africanus*; *Paranthropus robustus* was most likely capable of bipedal walking, but it was probably not an obligate biped. It has been suggested that the thumb of *P. robustus* would have been capable of the type of grip necessary for the manufacture of simple stone tool, but this claim has not been accepted by all researchers.

In 1959, Louis Leakey suggested that a new genus and species, *Zinjanthropus boisei* Leakey, 1959, was needed to accommodate OH 5, a subadult cranium recovered in 1959 from Bed I, Olduvai Gorge, Tanzania. A year later, John Robinson suggested that *Z. boisei* be subsumed into the genus *Paranthropus* as *Paranthropus boisei*, and in 1967, Phillip Tobias suggested that it should be subsumed into *Australopithecus*, as *Australopithecus boisei*; in this review, we refer to it as *Paranthropus boisei* (Leakey, 1959) Robinson, 1960. Additional fossils from Olduvai Gorge were subsequently added to the hypodigm, plus fossil evidence from Peninj, Omo Shungura, Konso, Koobi Fora, Chesowanja, and West Turkana, all of which are in East Africa. The temporal range of the taxon is *c*.2.3–*c*.1.4 Ma. *Paranthropus boisei* has a comprehensive craniodental fossil record, comprising several skulls and well-preserved crania, many mandibles, and isolated teeth. There is evidence of both large- and small-bodied individuals, and the range of the size difference suggests a substantial degree of body size sexual dimorphism, despite the evidence for modest canine sexual dimorphism. *Paranthropus boisei* is the only hominin to combine a wide, flat face, massive premolars and molars, small anterior teeth, and a modest endocranial volume (*c*.480 cm^3). The body of the mandibles of *P. boisei* is larger and wider than that of any other hominin, and the tooth crowns grow at a faster rate than has been recorded for any other early hominin. For a long time, there was no postcranial evidence that could, with certainty, be attributed to *P. boisei*, but a fragmentary associated upper limb skeleton from Olduvai Gorge (Domínguez-Rodrigo et al. 2013) and a better preserved associated upper limb skeleton from Koobi Fora (Richmond et al. 2011) almost certainly belong to that taxon. Some of the postcranial fossils from Bed I at Olduvai Gorge currently attributed to *Homo habilis* may belong to *P. boisei*.

The taxon *Paranthropus aethiopicus* (Arambourg and Coppens, 1968) Chamberlain and Wood 1985 was introduced as *Paraustralopithecus aethiopicus* to accommodate Omo 18.18 (or 18.1967.18), an edentulous adult mandible recovered in 1967 from Omo Shungura in Ethiopia. The hypodigm is small, but it includes a well-preserved adult cranium from West Turkana (KNM-WT 17000) together with mandibles (e.g., KNM-WT 16005) and isolated teeth from Omo Shungura (some also assign the Omo 338y-6 cranium to this taxon). No published postcranial fossils have been assigned to *P. aethiopicus*, but a proximal tibia from Laetoli may belong to it. The temporal range of *P. aethiopicus* is *c*.2.5–2.3 Ma. *Paranthropus aethiopicus* is similar to *P. boisei* except that the face is more prognathic, the cranial base is less flexed, the anterior teeth are larger, and the postcanine teeth are not so large or morphologically specialized.

The most recent addition to the hyper-megadont archaic hominin hypodigm is *Australopithecus garhi* Asfaw et al. 1999. It was introduced to accommodate specimens recovered in 1997 from Aramis in the Middle Awash study area, Ethiopia. The hypodigm is presently restricted to fossils recovered from the Hata Member in the Middle Awash study area, Ethiopia. The type specimen, the *c*.2.5 Ma BOU-VP-12/130, combines a primitive cranium with large-crowned postcanine teeth, and particularly large premolars. However, unlike *Paranthropus boisei*, the incisors are small and the canines are large and the enamel apparently lacks the extreme thickness seen in the latter taxon. A partial skeleton with a long femur and forearm was found nearby, but it is not associated with the type cranium and it has not been formerly assigned to *Au. garhi*. If the type specimen of *P. aethiopicus* (Omo 18.18) belongs to the same hypodigm as the mandibles that appear to match the *Au. garhi* cranium, then *P. aethiopicus* would have priority.

8 Tempo and Mode

The study of macroevolution in hominin evolution is complicated by a number of factors unrelated to taxonomy. First, hominin remains are extremely rare in the fossil record, most fossils are frustratingly incomplete, and because of these factors, the same morphological regions are not well enough represented in the fossil records of some taxa to allow meaningful comparisons to be made among taxa. Second, the evolutionary sequence for the majority of hominin lineages is unknown. Most hominin taxa, particularly early hominins, have no obvious ancestors, and in most cases, ancestor-descendent sequences (fossil time series) cannot be reliably constructed—two possible exceptions are mentioned below. Third, error from many sources—measurement, reconstruction, sampling, and dating—can lead to spurious conclusions about evolutionary patterns. Finally, differences in scale can lead to differences in interpretation of tempo and mode of macroevolutionary change. Depending on the time separating recovered fossils, gradual steady-rate evolutionary changes can appear to be punctuated, and punctuated changes can appear to be gradual. None of these complications are limited to studying evolution within the hominin clade, but many of these issues are exacerbated in paleoanthropology given the intense scrutiny that our own lineage receives.

At one time, or another, every early hominin discussed above has been presented as "the" ancestor of later hominins, but in our opinion, only two pairs of taxa, *Au. anamensis* and *Au. afarensis* (Kimbel et al. 2006), and *P. aethiopicus* and *P. boisei* (Wood and Schroer 2013), are plausible examples of ancestor/descendant relationships (i.e., are examples of anagenesis). In the case of the former pair, *Au. anamensis* and *Au. afarensis* are most likely time-successive taxa within a single lineage with the Laetoli hypodigm of the former taxon intermediate between *Au. anamensis* and the Hadar hypodigm of *Au. afarensis*. This hypothesis has been given support by the discovery of fossil evidence from Woranso-Mille in Ethiopia that is both temporally and morphologically intermediate between *Au. anamensis* and *Au. afarensis* (Haile-Selassie et al. 2010). As for *P. aethiopicus* and *P. boisei*, although there are differences between the taxa (Suwa 1988; Wood et al. 1994), they are consistent with the older, less derived taxon being the ancestor of the younger more derived taxon. Indeed, some researchers have taken the view that the hypodigms of the two taxa are so similar they should both be included in *P. boisei* (Walker and Leakey 1988).

Another complicating factor is the history and current status of the punctuated equilibrium (Eldredge and Gould 1972; Gould and Eldredge 1977; reviewed in Sepkoski 2012) model that has provided the context for many past discussions of tempo and mode in the hominin clade. This model suggests that while stasis (i.e., lack of morphological change) is the predominant pattern of evolution within species, bursts of rapid change occur at speciation events. Further, these events occur in small isolated populations where new selection pressures and genetic drift can have a more dramatic effect than in larger populations (e.g., as in Mayr's allopatric model of speciation) (Mayr 1942). In fact, Gould and Eldredge's (1977) classic paper on punctuated equilibrium used hominin evolution as one of the primary supporting examples (see also Eldredge and Tattersall 1975). Based on the known fossil record at the time, the authors argued that stasis was predominant within hominin taxa, with each species looking much the same at the beginning and at the end of its fossil record. This interpretation stimulated a rebuttal by Cronin et al. (1981), who suggested that the apparent evidence of stasis within hominins was the result of poor samples and uncertainties about dating and taxonomy. In a recent review of the status of current evolutionary biological thought on the punctuated equilibrium hypothesis, Pennell et al. (2014) argued that current thinking on punctuated equilibrium conflates four key questions that should be addressed independently for any group such as the hominins. First, what is the relative importance of gradualistic versus punctuated evolution? Second, what is the role of speciation events versus within-lineage evolution in the group? Third, when change is due to speciation, are these changes adaptive or driven by neutral processes? Finally, how important is species selection in shaping patterns of diversity? Lieberman and Eldredge (2014) countered by suggesting that Pennell et al. (2014) did not correctly define punctuated equilibrium (or define macroevolution at all, though this is included in the glossary on their first page). Lieberman and Eldredge also suggested that Pennell et al.'s four questions were about patterns of evolution rather than processes, or conflated the two, or do not directly relate to Lieberman and Eldredge's understanding of the punctuated equilibrium hypothesis. While we do not wish to get into a debate on the merits of the punctuated equilibrium model, or the views of either set of authors, we view Pennell et al. (2014) as an attempt to show that regardless of the original intentions of its authors (see Sepkoski 2012 for the history of this idea), all four questions are ways researchers have attempted to test the validity of the punctuated equilibrium model of evolution.

The dearth of unambiguous evidence for ancestor-descendent lineages within the hominin clade means that the hominin fossil record currently does not permit useful insights into Pennell et al.'s (2014) questions two, three, and four. However, a number of studies have addressed whether within hominin taxa, the dominant signal is one of stasis or gradual change, and most have done so with reference to testing a punctuated model of evolution. In order to investigate the tempo of evolution within an early hominin taxon, (a) the taxon needs to be distinctive; (b) it must have a good, well-dated fossil record, and (c) the sample needs to span enough time (at least several hundred thousand years) for any temporal trends to manifest. One, or more, of these criteria rules out most early hominin taxa, especially those only found in the southern African cave sites, and the only early hominin taxa that comply with these criteria are *Au. afarensis* and *P. boisei*. In both cases, when researchers tracked morphology that is well represented in the fossil record [Lockwood et al. (2000) for *Au. afarensis* and Wood et al. (1994) for *P. boisei*], the predominant signal across approximately a million years was stasis. In the case of a third taxon, *H. erectus*, researchers have reached different conclusions about the likelihood of stasis (Tobias 1985; Wood et al. 1994; Ruff et al. 1997; Lockwood et al. 2000), with Rightmire (1981), suggesting that there was no consistent evidence of directional change in skull and tooth dimensions, whereas when Wolpoff (1984) analyzed a more narrowly defined sample of *H. erectus's* mandibular, cranial, and dental features, he suggested that there was evidence for evolutionary change within that taxon.

Only a few traits are known from a wide enough range of fossil hominins to allow for quantitative, rather than qualitative, comparisons to be made across long periods of time. One of them, endocranial volume, has been regularly pointed out as a classic example of a macroevolutionary trend (Haldane 1949; Tobias 1971; Jerison 1973). Though current evidence on hominin brain size evolution points to an general increase over time (Holloway et al. 2004) and numerous hypotheses have been suggested as to the factors that caused this increase (e.g., Clutton-Brock and Harvey 1980; Martin 1996; Dunbar 1998), there is debate as to the tempo of brain size increase during hominin evolution. Some have argued for a gradual increase in hominin brain size over time (Lestrel and Read 1973; Lestrel 1976; but see Godfrey and Jacobs 1981; Lee and Wolpoff 2003) and others for an increasing rate of change (Tobias 1971; Holloway et al. 2004), and yet another sees evidence of stasis, followed by more rapid change (Ruff et al. 1997). A consistent theme of these interpretations is that there was a grade shift in endocranial volume *c.*1.8 Ma. One problem with past studies is that they present endocranial volume values as if they had neither dating nor measurement error. When Du et al. (in prep) analyzed a comprehensive data set on hominin brain size that took account of dating and measurement error, and which spanned the period from 3.5 to 0.5 Ma, they found that a gradual model of brain size received the strongest support.

9 Evolutionary Forces and Macroevolutionary Patterns

During the 1980s, there was a sizable backlash against portions of the punctuated equilibrium model of evolution (e.g., Lande 1980; Charlesworth et al. 1982). One of the chief reasons was because some versions of the model suggested that the causes of macroevolution are distinct from those leading to microevolution, thus suggesting that one of the central tenets of the modern synthesis is incorrect (e.g., Gould 1980). In a series of papers, population geneticists dissected various parts of the model, arguing that punctuated evolution was merely one end of a spectrum of evolutionary possibilities and likely did not play a major role in producing the patterns seen in the fossil record (Lande 1980; Charlesworth et al. 1982; Barton and Charlesworth 1984). While macroevolutionary patterns of evolution differ from microevolutionary patterns for some traits such as body mass (Uyeda et al. 2011), it is widely accepted that the idea that separate evolutionary processes are required to produce most macroevolutionary change is not supported by the evidence (reviewed in Charlesworth et al. 1982). The geological timescale is long enough that any variation in a trait, whether the result of many genes or a few, that increases fitness is likely to have been selected for and macromutations are extremely likely to be deleterious due to either their main effect or pleiotropic effects on other traits (Fisher 1930; Lande 1980; Charlesworth et al. 1982). Likewise, the recent suggestions that large morphological changes in hominin evolution were the result of a few simple changes in growth gradients or developmental fields (Lovejoy et al. 1999, 2003; Lovejoy and McCollum 2010) do not mean that one or just a few genes are at play (Lande 1980). Experimental attempts to identify the genes that control morphological changes suggest that variation in the vast majority of traits is controlled by multiple loci (i.e., it is polygenic) (Lynch and Walsh 1998).

Evolutionary forces—natural selection, random genetic drift, gene flow, and mutation—acting on heritable variation within populations, account for the majority of evolutionary change, but be that as it may, in paleoanthropology, it is nearly always assumed that macroevolutionary changes seen across the fossil record are the result of natural selection. While this is undoubtedly true for changes such as substantial increases in brain size between taxa, changes in morphology may also be due to the other three evolutionary forces, in addition to selection on other traits. Quantitative genetics (QG), which provides a mechanistic understanding of these evolutionary processes, began as an attempt to understand the inheritance of quantitative traits (height, weight, measurement of skeletal traits, etc.) (Provine 1971; Lynch and Walsh 1998), but its scope has expanded to include goals that range from understanding the nature of quantitative trait variation, the consequences of inbreeding, to developing predictive models for evolutionary change. Evolutionary quantitative genetics (EQG) takes concepts that were originally intended to look at changes in quantitative traits between generations, usually in association with livestock and crop improvement, and applies them to look at changes over evolutionary time (Roff 1997). A number of recent studies (Ackermann and Cheverud 2004; Rolian et al. 2010; Grabowski et al. 2011) have applied the theories and ideas of evolutionary quantitative genetics to the hominin fossil record in an attempt to provide insight into evolutionary forces that resulted in the patterns observed. In the next section, we focus on the first two forces, selection and drift, as these are likely the most relevant when applying evolutionary quantitative genetic methods to paleoanthropology. The results from these studies lead to the conclusion that morphological changes were likely due to a complex relationship between natural selection and random genetic drift.

9.1 Natural Selection

Most fossil analyses atomize the phenotype into a set of traits, observe how much a given trait differs between species, and then ascribe selection for a particular function as the cause of that change. But we know that organisms are integrated units, with many traits sharing some portion of their genetic background due to pleiotropy (i.e., the traits are integrated), and thus, natural selection on one trait leads to correlated responses in others (Olson and Miller 1958; Lande 1979; Gould and Lewontin 1979). This means that any change in morphology between fossil taxa may not be the result of direct selection for a particular trait or function, but is instead the result of a correlated response to selection on other integrated traits. Though this point is now generally accepted across much of biology, and some acceptance has occurred (Lovejoy et al. 1999, 2002, 2003; Strait 2001, Ackermann and Cheverud 2004), within paleoanthropology, atomization still reigns. Based on the work of Lande (1979), Lande and Arnold (1983), studies have reconstructed selection pressures that led to difference in morphology between species (Cheverud 1996, Marroig and Cheverud 2004, Rolian et al. 2010), including between fossil hominins (Ackermann and Cheverud 2004). The results of these studies suggest that at least some of these changes that were thought to be the result of natural selection driven by functional considerations were actually a correlated response to selection on other traits. For example, Rolian et al. (2010) found that the reduction in finger length seen during hominin evolution was likely a correlated response to selection to reduce toe length. In other words, the relatively short fingers that enable much of the manual dexterity that sets modern humans apart from other primates may not have been the result of selection on finger length. Instead, selection for shorter toes to permit habitual bipedalism led to changes in both homologous structures. Grabowski and Roseman (in press) tested the hypothesis that strong directional selection on many individual aspects of morphology was responsible for the large differences observed across a sample of fossil hominin hips spanning the Plio-Pleistocene. Their findings showed a complex and changing pattern of natural selection drove hominin hip evolution, and many, but not all, traits hypothesized to play functional roles in bipedalism evolved as a result of natural selection.

9.2 Random Genetic Drift

Tests for the roles of selection versus drift in producing the morphological diversity seen between worldwide modern human populations have become relatively common. For example, Betti et al. (2010) found that drift was much more important in shaping cranial diversity than selection due to climatic differences, with the exception of populations from extremely cold regions. Similarly, Betti et al. (2013) found that a combination of selection and drift explained variations in pelvic dimensions in modern human populations. With regard to fossil hominins, Ackermann and Cheverud (2004) tested whether random evolutionary processes alone could account for the morphological diversity seen in early hominin fossil crania. They also tested adaptive hypotheses about hominin facial diversity by estimating past selection pressures required to produce observed morphological change. Their results showed that though early hominin facial diversity exceeds levels expected if it had originated through random processes, diversity seen in early *Homo* did not. Weaver et al. (2007) tested the null hypothesis that morphological differences in the crania of Neanderthals and anatomically modern humans were the result of genetic drift. The researchers found that the null model could not be rejected, and thus, morphological differences between Neanderthals and modern humans could have been the result of genetic drift rather than selection.

10 Phylogenetic Relationships

For much of its history, hominin systematics was predicated on the assumption that there is a direct relationship between morphological similarity and genetic relatedness; the more hard tissue morphology two hominin taxa share, the closer their relationship. For extant taxa, this hypothesis can be tested against relationships based on molecular evidence. Such data, either on their own, or in combination with morphological evidence, have been used in efforts to try to resolve relationships among taxa, including those within large clades of medium- to large-sized mammals. But even at this "macro"-scale, it is apparent that a substantially similar skeletal phenotype does not always mean a shared recent evolutionary history. Long ago, Lankester (1870) suggested that the term homoplasy be used for morphology that is seen in what we now call sister taxa, but not in their most recent common ancestor. Such morphology gives the impression that the two taxa are more closely related than they really are. Because homoplasy can be mistaken for shared derived similarity (or synapomorphy), it complicates attempts to reconstruct phylogenetic relationships.

The confounding effects of homoplasy could be coped with if the "noise" generated by the latter was trivial compared to the strength of the phylogenetic "signal." But in some attempts to infer relationships among extant higher primates using skeletal data (in the form of either traditional non-metrical characters or characters generated from metrical data), the ratio of "noise" to "signal" was in the order of 1:2. The results of these analyses were not only frustratingly inconclusive, but when they were compared with the pattern of relationships generated using molecular data, some were misleading (Collard and Wood 2000). Other researchers suggested that this dismal performance was due to the exclusion of character-state data from fossil taxa (Strait and Grine 2004), but this argument is moot because soft tissue characters (for which there are no fossil data) are capable of recovering a pattern of relationships among extant higher primates that is consistent with the molecular evidence (Gibbs et al. 2000, 2002; Diogo and Wood 2011).

Therefore, something about the nature of hard tissue evidence may be problematic. Thankfully, hard tissue evidence *can* produce results congruent with the relationships generated from molecular data (e.g., Lockwood et al. 2004), as long as the anatomical regions targeted have a high enough signal-to-noise ratio. The research reviewed above suggests that the problem is with either, or both, the nature of the data or the scale of the enquiry, and not with cladistic methodology. However, that the type of data the fossil record provides (i.e., mostly craniodental hard tissue morphology) seems to be particularly prone to homoplasy when used at this relatively fine taxonomic level is not the best context for paleoanthropologists attempting to reconstruct phylogenies based on fossilized hard tissue remains.

There is also comparative evidence that homoplasy needs to be taken into account when generating hypotheses about the relationships among the taxa in the higher primate part of the tree of life. Although there is overwhelming molecular and morphological evidence for a (((*Pan*, *Homo*) *Gorilla*) *Pongo*) pattern of relationships among the extant hominids, selected morphological character states can be used to infer a (((*Pongo*, *Homo*) *Pan*) *Gorilla*) pattern of relationships, but these are almost certainly homoplasies. Similarly, homoplasy complicates attempts to resolve the relationships of fossil apes such as *Sivapithecus* (Young 2003), *Morotopithecus* (Nakatsukasa 2008), and *Chororapithecus* (Suwa et al. 2007). Moreover, studies of other mammalian clades evolving in Africa during the same time period as hominins and in similar paleoenvironments point to substantial and recurrent homoplasy [e.g., bovids (Gatesy et al. 1997), equids (Bernor et al. 2010), elephantids (Todd 2010), carnivores (Van Valkenburgh 2007), and Old World monkeys (Jablonski and Leakey 2008)]. There is no reason to assume that higher primate lineages were immune from the tendency to adapt morphologically in similar, and therefore phylogenetically confounding, ways to shared ecological challenges.

The important point is that shared similarities can only take one so far in determining phylogenetic relationships because homoplasy, as well as uncertainties in determining the polarity of character transformation, has the potential to generate substantial noise that serves to confound attempts to generate reliable hypotheses about phylogenetic relationships. These considerations have clear implications for generating hypotheses about phylogenetic relationships within the hominin clade and especially for the relationships of *Sahelanthropus*, *Orrorin*, and *Ardipithecus*. Even if these taxa share *some* derived features with later Pliocene hominins, it would be rash to simply presume that those features are immune from homoplasy, especially when other aspects of their respective phenotypes are consistent with a more distant relationship with the hominin clade.

11 Conclusions

It is difficult to believe, but the second, 1964, edition of Le Gros Clark's "*Fossil evidence for Human Evolution*" was the last time a review looked at the whole of what we now call the hominin fossil record (Clark 1964). Much has happened in the ensuing half century. We now have the advantage of new dating methods, new fossil evidence, new methods for capturing (e.g., 3D landmark data) and analyzing morphology (e.g., geometric morphometrics), and new methods for extracting data from fossils (e.g., imaging, molecular evidence). We also have the advantage of quantitative methods for alpha taxonomy, methods for generating hypotheses about relationships, and the emergence of functional morphological analysis. However, we need to test many of the assumptions (e.g., selection drives observed change, morphology is homologous, climate drives evolution, and all taxa are ancestors until proved otherwise) that underly attempts to improve our understanding of macroevolution within the hominin clade.

Much progress has been made, but many of the questions raised by Le Gros Clark are still with us. More fossil evidence is crucial, but an order of magnitude increase in the fossil evidence in the absence of equivalent progress in how we analyze the fossil record would not constitute an advance. Real progress will come when evidence and analysis move forward in tandem.

Acknowledgments BW was supported by the GW Provost and the GW Signature Program, and MG by the GW Signature Program.

References

Ackermann RR, Cheverud JM (2004) Detecting genetic drift versus selection in human evolution. Proc Natl Acad Sci USA 101:17946–17951. doi:10.1073/pnas.0405919102

Arambourg C, Coppens Y (1968) Découverte d'un australopithécien nouveau dans les gisements de l'Omo (Éthiopie). S Afr J Sci 64:58–59

Arnold C, Matthews LJ, Nunn CL (2010) The 10kTrees website: a new online resource for primate phylogeny. Evol Anthropol 19:114–118. doi:10.1002/evan.20251

Barton NH, Charlesworth B (1984) Genetic revolutions, founder effects, and speciation. Annu Rev Ecol Syst 133–164

Bernor RL, Armour-Chelu M (2010) Equidae. In: Werdelin L, Sanders B (eds) Cenozoic mammals of Africa. University of California Press, pp 685–721

Betti L, Balloux F, Hanihara T, Manica A (2010) The relative role of drift and selection in shaping the human skull. Am J Phys Anthropol 141:76–82. doi:10.1002/ajpa.21115

Betti L, von Cramon-Taubadel N, Manica A, Lycett SJ (2013) Global geometric morphometric analyses of the human pelvis reveal substantial neutral population history effects, even across sexes. PLoS ONE 8:e55909. doi:10.1371/journal.pone.0055909.t005

Bradley BJ (2008) Reconstructing phylogenies and phenotypes: a molecular view of human evolution. J Anat 212:337–353. doi:10.1111/j.1469-7580.2007.00840.x

Brandon-Jones, D, Eudey AA., Geissmann T, Groves CP, Melnick DJ, Morales JC, Shekelle M,Stewart CB (2004). Asian primate classification. International Journal of Primatology 25:97–164

Brunet M, Guy F, Pilbeam D et al (2002) A new hominid from the Upper Miocene of Chad, Central Africa. Nature 418:145–151. doi:10.1038/nature00879

Caccone A, Powell JR (1989) DNA divergence among hominoids. Evolution 43:925–942

Charlesworth B, Lande R, Slatkin M (1982) A neo-Darwinian commentary on macroevolution. Evolution 36:474–498

Cheverud JM (1996) Quantitative genetic analysis of cranial morphology in the cotton-top (*Saguinus oedipus*) and saddle-back (*S. fuscicollis*) tamarins. J Evol Biol 9:5–42

Le Gros Clark (1964) The fossil evidence for human evolution: an introduction to the study of paleoanthropology. University of Chicago Press, Chicago

Clutton-Brock TH, Harvey PH (1980) Primates, brains and ecology. J Zool 190:309–323
Collard M, Wood B (2000) How reliable are human phylogenetic hypotheses? Proc Natl Acad Sci USA 97:5003–5006
Cronin JE, Boaz NT, Stringer CB, Rak Y (1981) Tempo and mode in hominid evolution. Nature 292:113–122
Diogo R, Wood B (2011) Soft-tissue anatomy of the primates: phylogenetic analyses based on the muscles of the head, neck, pectoral region and upper limb, with notes on the evolution of these muscles. J Anat 219:273–359. doi:10.1111/j.1469-7580.2011.01403.x
Dobzhansky TG, Dobzhansky T (1937) Genetics and the Origin of Species. Columbia University Press, New York City
Domínguez-Rodrigo M, Pickering TR, Baquedano E et al (2013) First partial skeleton of a 1.34-million-year-old *Paranthropus boisei* from Bed II, Olduvai Gorge, Tanzania. PLoS ONE 8:e80347. doi:10.1371/journal.pone.0080347.s001
Dunbar RIM (1998) The social brain hypothesis. Evol Anthropol 6:178–190
Eldredge N (1989) Macroevolutionary dynamics: species, niches, and adaptive peaks. McGraw-Hill, New York City
Eldredge N (1993) What, if anything, is a species? In: Kimbel W, Martin L (eds) Advances in primatology. Springer, US, pp 3–20
Eldredge N, Gould SJ (1972) Punctuated equilibria: an alternative to phyletic gradualism. Models Paleobiol 82:115
Eldredge N, Tattersall I (1975) Evolutionary models, phylogenetic reconstruction, and another look at hominid phylogeny. Contrib Primatol 5:218–242
Fisher RA (1930) The genetical theory of natural selection, reprint, illustrated, annotated. Oxford University Press, Oxford
Gatesy J, Amato G, Vrba E et al (1997) A cladistic analysis of mitochondrial ribosomal DNA from the Bovidae. Mol Phylogenet Evol 7:303–319
Gibbs S, Collard M, Wood B (2002) Soft-tissue anatomy of the extant hominoids: a review and phylogenetic analysis. J Anat 200:3–49
Gibbs S, Collard M, Wood B (2000) Soft-tissue characters in higher primate phylogenetics. Proc Natl Acad Sci USA 97:11130–11132. doi:10.1073/pnas.190252697
Godfrey L, Jacobs KH (1981) Gradual, autocatalytic and punctuational models of hominid brain evolution: a cautionary tale. J Hum Evol 10:255–272
Goldschmidt R (1940) The material basis of evolution. Yale University Press, New Haven
Goodman M (1963) Classification and human evolution. Wenner-Gren Foundation, New York
Gould SJ (1980) Is a new and general theory of evolution emerging? Paleobiology 6:119–130
Gould SJ, Eldredge N (1977) Punctuated equilibria: the tempo and mode of evolution reconsidered. Paleobiology 115–151
Gould SJ, Lewontin RC (1979) The spandrels of San Marco and the Panglossian paradigm: a critique of the adaptationist programme. Proc R Soc Lond B Biol Sci 205:581–598
Grabowski MW, Polk JD, Roseman CC (2011) Divergent patterns of integration and reduced constraint in the human hip and the origins of bipedalism. Evolution 65:1336–1356. doi:10.1111/j.1558-5646.2011.01226.x
Grünbaum A (1902) Note on the blood relationship of man and the anthropoid apes. Lancet 159:143
Groves C (2001) Primate Taxonomy. Smithsonian Books
Groves CP (2005) Geographic variation within eastern chimpanzees (Pan troglodytes cf. schweinfurthiiGiglioli, 1872). Australasian Primatology. 17:19–46
Guy F, Lieberman DE, Pilbeam D et al (2005) Morphological affinities of the *Sahelanthropus tchadensis* (Late Miocene hominid from Chad) cranium. Proc Natl Acad Sci USA 102:18836–18841. doi:10.1073/pnas.0509564102
Haile-Selassie Y (2001) Late Miocene hominids from the Middle Awash, Ethiopia. Nature 412:178–181. doi:10.1038/35084063
Haile-Selassie Y (2004) Late Miocene teeth from Middle Awash, Ethiopia, and early hominid dental evolution. Science 303:1503–1505. doi:10.1126/science.1092978

Haldane JBS (1949) Suggestions as to quantitative measurement of rates of evolution. Evolution 3:51–56

Hallam A (1989) What can the fossil record tell us about macroevolution. In: Hecht MK (Ed) Evolutionary Biology at the Crossroads. Queens College Press, Flushing, NY

Henry, A.G., Ungar, P.S., Passey, B.H., Sponheimer, M., Rossouw, L., Bamford, M., Sandberg, P., de Ruiter,D.J., Berger, L., 2012. The diet of Australopithecus sediba. Nature

Holloway RL, Broadfield DC, Yuan MS (2004) Endocasts of early hominids. In: The human fossil record: brain endocasts-the paleoneurological evidence, vol 3, pp 39–107

Huxley TH (1863) Evidence as to mans place in nature by Thomas Henry Huxley. Williams and Norgate, London

Jablonski NG, Leakey MG (2008) Koobi Fora research project vol. 6. The fossil monkeys. California Academy of Sciences, San Francisco

Jerison HJ (1973) Evolution of the brain and intelligence. Academic Press, New York

Kimbel WH, Lockwood C, Ward C et al (2006) Was *Australopithecus anamensis* ancestral to *A. afarensis*? A case of anagenesis in the hominin fossil record. J Hum Evol 51:134–152. doi:10.1016/j.jhevol.2006.02.003

King M-C, Wilson AC (1975) Evolution at two levels in humans and chimpanzees. Science 188:107–116

Lande R (1979) Quantitative genetic analysis of multivariate evolution, applied to brain: body size allometry. Evolution 33:402–416

Lande R (1980) Microevolution in relation to macroevolution. Paleobiology 6:233–238

Lande R, Arnold SJ (1983) The measurement of selection on correlated characters. Evolution 37:1210–1226

Langergraber KE, Prufer K, Rowney C et al (2012) Generation times in wild chimpanzees and gorillas suggest earlier divergence times in great ape and human evolution. Proc Natl Acad Sci USA 109:15716–15721. doi:10.1073/pnas.1211740109

Lankester ER (1870) II.—On the use of the term homology in modern zoology, and the distinction between homogenetic and homoplastic agreements. The Annals and Magazine of Natural History 6:34–43

Lee SH, Wolpoff MH (2003) The pattern of evolution in Pleistocene human brain size. Paleobiology 29:186–196

Lestrel PE (1976) Hominid brain size versus time: revised regression estimates. J Hum Evol 5:207–212

Lestrel PE, Read DW (1973) Hominid cranial capacity versus time: a regression approach. J Hum Evol 2:405–411

Levinton JS (2001) Genetics, paleontology, and macroevolution. Cambridge University Press, Cambridge

Lieberman BS, Eldredge N (2014) What is punctuated equilibrium? What is macroevolution? A response to Pennell et al. Trends Ecol Evol 1–2. doi:10.1016/j.tree.2014.02.005

Locke DP, Hillier LW, Warren WC et al (2011) Comparative and demographic analysis of orangutan genomes. Nature 469:529–533. doi:10.1038/nature09687

Lockwood CA, Kimbel WH, Johanson DC (2000) Temporal trends and metric variation in the mandibles and dentition of *Australopithecus afarensis*. J Hum Evol 39:23–55. doi:10.1006/jhev.2000.0401

Lockwood CA, Kimbel WH, Lynch JM (2004) Morphometrics and hominoid phylogeny: support for a chimpanzee–human clade and differentiation among great ape subspecies. Proc Natl Acad Sci U S A 101:4356–4360

Lovejoy CO, Cohn MJ, White TD (1999) Morphological analysis of the mammalian postcranium: a developmental perspective. Proc Natl Acad Sci U S A 96:13247–13252

Lovejoy CO, McCollum MA (2010) Spinopelvic pathways to bipedality: why no hominids ever relied on a bent-hip-bent-knee gait. Philos Trans R Soc Lond B Biol Sci 365:3289–3299. doi:10.1098/rstb.2010.0112

Lovejoy CO, McCollum MA, Reno PL, Rosenman BA (2003) Developmental biology and human evolution. Annu Rev Anthropol 32:85–109. doi:10.1146/annurev.anthro.32.061002.093223

Lovejoy CO, Meindl RS, et al. (2002) The Maka femur and its bearing on the antiquity of human walking: applying contemporary concepts of morphogenesis to the human fossil record. Am J Phys Anthropol 119:97–133. doi:10.1002/ajpa.10111
Lynch M, Walsh B (1998) Genetics and analysis of quantitative traits, illustrated. Sinauer, Sunderland
Manaster BJ (1979) Locomotor adaptations within the *Cercopithecus* genus: a multivariate approach. Am J Phys Anthropol 50:169–182. doi:10.1002/ajpa.1330500205
Marroig G, Cheverud JM (2004) Did natural selection or genetic drift produce the cranial diversification of neotropical monkeys? Am Nat 163:417–428. doi:10.1086/381693
Martin RD (1996) Scaling of the mammalian brain: the maternal energy hypothesis. News Physiol Sci 11:149–156
Mayr E (1942) Systematics and the origin of species, from the viewpoint of a zoologist. Harvard University Press, Harvard
Mounier A, Marchal F, Condemi S (2009) Is *Homo heidelbergensis* a distinct species? New insight on the Mauer mandible. J Hum Evol 56:219–246. doi:10.1016/j.jhevol.2008.12.006
Nakatsukasa M (2008) Comparative study of Moroto vertebral specimens. J Hum Evol 55:581–588. doi:10.1016/j.jhevol.2008.04.009
Nixon KC, Wheeler QD (1990) An amplification of the phylogenetic species concept. Cladistics 6:211–223
Nuttall GHF, Graham-Smith GS, Pigg-Strangeways TSP (1904) Blood immunity and blood relationship. Cambridge University press, Boston
Olson EC, Miller RL (1958) Morphological integration. University of Chicago Press, Chicago
Paterson HE (1985) The recognition concept of species. In: Vrba ES (ed) Species and speciation. Transvaal Museum Monograph, pp 21–29
Pennell MW, Harmon LJ, Uyeda JC (2014) Is there room for punctuated equilibrium in macroevolution? Trends Ecol Evol 29:23–32. doi:10.1016/j.tree.2013.07.004
Perelman P, Johnson WE, Roos C et al (2011) A molecular phylogeny of living primates. PLoS Genet 7:e1001342. doi:10.1371/journal.pgen.1001342.s025
Prado-Martinez J, Sudmant PH, Kidd JM et al (2014) Great ape genetic diversity and population history. Nature 499:471–475
Provine WB (1971) The origins of theoretical population genetics, 3rd ed. University of Chicago Press, Chicago
Prufer K, Munch K, Hellmann I et al (2012) The bonobo genome compared with the chimpanzee and human genomes. Nature. doi:10.1038/nature11128
Richmond BG, Green DJ, Braun DR, et al (2011) New fossils from Ileret, Kenya, and the evolution of hominin hand function. In: Proceedings of American Association of Physical Anthropology Meetings. Wiley-Blackwell Commerce Place, 350 Main St, Malden 02148, MA USA, pp 253–253
Rightmire GP (1981) Patterns in the evolution of *Homo erectus*. Paleobiology 7:241–246
Roff DA (1997) Evolutionary quantitative genetics. Chapman & Hall, New York
Rolian C, Lieberman DE, Hallgrimsson B (2010) The coevolution of human hands and feet. Evolution 64:1558–1568. doi:10.1111/j.1558-5646.2010.00944.x
Ruff CB, Trinkaus E, Holliday TW (1997) Body mass and encephalization in Pleistocene *Homo*. Nature 387:173–176. doi:10.1038/387173a0
Sarich VM, Wilson AC (1967) Rates of albumin evolution in primates. Proc Natl Acad Sci USA 58:142–148
Scally A, Dutheil JY, Hillier LW et al (2012) Insights into hominid evolution from the gorilla genome sequence. Nature 483:169–175. doi:10.1038/nature10842
Senut B, Pickford M, Gommery D et al (2001) First hominid from the Miocene (Lukeino Formation, Kenya): premier hominidé du Mioccne (formation de Lukeino, Kenya). Comptes Rendus de l'Académie des Sciences-Series IIA-Earth and Planetary Science 332:137–144
Sepkoski D (2012) Rereading the fossil record. University of Chicago Press, Chicago
Simpson GG (1944) Tempo and mode in evolution. Columbia University Press, New York
Simpson GG (1953) The baldwin effect. Evolution 7:110–117

Simpson GG (1961) The major features of evolution. Columbia University Press, New-York
Smith AB (2009) Systematics and the fossil record: documenting evolutionary patterns. Wiley, New York
Singleton I, Wich S, Husson S, Stephens S, Utami-Atmoko SS, Leighton M, Rosen N, Traylor-Holzer K, Lacy R, Byers O (2004). Orangutan population and habitat viability assessment: final report.IUCN/SSC Conservation Breeding Specialist Group, Apple Valley, MN.
Stevens NJ, Seiffert ER, O'Connor PM et al (2014) Palaeontological evidence for an Oligocene divergence between Old World monkeys and apes. Nature 497:611–614. doi:10.1038/nature12161
Strait DS (2001) Integration, phylogeny, and the hominid cranial base. Am J Phys Anthropol 114:273–297. doi:10.1002/ajpa.1041
Strait DS, Grine FE (2004) Inferring hominoid and early hominid phylogeny using craniodental characters: the role of fossil taxa. J Hum Evol 47:399–452. doi:10.1016/j.jhevol.2004.08.008
Suwa G (1988) Evolution of the "robust" australopithecines in the Omo succession: evidence from mandibular premolar morphology. In: Grine FE (ed) Evolutionary history of the robust australopithecines, pp 199–222
Suwa G, Kono RT, Katoh S et al (2007) A new species of great ape from the late Miocene epoch in Ethiopia. Nature 448:921–924. doi:10.1038/nature06113
TCSAC (2005) Initial sequence of the chimpanzee genome and comparison with the human genome. Nature 437:69–87. doi:10.1038/nature04072
Tobias PV (1971) The brain in hominid evolution. Columbia University Press, New York
Tobias PV (1985) Punctuational and phyletic evolution in the hominids. In: Vrba ES (ed) Species and speciation. Transvaal Museum Pretoria, pp 131–141
Todd NE (2010) New phylogenetic analysis of the family Elephantidae based on cranial-dental morphology. Anat Rec 293:74–90. doi:10.1002/ar.21010
Uyeda JC, Hansen TF, Arnold SJ, Pienaar J (2011) The million-year wait for macroevolutionary bursts. Proc Natl Acad Sci USA. doi:10.1073/pnas.1014503108
Van Valkenburgh B (2007) Déjà vu: the evolution of feeding morphologies in the Carnivora. Am Zool 47:147–163
Walker A, Leakey RE (1988) The evolution of *Australopithecus boisei*. In: Grine FE (ed) Evolutionary history of the "robust" australopithecines. Transaction Publishers, New Jersey, pp 247–258
Weaver TD, Roseman CC, Stringer CB (2007) Were neandertal and modern human cranial differences produced by natural selection or genetic drift? J Hum Evol 53:135–145. doi:10.1016/j.jhevol.2007.03.001
White TD (2010) Human origins and evolution: Cold Spring Harbor, deja vu. Cold Spring Harb Symp Quant Biol 74:335–344. doi:10.1101/sqb.2009.74.016
White TD, Asfaw B, Beyene Y et al (2009) *Ardipithecus ramidus* and the paleobiology of early hominids. Science 326:64. doi:10.1126/science.1175802
White TD, Suwa G, Asfaw B (1994) *Australopithecus ramidus*, a new species of early hominid from Aramis, Ethiopia. Nature 371:306–312. doi:10.1038/371306a0
Wolpoff MH (1994) Yes it is, no it isn't: a reply to van Vark and Bilsborough. Am J Phys Anthropol 95:92–93
Wolpoff MH (1984) Evolution in *Homo erectus*: the question of stasis. Paleobiology 389–406
Wood B (2010) Reconstructing human evolution: achievements, challenges, and opportunities. Proc Natl Acad Sci USA 107(Suppl 2):8902–8909. doi:10.1073/pnas.1001649107
Wood B, Schroer K (2013) *Paranthropus*. In: A companion to paleoanthropology. Blackwell Publishing Ltd, New York, pp 457–478
Wood B, Wood C, Konigsberg L (1994) *Paranthropus boisei*: an example of evolutionary stasis? Am J Phys Anthropol 95:117–136. doi:10.1002/ajpa.1330950202
Young NM (2003) A reassessment of living hominoid postcranial variability: implications for ape evolution. J Hum Evol 45:441–464. doi:10.1016/j.jhevol.2003.09.001
Zuckerkandl E, Jones RT, Pauling L (1960) A comparison of animal hemoglobins by tryptic peptide pattern analysis. Proc Natl Acad Sci USA 46:1349

Facing the Big Sixth: From Prioritizing Species to Conserving Biodiversity

Elena Casetta and Jorge Marques da Silva

Abstract This chapter deals with biodiversity conservation efforts aimed at facing "the Big Sixth"—the new mass extinction we could be entering, the first one involving our own species as a primary cause. Following Michael Soulé (1985), we characterize conservation biology as a form of biodiversity surgery and illustrate the main difficulties that this very special kind of surgery has to meet. First of all, we briefly discuss the difficulties that arise in *declaring* the extinction of a species. We then focus on three challenges that *facing* extinction requires to take up: How to prioritize species; which conservation targets to focus on; and how to ethically justify species conservation. As we show, matters are complicated by the fact that although species continue to play a central role in policies aimed at preserving the variety of life, biodiversity is not just a matter of species preservation. Finally, based on the analysis of such challenges, we compare two legislations currently in force, namely the Endangered Species Act (the primary legislation providing federal legal protection to endangered species in the United States) and the Habitats Directive (which, together with the Birds Directive, forms the cornerstone of the European Union's nature conservation policy). We conclude that neither legislation is fully adequate for biodiversity conservation: The Endangered Species Act was not designed to preserve habitats, which constitute the higher level of biodiversity, and the Habitats Directive misses the objective

We are grateful to Michael Baumtrog, José Lino Costa, Luca Tambolo, and two anonymous referees for their precious comments and remarks on a previous version of this chapter.

E. Casetta (✉)
CFCUL, Centre for Philosophy of Science, University of Lisbon, Lisbon, Portugal
e-mail: elenattesac@gmail.com

E. Casetta
LabOnt, Laboratory for Ontology, University of Turin, Turin, Italy

J. Marques da Silva
Department of Plant Biology and Center for Biodiversity, Functional and Integrative Genomics, Faculty of Sciences, University of Lisbon, Lisbon, Portugal

E. Serrelli and N. Gontier (eds.), *Macroevolution*,
Interdisciplinary Evolution Research 2, DOI 10.1007/978-3-319-15045-1_12

of preserving genes, which constitute the lower level of biodiversity and thereby the evolutionary potential of populations. We therefore suggest that reforms are needed both in North American and European biodiversity conservation policies.

Keywords Conservation biology · Endangered Species Act · Evolutionary potential · Habitat Directive · Natura 2000

1 Introduction

Two macroevolutionary phenomena—speciation and extinction—govern the increase and decrease of biodiversity on the evolutionary timescale. *Background extinction*, namely the extinction that occurs naturally in the evolution process, just like speciation, is an evolutionary process that shapes life on Earth. Speciation and extinction can be considered two sides of the same biodiversity coin: New species are always on the rise, while others are on the way out. In anagenesis, an ancestral species—or, more generally, a single phylogenetic ancestral lineage—ceases to exist because after accumulating enough mutations, it becomes a "new" species. In cladogenesis, an ancestral species gives rise to more than one descendant species and the branching process involves, at least according to certain models, the extinction of the ancestral species. Background extinction, then, is mostly[1] compensated by speciation, and it is just part and parcel of the production of new diversity.

The way in which speciation and extinction have shaped biodiversity along the evolutionary timescale is particularly apparent in two peculiar phenomena, namely *mass extinction* and *adaptive radiation*. Adaptive radiations are ecological differentiations among a group of related populations that can lead to speciation (by cladogenesis). Since in adaptive radiations a taxon made of a small number of ancestral species diversifies into a larger number of descendant species, they help to account for the increase of diversity over time, in spite of mass extinctions. From the initial absence of life, the Earth runs up toward 5–10 million species,[2] notwithstanding setbacks. Fossil evidences suggest that five mass extinctions, known as "the Big Five," have punctuated the history of life on Earth: the Ordovician–Silurian, the Late Devonian, the Permian–Triassic, the End Triassic, and the Cretaceous–Tertiary.[3] Ninety-nine percent of the species that have inhabited Earth are now extinct and have been replaced by other species

[1] Not all background extinctions are compensated by speciation. In "ultimate extinction" a species becomes extinct when the last individual dies and no new species originate from it. Not unlike extinction by speciation, ultimate extinction is a process that naturally shapes life. A species usually becomes extinct when it is no more adapted (and not able to adapt) to a changed environment.

[2] For a discussion on the number of species, see May and Beverton (1990).

[3] Mass extinctions take place when extinction rates appear to be exceptionally high compared to background extinction rates. Notice that the number of mass extinctions is open to doubt, since the definition of what a mass extinction is and the criteria to classify an extinction event as a mass extinction are under discussion (Hallam and Wignall 1997).

(Barnosky et al. 2011). The historical relevance of mass extinctions (as well as the evolutionary recoveries which follow them) and the tenability of the distinction between background and mass extinction are subjects of heated controversy. According to one view, mass extinctions have far-reaching historical significance, mainly for two reasons. First, mass extinctions alter the normal rules of the evolutionary game, since species go extinct for reasons that are independent of their histories of selection. For example, if a meteorite is the cause of the extinction of a species, that species becomes extinct no matter how well adapted it is to its environment and clearly its extinction will not be compensated by a speciation. Second, mass extinctions usually corresponded to the extirpation of dominant groups and such extirpation allowed the variety of life to increase. For instance, it has been hypothesized that without the extinction of the dinosaurs at the end of the Cretaceous period, mammals would not have diversified. Still, the tenability of the categorical distinction between mass and background extinctions has been questioned since it could jeopardize theoretical extrapolationism, that is, the explanation of macroevolutionary results by means of microevolutionary causes only (Gould 2002, Chaps. 2 and 12; Jablonski 2005).

According to a different view, mass extinctions did not alter the course of evolution but rather just intensified its normal course, accelerating a process that was already under way and that was mainly due to factors internal to lineages. Paleontologist Jack Sepkoski[4] hypothesized that both the replacement of the Cambrian by the Paleozoic fauna and the replacement of the Paleozoic by the Modern fauna were accelerated by the mass extinction but not caused by it. The "new" faunas were already on their way, and they were becoming dominant before the mass extinction intervened. This chapter takes a perspective of this second type, in line with the approach outlined by Sterelny and Griffiths (1999). According to this approach, mass extinctions have historical importance since they fundamentally shape the diversity of life on Earth, but they do not do so by putting in place mechanisms other than microevolutionary ones. Rather,

> they do so by the operation of normal mechanisms in an abnormal world. So conceived, extrapolationism is consistent with the idea that mass extinctions fundamentally reshape the tree of life (Sterelny and Griffiths 1999: 306).

On the basis of projected extinctions for the next 50 years, it has been suggested that we could be experiencing a new mass extinction, the so-called "Big Sixth", which would also be the first one caused by our own species and its activities (Rosa 2004). The primary causes of this extinction include pollution, habitat alteration and transformation of the landscape, climate change resulting from greenhouse gas emissions, introduction of invasive species, and overexploitation of species (Wilson 1988; Pimm et al. 1995; Barnosky et al. 2011). The Big Sixth can be divided into two discrete phases. Phase one commenced about 100,000 years ago when our species began to disperse out of Africa to different parts of the world. Fossils attest that in this phase Neanderthals became extinct,

[4] See Sterelny and Griffiths (1999: 304).

surviving for less than 10,000 years leading up to our arrival. Phase two began about 10,000 years ago, with the starting of agriculture and the resulting increase of human population size.[5] As of today, the available evidence indicates that from 1970 to 2010 biodiversity has continued to decline, with eight out of ten of the relevant indicators showing declines in population trends. Decreases of the emblematic coral reefs, mangroves, and sea grass beds, as well as reductions in populations of several species of vertebrates, such as habitats specialist birds and shorebirds, have been documented (Butchart et al. 2010: 1165).

As stated, from the point of view of general diversity, species that are lost in extinction get replaced by new species, or because from an old species one or more "new" species originate (extinction by speciation), or because the extinction of a species opens ecological niches (ultimate extinction). The same holds for mass extinction, at least according to the perspective taken in this chapter. Unfortunately, with anthropogenic extinction, things go differently. Anthropogenic pressure over the ecosystems has endured since the Paleolithic era, and we cannot foresee an end to it. Thereby, the loss of hundreds of thousands of species caused by anthropogenic extinction is not—and cannot be—compensated for by speciation, not even, probably, by radiation (Rolston 1985: 724–725).

In this light, it seems plausible to claim that even though humans may have no duty toward avoiding natural extinctions, they do have duties toward mitigating anthropogenic extinctions—not only because of the potentially unprecedented magnitude of the Big Sixth, but also because mankind is its main propagator. Indeed, this is probably one of the reasons why the extinction of species and higher taxa is becoming more and more of a concern for conservation biologists and ecologists as well as stakeholders such as governments, decision makers, managers, and society at large. Biodiversity[6] is widely conceived as something to be protected, and this means, first of all, facing the present-day anthropogenic increase of the extinction rate. As conservation biologist Michael Soulé puts it in his foundational paper (1985), conservation biology is a *mission-* or *crisis-oriented discipline*, and its relation to biology can be compared to that of surgery to physiology, or of war to political science. How is this "biodiversity surgery" to be performed? After some preliminary remarks on the problems raised by declaring extinction, Sect. 2 focuses on three different challenges that facing extinction requires us to take up, viz., how to prioritize species, which conservation targets to focus on, and how to ethically justify biodiversity conservation. Then, in Sect. 3, based on the analysis of these three difficulties, we compare two concrete legislative acts currently in force, the Endangered Species Act (the primary legislation providing federal legal protections to endangered species in the United States) and the Habitats Directive (which, together with the Birds Directive, forms the cornerstone of European Union's nature conservation policy) arguing that reforms are needed in both legislative Acts and in conservation practices related to them.

[5] For this recontrustruction see Eldredge (2001).

[6] We make reference to a very broad definition of 'biodiversity' as the diversity of life on Earth. On the difficulties of defining biodiversity, see DeLong (1996).

2 Biodiversity Surgery

2.1 Declaring Extinction

When can a species be declared extinct? At a first glance, defining extinction and putting forward the criteria to establish when a species can be declared extinct may seem quite uncontroversial. Who would question that extinction is the end of a lineage and who would deny that dinosaurs are extinct? But things are not so easy.[7] In order to see why, we consider the following example—which, notice, does not involve controversial species concepts, or "nonstandard" organisms such as asexually reproducing ones, or the like.[8]

Assume Mayr's (1970) biological species concept, according to which a species is a group of interbreeding organisms producing fertile offspring. Assume, moreover, that the extinction of a sexual species amounts to its disappearance, or the ending of the reproductive lineage. It may seem quite natural to assume that the criterion for determining whether a species has become extinct is the death of its last member. Now, consider the Amur leopard subspecies,[9] which today, mainly due to habitat alterations caused by the exploitation of forests, only consists of about twenty individuals living in the southwestern Primorye region of Russia (IUCN 2012). Clearly, the Amur leopard is not extinct at present, since at least twenty Amur leopards inhabit the world. Equally clearly, in millions of years (to be optimistic), no Amur leopards will inhabit the world any longer—they will be extinct. But consider the time between today and that moment. When *exactly* will the Amur leopard be extinct? "With the death of the last individual" does not seem to be a thorough answer for at least two reasons.

1. Let us say that there is just one individual remaining. In order to declare the Amur leopard extinct, should we take into consideration its death, or rather its complete disappearance? Such a question is not simply a matter of splitting hairs. For instance, imagine that after the death of the last individual, a laboratory takes some genetic material from it and succeeds in cloning a new population of Amur leopards, or that a fertilized egg was present in the last individual.[10]
2. Let us suppose that there are two individuals remaining, both of them male (or female). In such a scenario, it would seem reasonable to declare the species extinct *before* the death of the last individual. Or maybe it should be said that a *sexual* species goes extinct after the disappearance of the last couple, even

[7] On the nature of extinction see also Delord (2007).

[8] Cf. Casetta (2013).

[9] As we will see shortly, from the point of view of conservation policy, there is no substantial difference between conserving species and subspecies.

[10] On this view of extinction and the scenarios that the age of biotechnologies can arise, see Gunn (1991).

if individuals of the same sex are still alive (Delord 2007). Yet, we know that several animals may hybridize, either spontaneously or by human intervention and that hybridizing becomes more likely as selective pressures increase. Thus, it would not be implausible to think that the Amur leopard, in a critical situation, might crossbreed with, say, the North Chinese leopard and produce fertile offspring. In this case, the genetic make-up of the Amur leopard would prosecute in a new lineage. Should we conclude, then, that it is extinct or not?

Declaring extinction is not a trivial matter[11]; decisions have to be made, and such decisions can have consequences on prioritization and, more generally, on conservation policies. Notice, moreover, that often there is no biological fact that could settle the issue. Conventions and procedures have to be agreed upon and established, and as mentioned above, facing the Big Sixth is not a matter for biologists only.

The International Union for Conservation of Nature (IUCN) Red List of Threatened Species, widely recognized as the most comprehensive inventory for evaluating the conservation status of plant and animal species, tries to set upon precise criteria to evaluate the extinction risk. According to the system of categories and criteria adopted by the IUCN for including and classifying species at risk of extinction into the Red List, a species or other taxon has to be declared extinct when

> there is no reasonable doubt that the last individual has died. A taxon is presumed Extinct when exhaustive surveys in known and/or expected habitat, at appropriate times (diurnal, seasonal, annual), throughout its historic range have failed to record an individual. Surveys should be over a time frame appropriate to the taxon's life cycle and life form (IUCN 2012: 14).

As mentioned in the Introduction, extinction concerns not only species, but also subspecies and higher taxa as well. More generally, extinction concerns *evolutionary lineages*. Nonetheless, from the point of view of conservation, the extinction of a subspecies can be treated as a species' extinction. This seems confirmed by the fact that IUCN explicitly states that its criteria "can be applied to any taxonomic unit at or below the species level" (IUCN 2012: 4).

"Mere" extinction has to be distinguished from *extinction in the wild.* A species (or a higher taxon) is declared *extinct in the wild* when

> it is known only to survive in cultivation, in captivity or as a naturalized population (or populations) well outside the past range. A taxon is presumed Extinct in the Wild when exhaustive surveys in known and/or expected habitat, at appropriate times (diurnal, seasonal, annual), throughout its historic range have failed to record an individual. Surveys should be over a time frame appropriate to the taxon's life cycle and life form (IUCN 2012: 14).

It is worthwhile to note that the survival of naturalized populations of a species "well outside the past range" is not a criterion of exemption from being classified as "extinct in the wild," although such species (their individual organisms) can thrive without the support of any anthropogenic action. In fact, naturalized

---

[11] An additional problem, which will be addressed in a short while, is whether species surviving ex situ only (for instance, in zoos) are to be considered extinct or not.

populations are populations of non-native organisms that in most cases are able to sustain themselves reproductively and that may even become invasive.

Moreover, the distinction between "extinct" and "extinct in the wild" correlates with a view of biodiversity that distinguishes between native and introduced species (NRE 1997). Introduced species may increase species diversity, as they increase species richness, and thereby may increase the mathematical indices broadly known as "diversity indices" (Hamilton 2005). However, introduced species can hardly be seen as positive elements in biodiversity evaluation and are often considered as a menace to native species. This is partly due to the potential invasive role of introduced species, which can decrease the population of native species and therefore decrease species abundance, negatively impacting diversity indices. According to Eldredge (2001), for instance, invasive species have contributed to 42 % of all threatened and endangered species in the USA. Even when competition with native species does not occur, however, introduced species are often seen as less valuable components of biodiversity. This is probably due mainly to the widespread positive connotation associated with the concept of *wilderness*. In spite of all the attempts to clearly distinguish between biodiversity conservation and wilderness preservation (Sarkar 2005: 14), the idea of wilderness plays a foundational role in environmental ethics and is therefore largely embedded in the scientific ecology community. Thus, extinction in the wild is considered as a major loss, with (full) extinction only adding a minor incremental loss.[12]

Within such a traditional approach to biodiversity, we are usually led to consider in situ conservation as a better conservation strategy than ex situ conservation.[13] But, of course, in situ conservation requires more structured strategies. This raises some obvious, but difficult questions: What should our first and primary conservation target be if we want to conserve species in their environment? Does conserving biodiversity reduce to stopping extinction? Is there a hierarchy of species importance? An attempt to address such questions will be presented below.

2.2 Prioritizing Species

Since the material impossibility (because of the scarcity of time and financial resources) of saving all endangered species, to proceed in conservation efforts, one needs to place priorities among species that will benefit from conservation measures.[14]

[12] Besides this "reverence for wilderness," the fact that extinction in the wild terminates the ecological role of a species is also of key importance in its being considered a major loss.

[13] In situ conservation is usually defined as the conservation of species in their natural surroundings, while ex situ conservation is conservation of species outside their natural habitats, in captive breeding programs. For a discussion of the dichotomy see Braverman (2014).

[14] See also Casetta, Marques da Silva (2014).

IUCN, for instance, distinguishes three categories of threatened species: Vulnerable (V), Endangered (E), and critically endangered (CE). Such categories are nested: Critically Endangered species are also Endangered and Vulnerable, but not vice versa. Clearly, the main priority is the category including Critically Endangered species. A species (or a subspecies, or a higher taxon) is CE "when the best available evidence indicates that it *is* … facing an extremely high risk of extinction in the wild" (IUCN 2012: 14).

A range of tentatively quantitative criteria, mainly directly or indirectly concerning the size of populations, is used to include a taxon in one of the above categories. Examples are the population size itself, the reduction rate in population size, the geographic range in the form of extent of occurrence or the area of occupancy, or both.

The number of remaining individuals in a species, however, is only a prima facie thermometer to establish the urgency of first aid intervention. The assessment of species relative to differential urgency is not enough for at least two reasons. First, stopping extinction is probably not sufficient to conserve biodiversity. As will become clear in the remaining chapter, biodiversity cannot be reduced to a mere list of species. Second, as we will see shortly and as the IUCN recognizes, is because the category of threat is neither a necessary nor a sufficient condition to determine priorities for conservation actions. Rather, it provides a mere assessment of the extinction risk based on criteria that are as quantitative as possible.

Nevertheless, within IUCN criteria, it is possible to discern a path for proceeding from prioritization to conservation. In biodiversity conservation, the aim is not just to save endangered species, but also to allow the continued production of novel diversity, i.e., to improve the conditions and mechanisms that help the diversity to be generated. To put it differently, although conservation biologists usually focus on the *effects* of biodiversity and of their decline—namely the differing entities and their loss—a new approach is being promoted. In the new approach, the perspectives of conservation biology and evolutionary biology would converge and become integrated (Maclaurin and Sterelny 2008: 20 ff.; Forest et al. 2007). This approach would be centered on the processes that *cause* biodiversity and on how to maintain and improve them. Within such an integrated perspective, we suggest evolutionary potential should take a central role in conservation efforts, starting from species prioritization. Evolutionary potential can be broadly defined as the short-term ability of a species or a lineage to respond to selection without going extinct in the process. The significance of a *short-term* capability can hardly be underestimated, in particular in dealing with anthropogenic extinction caused by the consequences of climate change and ongoing habitat fragmentation. Indeed, evidence is already available which suggests that many short-lived species' organisms have the ability to evolve rapidly in response to climate change, fisheries, hunting, and other anthropogenic pressures (Reusch and Wood 2007; Santamaría and Méndez 2012).

That evolutionary potential should take a central role in conservation efforts is not a new idea, of course, nor an original one. Unfortunately, managers and policy makers tend to neglect evolutionary potential and evolutionary processes more generally, in developing criteria for prioritization and conservation (Mace and Purvis 2008).

Also IUCN prioritization criteria do not explicitly mention evolutionary processes or evolutionary potential. Nonetheless, their taking not only the size of the remaining population of a species but also its *reduction rate* into account can be seen as a step in the direction of the integrated perspective that we advocate here. As the population ecologist and conservationist Graeme Caughley argued, the "small population paradigm," which considers smallness as a cause when it should rather be considered as an effect, "provides an answer only to a trivial question: how long will the population persist if nothing unusual happens?" On the contrary, the "declining population paradigm," albeit in need of more theoretical elaboration, "summons an investigation to discover the cause of the decline and to prescribe its antidote" (1994: 215).

2.3 From Species to Biodiversity

While it is apparent that species are the currency of biodiversity, biodiversity is not just a matter of species. Accordingly, conserving biodiversity is probably not just a matter of stopping extinction. Keeping an eye to the importance of conserving evolutionary potential, which objects are and/or should be targets of our conservation policies?

Although the term "biodiversity" has a well-known history, it is useful to sketch it very briefly in order to trace the main phases the concept passed through and the corresponding differences in conservation targeting. The term "biodiversity" is of rather recent coinage. Biologist Walter G. Rosen blended the expression "biological diversity" to label the National Forum on BioDiversity organized in 1986 by the National Research Council under the patronage of the Smithsonian Institution and the National Academy of Sciences (Takacs 1996). The name was intended as a slogan to draw attention and financial support to combat the rapid decrease in the number of species and to raise political and academic awareness of species loss and decline caused by human activities. When it made its appearance, the term "biodiversity" was, implicitly or explicitly, intended to refer to the variety of species. In his contribution to the proceedings of the National Forum, Wilson (1988) explicitly equated the amount of biological diversity with the number of species and the loss of diversity with their extinction. It therefore comes as no surprise that during this early phase the targets of biodiversity conservation were first and mainly the species. Assessing biodiversity was considered as one and the same thing as inventorying species, and conserving biodiversity consisted in maintaining—and perhaps improving—the inventory. Hotspots prioritization and efforts to produce species catalogs such as the IUCN Red List, or the EOL (the Encyclopedia of Life, an online species database started by E.O. Wilson in 2007)[15] are emblematic of this approach.

Besides historical considerations, targeting species seems to be a good move for several reasons. Even though they are affected by the so-called Species Problem

[15] http://eol.org.

(addressed below), species are nonetheless theoretically well-founded entities. In fact, there is a widespread agreement that the concept of species is well entrenched in the framework of evolutionary biology. Moreover, we already possess good—even if not complete or fully coherent—species inventories, such as species collections in natural museums, or online species databases such as the previously mentioned EOL. These inventories are accompanied by some fairly reliable ways to delimit and recognize species in practice, by means of both traditional taxonomic tools and molecular techniques such as DNA barcoding (Barberousse and Samadi 2014; Hebert et al. 2003). Finally, phrasing the biodiversity crisis issue in terms of species has the advantage of making it understandable for the general public, which is important given that public concern has a great influence over the allocation of resources for conservation.

Although the number of species played a leading role, at least during this first historical phase, it soon became clear that biodiversity could not simply be reduced to a mere inventory of species and that several other aspects must be taken into account as well.

Criticisms of conservation policies strictly based on species diversity have appeared since the late 1980s. In particular, three types of criticisms have been advanced. The first is that mainly because of the Species Problem[16]—that is the connected issues of defining, delimiting, and identifying species—species are not good conservation targets. At present, more than twenty different concepts of species are in circulation (Richards 2010) and the argument over how a species should be defined appears to be an endless story. One cause of disagreement can be traced back (Mayr 1996) to the opposition between a static view of species and an evolutionary one. According to what Mayr (1959) calls "typological thinking," species are classes of organisms which share one or more essential properties—like in Linnaeus' *Systema naturae*. Such a view has been challenged and defeated during the post-Darwinian era, in which species have come to be viewed as ever-changing evolutionary units. Still, the disagreement on how to define and delineate them remains. Although no conclusive solution to the Species Problem has thus far been offered, it seems that a certain agreement has been reached—at least operationally—on a general definition or characterization of a species as a lineage or a segment of a lineage. For instance, LaPorte (2007) defines a general concept of species as "the least inclusive salient and stable lineage to which an organism belongs," and de Queiroz (1998) talks of species as "separately evolving metapopulation lineages."

The second criticism of species-based conservation policies is that focusing on species, even though it has been a strategy of great value in protecting them, is too narrow a strategy for biodiversity conservation: Taking into account species diversity is simply not representative enough of biodiversity as a whole.

> Biodiversity isn't species—biodiversity is the whole tree of life … only someone sharing the BSC [Biological Species Concept] view that species are fundamental … should think that species are the basic units of biodiversity, or that a list of currently named species in some way provides an inventory of biodiversity (Mishler 2010: 118).

[16] For an overview on the Species Problem see: Stamos (2004), Wilkins (2009), Richards (2010). For its cognitive causes, see Hey (2001).

In the 1987 report on "Technologies to Maintain Biological Diversity," the Office of Technology Assessment argues that, in addition to the diversity of species, the diversity of patterns and processes at all levels of living matter organization must be taken into account in order to assure the conservation of biological diversity (OTA 1987). However, while this need was in principle recognized by federal agencies, the bulk of conservation policies still remains centered on what biologist Hutto and his colleagues (1987: 3) designated as "the species approach," based on the maintenance of viable populations of selected species.

This leads us to the third type of criticism, which focuses on the alleged objectivity of the quantitative criteria on which the species approach is based. The selection of the so-called indicator species may play a role in some applications of the species approach. In these cases, populations of those species are chosen to be monitored and it is assumed that the populations of the other species of the same system will follow the same trend as indicator species. However, "no biologist would argue that a single species can be chosen to speak for the welfare of all others" (Hutto et al. 1987: 1). The idea that a chosen group of species is a reliable source of information concerning the welfare of all the other species in the same system is debatable. Moreover, it is not at all clear how many indicator species are needed in order to monitor the health of a system and it is far from obvious which species should be chosen as indicators. Finally, even if these issues could be satisfactorily handled, the species approach runs the risk of circularity, since

> we must realize that the only species whose welfare will be assured by that of an indicator species will be those whose niches are entirely subsumed by, or included within, that of an indicator species (Hutto et al. 1987: 1).

The problems with the species approach are not restricted to the selection of indicator species; they also occur when species are chosen on the basis of their risk of extinction. In spite of the Fish and Wildlife Service and Forest Service biologists' attempt to make use of objective selection criteria, the choice of species seems to be anything but neutral. This is indicated by the fact that many of the species listed by both agencies are those that are considered useful for humans as food, for sport, or hides, or even mere aesthetic enjoyment, and not necessarily those that are more threatened or endangered (Hutto et al. 1987).

More recently, different approaches have emerged which tend to identify conservation targets other than species. In particular, a transition from a static to a more dynamic view of conservation targets can be traced (Norton 2001) and recent conservation policies, focused on evolving systems and ecosystem processes, seem to target not only entities but also processes:

> We do know that the full absorption of evolving systems thinking into environmental management will have a far-reaching impact on advocated policies, and will almost certainly require more attention to interspecific relationship and system-level characteristic (Norton 2001: 77).

Of course, here we are dealing with an in progress phase, whose features are more to be constructed than they are to be described. Nevertheless, it seems that an increasing attention to evolutionary processes and evolutionary potential that we mentioned above clearly distinguishes the present phase from the previous ones. More generally, it seems that current conservation paradigms—aimed at facing not only "traditional" causes of diversity loss, but also the "new" challenges posed by climate change, invasive species, GMOs introduction, and rapid urbanization as well—are theoretically evolving toward a greater consideration of processes and dynamism. To what extent such a theoretical shift is also adequately reflected in our conservation practices remains a different matter. It should not be forgotten that conservation policies and practices are shaped not only by scientists, but also by a wide range of actors, including decision makers, governments, and NGOs, besides being constrained by political and economic limits.

Alternatives to the species approach can either focus on entities at a lower (or more fine-grained) level than species, or on a higher and more comprehensive level. In the first case, conservation will be guided mainly by the aim of preserving the inventory. Unlike the species approach, however, the inventory will be made of entities other than species and preserving the inventory will often be seen as a means to an end, namely to conserve processes in which the entities of the inventory are involved by preserving the entities themselves. In the second case, conservation efforts will take a different, more holistic form and conservation of biodiversity will be guided not by the aim of preserving the integrity of the inventory, but rather by the aim of maintaining the health of ecosystems' functions (and services).

An example of the first type of a approach is the evolutionarily significant units (ESUs) approach. It targets evolutionary processes, taking into account the evolutionary potential of populations and not only their size. Accordingly, it develops "strategies that will ensure that sufficient populations are viable to enable the species to survive in the short-term and to diversify in the future" (Moritz 1999: 217). The concept of an Evolutionary Significant Unit was originally introduced by geneticist Ryder (1986) in order to identify populations with significant genetic attributes for present and future generations and was rapidly imported to the administrative context (Waples 1991). The basic idea is that rather than preserving all phenotypic variants, it would be worthwhile to preserve those populations, or sets of populations, which "shows evidence of being genetically separate from other populations, and contributes substantially to the ecological or genetic diversity found within the species taxon as a whole" (Hey et al. 2003: 600). The premise here is that as long as evolutionary processes are able to operate, their products, in particular specific adaptive phenotypes, can be replaced or recreated.[17]

A different approach to conservation focuses, as said, on ecosystem functions, and in its more recent variants, on the products of those functions, such as ecosystem services.[18] The enormous efforts expended in carrying out the MEA, Millennium

[17] This perspective can be traced back to Frankel (1974). Unfortunately, however, appealing the ESU's concept may seem, it has to be noticed that the implementation of the concept—whose formulation has evolved over time—has been often inconsistent across different cases (Crandall et al. 2000).

[18] Ecosystem services "represent the benefits human population derive, directly or indirectly, from ecosystem functions" (Costanza et al. 1997: 253) such as food production, water supply, climate regulation, soil formation, pollination, as well as recreation and cultural services.

Ecosystem Assessment (Pereira et al. 2005; Carpenter et al. 2009), initiated in 2001 and involving the work of more than 1,360 experts worldwide, can be viewed as emblematic of the ecosystem functions approach. The aim of the MEA was to provide a state-of-the-art appraisal of the conditions and trends in the world's ecosystems and ecosystem services, to assess the consequences of changes of ecosystems for human well-being, and to put forward the scientific basis for action required to enhance the conservation and sustainable use of those systems.

Notice that in both the ESUs approach and the ecosystem functions approach, which both attempt to go beyond the species approach, the importance of species remains fundamental. Though the ESUs approach targets entities other than species, thereby enlarging the inventory to the genetic level and including evolutionary potential, it still remains connected, according to some, to the species approach. Joseph LaPorte, for instance, writes that ESUs

> provides a finer and more multifaceted gauge than species according to coarser conceptions, but it seems either to describe species according to finer conceptions or else to incorporate the relevant criteria (2007: 266).

As for the role of species in the Ecosystems Functions approach, things are no less problematic. As E.O. Wilson puts it:

> Eliminate one species, and another increases in number to take its place. Eliminate a great many species, and the local ecosystem starts to decay visibly. Productivity drops as the channels of the nutrient cycles are clogged. More of the biomass is sequestered in the form of dead vegetation and slowly metabolizing, oxygen starved mud, or is simply washed away … Fewer seeds fall, fewer seedlings sprout. Herbivore decline, and their predators die away in close concert (1992: 14).

Although the richness and evenness (relative abundance) of species of an ecological system are two fundamental parameters of its structure, the correlation between them and the well-being of the system is not trivial, since the relationship between species diversity/richness and the stability of the system (its ability to recover to an equilibrium state after disturbance, in other words, its resilience) is a controversial one (McCann 2000). According to the diversity–stability hypothesis, the richer a system is in species diversity, the higher is its stability. This is because a greater diversity means there is a higher probability that after disturbance there will be some species able to compensate for the loss of others (MacArthur 1955; Elton 1958). While fairly intuitive, this hypothesis was challenged as early as the 1970s. May (1973), for instance, using mathematical models, found out that diversity tends instead to destabilize the dynamics of a system. More precisely, the stability of a system increases with the number of its components, but only until a certain point, after which the system becomes unstable. Nevertheless, recent empirical evidence, gathered mostly in microorganisms, highlights the positive relation between diversity, community stability, and functional resilience after perturbation (e.g., Girvan et al. 2005). A different hypothesis that has been proposed to explain the relation between the richness of species in an ecosystem and their role in its functioning—the redundancy hypothesis (Walker 1992)—takes into account the differential role that species can play. The redundancy hypothesis predicts that, since in most ecosystems several species accomplish very similar functions, the loss of some of them will have little impact on the system as a whole.

As we have seen, different approaches to conservation have been espoused. Each of them, in prioritizing or targeting some entities or processes at the expenses of others, makes choices concerning the entities or processes that are deemed worthy of conservation. On the one hand, such choices are the result of empirical considerations as well as political and economical constraints. On the other hand, they display—or perhaps are partially the result of—our environmental value commitments. Put differently, our ethical approach to biodiversity and species contributes to shape our conservation policies and practices.

2.4 Paving the Way for Action

Imagine the following scenario, again involving our Amur leopards. After a sudden, bizarre, climate change, *Plasmodium falciparum*—a protozoan parasite that causes malaria in humans—is almost extinct, and it is now listed among Critically Endangered species. The number of Amur leopards, however, has started to increase to the point that they are no more listed as CE but just as a Vulnerable species. Are you certain that you would like to focus your conservation efforts on *Plasmodium falciparum*? Whatever answer one favors, it should be clear by now that mere empirical considerations are not enough to settle such issues. Moreover, notice that analogous scenarios can be envisaged for habitats as well as for ecosystems. Think of the stress that traditional approaches to conservation put on wilderness (Sect. 2.1): Why should preserving allegedly "pristine" environments be a better choice than focusing the attention on urbanized landscapes? To explore such issue—we would like to argue—one has to pay attention to the connection between different understandings of the importance of biodiversity and species, on the one hand, and conservation practices, on the other hand. The bridge between the two, we suggest, is unraveled by environmental ethics.

As already mentioned, conserving biodiversity is a sort of surgery; it is always performed in a state of emergency and without knowing all the relevant facts. Decisions have to be made before having all the empirical data at hand and before a full theoretical analysis of the data is available. In such a situation of uncertainty, ethical norms that guide conservation are, at least according to Soulé (1985), a genuine part of the discipline, as is always the case with crisis-oriented disciplines.

To stress the surgery metaphor, a surgeon cannot just operate on the sickest organ in a body; he or she also has to decide whether operating is the most appropriate course of action, keeping in mind the general well-being of the patient. The surgeon must check whether it is really *that* organ that should be operated on, or maybe that one plus another one which is connected to it, or maybe two unrelated organs because surgery would be useless otherwise. In other words, even in a crisis, some questions must be answered in order to put forward the most effective conservation policies and practices.

It is clear that we want to protect species—and biodiversity—because we think that they have value. Still, admitting that species are valuable, which type of value are we dealing with? Is our attitude to favor the Amur leopards to the detriment of parasitic protozoans somehow justified, or justifiable? Moreover, as mentioned in the Introduction, background extinction as well as the Big Five are caused by events that are completely independent of our species. In the case of the Big Sixth and the anthropogenic biodiversity crisis, things are different—*Homo sapiens* seem to have a moral responsibility toward other species and biodiversity at large *in themselves* as well as toward its own future generations.

First of all, a distinction can be drawn between something having an *instrumental value*, because it fulfills some material or immaterial need of someone/something else, and something having an *intrinsic value*—a value per se. Then, among instrumental values, we may distinguish between *commodity value*, *amenity value*, *existence value,* and *option value* of a species, or subspecies, or any other biological entity.[19] A species' commodity value, if any, is its market value. The Amur leopard's commodity value may be identified, for instance, with the marketplace value of the fur coat that can be made from it, or with the price of the zoo ticket that people would be willing to pay to view it. A species' *amenity value* is generated by the possible improvement that our intangible life quality can enjoy, thanks to it. Think, for instance, of the aesthetic excitement that the view of an Amur leopard may elicit in us. *Existence value* reflects the public's willingness to contribute to its preservation. Few people will have the opportunity to observe an Amur leopard and enjoy its beauty, but many more feel satisfaction from knowing that Amur leopards still exist in remote places and therefore might be willing to preserve them, contributing with donations and volunteer work for public campaigns. Finally, a species' *option value* is a potential value. Some species could have no *present instrumental value*—having no utility at all as a mean to a known end. Yet, such species can be a means to a potential, still unknown end and we may then decide to conserve some of them because we believe that although we do not need them now, we will perhaps need them in the future, for reasons that we are currently unable to envisage. Instrumental values such as the ones sketched above are the subject matter of environmental economics; intrinsic value is the main focus of environmental ethics. A classic example of intrinsic value is happiness: We pursue it because we think it is good in itself, and not in view of its use in obtaining something else.

Within environmental ethics, some advocate environmental *intrinsic value theories*, according to which biodiversity and particularly species have a value in themselves, for what they are, and not as a means to obtain some other goal. Advocates of *instrumental value theories* claim instead that species have an

[19] Several classifications of species values are available. Here, we follow, at large, the ones sketched by Norton (1988) and the comprehensive overview offered by Sandler (2012).

instrumental value in that they do not have a value per se but only insofar as they allow us to reach some other end that we find desirable. The desired end does not necessarily need to be for our species, as in the case of commodity or amenity value, it can also be for some other intrinsically valuable entity. Species, for instance, are typically considered as instrumental to the end of ecological resilience (species richness is considered important because the richer in species a system is, the higher is its integrity—recall, however, that such a claim is a debated one, as we have seen in Sect. 2.3). Notice, by the way, that *having a value* is often not a matter of "yes" or "no" but rather a matter of degree and that values are not mutually exclusive. It is also for these reasons that prioritization usually requires difficult choices.

The idea that species or some other biological entities or processes or even overall biodiversity have *only* an instrumental value typically characterizes *anthropocentric environmental ethics theories*. Anthropocentrism can be seen as the view that the only intrinsically valuable entities are human beings, and therefore, nature is instrumentally valuable just to the extent that it fulfills human needs. A strong and a weak version of anthropocentrism can also be distinguished. Strong anthropocentrism aims at accounting for the needs of the current generation *only*—therefore, unlimited resource exploitation is admissible. Weak anthropocentrism, on the other hand, aims at accounting for the needs of the present generation *and* future human generations. It is thus commonly referred to as intergenerational anthropocentrism, and it stands as the ethical foundation of the political concept of *sustainable development*.

Non-anthropocentric environmental ethics theories acknowledge that, besides human beings, other natural entities possess intrinsic value. Two perspectives, biocentrism and ecocentrism, acknowledge intrinsic value for different types of non-human entities: non-human *individuals* (individual animals, plants, etc.) in the case of biocentrism and environmental *collective* entities (species, ecosystems, etc.) in the case of ecocentrism. (For this reason, ecocentrism is considered a *holistic* environmental ethics theory). The idea that species or some other collective biological entities or processes—or even biodiversity understood as a property—have an intrinsic value typically characterizes *ecocentric environmental ethic theories*. Stenmark (2002) recognizes a strong and a weak version of both biocentrism and ecocentrism. Strong biocentrism includes those theories that attribute as much or more value to non-human individuals than to human beings, while weak biocentrism includes those theories that attribute less value to non-human individuals than to human beings. Strong ecocentrism includes those theories that attribute the same or even more value to environmental collectives than to individuals, and finally, weak ecocentrism attributes less value to environmental collectives than to individuals. Although somehow schematic, this sixfold classification of environmental ethics theories provides a useful tool for analyzing the foundation of environmental policies.

3 The US Endangered Species Act and the EU Habitats Directive: A Functional and Ethical Comparative Analysis

Two of the main legislative acts on environment conservation are the 1973 Endangered Species Act, currently in force in the United States, and the 1992 Habitats Directive, currently in force in the European Union.[20] They are probably the most relevant pieces of legislation concerned with biodiversity conservation—with the only possible exception being the global international treatises: the Convention on International Trade in Endangered Species of Wild Fauna and Flora (CITES) and the Convention on Biological Diversity (CBD). The main objective of the Habitats Directive was to institute Natura 2000, a network of natural sites within the European Union whose aim is to preserve some of the most representative habitats and species of Europe's fauna and flora (European Council 1992). In its turn, the Endangered Species Act was designed to give federal protection from extinction to all endangered species of fauna and flora (93rd United States Congress 1973).

The two legislations and the policies they encompass arose in different historical moments (the ESA in 1973 and the Habitats Directive in 1992) and were conceived in and for two different political scenarios: the ESA for a more politically integrated territory (USA), the Habitats Directive for a recent, in progress, political union (the European Union). Do these differences in political context translate into fundamental differences in the conservation policies envisaged by the ESA and the Habitats directive? In particular, do the two legislations rest on a different ethical evaluation of biodiversity? Do they target different objects for preservation? Do they integrate the evolutionary potential in conservation policies differently? To answer these questions, a thorough comparative analysis of the ESA and the Habitats Directive is presented in the remaining parts of this work.

3.1 Protecting the Evolutionary Potential

In Sect. 2.2, we suggested that taking the evolutionary potential of species or populations into account would help prioritization choices and, more broadly, that in order to go beyond the species approach, a consideration of the evolutionary process is in order. Do the Habitats Directive and the ESA differ in this respect?

---

[20] The texts of the two legislations to which we make reference in the remainder of this work can be downloaded, respectively, from http://www.nmfs.noaa.gov/pr/pdfs/laws/esa.pdfand http://eur-lex.europa.eu/LexUriServ/LexUriServ.do?uri=OJ:L:1992:206:0007:0050:EN:PDF.

Albeit the terms "evolution" and "evolutionary" are absent from the letter of the ESA, it is possible to argue that the ESA pays strong attention to the evolutionary potential of species, as witnessed by the presence of the concept of an Evolutionary Significant Unit in the implementation—though not in the text—of the Act. Let us review this point in more detail.

The concept of an Evolutionary Significant Unit made its appearance in the ESA in order to clarify the concept of distinct population segment (DPS) and evolved as a practical response to the specific challenges of conservation. Distinct population segments are those populations, or groups of populations, that are significant in relation to the entire species and discrete from other populations of the species. In order to clarify when a certain population has to be considered significant and discrete for the purposes of the ESA, looking at whether it represents an evolutionarily significant unit or not has been a suggested approach. This approach took place in the first petitions for listing the Pacific Northwest salmon among endangered species, filed in 1990, that asked to include distinct population segments. At that time, however, the agencies in charge of implementing the ESA lacked any formal guidance for how to interpret the distinct population segments provision in the Act because the provision as defined for terrestrial species, such as grizzly bears, bald eagles, and alligators, during the 1990s used a wide variety of criteria. Therefore, Robert Waples (1991), an official of the National Marine Fisheries Service, developed a conceptual framework laying down that a salmon population (or a group of salmon populations) should be considered a DPS if it represents an Evolutionary Significant Unit of that species.

> A vertebrate population will be considered distinct (and hence a 'species') for purposes of conservation under the Act if the population represents an evolutionarily significant unit (ESU) of the biological species. An ESU is a population (or group of populations) that (1) is substantially reproductively isolated from other conspecific population units, and (2) represents an important component in the evolutionary legacy of the species.

To fulfill the ESU requirements, a population unit should both be under reproductive isolation and show evolutionary significance. The second condition is prevalent, as isolation is not demanded to be complete, but is required only to the extent needed to allow evolutionarily important differences to arise in different units. The policy based on ESUs has only been applied to the Pacific salmon by the National Marine Fisheries Service. The desire for guidelines that would produce more reliable and consistent results motivated the development of a policy, common to the US Fish and Wildlife Service and the National Marine Fisheries Service, on how to interpret the concept of distinct population segments in the ESA, finalized in 1996. The joint policy is based on two criteria—discreteness and significance—that closely parallel those in the ESU salmon policy (Waples 2006).[21] The two

[21] Alternatives to the ESU worth mentioning are the designatable units (DUs) employed by the Committee on the Status of Endangered Wildlife in Canada for status assessment of species under the Canadian Species at Risk Act. According to (Green 2005), DUs would fulfill "the need for a practicable procedure to identify infraspecific entities for status assignment" avoiding at the same time "value judgments concerning evolutionary importance".

agencies have used the joint policy to make DPS determinations for a number of species, including the gray wolf, Sierra Nevada bighorn sheep, cactus ferruginous pygmy owl, white marlin, and Puget Sound killer whale (Waples 2006). Therefore, albeit the Endangered Species Act does not explicitly mention evolutionary processes and potential, if we look at its practical implementation, there are little doubts that the preservation of evolutionary processes and potential is of central concern.[22]

In sharp contrast with ESA, genetic studies of local populations have seldom been used to justify the inclusion of new conservation areas in the Natura 2000 network (Santamaría and Méndez 2012), meaning that evolutionary processes and potential are of minor concern. Furthermore, only one reference to genetics is made in the text of the Habitats Directive. It is found in a passage in which the need to conserve landscape "stepping stones" (such as small woods or ponds) essential for dispersal and genetic exchange of wild species is recalled. However, Annex III requires that for the assessment of the Natura 2000 sites, the "global ecological value of the site for the biogeographical regions concerned" and the degree of isolation of priority species be taken into account. This apparently hosts, to a certain extent, the conservation status of genetically distinct populations and their expected contribution to the species' evolutionary potential as criteria for asserting the conservation status of species and justifying new classified areas. Nonetheless, the conservation of genetically distinct populations, races, or subspecies is not specifically required by the Directive.

In sum, the ESA might have the capability of addressing the conservation of evolutionary mechanisms and potential in spite of the biases introduced by interest groups, political pressures, and bureaucratic constraints in the selection of protected species. However, the historical evidence of the small number of species and populations that enhanced their conservation status, thanks to measures inspired by the ESA, casts doubts on this possibility (Langpap and Kerkvliet 2002). In any case, the capability of the Natura 2000 network to address the conservation of evolutionary potential seems even lower: Even though some references to the conservation of genetic diversity are traceable in the Habitats Directive (mainly in Annex III), they are not mandatory and, in practice, there is an almost total absence of compliance of member states with these aims. Neither of the two legislative acts adequately targets the objective of conserving evolutionary potential—even though the ESA seems closer to achieving this objective.

This remains, however, a central goal. On the one hand, at large, biodiversity is the product of evolution and thereby its long-term maintenance is dependent on evolutionary processes and on conserving and improving evolutionary potential. On the other hand, there is compelling evidence of an acceleration of microevolutionary changes in contemporary populations due to human disturbance or activity,

[22] To identify and characterize ESUs and DPSs, ESA increasingly relies on the use of molecular and population genetics, which have raised concerns regarding the limitation of lay public participation in listing decisions (Kelly 2010).

including, somehow ironically, changes elicited by the application of conservation programs (Santamaría and Méndez 2012). Those anthropogenic microevolutionary changes play a crucial role in allowing the populations of threatened species to adapt to rapidly changing environments and can eventually lead, in the long term, to macroevolutionary speciation. Protecting evolutionary processes is, therefore, of critical importance.[23]

3.2 Biodiversity Concepts and Conservation Targets

In Sect. 2.3, we argued that the different views of biodiversity underlying distinct conservation policies might influence the choice of conservation targets. Now, we ask whether this is the case with the two legislative acts under review. Ideally biodiversity conservation policies should recognize the importance of the different levels of biodiversity and be designed and implemented to conserve the entire spectrum that goes from genes to ecosystems, going beyond a mere species-centered approach. But, do they do it?

The network based on Habitats Directive—Natura 2000—is probably the most ambitious supranational policy measure aiming at preserving biodiversity through land use regulation. In comparison with nature conservation policies targeting only species, Natura 2000 goes one step further by protecting not only endangered species, but also natural and semi-natural habitats. The network is therefore concerned with biodiversity at the level of both species and ecosystems (Rosa and Marques da Silva 2005). Furthermore, the explicit reference to semi-natural habitats denotes that the concept of wilderness does not play a core role, suggesting a conception of biodiversity quite different from the one embraced by IUCN (see Sect. 2.1).

Albeit the main modus operandi of the ESA is also based on the classification of protected areas, these are not taken as ends in themselves but rather as instruments to the ultimate goal of endangered species preservation. Protected areas are declared to be such because they encompass habitats that are critical for the in situ conservation of endangered species. The value that the ESA accords to habitats is, therefore, just an instrumental value. This claim is further confirmed by the transitory character of the protection measures prescribed by the ESA, among which we mention the rule that established that protected areas shall be declassified if the species they contain are no longer threatened (93rd United States Congress 1973).[24]

> The Secretary shall implement a system in cooperation with the States to monitor effectively for not less than 5 years the status of all species which have recovered to the point at which the measures provided pursuant to this Act are no longer necessary and which, in accordance with the provisions of this section, have been removed from either of the lists published under subsection.

---

[23] Ashley et al. (2003) called, for instance, for an "Evolutionary Enlightened Management".

[24] Notice also that Natura 2000 contemplates the possibility of classified sites being delisted once recovered, but the entire network was conceived as a permanent structure.

It should be noted, however, that although the ESA's measures were conceived more as an emergency surgery than as long-term therapy, the emergency usually turns out to be longer than expected. For example, by 1999, of 1,746 species listed (FWS 1999), only 27 had been delisted [8 because they actually recovered, 7 because they went extinct, and 11 because additional information revealed that the listing was unwarranted (Abbitt and Scott 2001)]. Such a focus on species seems to show that the ESA is mainly concerned with biodiversity conservation at the level of species, with minor or no interest in the conservation of biological diversity at the ecosystems level. So, what about genetic diversity?

The complete absence of the terms "genetic" and "genetics" in the text of the ESA could lead the reader to conclude that biodiversity at the level of genes is also of minor or no interest to the ESA. That would, however, be an inaccurate conclusion. Genetic information is commonly used by the two federal agencies in charge of implementing the ESA: the Fish and Wildlife Services, under the Department of the Interior, and the National Marine Fisheries Services, within the Department of Commerce. Both agencies allow for three classes of entities to be listed as endangered or threatened: species, subspecies, and DPSs [and, for the purposes of the ESA, an entity, once listed, "becomes" a species, regardless of whether it was originally listed as a species, a subspecies, or a DPS (Kelly 2010)].

Subspecies and species are common taxonomic concepts, although not without difficulties, as the Species Problem attests (Sect. 2.3). For instance, regarding the absence of a shared definition of species, the ESA adopted the previously mentioned Mayr's biological species concept—one of the most commonly used in the scientific community, in particular by zoologists. Indeed, in the text of the ESA, it is specified that "The term 'species' includes any subspecies of fish or wildlife or plants, and any distinct population segment of any species of vertebrate fish or wildlife which interbreeds when mature."

By contrast, the notion of a "distinct population segment" is a less standard biological concept. Although it may be applied to any species—and it seems particularly fitted to sessile organisms like vascular plants—under the ESA, the concept is currently restricted to vertebrates. Though a certain species as a whole may not be in danger of extinction, designation as DPS offers protection to imperilled, isolated populations of a species (Kelly 2010). Population genetics operates both as a tool to identify DPSs and as the motivation to protect them. DPSs are preserved due to their unique gene pool, implicitly assumed (when no genetic studies have been performed and classification is based, for instance, on geographic analysis only) or explicitly demonstrated (by means of molecular biology techniques) to be significantly different from the gene pools of other populations of the same species. Thereby, the designation of DPSs as protected species highlights the concern of the ESA with the genetic level of biodiversity.

In contrast, as mentioned in Sect. 3.1, the term "genetic" is present in the text of the Habitats Directive, in a reference to the importance of preserving landscape features such as ponds or small woods that, by virtue of their role as "stepping stones," are essential for the migration, dispersal, and genetic exchange of wild species. However, there are no explicit references to genetic evaluation and preservation in the process of selecting sites to be listed. Therefore, the genetic level of biodiversity seems to be of minor concern in the Habitat Directive—and, accordingly, in the Natura 2000 network—compared to the ESA. In summary, both the ESA and the Habitats Directive are concerned with conserving biodiversity at the species level, but their interests diverge regarding other organizational levels of biodiversity. Whereas the ESA has a strong focus on the conservation of genetic diversity, the emphasis of the Habitats Directive is, as its designation explicitly recalls, is on the higher organizational level of habitats.

3.3 Ethical Frameworks

Identifying the reasons for conserving biodiversity is not only valuable from a theoretical perspective but is also useful from a practical point of view. A correct assessment of the value system is essential to designing conservation policies that adequately cope with the trade-off between biodiversity and other societal values. Is the ethical motivation underlying the ESA and the Habitats Directive the same? Do they share an anthropocentric, biocentric, or ecocentric view of biodiversity (see Sect. 2.4)? An abundant literature on the ethics of biodiversity is available, but, to our knowledge, no one has offered any comparison of the ethical frameworks of the ESA and the Habitats Directive. In what follows, we will provide some remarks on this issue.

The ESA does not explicitly explain its motivations for biodiversity conservation. Nevertheless, some insights into the ethical framework underlying it can be traced. First of all, there is a commitment to comply with the international obligations of the United States. Indeed, the ESA expresses that "develop[ing] and maintain[ing] conservation programs which meet national and international standards is a key to meeting the Nation's international commitments."

This concern can be seen as expressing both indirect duties toward biodiversity and an anthropocentric perspective on its value given that the value of biodiversity rests in its potential to fulfill international agreements. It should be noted, however, that this is only a provisional evaluation of the ethical framework of the ESA, since the ethical motivations for engaging in the international agreements will ultimately define the ethical perspective of this legislative act. Still, in the same sentence, the ESA presents other motivations for biodiversity conservation, such as "better safeguarding, for the benefit of all citizens, the Nation's heritage in fish, wildlife, and plants."

This reinforces the anthropocentric perspective, explicitly addressing "the benefit for all citizens." There are no direct and explicit references to future generations suggesting a strong anthropocentric position. However, the term "heritage" denotes a temporal perspective, something that was received from ancestors and (presumably) should be given to the forthcoming generations, thus revealing a position characteristic of weak or intergenerational anthropocentrism. The biocentric perspective, as expected, can be ruled out, since the ESA explicitly admits the killing of individual organisms when required to spare collective entities.

Interestingly, the ESA recurrently refers to "the well-being" of species (and to the need to preserve them), since these collective entities could share the subjective feeling of their constituent individual organisms—but no references to the well-being of the latter are made. In summary, the scarce indications available regarding the values behind the ESA suggest a strong anthropocentric ethics, mitigated with some intergenerational elements.

Let us move now to the Habitats Directive. Prima facie reasons allow us to presume that its underlying ethics may align it with an ecocentric paradigm since it targets species and habitats which are central entities to holistic environmental ethics systems (Rosa and Marques da Silva 2005). Furthermore, Natura 2000 is supported by environmental NGOs whose members, or at least part of them, presumably hold non-anthropocentric ethical assumptions (Dobson 2000). However, the Habitats Directive's preamble explicitly states that Natura 2000 was conceived as a contribution to sustainable development:

> Whereas, the main aim of this Directive being to promote the maintenance of biodiversity, taking account of economic, social, cultural and regional requirements, this Directive makes a contribution to the general objective of sustainable development; whereas the maintenance of such biodiversity may in certain cases require the maintenance, or indeed the encouragement, of human activities.

In its current definition and understanding, sustainable development is a paradigmatic mark of weak (i.e., intergenerational) anthropocentrism. In fact, the Habitats Directive deals directly or indirectly with many of the concepts typically associated with weak anthropocentrism. It refers, for example, to asset value, cultural values, natural heritage, and natural resources. Furthermore, the Habitats Directive admits that many human activities (implicitly including productive and economic activities such as forestry and agriculture) need not be prohibited in order to conserve biodiversity. Nevertheless, it would be a mistake to connect the ethical background of the Habitats Directive with strong anthropocentrism because the relevance given to sustainability and stewardship points elsewhere. Furthermore, the greater importance explicitly conferred by the Habitats Directive to biological over socioeconomic criteria in the selection, management, and protection of conservation sites would be barely allowable under typical strong anthropocentrism (Rosa and Marques da Silva 2005). Also, like in the case of the ESA, no direct or indirect reference to values connected to a biocentric or ecocentric ethical system is present in the Habitats Directive. Thus, it seems that the underlying ethics of the Habitats Directive and the corresponding Natura 2000 network is that of weak (intergenerational) ethics. In summary, the underlying ethics of both the ESA and the Habitats Directive are anthropocentric, clearly intergenerational in the case of the latter, and closer to the strong version in the case of the former.

4 Concluding Remarks

Abundant evidence shows that we may be entering a new mass extinction, the first one to be anthropogenic. Since our species and its activities are the principal cause of the alleged Big Sixth, it seems reasonable to think that, insofar as we are moral agents, we have responsibilities toward other people and future generations (if not directly toward species other than ours). These responsibilities include preserving biodiversity as a source of well-being. Current conservation policies, such the North American ESA and the European Habitats Directive, were created and are still running under an anthropocentric ethical paradigm, in line with the dominant political concept of sustainable development. In spite of the many criticisms that have been raised against both legislations, and the difficulties in comparing their effectiveness,[25] the anthropocentric ethical framework, per se, is not hindering their conservation effectiveness. On the contrary, the prevalence of a static view of species, characteristic of typological thinking, is of greater concern. Preserving biodiversity is not only a matter of stopping species extinction: conservation of all levels of biodiversity, under (genes) and above (habitats) species, should be directly addressed. Furthermore, conserving the evolutionary processes is of focal importance. The conceptual architecture of both the ESA and the Habitats Directive is not fully capable of fulfilling these needs: Whereas the ESA was not designed to preserve the higher level of biodiversity (habitats), the Habitats Directive misses the objective of preserving the lower level (genes), and thereby the evolutionary potential of populations. Therefore, reforms in both North American and European biodiversity conservation policies are needed.

References

93rd United States Congress (1973) Endangered species act (ESA): an act to provide for the conservation of endangered and threatened species of fish, wildlife, and plants, and for other purposes. Public Law 93–205

Abbitt RJF, Scott JM (2001) Examining differences between recovered and declining endangered species. Conserv Biol 15:1274–1284

Ashley MV, Willson MF, Pergams ORW, O'Dowd DJ, Gende SM (2003) Evolutionarily enlightened management. Biol Conserv 111:115–123

[25] The effectiveness of legislation is usually difficult to measure. In the case at issue, a comparative evaluation of results is made even more difficult for a series of reasons; for instance, the fact that the ESA became effective 20 years before the Habitat Directive, that they target slightly different objectives, and that—as repetitively stressed—conserving biodiversity is not just saving species. In the case of the CBD, for instance, more than twenty indicators have been monitored, showing that the 2010 Biodiversity Target ("to achieve by 2010 a significant reduction of the current rate of biodiversity loss at the global, regional and national level") failed. But to our knowledge, no similar monitoring has been done for the ESA and the Habitat Directive. For an attempt of comparing the effectiveness of the two laws, and a critical discussion of its difficulties, see Verschuuren (2004).

Barberousse A, Samadi S (2014 forthcoming) La taxonomie et les collections d'histoire naturelle à l'heure de la sixième extinction. In: Casetta E, Delord J (eds) Versatile biodiversité. Enjeux philosophiques et scientifiques. Édition matériologiques, Paris
Barnosky AD et al (2011) Has the Earth's sixth mass extinction already arrived? Nature 471:51–57
Braverman I (2014) Conservation without nature. The trouble with in situ versus ex situ conservation (18 Nov 2013). Geoforum 51:47–57. Available at SSRN:http://ssrn.com/abstract=2356563(Forthcoming; SUNY Buffalo Legal Studies Research Paper No. 2014-005)
Butchart SHM et al (2010) Global biodiversity: indicators of recent declines. Science 328:1164–1168
Carpenter SR et al (2009) Science for managing ecosystem services: beyond the millennium ecosystem assessment. Proc Natl Acad Sci USA 106:305–312
Casetta E (2013) The temporal boundaries of biological species. In: Graziani P, Grimaldi G (eds) Open problems in philosophy of sciences. College Publications, London, pp 35–45
Casetta E, Marques da Silva J (2014) Biodiversity surgery. Some epistemological challenges in facing extinction. Axiomathes. doi:10.1007/s10516-014-9244-9
Caughley G (1994) Directions in conservation biology. J Anim Ecol 63(2):215–244
Costanza R et al (1997) The value of the world's ecosystem services and natural capital. Nature 387:253–260
Crandall KA et al (2000) Considering evolutionary processes in conservation biology. Tree 15(7):290–295
DeLong DC Jr (1996) Defining biodiversity. Wildl Soc Bull 24:738–749
Delord J (2007) The nature of extinction. Stud Hist Philos Biol Biomed Sci 38:656–667
de Queiroz K (1998) The general lineage concept of species, species criteria, and the process of speciation. In: Howard DJ, Berlocher SH (eds) Endless forms: species and speciation. Oxford University Press, Oxford, pp 57–75
Dobson A (2000) Green political thought, 3rd edn. Routledge, London
Eldredge N (2001) The sixth extinction. ActionBioscience.org. URL http://www.actionbioscience.org/evolution/eldredge2.html
Elton CS (1958) Ecology of invasions by animals and plants. Chapman & Hall, London
European Council (1992) Council directive 92/43/EEC of 21 May 1992 on the conservation of natural habitats and of wild fauna and flora. Retrieved 7 Nov 2002 (Official Journal of the European Communities L 206, 22.07.1992)
Forest F et al (2007) Preserving the evolutionary potential of floras in biodiversity hotspots. Nature 445(7129):757–760
Frankel OH (1974) Genetic conservation: our evolutionary responsibility. Genetics 78:53–65
Girvan MS et al (2005) Bacterial diversity promotes community stability and functional resilience after perturbation. Environ Microbiol 7(1):301–313
Gould SJ (2002) The structure of evolutionary theory. Belknap Press of Harvard University Press, Cambridge
Green DM (2005) Designatable units for status assessment of endangered species. Conserv Biol (19)6:1813–1820
Gunn AS (1991) The restoration of species and natural environment. Environ Ethics 13:291–310
Hallam A, Wignall PB (1997) Mass extinctions and their aftermath. Oxford University Press, Oxford
Hamilton J (2005) Species diversity or biodiversity? J Environ Manage 75:89–92
Hebert PDN et al (2003) Biological identifications through DNA barcodes. Proc Roy Soc Lond 270:313–321
Hey J (2001) The mind of the species problem. Trends Ecol Evol 16:326–329
Hey J et al (2003) Understanding and confronting species uncertainty in biology and conservation. Trends Ecol Evol 18:597–603
Hutto RL, Reel S, Landres PB (1987) A critical evaluation of the species approach to biological conservation. Endangered Species Update 4:1–4

IUCN (2012) IUCN red list: categories and criteria, version 3.1, 2nd edn. URL http://www.iucnredlist.org/static/categories_criteria_3_1
Jablonski D (2005) Mass extinction and macroevolution. Paleobiology 31(2):192–210
Kelly RP (2010) The use of population genetics in endangered species act listing decisions. Ecol Law Q 37:1107–1159
Langpap C, Kerkvliet J (2002) Success or failure? Ordered probit approaches to measuring the effectiveness of the endangered species act. 2002 annual meeting, 28–31 July, Long Beach, 19713, American Agricultural Economics Association (New Name 2008: Agricultural and Applied Economics Association)
LaPorte J (2007) In defense of species. Stud Hist Philos Biol Biomed Sci 38:255–269
MacArthur RH (1955) Fluctuations of animal populations and a measure of community stability. Ecology 36:533–536
Mace GM, Purvis A (2008) Evolutionary biology and practical conservation: bridging a widening gap. Mol Ecol 17:9–19
Maclaurin J, Sterelny K (2008) What is biodiversity. University of Chicago Press, Chicago
May RM (1973) Stability and complexity in model ecosystems. Princeton University Press, Princeton
May RM, Beverton RJH (1990) How many species? [and discussion]. Phil Trans Biol Sci 330(1257):293–304
Mayr E (1959) Darwin and the evolutionary theory in biology. In: Meggers J (ed) Evolution and anthropology: a centennial appraisal. Anthropological Society of Washington, Washington, D.C., pp 1–10
Mayr E (1970) Populations, species, and evolution. Harvard University Press, Cambridge
Mayr E (1996) What is a species and what is not? Philos Sci 63:262–277
McCann KS (2000) The diversity-stability debate. Nature 405:228–233
Mishler BD (2010) Species are not uniquely real biological entities. In: Ayala FJ, Arp R (eds) Contemporary debates in philosopphy of biology. Wiley-Blackwell, Singapore
Moritz C (1999) Conservation units and translocations: strategies for conserving evolutionary processes. Hereditas 130:217–228
Norton BG (1988) Commodity, amenity, and morality. In: Wilson, pp 200–205
Norton BG (2001) Conservation biology and environmental values: can there be a universal earth ethic? In: Potvin C et al (eds) Protecting biological diversity: roles and responsibilities. McGill-Queen's University Press, Montreal
NRE (1997) Victoria's biodiversity–our living wealth. Department of Natural Resources and Environment, Melbourne, Victoria
OTA (1987) Technologies to maintain biological diversity. Congress of the United States, Office of Technology Assessment, Washington, D.C.
Pereira H et al (2005) Ecosystem services and human well-being: a participatory study in a mountain community in Northern Portugal. Ecol Soc 10(2):14
Pimm SL et al (1995) The future of biodiversity. Science 269:347–350
Reusch TB, Wood TE (2007) Molecular ecology of global change. Mol Ecol 16(19):3973–3992
Richards R (2010) The species problem. Cambridge University Press, Cambridge
Rolston H (1985) Duties to endangered species. Bioscience 35(11):718–726
Rosa HD (2004) The bioethics of biodiversity. In: Susanne C (ed) Societal responsibilities in life sciences. Kamla-Raj, Delhi
Rosa HD, Marques da Silva J (2005) From environmental ethics to nature conservation policy: Natura 2000 and the burden of proof. J Agric Environ Ethics 18:107–130
Ryder OA (1986) Species conservation and systematics: the Dilemma of subspecies. Trends Ecol Evol 1:9–10
Sandler RL (2012) The ethics of species. Cambridge University Press, Cambridge
Santamaría L, Méndez PF (2012) Evolution in biodiversity policy. Current gaps and future needs. Evol Appl 5:212–218
Sarkar S (2005) Biodiversity and environmental philosophy: an introduction. Cambridge University Press, New York

Soulé ME (1985) What is conservation biology? Bioscience 35:727–734
Stamos D (2004) The species problem. Lexington Books, Lanham
Stenmark M (2002) The relevance of environmental ethical theories for policy making. Environ Ethics 24:135–148
Sterelny K, Griffiths P (1999) Sex and death. University of Chicago Press, Chicago
Takacs D (1996) The idea of biodiversity. Philosophies of life. John Hopkins University Press, Baltimore
Verschuuren J (2004) Effectiveness of nature protection legislation in the European Union and the United States: the habitats directive and the endangered species act. In: Dieterich M, van der Straaten J (eds) Cultural landscapes and land use: the nature conservation-society interface. Kluwer Academic Publishers, Dordrecht, pp 39–67
Walker BH (1992) Biodiversity and ecological redundancy. Conserv Biol 6:18–23
Waples RS (1991) Pacific Salmon, Oncorhynchus spp., and the definition of "species" under the endangered species act. Mar Fish Rev 53:11–22
Waples RS (2006) Distinct population segments. In: Scott JM, Goble DD, Davis FW (eds) The endangered species act at thirty: conserving biodiversity in human-dominated landscapes. Island Press, Washington, D.C., pp 127–149
Wilkins JS (2009) Defining species. A sourcebook from antiquity to today. Peter Lang, New York
Wilson EO (ed) (1988) Biodiversity. National Academy Press, Washington, D.C.

MIX
Papier aus verantwortungsvollen Quellen
Paper from responsible sources
FSC® C105338

If you have any concerns about our products,
you can contact us on
ProductSafety@springernature.com

In case Publisher is established outside the EU,
the EU authorized representative is:
Springer Nature Customer Service Center GmbH
Europaplatz 3, 69115 Heidelberg, Germany

Printed by Libri Plureos GmbH
in Hamburg, Germany